T0138389

Foraging

Foraging

Behavior and *Ecology*

Edited by David W. Stephens,
Joel S. Brown, and
Ronald C. Ydenberg

The University of Chicago Press
Chicago & London

DAVID W. STEPHENS is Professor of Ecology, Evolution, and Behavior at the University of Minnesota and author, with J. R. Krebs, of *Foraging Theory*.
JOEL S. BROWN is Professor of Biology at the University of Illinois at Chicago and author, with T. L. Vincent, of *Evolutionary Game Theory, Natural Selection, and Darwinian Dynamics*.
RONALD C. YDENBERG is Professor in the Behavioral Ecology Research Group and Director of the Centre for Wildlife Ecology at Simon Fraser University.

The University of Chicago Press, Chicago 60637
The University of Chicago Press, Ltd., London
© 2007 by The University of Chicago
All rights reserved. Published 2007
Printed in the United States of America
16 15 14 13 12 11 10 09 08 07   1 2 3 4 5

ISBN-13: 978-0-226-77263-9 (cloth)
ISBN-13: 978-0-226-77264-6 (paper)
ISBN-10: 0-226-77263-2 (cloth)
ISBN-10: 0-226-77264-0 (paper)

Library of Congress Cataloging-in-Publication Data

Foraging : behavior and ecology / [edited by] David W. Stephens, Joel S. Brown & Ronald C. Ydenberg.
     p.   cm.
   ISBN-13: 978-0-226-77263-9 (cloth : alk. paper)
   ISBN-13: 978-0-226-77264-6 (pbk. : alk. paper)
   ISBN-10: 0-226-77263-2 (cloth : alk. paper)
   ISBN-10: 0-226-77264-0 (pbk. : alk. paper)
   1. Animals—Food.   I. Stephens, David W., 1955–   II. Brown, Joel S. (Joel Steven),
1959–   III. Ydenberg, Ronald C.
   QL756.5.F665 2007
   591.5′3—dc22

                                                                      2006038724

⊚ The paper used in this publication meets the minimum requirements of the American National Standard for Information Sciences—Permanence of Paper for Printed Library Materials, ANSI Z39.48-1992.

Contents

II

Foreword

On October 1, 1975, JK wrote the following in a letter (no email in those days!) to Ric Charnov to report the pilot results of the first experimental test of the "classic" diet model under properly controlled conditions of encounter rate, handling time, and prey energy content:

Here are the results—read 'em and gloat:

	percentage small prey in the diet predicted by:		
treatment	random foraging	prey model	observed
1	50	50	47
2	50	50	37
3	25	0	0
4	50	0	2
5	67	0	9

The last three rows demonstrated the crucial counterintuitive prediction that small prey would be excluded from the diet, independently of their encounter rate, if the encounter rate with large prey were above a certain quantifiable threshold.

Those were heady days! Setting aside the fact that the small prey were not *totally* ignored, it seemed as though a very simple, testable model, derived from a few starting assumptions about rate maximization and constraints on foraging, could actually predict how an animal responded in an experiment. It's hard to overstate the excitement at the time.

Shortly afterward, Richard Cowie's quantitative test of the patch model appeared (Cowie 1977), and the first use of stochastic dynamic modeling predicted the trade-off between sampling and exploitation of a new environment (Krebs et al. 1978). It really looked as though a new quantitative theoretical framework for behavioral ecology had been born out of the ideas of MacArthur and Pianka (1966), Emlen (1966), Charnov (1976a, 1976b), and Parker (1978). By the time Stephens and Krebs published their monograph on foraging theory in 1986, the optimal foraging industry had been in full swing for a nearly a decade, and large numbers of laboratory and field studies seemed to underline the power of the theory.

But by no means everyone was convinced. At the Animal Behavior Society symposium held in Seattle in 1978 (Kamil and Sargent 1981), Reto Zach and Jamie Smith concluded their article "Optimal Foraging in Wild Birds" as follows: "Most feeding problems in the wild are complex and it is therefore difficult to define optima. Furthermore, optimal foraging theory cannot be tested conclusively. Optimal foraging theory is thus of limited use only. Fortunately there are other promising approaches to the developmental and comparative analysis of foraging skills."

By 1984, the time of the seminal "Brown Symposium" (Kamil et al. 1987) (referring of course to the eponymous university, not the color of the resulting book—which was green), not only had the field of optimal foraging theory become broader, but Russell Gray and John Ollason had developed excoriating critiques of the whole enterprise. Russell Gray summarized his views in these terms: "Despite its popularity, OFT faces a long list of serious problems. . . . These problems are generally downplayed within the OFT literature and the validity of the optimality assumption is taken on faith. This faith does not seem to be particularly useful." John Ollason was equally, if not more, astringent, commenting that when predictions of OFT and data coincide, "a labyrinthine tautology has been constructed that is based on assumption piled on assumption."

With the benefit of twenty years' hindsight, who was right? Was it the enthusiastic optimists or the cynical critics? The answer is, "a bit of both." On one hand, there is no doubt that the initial hopes for a simple, all-embracing theory that paid little attention to behavioral mechanisms were soon dashed. On the other hand, as the research has matured, important insights into behavior and ecology have been fostered by optimal foraging theory. Indeed, many important questions have been asked because of optimality thinking, and asking the right questions is the basis of successful science. Furthermore, the breadth of impact of foraging theory across many disciplines is remarkable.

This book shows how the field has broadened and deepened. Simplicity and coherence have been left behind, but diversity, richness of texture, and understanding have been gained. The tentacles of foraging theory, in its broadest sense, have extended to form links with neuroethology, behavioral economics, life histories, animal learning, game theory, and conservation biology.

Perhaps most important of all, the simplistic approach to building and testing models of behavior that characterized some of the early foraging literature has been replaced by a more sophisticated comparative analysis of models. Take risk sensitivity as an example. Caraco's early experimental work (Caraco et al. 1980) and Stephens's theoretical formulation (Stephens 1981) provided a beguilingly simple combination of theory and data: animals should be risk prone when their expected energy budget is negative and risk averse when it is positive. Houston and McNamara (1982) subsequently extended Stephens's idea, using stochastic dynamic models, to predict changes in risk sensitivity depending on both energetic state and time horizon. The theory became more sophisticated, but did not encompass mechanisms of decision making: its predictions were based on arguments about adaptation. But when mechanisms were considered, it turned out that the purely functional approach embodied in risk sensitivity theory was not the one that most successfully accounted for the experimental data.

Kacelnik and Bateson (1997) compared the predictions of four kinds of models: risk sensitivity theory, short-term rate maximization, scalar utility theory, and associative learning theory. The first kind of model is based on functional arguments; the second is descriptive, predicting choices from regularities previously observed in data; the third derives from the psychophysics of perception; and the fourth examines the consequences of established principles of animal learning.

Although some of the early studies seemed to confirm the predictions of risk sensitivity theory (namely, experimental animals reversed their preference for variance depending on manipulations of their energy reserves), this result was not robust. The single most reliable phenomenon is that, when averages are equal, animals prefer variable over fixed *delays* to food and fixed over variable *amounts* of food. In other words, they are risk prone for delay to reward and risk averse for amount. Risk sensitivity theory does not explain or predict this observation, while scalar utility theory predicts both effects at a qualitative level. None of the theories is fully successful in terms of quantitative predictions: each predicts some results and fails to predict others. Furthermore, the different models are as interesting in the ways in which they fail as they are in their successes.

This example illustrates several points. First, in its more mature phase, foraging theory has moved from simply testing the predictions of one kind of model to comparing the ability of a range of models to explain the data. Second, while it is still important conceptually to distinguish accounts of behavior based on functional arguments from those based on causal mechanisms, the interplay between these two kinds of explanations benefits both approaches. On one hand, without the input from functional modeling (as embedded in risk sensitivity theory), the question of preference for variance would not have been examined in the light of mechanisms. But on the other hand, if one of these mechanistic models turns out to be better at predicting behavior, the functional theory needs to be reexamined. For instance, earlier risk sensitivity models may have incorrectly identified the selective forces that act on animal risk taking. The success of scalar utility theory suggests that selection may have favored a logarithmic encoding of stimulus intensity to allow the animal to cope with a wide range of stimuli, which leads automatically to preference for variable delays and fixed amounts.

This excellent volume sets the stage for the next decade of research, as a result of which the field of foraging will no doubt have evolved and been transformed again.

John Krebs
Alex Kacelnik
Oxford, June 2006

Acknowledgments

From the editors:
The editors thank the authors for their patience and good will throughout the production of this volume. We thank Christie Henry for her advice and assistance, which—quite literally—made this project possible. We thank Todd Telander, who translated our figures from meaningless scrawls to a coherent and aesthetically pleasing whole. Finally, we thank Norma Roche for her careful and competent copyediting.

Chapter 1
The overview we present here has been shaped by discussions with many colleagues over the past several decades. Joel Heath and Grant Gilchrist took the eider videos to which the reader is referred in the opening passage, and Joel Heath maintains the Web site on which they are displayed. We thank Dave Moore and Jon Wright for discussion on particular points.

Chapter 2
I thank Tom Getty, Colleen McLinn, and Ron Ydenberg for comments on the manuscript. The National Science Foundation (IBN-0235261) and the National Institute of Mental Health (RO1-MH64151) supported my research during the preparation of this manuscript.

Chapter 3
We would like to thank Robert Gegear and Peter Cain for their many helpful comments on the manuscript and Jennifer Hoshooley for valuable discussion. Preparation of this chapter was supported by grants from the Natural Sciences and Engineering Research Council of Canada.

Chapter 4
We would like to thank Al Riley, George Barlow, Seth Roberts, Andy Suarez, Karen Nutt, and the Animal Behavior lunch group for helpful comments and discussions on early drafts of this manuscript. Thank you also to the editors of this volume for many helpful comments.

Chapter 5
We thank the editors for inviting our participation and for their constructive criticisms. Many individuals helped shape our current views on foraging, especially Joel Brown, Richard Holmes, Robert Holt, William Karasov, Burt Kotler, Carlos Martínez del Rio, Douglas Levey, Timothy Moermond, and Mary Willson. We especially thank Carlos Martínez del Rio, Brenda Molano-Flores, Dennis Whelan, and Mary Willson for reviewing previous drafts.

Chapter 6
I am deeply grateful to three scientists for all that they have taught me: Tom Caraco, John Krebs, and Tony Parsons.

Chapter 7
We thank Dave Stephens and Ron Ydenberg for valuable comments and editing help. AB was supported by a grant from the Swedish Research Council, VR. ÅL warmly acknowledges the ever-inspiring collaboration with Thomas Alerstam and the Migration Ecology Research Group in Lund.

Chapter 8
I am deeply grateful to colleagues, friends, and students in the Behavioral Ecology Research Group and the Centre for Wildlife Ecology at Simon Fraser University for their commitment to collegial scientific inquiry and for their parts in the parade of ideas, discovery, and natural history that makes working there so endlessly fascinating.

Chapter 9
I thank Earl Werner, Shannon McCauley, Luis Schiesari, Mara Savacool Zimmerman, Mike Fraker, Kerry Yurewicz, Steve Lima, Annie Hannan, Uli Reinhardt, Graeme Ruxton, Tim Caro, Dan Blumstein, Anders Brodin, Will Cresswell, Ron Ydenberg, and Dave Stephens for helpful comments on the manuscript, and Robert Gibson and David McDonald for help in locating a reference.

Chapter 10
We thank E. A. Marschall, K. M. Passino, R. Ydenberg, D. Stephens, and students in our graduate course in behavioral ecology for comments on the manuscript.

Chapter 11
We thank Chris Whelan and the editors for very helpful comments on the chapter. RDH thanks NSF, NIH, and the University of Florida Foundation for support, and Burt Kotler, Joel Brown, Tom Schoener, Doug Morris, Per Lundberg, and John Fryxell for stimulating conversations on foraging. TK thanks NSF for a graduate research fellowship

Chapters 12 and 13
We are grateful to our many colleagues and students over the years whose discussions, ideas, and insights contributed so very much to our own ideas and worldview. These include Zvika Abramsky, Leon Blaustein, Sasha Dall, Mike Gaines, Bob Holt, Bill Mitchell, Doug Morris, Ken Schmidt, and Tom Vincent. We are especially grateful to our teacher and mentor, Mike Rosenzweig.

Chapter 14
Thanks to Joel Brown and Dave Stephens for substantive help with the manuscript. Thanks to Peter Raven and the Missouri Botanical Garden for sabbatical year hospitality. Thanks to my colleagues Zvika Abramsky, Burt Kotler, and Yaron Ziv for continual intellectual stimulation.

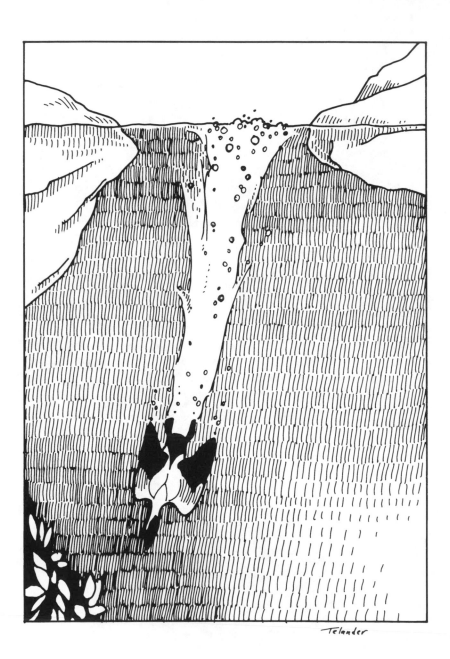

Foraging: An Overview

Ronald C. Ydenberg, Joel S. Brown, and David W. Stephens

1.1 Prologue

Hudson Bay in winter is frozen and forbidding. But, at a few special places where strong tidal currents are deflected to the surface by ridges on the seafloor, there are permanent openings in the ice, called polynyas, that serve as the Arctic equivalent of desert oases. Many polynyas are occupied by groups of common eiders. When the current in the polynya slackens between tide changes, these sea ducks can forage, and they take advantage of the opportunity by diving many times. With vigorous wing strokes they descend to the bottom, where they search though the jumbled debris, finding and swallowing small items, and occasionally bringing a large item such as an urchin or a mussel clump to the surface, where they handle it extensively before eating or discarding it. (Readers can take an underwater look at a common eider diving in a polynya at www.sfu.ca/eidervideo/. These videos were made by Joel Heath and Grant Gilchrist at the Belcher Islands in Hudson Bay.)

This foraging situation presents many challenges. Eiders must consume a lot of prey during a short period to meet the high energy demand of a very cold climate. Most available prey are bulky and of low quality, and the ducks must process a tremendous volume of material to extract the energy and nutrients they need. They must also keep an eye on the clock, for the strong currents limit the available foraging time.

Throughout the winter, individual ducks may move among several widely separated polynyas or visit leads in the pack ice when the wind creates openings. Foxes haunting the rim of the polynyas and seals in the water below create dangers that require constant wariness. In this unforgiving environment, the eider must meet all these challenges, for in the Arctic winter, a hungry eider is very soon a dead eider.

1.2   Introduction

Twenty years ago, Dave Stephens and John Krebs opened their book *Foraging Theory* (1986) with an example detailing the structure of a caddisfly web. The example showed how the web could be analyzed as a trap carefully constructed to capture prey. The theme of the book was that foraging behavior could also be looked at as "well-designed." In it, they reviewed the basic theoretical models and quantitative evidence that had been published since 1966. In that year, a single issue of *The American Naturalist* carried back-to-back papers that may fairly be regarded as launching "optimal foraging theory." The first, by Robert MacArthur and Eric Pianka, explored prey selection as a phenomenon in its own right, while the second, by John Merritt Emlen, was focused on the population and community consequences of such foraging decisions. This book gives an overview of current research into foraging, including the offspring of both these lines of investigation.

The reader will discover that foraging research has expanded and matured over the past twenty years. The challenges facing common eiders in Hudson Bay symbolize how the study of foraging has progressed. Some of these problems will be familiar to readers of *Foraging Theory* (which items to eat?), but their context (diving) requires techniques that have been developed since 1986. Eiders work harder when they are hungry, so their foraging is state-dependent. The digestive demand created by bulky prey and the periodicity in prey availability mean that their foraging decisions are time-dependent (dynamic). Predators are an ever-present menace, and eiders may employ variance-sensitive tactics to help meet demand. Furthermore, the intense foraging of a hundred eiders throughout an Arctic winter in a small polynya must have a strong influence on the benthic community as these prey organisms employ their own strategies to avoid becoming food for eiders.

All these topics have been developed greatly since 1986. This book argues that foraging has grown into a basic topic in biology, worthy of investigation in its own right. Emphatically, it is not a work of advocacy for a particular approach or set of models. The enormous diversity of interesting foraging problems across all levels of biological organization demands many different approaches, and our aim here is to articulate a pluralistic view. However, foraging research was originally motivated by and organized around optimality models and the ideas of behavioral ecology, and for that reason, we take Stephens and Krebs's 1986 book as our starting point. We aim to show that the field has diversified enormously, expanding its purview to look at topics ranging from lipids to landscapes.

A colleague recently asked when we would finally be able to stop testing the patch model. Our answer was that there is no longer a single patch model, any more than there is a single model of enzyme kinetics. The patch model and the way it expresses the concept of diminishing returns is so useful that it plays a role in working through the logic of countless foraging contexts. Hence, it often helps in developing hypotheses—which is what we are really interested in testing. In exactly analogous ways, working scientists everywhere use the conceptual structure of their discipline to develop and test hypotheses. If their discipline is healthy, it expands the concepts and methods it uses, just as we feel has been happening in foraging research.

We have aimed the text at a hypothetical graduate student at the outset of her career, someone reading widely to choose and develop a research topic. This book is best used in an introductory graduate seminar or advanced undergraduate reading course, but should be useful to any biologist aiming to increase his familiarity with topics in which foraging research now plays a role. We begin with a chapter-by-chapter comparison with Stephens and Krebs (1986) to give a brief overview of how the field of foraging research has developed over the past two decades, identify the main advances, and introduce students to the basics.

1.3   A Brief History of Optimal Foraging Theory

Interest by ecologists in foraging grew rapidly after the mid-1960s. Scientists in areas such as agricultural and range research already had long-standing interests in the subject (see chap. 6 in this volume). Entomologists, wildlife biologists, naturalists, and others had long been describing animal diets. So what was new? What generated the excitement and interest among ecologists?

We believe that the answer to this question is symbolized by a paper published by the economist Gordon Tullock in 1971, entitled "The coal tit as a careful shopper." Tullock had read the studies of Gibb (1966) on foraging by small woodland birds on insects, and he suggested in his paper that one could apply microeconomic principles to understand what they were doing. (We do not mean to suggest that Tullock originated this approach, merely that his paper clearly expressed what many ecologists were thinking.) The idea of using an established concept set to investigate the foraging process from first principles animated many ecologists. This motivation fused with developing notions about natural selection (Williams 1966) and the importance of energy in ecological systems to give birth to "optimal foraging theory" (OFT). The new idea of optimal foraging theory was that feeding strategies evolved by natural selection, and it was a natural next step to use the techniques of optimization models.

Although the terminology differs somewhat among authors, the elements of a foraging model have remained the same since the publication of Stephens and Krebs's book. At their core, models based on optimal foraging theory possess (1) an objective function or goal (e.g., energy maximization or starvation minimization), (2) a set of choice variables or options under the control of the organism, and (3) constraints on the set of choices available to the organism (set by limitations based on genetics, physiology, neurology, morphology, and the laws of chemistry and physics). In short, foraging models generally take the form, "Choose the option that maximizes the objective, subject to constraints." A specific case may be matched with a detailed model (e.g., Beauchamp et al. 1992), or a model may conceptualize general principles to investigate the logic underlying foraging decisions, such as whether an encountered item should be eaten or passed over in favor of searching for a better item.

We now regard the rubric "optimal foraging theory," used until the mid-1980s, as unfortunate. Although optimality models were important, they were not the only component of foraging theory, and the term emphasized the wrong aspects of the problem. "Optimality" became a major focus and entangled those interested in the science of foraging in debates on philosophical perspectives and even political stances, which, needless to say, did more to obscure than to illuminate the scientific questions. A few key publications will enable the reader to appreciate this history and the intensity of debate. Stephens and Krebs (1986) reviewed the issues up to 1986 (see Pyke et al. 1977; Kamil and Sargent 1981; and Krebs et al. 1983 for earlier reviews). Perry and Pianka (1997) provided a more recent review, and showed that while the titles of published papers dropped the words "optimal" and "theory" after the mid-1980s, foraging remained an active area of research. Sensing opprobrium from their colleagues, scientists evidently began to shy away from identifying with optimal foraging theory. If the reader doubts that this was a real factor, he or she should read the article by Pierce and Ollason (1987) entitled "Eight reasons why optimal foraging theory is a complete waste of time." In a more classic (and subtle) vein, Gould and Lewontin (1979) criticized the general idea of optimality in their famous paper entitled "The spandrels of San Marco and the Panglossian paradigm: A critique of the adaptationist programme" (later lampooned by Queller [1995] in a piece entitled "The spaniels of St. Marx"). Many other publications have addressed these and related themes.

A persistent source of confusion has been just what "optimality" refers to. Critics assert that it is unreasonable to view organisms as "optimal," using biological arguments such as the claim that natural selection is a coarse mechanism that rarely has enough time to perfect traits, or that important features of organisms may originate as by-products of selection for other traits. These arguments graded into ideological stances, such as claims that use of "optimality" promotes a worldview that justifies profound socioeconomic inequalities. It is difficult to disentangle useful views in this literature from overheated rhetoric, a problem exacerbated by careless terminology and glib applications on both sides. Our view is that most of this debate misses the point that "optimality" should not be taken to describe the organisms or systems investigated. "Optimality" is properly viewed as an investigative technique that makes use of an established set of mathematical procedures. Foraging research uses this and many other experimental, observational, and modeling techniques.

Nor does optimality reasoning require that animals perform advanced mathematics. As an analogue, a physicist can use optimality models to analyze the trajectories that athletes use to catch a pass or throw to a target. However, no one supposes that any athlete is performing calculus as he runs down a well-hit ball (see section 1.10 below).

The word "theory" was also a stumbling block for many ecologists, who regarded it as a sterile pursuit with little relevance to the rough-and-tumble reality of the field. Early foraging models were very simple, and their explanatory power in field situations may have been oversold (see, e.g., Schluter 1981). Ydenberg (chap. 8 in this volume), for example, makes clear the limitations of the basic central place foraging model put forward in 1979. But, informed by solid field studies (e.g., Brooke 1981), researchers identified the holes in the model and developed theoretical constructs to address them (e.g., Houston 1987). Errors in the formulation of the basic model were soon corrected (Lessells and Stephens 1983; Houston and McNamara 1985). This historical perspective shows how misrepresentative are oft-repeated claims such as, "Empirical studies of animal foraging developed more slowly than theory" (Perry and Pianka 1997). As in most other branches of scientific inquiry, theory and empirical studies proved, in practice, to be synergistic partners. Their partnership is flourishing in foraging research, and theory and empiricism in both laboratory and field are important parts of this volume.

If the basics of foraging models have remained unchanged since the publication of Stephens and Krebs's book (1986), the range and sophistication of objective functions, choice variables, and constraint sets has expanded. Mathematics has spawned new tools for formulating and solving foraging models. And advances in computing have permitted ever more computationally intensive models. The emphasis of modeling has expanded from analytic solutions to include numerical and simulation techniques that require mind-boggling numbers of computations. The last two decades have seen a pleasing lockstep among empirical, modeling, mathematical, and computational advances.

New concepts have also emerged. Some of the biggest conceptual advances in foraging theory have come from the realization that foragers must balance food and safety (see chaps. 9, 12, and 13 in this volume), an idea that ecologists had just begun to consider when Stephens and Krebs published their book in 1986. Box 1.1 outlines the history of this important idea.

BOX 1.1   Prehistory: Before Foraging Met Danger
Peter A. Bednekoff

The theory of foraging under predation danger took time to formulate. Broadly speaking, students of foraging hardly ever addressed the effects of predation during the 1970s, but they gave increasing attention to predation in the 1980s, and predation enjoyed unflagging interest through the 1990s. From the start, behavioral ecologists took the danger of predation seriously; but they treated foraging and danger separately. In the first edition of *Behavioral Ecology* (Krebs and Davies 1978), the chapter on foraging (Krebs 1978) is immediately followed by one dealing with predators and prey (Bertram 1978), with another chapter on antipredator defense strategies not far behind (Harvey and Greenwood 1978). The thinking seems to have been that these phenomena operated on different scales, such that danger might determine where and when animals fed, but energy maximization ruled how they fed (Charnov and Orians 1973; Charnov 1976a, 1976b). This was a useful scientific strategy: it was important to test whether energetic gain affected foraging decisions before testing whether energetic gain and danger jointly affected foraging decisions. We probably can separate foraging from some kinds of activities. For example, male manakins may spend about 80% of their time at their display courts on leks (Théry 1992). Male manakins probably need to secure food as rapidly as possible when off the lek and to display as much as possible when on the lek. Therefore, foraging and displaying are separate activities. Survival, however, is a full-time job. Animals cannot afford to switch off their antipredator behavior. Because

(Box 1.1 continued)

trade-offs between danger and foraging gain can occur at all times and on all scales, the effects of danger can enrich all types of foraging problems.

A more subtle difficulty may have delayed the integration of foraging and danger: the two models that dominated early tests of foraging theory, the diet and patch models, do not readily suggest ways to integrate danger (see Lima 1988b; Gilliam 1990; Houston and McNamara 1999 for later treatments). Several graphical models dealt with predation and other aspects of foraging (Rosenzweig 1974; Covich 1976) and one chapter juxtaposed diet choice and antipredator vigilance models, both important contributions made by Pulliam (1976). Although the pieces seem to have been available, integration did not happen quickly. Even the early experimental tests treated danger as a distraction rather than a matter of life and death (Milinski and Heller 1978; Sih 1980). These studies would have reached similar conclusions if they had considered competitors rather than predators.

The first mature theory of foraging and predation concentrated on habitat choice and did not consider the details of foraging within habitats (Gilliam 1982). This theory assumed that animals grew toward a set size with no time limit. It showed that animals should always choose the habitat that offers the highest ratio of growth rate, g, to mortality rate, M. In order to avoid potentially dividing by zero, Gilliam expressed his solution in terms of minimizing the mortality per unit of growth, so we call this important result the mu-over-g rule. Departures from the basic assumptions lead to modifications of the M/g rule. This rule is a special case of a more general minimization of

$$\frac{M + r - \frac{b}{v}}{g},$$

where r is the intrinsic rate of growth for the population, b is current reproduction, and V is expected future reproduction (Gilliam 1982; Werner and Gilliam 1984). The familiar special case applies to juveniles in a stable population: juveniles are not yet reproducing, so b is zero, and the population is stable, so its growth rate, r, is also zero (Gilliam 1982; Werner and Gilliam 1984). Gilliam never published this work from his dissertation, but Stephens and Krebs (1986) cogently summarized the special case. Although the M/g rule is incomplete for various situations (Ludwig and Rowe 1990; Houston et al. 1993), it is surprisingly robust (see Werner and Anholt 1993). Modified versions may be solutions for problems that do not superficially

(Box 1.1 continued)

resemble the one analyzed by Gilliam (Houston et al. 1993), and Gilliam's M/g criterion may reappear from analysis of specific problems (e.g., Clark and Dukas 1994; see also Lima 1998, 221–222, and chap. 9 in this volume).

In hindsight, we can see that various studies in the early 1980s pointed to the pervasive effects of danger on foraging (e.g., Mittelbach 1981; Dill and Fraser 1984; Kotler 1984), but these effects were not immediately integrated into the body of literature on foraging. Besides Gilliam's studies, Stephens and Krebs mentioned only one other study of foraging under predation danger, which found that black-capped chickadees sacrifice their rate of energetic gain in order to reduce the amount of time spent exposed at a feeder (Lima 1985a). This influential book seems to have just preceded a flood of results. In the mid-1980s, students of foraging found that danger influences many details of foraging and other decisions made by animals (Lima and Dill 1990). The general framework has continued to be productive and currently shows no sign of slowing its expansion (see Lima 1998).

A second profoundly important concept is "state dependence," the idea that the tactical choices of a forager might depend on state variables, such as hunger or fat reserves. This concept developed in ecology in the late 1970s and 1980s and is described in sections 1.8 and 1.9 below. Stephens and Krebs (1986) used the idea of state dependence in two chapters and anticipated the still-growing impact of this concept.

A third important conceptual advance not considered at all in Stephens and Krebs (1986) lies in social foraging games and the consequences of foraging as a group. Foraging games between predator and prey represent an extension of both game theory and foraging theory. Here the objective function of the prey takes into account its own behavior as well as that of the predator, and the predator's objective function considers the consequences of its behavior and that of its prey. We anticipate that these models will find application in a variety of basic and applied settings.

1.4   Attack and Exploitation Models

The second chapter of Stephens and Krebs (1986) develops the foundational models of foraging, the so-called "diet" and "patch" models. The treatment is clear and rigorous, and the beginning student is encouraged to use their chapter as an excellent starting point. In addition to the classic review articles listed above, one can find recent reviews of the published tests of these models in Sih and Christensen (2001; 134 published studies of the diet model) and Nonacs (2001; 26 studies of the patch model).

The significance of these two models lies in the types of decisions analyzed. The terms "diet" and "patch" are misnomers in the sense that the decisions are more general than choices about food items or patch residence time. Stephens and Krebs (1986) termed these models the "attack" and "exploitation" models to underscore this point, but these terms have never caught on.

The diet model analyzes the decision to attack or not to attack. The items attacked are types of prey items, and the forager decides whether to spend the necessary time "handling" and eating an item or to pass it over to search for something else. The model identifies the rules for attack that maximize the long-term rate of energy gain. Specifically, the model predicts that foragers should ignore low-profitability prey types when more profitable items are sufficiently common, because using the time that would be spent handling low-profitability items to search for more profitable items gives a higher rate of energy gain. The diet model introduced the principle of lost opportunity to ecologists, who have since used the concept in many other settings (e.g., "optimal escape"; Ydenberg and Dill 1986). The diet model considers energy gain, but the same rules apply in non-foraging situations of choice among items that vary in value and involvement time.

The patch model asks how much time a forager should invest in exploiting a resource that offers diminishing returns before moving on to find and exploit the next such resource. The "patches" are localized concentrations of prey between which the predator must travel, and the rule that maximizes the overall rate of energy gain is to depart when more can be obtained by moving on. In this sense, the patch model also considers lost opportunity, but its real value was to introduce the notion of diminishing returns. If the capture rate in a patch falls as the predator exploits it—a general property of patches— then the maximum "long-term" rate of gain (i.e., over many patch visits) is that patch residence time at which the "marginal value" (i.e., the intake rate expected over the next instant) is equal to the long-term rate of gain using that patch residence rule. Because diminishing returns are ubiquitous, this so-called "marginal value theorem" (Charnov 1976b) can be used in many situations. For example, we can think of eiders as "loading" oxygen into their tissues prior to a dive. The rate at which they can do so depends on the difference in partial pressure between the tissues and the atmosphere, and hence the process must involve diminishing returns. How much oxygen they should load depends on the situation, and the "patch" model gives us a way to analyze the problem (Box 1.2).

BOX 1.2 **Diving and Foraging by the Common Eider**
Colin W. Clark

Common eiders and other diving birds capture prey underwater during "breath-hold" diving. During pauses on the surface between dives, they "dump" the carbon dioxide that has accumulated in their tissues and "load" oxygen in preparation for the next dive. (Heat loss may also be a significant factor in some systems, but is not considered here.) Figure 1.2.1 schematically portrays a complete dive cycle. This graph shows a slightly offbeat version of the marginal value theorem.

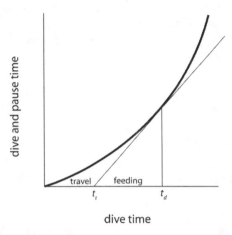

Figure 1.2.1. The relationship between dive time (composed of round-trip travel time to the bottom plus feeding time on the bottom) and the total amount of time required for a dive plus subsequent full recovery (pause time). The relationship accelerates because increasingly lengthy pauses are required to recover after longer dives. Small prey are consumed at rate c during the feeding portion of the dive. The problem is to adjust feeding time ($t_d - t_t$) to maximize the rate of intake over the dive as a whole. The tangent construction in the figure shows the solution. The reader can check the central prediction of this model by redrawing the graph to portray dives in deeper water (i.e., make travel time longer). The repositioned tangent will show that dives should increase in length if energy intake is to be maximized.

A dive consists of round-trip travel time to the bottom (t_t) and time on the bottom spent finding and consuming small mussels (feeding time). Travel time is a constraint, and it is longer in deeper water or, as in the eider example in the prologue, faster currents. Dive time (t_d) consists of travel time plus feeding time. Dive-cycle time consists of dive time plus the pause time on the surface between dives (t_s). How should an eider organize its dives to maximize the feeding rate?

(Box 1.2 continued)

Let

> $F_s(t_s) = O_2$ intake from a pause of length t_s,
> $F_d(t_d) = O_2$ depletion from a dive of length t_d,
> $Y(t_d)$ = energy intake (number of mussels times energy per mussel)
> from a dive of length t_d.

The average rate of food intake is thus

$$\frac{Y(t_d)}{(t_d + t_s)} \qquad\qquad (1.2.1)$$

which is maximized subject to the condition that oxygen intake must equal oxygen usage, so

$$F_d(t_d) = F_s(t_s). \qquad (1.2.2)$$

To solve this problem graphically, first solve equation (1.2.2) for t_s as a function of t_d:

$$t_s = \Phi(t_d). \qquad (1.2.3)$$

Here $\Phi(t_d)$ represents the pause time required to recover oxygen reserves after a dive of length t_d. One would expect that $\Phi'(t_d)$ would increase with t_d. This is the source of the diminishing returns in this model—increasingly longer times are required to recover after longer dives. An attractive feature of this model is that it requires an estimate of $\Phi(t_d)$, which can be obtained from observational data, rather than the separate functions F_s and F_d.

Suppose that $Y(t_d) = 0$ if $t_d < t_t$ (no food can be consumed if the dive is not long enough to travel to the bottom and back), and that if $t_d > t_t$, then

$$Y(t_d) = c \cdot (t_d - t_t), \qquad (1.2.4)$$

meaning that energy is ingested at the rate c during the portion of the dive spent feeding on the bottom. The optimization problem is to adjust the length of the dive $(t_d; t_d > t_t)$ to maximize the rate of energy gain, which is

$$\frac{c \cdot (t_d - t_t)}{(t_d + \Phi(t_d))}. \qquad (1.2.5)$$

Write $\psi(t_d) = t_d + \Phi(t_d) =$ total dive time plus pause time. Then maximizing equation (1.2.5) is equivalent to adjusting t_d to minimize $\psi(t_d)/(t_d - t_t)$. This is shown in the graph, and the optimal dive time is easily found.

(Box 1.2 continued)

The model predicts that dive and surface time both increase with travel time (dive depth), that the level of oxygen loading increases with depth, and that the optimal dive length is independent of resource quality (c).

While these simple models do not apply universally like Newton's laws, they are foundational, and it is hard to overstate their importance in the logical development of foraging theory. The patch model may in fact be the most successful empirical model in behavioral ecology; its basic predictions have been widely confirmed, at least qualitatively, although it is not always clear that the logic of the patch model correctly describes the situation being modeled. Stephens and Krebs (1986) considered mainly long-term average rate maximizing, but investigators have since shown that animals sometimes behave as "efficiency" maximizers (Ydenberg 1998). The links between efficiency maximizing and rate-maximizing currencies have interesting implications for energy metabolism and workloads (chap. 8 in this volume explores this topic further).

The simplicity of both the diet and patch models is deceptive, and the beginning student will have to work hard to master their subtleties. They show that the modeler's real art is not mathematics per se (after all, the math is elementary), but rather in distilling the essentials from so many and such varied biological situations.

1.5   Changed Constraints

Stephens and Krebs devoted their third chapter to what they called "changed constraints": relatively minor modifications of the basic models, including simultaneous prey encounter, central place foraging, nutrient constraints, and discrimination constraints. They could devote an entire chapter to minor modifications because, at the time, foraging theory was a fairly unitary field. Contemporary foraging research, as this volume demonstrates, finds itself addressing areas from neurobiology to community ecology, and it is no longer possible to imagine a cohesive chapter on minor modifications. Nonetheless, many of the issues raised in that chapter are important in other ways. To illustrate this point, we discuss the problem of sequential versus simultaneous prey encounter in some detail here.

Animals frequently encounter food items simultaneously: bees encounter groups of flowers, monkeys encounter many fruits on a tree, and so on. Such situations have elements of both patch and diet problems, and disentangling the two was the early focus of "simultaneous encounter" models.

Although the simultaneous encounter problem arose as a complaint about the simplistic assumptions of early foraging models, it has developed deep connections with other approaches to behavior. When animals in experiments must choose between simultaneously presented options that differ in delay and amount, psychologists have found a strong preference for immediacy that appears to fly in the face of long-term energy maximization (Ainslie 1974; Green et al. 1981; Mazur and Logue 1978; Rachlin and Green 1972). An intriguing aside is that psychologists view this impulsiveness as a model of several important problems in human behavioral control (Rachlin 2000). For example, children who are better at waiting for benefits perform better in school (Mischel et al. 1989); while phenomena such as addiction and suicide are seen as failures of impulse control.

Foraging theorists have reasoned that delayed food is worth less than immediate food because (for example) an interruption might prevent an animal from collecting a delayed food item (Benson and Stephens 1996; McNamara and Houston 1987a); in other words, delayed food items are "discounted." The difficulty with this approach is that there is a wide gulf between plausible discounting rates and observed animal impulsiveness. Reasoning from first principles analogous to the arguments for animal discounting, economists assume that human monetary discounting hovers in the neighborhood of 4% *per year* (Weitzman 2001). Experimental studies of impulsivity with pigeons, however, require a discount rate of up 50% *per second*. This large difference (8 orders of magnitude!) makes discounting unlikely to be a general explanation for animals' strong preference for immediacy.

In an alternative approach, Stephens and colleagues (Stephens 2002; Stephens and Anderson 2001; Stephens and McLinn 2003) have argued that impulsive choice rules exist because they perform well (that is, achieve high long-term intake rates) in sequential choice situations. This idea is called the ecological rationality hypothesis. According to this view, animals perform poorly when we test them in simultaneous choice situations because they misapply rules that are more appropriate for sequential choice problems. Impulsiveness is not a consequence of economic forces that discount delayed benefits, but a consequence of a rule that achieves high long-term gains in naturally occurring choice situations.

The simultaneous encounter problem is also linked to the problem of understanding the value of information in foraging (Mitchell 1989; see also chap. 2 in this volume). A forager can exploit a simultaneously encountered set of resources in several ways, in the same way that the famous traveling salesman of operations research can choose several routes through a collection of cities, only one of which maximizes the profitability of the trip. A nectar-collecting bee may use the same flower patches every day, and we would expect it to use them in a consistent order that is sensitive to both their relative qualities and their arrangement in space. Within foraging theory, this orderly use and reuse of a spatial array of resources is known as "traplining" and has been studied in nectivorous birds (Gass and Garrison 1999; Kamil 1978), bees (Thomson et al. 1997; Williams and Thomson 1998), and frugivorous monkeys (Janson 1998). However, because the world changes continually, unpredictably, and subtly, we can be sure that a traplining forager is obtaining not only food, but also information about the current state of the world. What is not understood is whether this information potential should affect the route. Understanding how animals collect and use information about resources in this and other foraging situations is a fundamental problem in foraging behavior.

1.6   Information

The classic diet and patch models assumed that foragers had perfect knowledge of the model's parameters. Stephens and Krebs (1986) called this the "complete information assumption." While useful as an analytic simplification, it clearly cannot be generally true. Foraging theorists first developed incomplete information models for patch exploitation scenarios (Green 1980; Iwasa et al. 1981; McNamara 1982; Oaten 1977), the idea being that experience in the first few moments of a patch visit can provide information about patch quality. The general mathematical problem of optimal behavior when patch quality is uncertain is difficult (McNamara 1982), but modelers have made progress by considering simpler special cases. For example, in a series of elegant experiments, Lima (Lima 1983, 1985b) considered a case in which patches were either completely empty or completely full. In this case, the first prey capture within the patch tells the forager that this patch is one of the better, full types. Another information problem concerns foragers that use a number of habitats whose qualities vary so that the forager cannot be sure at any given time which is best; sampling (i.e., making a visit) is required (Devenport et al. 1997; Krebs and Inman 1992; Shettleworth et al. 1988; Tamm 1987).

At first, the "problem" of incomplete information seems straightforward and unitary (animals can't possibly have complete information about every relevant feature of their environment). But the foraging environment can be uncertain in a virtually unlimited number of ways. Moreover, animals can acquire information via many channels and methods. This complexity means that there is no single solution to "the problem of incomplete information."

There is, however, a common approach to all incomplete information problems (statistical decision theory; DeGroot 1970). In this approach, a statistical distribution of states represents the forager's *prior* information. The forager's actions and subsequent experience provide an updated distribution of states via Bayes's theorem (called the *posterior* distribution), which can be used to choose better behavioral alternatives. The central questions are (1) whether, and how, the forager should change its behavior to obtain an updated distribution of states, and (2) how the forager should act in response to changes in updated information about states. An answer to question 2—what would you do with the information if you had it?—is required before we can answer question 1—should you change your behavior to obtain information? Stephens reviews several examples of this approach in chapter 2 of this volume.

Although the basic theoretical issues surrounding information problems are clear, much remains to be done. Empiricists need to follow up the early experimental studies of tracking and patch sampling, and modelers need to incorporate empirical insights into new models. Within the field of foraging, workers with interests in information have been attracted to related problems such as learning, memory, and perception (see chaps. 3 and 4 in this volume), and it seems likely that we will have to look to these areas for progress in information problems. And there is a growing interest in information problems within behavioral ecology, spurred on by a long-standing interest in sexual signaling and other forms of communication (Dall and Johnstone 2002), that may reinvigorate interest in foraging information problems.

1.7 Consumer Choice

Stephens and Krebs's chapter 5, entitled "The economics of choice," considered situations in which foragers face trade-offs. In such situations, increasing the gain of one thing important for fitness (say, food intake) compromises the attainment of another (say, safety). The chapter provided a brief introduction to microeconomic consumer choice theory, which provides a framework for analyzing trade-off problems by assigning "utility" so that their value can be measured on a common scale. Animal psychologists had used this approach in operant conditioning experiments, with some success, to study the choices made by animals between, for example, different food types obtainable by varying amounts of bar pressing, or different delays to reinforcements of different sizes. Behavioral ecologists had far less success with this theoretical structure because it was difficult to express the fitness value of very different things (e.g., food and safety) in a common currency. When *Foraging Theory* was published in 1986, the "differing currencies" problem seemed formidable indeed.

The most satisfactory solution to the trade-off problem came from re-thinking the structure of optimization problems. In fact, Stephens and Krebs hinted at this solution in a section entitled "Trade-offs and dynamic optimization" (1986, 161; see also section 1.8 below), explaining that one can use dynamic optimization to study trade-offs because "it seems natural" to formulate trade-off decisions as functions of state variables.

A state variable describes a property or trait of a system, such as an organism or a social insect colony. The state might be hunger, size, or temperature, but it could be anything. The key is that behavior alters the future value of the state variable. The organism has a number of behavioral options, each of which has consequences for the state. These consequences are more easily measurable than the fitness value (i.e., cost or benefit) of a behavior. It is the state of the organism that is (eventually; see below) evaluated in fitness terms. State variable models provide the best means to resolve "differing currencies" problems and have been widely applied since 1986 (Houston and McNamara 1999).

1.8   Dynamic Optimization

Real-world foraging problems not only include uncertainties and trade-offs, but are also dynamic. For example, eiders in polynyas may accelerate their rate of work as the end of a foraging period approaches, and they may postpone recovery from diving in order to continue feeding while it is possible (e.g., Ydenberg and Clark 1989). Theorists have long recognized that analyzing such strategic options requires a dynamic model. The first edition of the authoritative compilation entitled *Behavioural Ecology* (Krebs and Davies 1978) devoted an entire chapter to dynamic optimization (McCleery 1978), as did Stephens and Krebs (1986), but behavioral ecologists avoided or ignored dynamic optimization because of the difficulty and mathematical abstruseness of the subject.

This all changed quite suddenly in the mid-1980s with the development of what are now called "dynamic state variable models," pioneered by Mark Mangel, Colin Clark, John McNamara, and Alasdair Houston (see Mangel and Clark 1988; Clark and Mangel 2000; Houston and McNamara 1999). Even nonmathematical biologists can easily understand dynamic state variable models and implement them (in principle at least) on small computers. In addition, dynamic state variable models solved the "differing currencies" problem described in the previous section.

A dynamic state variable model uses one or more state variables to describe a system at time t. For example, in a model by Beauchamp (1992), the state variables *w* and *n* represent the number of foragers and the size of the nectar reserve in a honeybee colony. The state variables change in value from one period to the next according to specified "dynamics": the nectar reserve increases as nectar is delivered and decreases as nectar is used by the hive for production. Colony reproduction and mortality determine the dynamics of forager number. The decision variable—in this case, the number of flowers visited per foraging trip—affects the change in the value of the state variables. As the bees visit more flowers, they deliver more nectar, but foragers also die at a higher rate. The objective is to calculate the strategy (the number of flowers the bees should visit as a function of *t*, *n*, and *w*) that maximizes colony size at the end of the summer foraging period, subject to the condition that the honey store is large enough to survive the winter

Dynamic state variable models accomplish this using the following algorithm. Computations begin in the last period, *T* (called "big T"). We can use the "terminal fitness function," the empirical relation between the values of the state variables and fitness, to assign a value to every possible outcome in the last period. Next, we can use the results from period *T* to find analogous values for the second to last period ($T - 1$). These calculations determine for every value of the state variables the decision that leads to highest expected fitness in the final period (T). The fitness value of that choice is calculated. Next, we use the results from period $T - 1$ to make the same calculations for period $T - 2$. We can use this backward induction method to derive the entire strategy—that is, the fitness-maximizing behavioral choice for every value of the state variable at every time. Small computers can solve even large problems quickly using this scheme. While the practicality of such extensive computation no longer poses a barrier, its interpretation does. In common with other numerical techniques, such as genetic algorithms, the solutions are specific to particular models. Generality is elusive, but does come with wide application and testing.

Dynamic state variable models represent an invaluable addition to foraging theory's toolkit, and they have already contributed to two fundamentally important advances. First, they have established the widely applicable notion of state dependence (Houston and McNamara 1999). Dynamic state variable models formalize the interaction between state and action, and thus connect short-term behavioral decisions to long-term fitness consequences. They also provide deep insights into the trade-off between food and safety because the differing effects of feeding and predation are accommodated within a conceptually unified framework. As this book shows, investigations of this classic trade-off represent one of the biggest advances that the field has made over the past twenty years.

1.9   Variance-Sensitive Foraging

In nature, random variation in prey size, handling time, the time between successive encounters with prey, and other components of the foraging process combine to create variance around the expected return of a particular foraging strategy. Stephens and Krebs treated this concept in their chapter 7. Naively, one might think that many small sources of variation would cancel each other out, but in fact, their combined effect is additive and can be quite large. For example, Guillemette et al. (1993) computed that the total daily intake of a wintering eider when feeding on small mussels could vary between about 800 and 1,800 kJ (coefficient of variation 12%). Eiders experience even more variance when they feed on large crabs (coefficient of variation about 23%). The theory tells us that foragers ought to be "sensitive" to this variance. Consider a situation in which a forager will starve if it gains less than some threshold amount. If the forager expects to gain more than required, it should prefer foraging choices that offer low variance because this strategy minimizes the probability of a shortfall. On the other hand, if the forager expects to gain less than it needs, a high-variance choice will increase the probability of survival. In general, variance sensitivity is expected whenever the (absolute value of) fitness effects of returns above and below the mean gain are unequal.

Variance sensitivity first came to the attention of foraging ecologists through an experiment carried out by Tom Caraco, Steve Martindale, and Tom Whitham and published in 1980. By 1986, several other ecologists had documented its occurrence, and theorists had begun to flesh out its theoretical basis. Experimental psychologists had long known of apparently similar phenomena from conditioning experiments in which animals choose between constant and variable rewards. Work on these issues since the publication of *Foraging Theory* in 1986 (see summary by Houston and McNamara 1999) has been steady, and a coherent framework has begun to emerge that makes sense of many of the experimental results. Major puzzles remain, however, such as the strong preference experimental animals show for "immediacy" (see section 1.5 above), but here a recent paradigm called "ecological rationality" (Stephens 2002; Stephens and Anderson 2001) suggests a way of looking at the problem that promises a solution with broad implications for the way that animals view their world.

In contrast to the attention that theorists and laboratory experimentalists have given to variance sensitivity, field ecologists have virtually ignored it. In general, they seem suspicious of the theory as somewhat contrived and have doubts about its applicability or relevance in nature. Clearly, we believe they are wrong. The growing strength of this approach suggests that fieldworkers should begin to examine its role in nature.

1.10   Rules of Thumb

Foraging researchers have long distinguished between the methods theoreticians use and the mechanisms animals use to make foraging decisions. For example, the patch model tells us that animals should leave patches when the derivative of the gain function equals the overall habitat rate of intake, but as we explained above, foragers do not determine their actions using higher mathematics. But if not, how do they do it? Animals could achieve the *behavior* predicted by the marginal value theorem in any of several ways that do not involve calculating derivatives. Students of foraging recognized this as the "rule of thumb" problem: modelers predict behavior with calculus and algebra, but animals use "rules of thumb" to make their foraging decisions. The idea is that the cost of more complex mechanisms means that a rule of thumb is better than a direct neurophysiological implementation of the theoretician's solution method: a rule of thumb is simpler, cheaper, and faster.

"Rules of thumb" research offers an apparently appealing connection between adaptationist models of traditional foraging and mechanistic studies of choice and decision making. In practice, this research program has not advanced very far over the last twenty years; after an early flurry (e.g., Cheverton et al. 1985; Kareiva et al. 1989), interest in rules of thumb has all but vanished among behavioral ecologists. We believe that this is because the paradigm—except for the basic notion that animals do not use the diet, patch, or other models to solve foraging problems—is fundamentally flawed. We think it unlikely that animals use simple rules to approximate fitness-maximizing solutions to foraging problems. They use intricate and sophisticated mechanisms involving sensory, neural, endocrine, and cognitive structures and active interactions with genes. Sherry and Mitchell's description of the honeybee proboscis extension response in chapter 3 is an example that hints at the complexity of the underlying mechanisms. In this volume, we have highlighted the increasing attention that foraging research pays to mechanisms with three chapters (3, 4, and 5) and seven text boxes devoted to mechanisms. This information will provide a firmer foundation for meaningful predictions about the costs and complexity of rules.

These mechanisms have surely been shaped by natural selection over each species' long history and have evolved to function in the environmental situations that an animal's ancestors experienced. Hence, they must be rational in that context, and they may perform poorly in other contexts. Students of foraging (e.g., Stephens and Anderson 2001) offer a view of rationality that is based on evolution and plausible natural decision problems faced by foraging animals (see section 1.5 above). Economists, psychologists, and cognitive scientists (Gigerenzer and Selten 2001; Simon 1956; Tversky and Kahneman

1974) have pursued a related program of research under the heading of "bounded rationality" (or "decision-making heuristics"). For example, in a phenomenon known as the "base-rate fallacy," human decision makers typically overestimate the reliability of information about rare events. If, for example, a test for a rare disease is 90% accurate, people tend to assume that a positive test means there is a 90% probability that you have the disease. This assumption is wrong because it neglects the fact that there will be many false positives for rare diseases; the true probability of disease, given a positive test, is typically much lower than 90%. These studies show that human decision makers make systematic mistakes in comparison to globally optimal solutions. Advocates of bounded rationality see their approach as distinct from (and an important alternative to) traditional optimality, and they have spilled a great deal of ink in disputes about whether optimization can accommodate the empirical results of bounded rationality.

After a long absence from the scene, "rules of thumb," based on a deeper appreciation of mechanisms, are poised for a reemergence.

1.11   Foraging Games

The traditional patch and diet models consider solitary foragers facing an unresponsive environment, but real life is more complicated. Foragers respond to their predators, and their prey responds to their presence. Animals may forage in groups, and so competitors also form a responding part of their environment. These problems, and many others, require a game theoretical approach. Curiously, Stephens and Krebs (1986) said nothing about game theory, even though it was a burgeoning topic in behavioral ecology at the time. However, game theoretical studies of foraging have since blossomed, and they appear in several chapters of this book. Giraldeau and Caraco (2000) provide a modern synthesis of the relevant concepts.

Games have players, strategies, rules, and payoffs. Their essential property is that a player's choice of strategy influences not only its own payoff, but also the payoffs of others. A player's actions will rarely maximize the payoffs of other players, and hence players commonly face conflicts of interest. Zero-sum games (in which the sum of payoffs to all players is a constant) always present conflicts of interest because one player's gain necessarily comes at the expense of other players. Even in non-zero-sum games (in which the sum of payoffs varies with players' strategies), players typically choose strategies that enhance their own pieces of the pie without necessarily maximizing the size of the collective pie. The tragedy of the commons, in which a private gain occurs at public expense (Hardin 1968), encapsulates this phenomenon of game theory.

In foraging games, the players can be individuals of the same species (school of fish), individuals from different species (mixed-species flocks of birds; lions and hyenas stealing each other's kills), or predator and prey (a stealthy mountain lion and a wary mule deer). Each player chooses from a list of available strategies. These strategies can include behavioral options (patch residence time, schedules of activity, and so on), but can also include physical and morphological traits. A foraging game has objective functions (one for each player) that determine payoffs, strategies for each player, and constraints that determine the array or range of choices available to each player. In a symmetric game, each player chooses from the same set of strategies and experiences the same consequences—each player has the same objective function and strategy set. When strategies are discrete and finite we can use a matrix to show the payoffs of strategic choices in the game. For example, players in the producer-scrounger game have two choices: "find food" or "share in the food that someone else has found." Thus, we can use a two-by-two table (or game matrix) to show the consequences associated with all combinations of actions. The matrix presentation works particularly well for two-player "contests" in which pairwise interactions determine payoffs. In other situations, a matrix representation of the game is not helpful, or even possible. For example, in games of vigilance or time allocation, the strategies are continuous and quantitative. In these continuous games, the objective function takes the form of a function that includes a variable for the individual's strategy, variables for the strategies of others, and possibly a variable for the population sizes of individuals with each of the respective strategies (e.g., the fitness generating function; Vincent and Brown 1988). Boxes 1.3 and 1.4 give examples.

Game theorists apply two similar solution concepts to foraging games: Nash equilibrium and the evolutionarily stable strategy (ESS). A set of strategy choices among the different players is a Nash equilibrium if no individual can improve its payoff by unilaterally changing its strategy. (For this reason, a Nash equilibrium is called a "no-regret" strategy.) An ESS is a strategy or set of strategies which, when common in the population, cannot be invaded by a rare alternative strategy. The two concepts are related: an ESS is always a Nash equilibrium but not vice versa (Vincent and Brown 1988).

1.12 An Overview of This Book

This brief review shows that research into foraging has expanded and advanced at a steady pace since the 1986 publication of *Foraging Theory*. Advances have been recorded in most, but not all, of the topic areas covered by its chapters 2–8. The major point of contention with critics—whether organisms are

BOX 1.3   A Two–Player, Symmetric, Matrix Game

Consider the following payoff matrix:

	A	B	C	D
A	3	6	5	3
B	5	9	1	2
C	3	11	5	1
D	2	7	6	4

Each player has four strategy choices (choose row A, B, C, or D). A player determines its payoff by matching its strategy as a row with its opponent's strategy as a column. Neither player knows what strategy its opponent will play, and it must choose its own strategy in advance. The player's payoff is the intersection of the appropriate row (the focal individual's strategy) with the appropriate column (the opponent's strategy). Each strategy has certain merits. Strategy A is a max-min strategy. It is the pessimist's strategy: "Since I am not sure what my opponent is going to play, I am going to assume that it will be the strategy that minimizes my payoff!" It maximizes the lowest payoff that an individual can receive from playing an opponent that happens to play the least desirable strategy for that individual. The max-min strategy maximizes the row minima. However, if everyone plays strategy A, the focal individual would do well to use another strategy, such as B.

Strategy B is a group-optimal strategy. It is attractive in that it provides the highest overall payoff given that all individuals use the same strategy. As such, strategy B represents the maximum of the diagonal elements. However, if everyone plays B, a focal individual would be tempted to use strategy C.

Strategy C, the max-max strategy, is attractive for several reasons. It represents the optimistic assumption that the opponent is willing to play the most desirable strategy for the focal individual. Also, since row C has the highest average payoff, strategy C maximizes a player's expected payoff under the assumption that the other player selects its strategy at random. However, if everyone plays strategy C, it behooves the focal individual to play strategy D.

Strategy D, at first glance, may have little to commend it. It is not max-min or max-max, nor does it maximize the value of the diagonal elements when played against itself. However, if all individuals use strategy D, then a focal individual has no incentive to unilaterally change its strategy, because no other strategy offers a higher payoff. It is this property that makes strategy D a Nash equilibrium.

BOX 1.4   A Two-Player Continuous Game

The game we describe here is a type of producer-scrounger game. We imagine a pair of foragers, each with a strategy that influences its harvesting of resources and its share of the total resources harvested. We will let the strategy u take on any value between 0 and 1: $u \in [0,1]$. We imagine that each forager harvests resources at a rate $(1 - u)f$. Hence, resource harvest is maximized when each forager selects a strategy of $u = 0$. But a forager's share of the combined harvest is determined by the effort it devotes to bullying (u) relative to its opponent's bullying. Specifically, we assume that the first player's share of the combined harvest is $u_1/(u_1 + u_2)$; the rest goes to player 2. We assume that a player's share is 0.5 when both players use strategy $u_1 = u_2 = 0$. We can write a fitness generating function for this game as

$$G(v, u) = \left[\frac{vf}{(v + u)} \right] [2 - (v + u)]$$

or

$$G(v, u) = f \quad \text{if } u_1 = u_2 = 0.$$

In this formulation, v is the strategy of the focal individual and u is the strategy of the other individual or opponent. For instance, to generate the payoff function for player 2, we would set $v = u_2$ and $u = u_1$.

We can seek an ESS solution by maximizing G with respect to v and finding a solution where $v = u$. To do this, we start with the partial derivative of G with respect to v:

$$\frac{\partial G}{\partial v} = \left[\frac{fu}{(v + u)^2} \right] [2 - (v + u)] - \frac{vf}{(v + u)},$$

where the first term on the right-hand side represents the benefits of bullying (a higher share in the collective harvest) and the second term represents the cost of bullying (less collective harvest to bully for).

To find a candidate ESS solution, we set each individual's strategy equal $(v = u)$, and then set the expression equal to 0:

$$\left. \frac{\partial G}{\partial v} \right|_{v=u} = \frac{f(1 - 2u)}{2u} = 0,$$

which implies that $u^* = 0.5$. With further evaluation, it can be shown that this candidate solution is an ESS. At $u^* = 0.5$, neither forager gains from unilaterally changing its strategy (satisfying conditions for a Nash

(Box 1.4 continued)

equilibrium), and if a player has a strategy slightly away from 0.5, it will benefit from a unilateral change back toward 0.5 (this phenomenon is referred to as convergent stability; Cohen et al. 1999).

At this ESS, each forager splits its effort between procuring resources and haggling over its share of the collective harvest. Note that both foragers would be better off if they could agree to shift their strategies to values less than 0.5. In fact, a strategy of $u = 0$ would maximize collective gain. However, this situation would not be stable, because both players would be tempted to shift some of their effort from harvesting to bullying. Besides mimicking aspects of a producer-scrounger game, this game also illustrates what happens when individuals can contribute to a public good (in this case by harvesting resources) but pay a private cost (inability to haggle over one's share of the harvested resources).

optimal—turned out not to be a fundamental flaw, but simply a misinterpretation of what an optimality model means. The limitations and shortfalls of the basic models have been recognized and left behind, and students of foraging have developed new ideas and techniques to conquer problems that seemed very thorny in 1986. The field has matured and expanded beyond the set of topics Stephens and Krebs considered in 1986. To paraphrase Mark Twain, reports of the death of foraging theory have been greatly exaggerated!

In the remainder of this section we give an overview of this book, placing the successive chapters in perspective. Part 1 (chapters 2, 3, and 4) deals with information, neuroethology, and cognition. Animals respond to their environment at the speed of neural transmission. Quick, coordinated movement is a hallmark of animals, and of course, animals come equipped with senses and the neural machinery that connects these senses to muscular output, with often amazing specializations and elaborations. This part of the book explores the connection between foraging and the information processing systems of animals at several levels.

In chapter 2, David Stephens considers the economics of information use. Starting with first principles, he asks what kinds of information should be important to a foraging animal and what constrains animal information-collecting abilities. The first model in this chapter develops the link between movement (or action) and the value of information. The model shows that the potential to direct actions is fundamentally what makes information valuable. A complication arises, however, because the world is often an ambiguous place, in which the relation between stimulus and information is not clear-cut. The theory of signal detection illustrates the interplay between economics and constraint in animal information gathering. Students interested in how sensory and neural systems can contribute to efficient foraging will want to pay close attention to this chapter.

In chapter 3, David Sherry and John Mitchell provide a gentle introduction to the "wetware" that underlies the mechanisms for the information-gathering tasks outlined in chapter 2, especially the classic psychological phenomena of learning and memory (which are, of course, fundamentally information processing phenomena). The chapter outlines the basic properties of a simple neural system involved in foraging (the antennal lobes and mushroom bodies of the honeybee brain) and explains important new discoveries about the cellular and molecular basis of learning. Food caching and recovery (see chap. 7 in this volume) is a foraging phenomenon that has become an important model system in the neuroethology of memory. The chapter uses this system to introduce basic ideas about memory, including types of memories and current thinking about the neural structures that form and store these memories.

In chapter 4, Melissa Adams-Hunt and Lucia Jacobs address a "higher" level of mechanistic thinking, reviewing the cognitive phenomena involved in foraging. Readers without a background in this area will be surprised at their number and complexity. Even the apparently simple act of perceiving a potential food item involves cognitive concepts unfamiliar to most behavioral ecologists: sensory transduction, attention, categorization, generalization, search image, and so on. In addition to exploring perception, the chapter outlines basic ideas about memory, learning, and spatial orientation.

In the early days of foraging theory and behavioral ecology, a wall separated ultimate (or evolutionary) explanations from proximate (or mechanistic) explanations. Strong proponents of this separation held that these two approaches were different levels of analysis, each of which could be successfully pursued without knowledge of the other. But a growing number of behavioral ecologists, neuroethologists, and psychologists are taking down this wall. While many questions can be asked and answered satisfactorily at one level of analysis or the other, a more complete understanding results when we combine levels of analysis. This part of the book challenges the reader to ask how mechanistic and evolutionary thinking can be profitability combined, perhaps producing an entirely new perspective on foraging behavior.

When a snake strikes and kills a kangaroo rat—a common event in desert landscapes (see chaps. 12 and 13 in this volume)—much of the action of classic foraging models ends, but in fact the snake's job has just begun. The kill begins an elaborate and time-consuming process of consumption and processing. The snake, as many will know, must manage to swallow its prey whole and uses several special features of its jaw and musculature to accomplish this. Once the kangaroo rat enters the snake's digestive system, an impressive physiological up-regulation begins. The snake is not a frequent eater, and most of the time its digestive machinery is quiescent, waiting to be turned on only when it is needed. Like any other forager, the snake can use the energy and nutrients it acquires from a meal in any of several ways: it can store energy as fat or use it immediately for maintenance or reproduction. None of these "post-kill" phenomena are well integrated with conventional foraging theory, yet they are surely important to any complete understanding of foraging. Part 2 of this book (chapters 5, 6, and 7) deals with three themes that begin where conventional foraging models end.

In chapter 5, Chris Whelan and Ken Schmidt review issues of food acquisition, processing, and digestion. The chapter explains chemical reactor models of digestion and reviews evidence for the adaptive control of digestive processes. Foraging theorists have viewed digestion as a black box with fixed properties (i.e., so much in yields so much out), but the gut is an active partner with foraging behavior. It may be adjusted seasonally, daily, or even with the mix of foods in a particular meal. Of course, this fact has consequences for the foraging models and for ecological interactions, as Whelan and Schmidt explain.

In chapter 6, Jonathan Newman presents a novel synthesis of herbivory that views the foraging problems of elephants and grasshoppers as essentially similar. Newman focuses on four issues—where to eat, what to eat, how fast to eat, and how long to eat—to highlight the special problems of herbivores. He outlines theoretical and empirical progress in studies of the effects of complementary nutrients, a topic that received little more than a hand-waving style comment in Stephens and Krebs (1986). In addition, the chapter reviews new models of the encounter process that have emerged from considering the special problems of mammalian herbivores.

Conditions can always worsen, and when they do, an animal must either reduce its energy demands (e.g., hibernate) or rely on food stored as fat or cached in the environment. Anders Brodin and Colin Clark consider this active area of modeling and empiricism in chapter 7. As they explain, a dynamic approach is essential because one is fundamentally concerned with how food collected now will affect fitness in the future. In addition, hoarded food and food stored as fat are not the same: each option has advantages and disadvantages. Fat is readily available, but may limit mobility and make the fat individual more susceptible to predation. A hoarding animal can cache large quantities of food, but cached food is susceptible to spoiling, pilfering, and retrieval costs.

Each of these themes represents an area that was in a relatively primitive state of knowledge in 1986. Now each is a growth industry in its own right.

Seven third-party text boxes convey the level of activity in this area. In chapter 5, Fred Provenza gives his perspective on how learning and taste affect the behavior of herbivores, which may in turn influence plant diversity. In chapter 7, Alasdair Houston and John McNamara provide an authoritative essay on the strengths and weaknesses of current modeling trends, while Stephen Woods and Thomas Castonguay introduce us to the neuroendocrine pathways of meal regulation.

Part 3 of this book (chapters 8, 9, and 10) offers three growing points in the direct analysis of foraging behavior. In chapter 8, Ron Ydenberg considers provisioning, the delivery of food and materials to other places (e.g., nest sites) or individuals (e.g., young). Provisioning theory derives directly from central place foraging models, but incorporates some essential differences. In particular, the theory recognizes that provisioners deliver food, but do not consume it themselves, at least not immediately. They must power this delivery with other food consumed in "self-feeding," and models must avoid mixing the energetic accounts of those who pay for the delivery (the provisioner) with those of the recipients (e.g., young in the nest). The chapter uses this approach to ask questions about the evolution of metabolic capacity and the rate of work and about the effects of demand on foraging strategies.

Chapter 9 investigates the effects of predation danger on foraging behavior. That foraging exposes an animal to predators is not a new idea; protective cover has long been a central concept in wildlife science, for example. However, the idea that foragers mitigate the danger with modifications to their foraging strategies had just begun to take hold in 1986. The idea that these modifications must be traded off against the foraging rate was developed once dynamic state variable models became available to help with the analysis of such trade-offs, and the implications of this notion form one of the major recent trends in foraging research. Peter Bednekoff traces these developments and surveys the current state of affairs in this chapter. Currently, the range and impact of these behavioral modifications continues to expand as ecologists actively investigate their effects on populations and communities.

We noted above that game theory played a limited role in early foraging research. This is mysterious, since the essential theoretical apparatus was in active use in other areas of behavioral ecology, and since topics that we now know require its use (e.g., information centers) were being actively researched. This situation has changed greatly in recent years through the integration of established paradigms, such as stable group size and the Prisoner's Dilemma, into foraging contexts. Tom Waite and Kristin Field review the current state of this area of research in chapter 10. They point out that the conditions for foragers to be considered "social"—whenever the payoff of their foraging strategy depends on the strategies used by other foragers—are probably quite general, even in cases in which foragers never interact directly. Thus, game theory is essential to the analysis of a very broad array of foraging situations.

When the young Charles Elton visited Spitzbergen in the early 1900s, his observations of who ate whom led him to imagine a community organized in terms of what he called a food chain. This early insight into the role of foraging in the organization of communities foretells the developments outlined in part 4 of this book (chapters 11, 12, 13, and 14), in which we consider foraging and its relation to ecological communities. An organism needs energy and materials to reproduce, grow, and stave off death. In turn, these processes influence the distribution and abundance of species (chapter 11). Population interactions and species composition and diversity form the core of community ecology. Foraging shapes the intensity, quality, and form of community interactions and consequent opportunities for species coexistence and diversity (chapter 12). Foraging often is a form of predation in which foragers exert mortality on their prey, compete for resources with one another, and provide opportunities for their own predators (chapter 13). Foraging behaviors, by shaping the experiences of organisms, directly determine many of the environmental circumstances that shape the evolution and coadaptation of other, less plastic physiological and morphological traits. This, then, is the domain of foraging ecology. Foraging ecology considers the population, community, and evolutionary consequences of animals' feeding behaviors.

Feeding behaviors are central to ecological and evolutionary feedbacks between an individual and its environment. As such, they can offer behavioral indicators of a species' prospects. In chapter 14, Mike Rosenzweig considers feeding behaviors as valuable indicators for conservation. We often use population size as an indicator of conservation status, but the population itself is often the valued ecological component under protection. Unfortunately, changes in population size become a trailing indicator. By the time one notes a substantial decline in population size, it has already happened. On the other hand, flexible feeding behaviors of organisms should indicate the animals' current prospects and their perceptions of prospects in the future. A shift of feeding behavior in response to changes in the environment should provide a leading indicator of changes in population size. Furthermore, the behaviors themselves may be a valued component of the ecosystem. For example, responsiveness to predators is an integral part of being an elk, and loss of such antipredator behaviors results in significant changes in the elk's feeding behaviors and ecology (Laundre et al. 2001). Chapter 14 goes beyond the utility of foraging behavior for conservation. It proposes using the principles of foraging theory to understand human perspectives and goals. In this light, foraging theory provides a framework for incorporating human resource acquisition activities and their ecological consequences.

Part I

Foraging and Information Processing

Telander

Models of Information Use

David W. Stephens

2.1 Prologue

A rufous hummingbird perches on a prominent branch and surveys a flower-covered slope. Most of the time, it waits and watches. Occasionally, it flies off its perch to probe the hanging flowers of scarlet gilia within its territory. Scarlet gilia is a classic hummingbird flower. An inflorescence consists of six to twenty flowers, each of which is a long scarlet tube with a pool of nectar at the base.

Each inflorescence makes up a clearly defined patch in the sense of classic foraging theory—even more so than most patches because it consists of discrete, visitable entities; i.e., flowers. In applying the classic models of patch exploitation to this situation, we naturally think of the time taken to fly between inflorescences (travel time) and the obvious patch depletion that a hummingbird will experience when it revisits flowers. But our hummingbird's problem isn't quite so simple. Inflorescences vary: some consist of mostly empty flowers, while others have mostly full flowers. Our hummingbird's own behavior partially creates this pattern, but some other actors are involved as well. Robber bees move methodically from one flower to the next, making neat incisions in the corolla that allow their short tongues access to the nectar.

This variation means that while our hummingbird obtains food each time it probes a flower, it also obtains information: it finds out something about the quality of this inflorescence and possibly about neighboring inflorescences. If *this* flower is full, then the neighboring flowers may also be full. Does this information value of a flower visit change our thinking about this otherwise straightforward patch exploitation problem? Surely it must. In this scenario, the information value of that "next flower" is an important component of the economics of patch departure decisions.

Foraging animals obtain food *and* information about food resources as they go about the business of feeding. Of course, animals also acquire and use information when they choose mates, defend territories, or avoid predators. Foraging has, however, served as a productive model for the study of information problems in behavioral ecology. The idea that animals may act on and seek to obtain information about food resources connects foraging models to central questions in psychology. One can think of Pavlov's dogs as responding to information about a new environmental relationship between a metronome and food.

2.2   The Basic Problem: Incomplete Information

Let's simplify the hummingbird example to illustrate the basic properties of foraging information problems. Suppose that there are two types of inflorescences: one type consists entirely of FULL flowers, and the other consists entirely of EMPTY flowers. The hummingbird's problem has two components. First, our hummingbird cannot know whether any particular inflorescence is FULL or EMPTY. However, it can reduce this uncertainty by probing a single flower. In this simple FULL/EMPTY scenario, a single flower tells all about the inflorescence (we'll consider a more complex situation shortly).

In general, when we say that a forager faces an incomplete information problem, we mean that it is uncertain about some relevant feature of the environment, and that it can take some action that will reduce this uncertainty (i.e., obtain information), typically at some cost in time or energy.

Consider again a hummingbird faced with different types of inflorescences. Here we'll consider a broader range of inflorescence types than just FULL and EMPTY. Some inflorescences are very good food sources, some mediocre, and some poor. How valuable is it for our hummingbird to know which type of inflorescence it's dealing with? Several authors have dealt with this question theoretically (Gould 1974; Stephens 1989; Stephens and Krebs 1986), and the results provide general insights into the nature of incomplete information problems. If the hummingbird knows that it's facing a "very good" inflorescence, then it can implement a behavior that is appropriate for "very good" inflorescences. A best response exists for each type of inflorescence; the best response may be a long patch residence time in a very good inflorescence and a short patch residence time in a poor inflorescence. Mathematically, we can imagine of list of the "best responses" associated with each of several possible states, and a "well-informed" hummingbird will be able to use information to adopt the best response for each state (inflorescence type, in our example).

In contrast, if our hummingbird must act in ignorance, then it must pick a single response that does best on average (that is, averaging across all possible inflorescence types). This "single best response" will typically represent a compromise: it's an acceptable response overall, but it isn't the best response for any given inflorescence type.

In principle, we can calculate the average benefit that a well-informed hummingbird obtains by calculating the average payoff that a hummingbird adopting the best response for each state obtains. We can also calculate the average benefit that an uninformed hummingbird obtains by calculating the average payoff for a hummingbird that adopts the same behavior in all inflorescence types. The value of being informed is the difference between these two averages—specifically, the difference between (1) the expected value of adopting a behavior that matches each possible state and (2) the expected value of treating each state in the same way. What's the difference between being informed and uninformed? An uninformed forager must choose a single action representing a compromise solution for all possible states, while an informed forager can tailor its action to each possible state, and this *is* the advantage that information confers.

To push this point a little further, consider the following odd situation. Suppose that five inflorescence types exist, but the best action is the same for all five types. What's the value of information now? A moment's reflection will tell you that it's zero: if the best possible action is the same for all states, then the single best action must also be the same, the two averages that we use to calculate the value of information must be the same, and their difference must be zero. If states don't affect action, then information has no value—again, the potential to change *actions* makes information valuable.

The reader may find this boringly obvious: surely everyone knows that information matters only when it makes a difference. But the interesting observation here is that some differences matter more than others. Our premise that the best actions are the same doesn't mean that states don't make any difference to the animal. They could make a big difference: some states might signify a big payoff, while others might result in a loss. Information has no value because these differences don't affect action: even though things may change from one state to the next, the best action is always the same. The take-home lesson is simple but important: *information is valuable when it can tell you something that changes your behavior.*

2.3   Information in Prey Choice: Signal Detection

Consider an insectivorous bird that eats greenish black beetles. Some beetles taste good, but others taste bad because they contain noxious secondary plant compounds. Greenish beetles tend to be noxious, but the situation is fuzzy: some black beetles are noxious, and some greenish ones are not noxious. A greenish beetle is just a bit more likely to be noxious. While color provides only fuzzy information, the forager must still make a "crisp" decision to attack or ignore an encountered beetle. (In theory, of course, the forager could make a halfway decision, such as "investigate further"; this possibility raises several interesting problems, which come under the heading of sequential decision making.) In many discrimination problems, a forager cannot "know" exactly which state is true. Instead, it has information about the relative likelihood of states.

The bird's problem resembles the classic "signal detection" problem that students of perception and sensation have long studied (Egan 1975; Swets 1996). In such a problem, we typically call one possible state "True" (say, finding a tasty beetle) and the other "False" (finding a noxious beetle), and we describe the alternative actions as "Yes" (attack the beetle) and "No" (ignore the beetle). Now we've imposed considerable structure on our general problem, as the following table shows:

	True	False
Yes	Correct Acceptance V_{CA}	False Alarm V_{FA}
No	Miss V_M	Correct Rejection V_{CR}

The table introduces some useful terminology and new notation. If the forager chooses the "Yes" action when the state is true, we call this a "correct acceptance" and say that the value of a correct acceptance is V_{CA}. If the forager chooses "Yes" and the state is false, we call this a "false alarm" and say that value of a false alarm is V_{FA}. If the forager chooses "No" and the state is true, we call this a "miss" and say that the cost of a miss is V_M. Finally, if the forager chooses "No" and the state is false, we call this a "correct rejection" and say that the value of a correct rejection is V_{CR}. Notice that there are two "correct" combinations and two types of errors. (This so-called "truth table" arises in many guises, and the student of information will do well to recognize its various forms. In statistics, the "miss" cell corresponds to the event measured in statistical significance—rejecting a true hypothesis—and the "false alarm" cell corresponds to "power." Truth tables also arise frequently in analyses of animal communication; see Bradbury and Vehrencamp 1998.)

Now we can solve this problem easily if the forager can know which state is true: it should choose "Yes" if the state is true and choose "No" if the state is false. We have erected this two-state/two-action framework to help us understand the complicated situation in which a decision maker must act with partially informative experience. Let's reconsider our green-to-black, noxious-to-tasty beetles. Suppose that our hypothetical forager observes the color of a given beetle, represented by a variable X (high X means that the beetle is blacker than green—and more likely to be tasty). So our forager might adopt a rule, such as "attack beetles when $X > a$." How should a be set?

Mathematically, any rule determines four conditional probabilities that correspond to the cells of our truth table:

1. the probability of a "Yes" response given that the state is "true" and the rule parameter equals a; in symbols, $P(\text{Yes}|\text{True} \& a)$, i.e., a correct acceptance
2. the probability of a "No" response given that the state is "true" and the rule parameter equals a; in symbols, $P(\text{No}|\text{True} \& a) = 1 - p(\text{Yes}|\text{True} \& a)$, i.e., a miss
3. the probability of a "Yes" response given that the state is "false" and the rule parameter equals a; in symbols, $P(\text{Yes}|\text{False} \& a)$, i.e., a false alarm
4. the probability of a "No" response given that the state is false and the rule parameter equals a; in symbols, $P(\text{No}|\text{False} \& a) = 1 - p(\text{Yes}|\text{False} \& a)$, i.e., a correct rejection

Notice also that these four probabilities are really two pairs of complementary probabilities $[P(\text{No}|\text{True} \& a) = 1 - P(\text{Yes}|\text{True} \& a)$ and $P(\text{No}|\text{False} \& a) = 1 - P(\text{Yes}|\text{False} \& a)]$, so we can simplify the mathematical problem by focusing on only two of them, but which two? By convention, we consider the two probabilities of acceptance, $P(\text{Yes}|\text{True} \& a) = P(\text{Correct Acceptance}|a)$ and $P(\text{Yes}|\text{False} \& a) = P(\text{False Alarm}|a)$.

The Receiver Operating Characteristic Curve

Now, consider how $P(\text{Correct Acceptance}|a)$ and $P(\text{False Alarm}|a)$ change as the forager changes the decision threshold a. Suppose that our insectivorous bird picks a threshold a value—say, $\breve{a}$—that leads it to always accept beetles regardless of their color. In this case, our forager will never miss a truly tasty beetle $[P(\text{Correct Acceptance}|\breve{a}) = 1]$, but the price of this advantage is that it always incorrectly accepts noxious beetles $[P(\text{False Alarm}|\breve{a}) = 1]$. At the other end of the spectrum, imagine that our insectivorous bird picks an a value—say, $\hat{a}$—that causes it to reject everything. Then the forager will never accept a noxious beetle $[P(\text{False Alarm}|\hat{a}) = 0]$, but it will always reject tasty beetles $[P(\text{Correct Acceptance}|\hat{a}) = 0]$. As the parameter a changes from values specifying "always accept" to values specifying "always reject," it determines

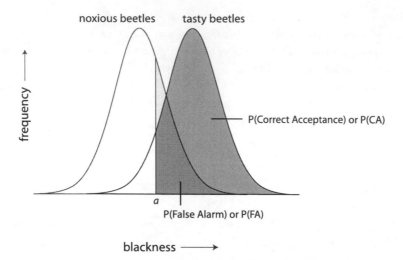

Figure 2.1. The relationship between P(False Alarm) and P(Correct Acceptance). P(False Alarm) is the area under the lower (noxious beetle) curve that is also above *a* (light shading). P(Correct Acceptance is the area under the higher (tasty beetle) curve that is above *a* (darker shading)

a relationship between P(False Alarm$|a$) and P(Correct Acceptance$|a$). This relationship, called the receiver operating characteristic (ROC) curve, is a fundamental part of our analysis because it gives a powerful and concise summary of the constraint imposed by imperfect discrimination. The receiver operating characteristic curve focuses our attention on the trade-off between high acceptance rates that lead to few misses but frequent false alarms, and high rejection rates that lead to few false alarms but frequent misses.

We can take the logic above a bit further to show how the entire receiver operating characteristic curve can be constructed. Figure 2.1 shows two overlapping color (green-to-black) distributions. The distribution on the right shows the (blacker) colors of tasty beetles, and the distribution on the left shows the (greener) colors of noxious beetles. If we choose an acceptance threshold *a*, the probabilities of acceptance are the areas under the curves above *a*, as indicated in the figure. P(Correct Acceptance$|a$) is the area above *a* under the upper "tasty beetle" curve, and P(False Alarm$|a$) is the analogous area above *a* under the lower "noxious beetle" curve. As *a* increases, the two probabilities of acceptance move in concert, tracing out a receiver operating characteristic curve, as figure 2.2 shows.

A comparison of figures 2.2A and 2.2B shows how receiver operating characteristic curves differ between easy and difficult discrimination problems. Part A shows a case in which the two distributions are well separated, making this an easy discrimination problem, because we can easily choose an *a* value that rejects most noxious beetles *and* accepts most tasty beetles. The figure shows how this situation leads to a strongly "bowed out" receiver

A. Easy Discrimination Problem

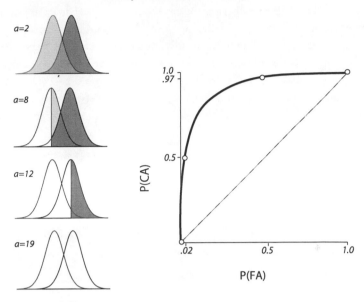

B. Difficult Discrimination Problem

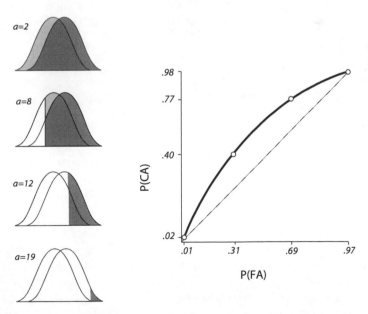

Figure 2.2. Two examples showing how receiver operating characteristic (ROC) curves are derived from noxious and tasty beetle distributions. (A) When the two overlapping distributions are well separated, the receiver operating characteristic (ROC) is bowed toward the ideal [P(FA) = 0, P(CA) = 1] point. (B) When the two distributions are close together, the curve is less bowed out and more linear.

operating characteristic curve. Part B shows a more difficult discrimination problem in which the two distributions overlap more, so that a forager finds it difficult to reject noxious beetles without also rejecting tasty ones. This situation leads to a much flatter receiver operating characteristic curve. In the limiting case, in which the two distributions are exactly the same (complete overlap), the receiver operating characteristic curve would be a straight line connecting (0,0) and (1,1). The extent to which the receiver operating characteristic curve bows out away from linearity is, therefore, a measure of the "discriminability" of the situation.

Finding the Optimal Discrimination Strategy

Now that we have the machinery of the receiver operating characteristic curve, we can find the "optimal" threshold a. We will simply quote the result here (Commons et al. 1991; Egan 1975; Gescheider 1985; Green and Swets 1966; Wiley 1994). We established above that the chosen value of the threshold a implicitly determines a point on the receiver operating characteristic curve. Of course, the reverse applies as well: for a given point on the receiver operating characteristic curve, we can find the corresponding a (doing this requires some very laborious algebra, but it is logically straightforward). So we will state our "solution" in terms of the receiver operating characteristic curve. The optimal point on the receiver operating characteristic is the point that has a slope equal to

$$ m^* = \frac{1-p}{p} \left[\left(\frac{V_{\mathrm{CR}} - V_{\mathrm{FA}}}{V_{\mathrm{CA}} - V_{\mathrm{M}}} \right) \right], \tag{2.1} $$

where p is the proportion of beetles that are tasty (so $1 - p$ are noxious), and the V terms come from the payoff table given above. This term, m^*, will be a large number if noxious beetles are much more common than tasty beetles (p near zero), predicting that the solution should be on a steep part of the receiver operating characteristic curve (implying a high, generally "unaccepting," a value; fig. 2.3). If, instead, tasty beetles are more common (p near 1), then m^* will be small, and the solution will be on the shallower (upper) portion of the receiver operating characteristic curve (implying a small, generally "accepting," a value). We can make similar predictions about the effect of the quotient $(V_{\mathrm{CR}} - V_{\mathrm{FA}})/(V_{\mathrm{CA}} - V_{\mathrm{M}})$: a large value pushes the optimal threshold toward rejection (the steep part of the receiver operating characteristic curve), and a small value shifts it toward acceptance (the shallow part of the receiver operating characteristic curve). This result agrees with intuition because a large $(V_{\mathrm{CR}} - V_{\mathrm{FA}})/(V_{\mathrm{CA}} - V_{\mathrm{M}})$ value means that the premium for correct be-

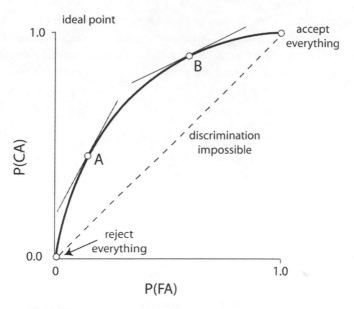

Figure 2.3. An annotated receiver operating characteristic (ROC) curve. Signal detection theory gives the optimal behavior in terms of a critical likelihood ratio that we can visualize as the slope of the receiver operating characteristic (ROC). For example, if true states are rare, then we expect a high critical likelihood ratio that corresponds to a point on the steep portion of the receiver operating characteristic curve, as in point A. If, on the other hand, true states are common, then we expect a lower critical likelihood ratio that corresponds to a point on the shallower portion of the receiver operating characteristic curve, as in point B.

havior is greater in the "false" state than in the "true" state (i.e., $V_{CR} - V_{FA} > V_{CA} - V_M$).

Signal Detection: A Summary

Now we have a fairly complete picture of optimal behavior in the face of an ambiguous signal. Figure 2.3 shows our model and its interpretation. The receiver operating characteristic curve shows us how difficult the discrimination problem is (in terms of where it lies between the ideal $P(FA) = 0$, $P(CA) = 1$ point and the discrimination-impossible $P(CA) = P(FA)$ line). Mathematically, the receiver operating characteristic curve shows us the achievable $P(CA) - P(FA)$ combinations (technically, everything beneath the curve is achievable, but we are not interested in points below the curve), and each combination has a corresponding likelihood ratio that we visualize as the slope of a tangent line. Finally, the term m^* [see eq. (2.1)],

$$\frac{1 - p}{p} \frac{V_{CR} - V_{FA}}{V_{CA} - V_M},$$

which compares the commonness of the "true" and "false" states with the economic consequences of actions in those states, specifies a critical likelihood ratio that we can superimpose on our receiver operating characteristic curve to determine which of its feasible combinations is best (Getty et al. 1987).

Two Basic Ideas

Taken together, these two ideas—the value of information and the problem of signal detection—offer basic lessons in the economics of animal information use. An observant student will notice that these ideas come up repeatedly, in various guises, in many treatments of information use, learning, communication, and cognitive processing. The remaining sections of this chapter consider specific information problems (namely, patch use and environmental tracking). In each case, I comment about the relevance of these two ideas.

2.4   Information in Patch Use

In this section we return to our rufous hummingbird and consider how incomplete information can influence patterns of patch exploitation. We apply the two basic ideas developed above to patch use, and we find that we need to consider sequential sampling problems to understand the role of information in patch use.

According to the classic models of patch leaving, foragers leave patches when within-patch gain rates decline to the point that the forager can do better elsewhere. While students of foraging will recognize the importance of this effect, early critics (Green 1980; Oaten 1977) of patch models recognized that information about patch quality might add an important dimension to these models. The idea is straightforward: as the animal forages in the patch, it might *discover* that the patch is especially good or especially bad, and this *discovery* may tip the balance between leaving and staying.

The simplest models of this type imagine egg-carton-like patches (Green 1980; Lima 1983, 1985), like the inflorescences visited by our hummingbird, in which a forager checks discrete sites within a patch, and each site can be full or empty. As the forager exploits a patch, it "checks" each site for food and obtains information about the relative frequency of full and empty sites within that patch.

Imagine a world in which inflorescences have a fixed number of flowers (say, s, for patch size), that each flower can be either full or empty, and finally, that only two types of inflorescences (patches) exist: either completely empty or partially full. Let q represent the relative frequency of partially full patches (so $1 - q$ is the frequency of empty patches). In partially full inflorescences, p

of the flowers have some nectar, and $1 - p$ have none. (Notice that p and q represent different proportions; specifically, it is not true that $p = 1 - q$.) These assumptions create a relatively simple information problem because finding a single "full" flower means that this inflorescence is of the partially full type. On the other hand, sampling a string of n empty flowers provides ambiguous information because it may indicate that this inflorescence is empty, or it may just be a run of bad luck in a partially full patch.

Suppose that our hummingbird adopts a rule: leave after n empties, but visit all s flowers if you discover any full flowers in the first n visits. Figure 2.4 shows the optimal giving-up time, n, as a function of p (the fullness of partially full patches) for two levels of q (the relative frequency of partially full patches) We can see several intuitively appealing results. First, the optimal giving-up time n^* decreases with p, this makes sense because when p is high, the forager can more easily discriminate partially full and empty patches. Second, n^* decreases with q (the prior, or environmental, probability of empty patches). This is a signal detection effect: decision makers should set a "pickier" threshold when true states are rare. Finally, n^* increases with the travel time τ. This is the classic "options elsewhere" effect: when a forager can quickly find a fresh patch, it should spend less time checking the current patch.

While these results agree with our expectations, we can learn a bit more by applying our two basic ideas about the value of information and the problem of signal detection to this basic foraging problem.

The Value of Information

A forager with perfect information would spend s time units in each partially full patch and no time exploiting empty patches. An omniscient forager, therefore, would obtain a rate of

$$\frac{(1 - q)sp}{\tau + s(1 - q)}. \tag{2.2}$$

In contrast, a forager that must act without information would have to spend s time units in all patches (assuming that our patches are the only food resource in the environment). This gives a rate of

$$\frac{q \cdot 0 + (1 - q)sp}{\tau + s}. \tag{2.3}$$

The value of information is therefore

$$\frac{(1 - q)sp}{\tau + s(1 - q)} - \frac{(1 - q)sp}{\tau + s}, \tag{2.4}$$

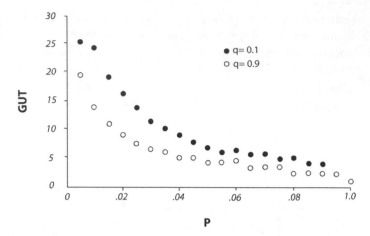

Figure 2.4. The relationship between p (probability of food in partially full patches) and optimal giving-up time (GUT). As p increases, the optimal giving-up time decreases. This is a discrimination effect; when p is near zero, a larger sample is required to discriminate empty patches from partially full patches. Predicted giving-up times are also generally longer when q (the probability of partially full patches in the environment) is small.

or

$$\frac{q(1-q)s^2p}{(\tau+s)(\tau+s-sq)}. \tag{2.5}$$

In agreement with the general development of our model, the value of information is (approximately) proportional to the variance in ideal behaviors ($s^2q(1-q)$, which is the variance of the random process in which a forager spends either s or 0 time units in a patch). Notice especially that the value of information peaks at intermediate q values (i.e., $q \approx 1/2$). On the other hand, information has less value when q takes extreme values. For example, if we assume that a forager must pay a cost to implement a giving-up time, then we might predict that a forager will adopt a fixed, non-information-gathering strategy when q is near 0 or 1.

Signal Detection: Sampling versus Deciding

Although the basic principles of signal detection apply here, the specific predictions of elementary signal detection theory do not transfer cleanly to the patch sampling problem. Signal detection theory tells us how a decision maker should act in response to a sample: say Yes if the sample X exceeds the threshold a. The patch sampling problem focuses on the intensity or level of sampling: how many sites you must check before concluding that this patch isn't worth further exploration. The question of when a forager should stop sampling raises questions of general significance. In this chapter, however, I can only comment on two basic effects. First, things that increase the value of information (as described in section 2.2) will tend to increase the number of samples taken. Second, the future value of a hypothetical next sample plays a key role in deciding whether to take that next sample. For example, in a large patch, the next sample may reveal that this patch contains many more prey, while in a small patch, the same piece of good news is simply not as significant because the smaller patch contains less food, even if it is full. The future value of the sample plays a key role in models of sampling intensity.

Patch Potential

The discussion above illustrates how information can influence patch departure decisions using a very simple example. The reader may have already thought of many possible complications: real environments may contain many more patch types beyond the partially full/empty dichotomy used in our example; foragers may be able to recognize some patch qualities without direct sampling. McNamara (1982) has offered a useful graphical method that can simplify our thinking about these complications. In this technique, we suppose that the forager keeps a running account of the quality, or *potential*, of the current patch. Typically, we suppose that there is a potential function $H(t, x)$ that is a function of time in the current patch, t, and the number of prey so far obtained in the current patch, x. The potential function gives an estimate of patch value as the forager exploits a patch (spending time and collecting prey), and the forager will leave the patch when $H(t, x)$ falls below some critical value. The potential function provides a helpful framework because it reduces a nearly infinite array of possible within-patch experiences to a single value, and in doing so, it gives us a general way to represent a forager's patch-leaving rule.

One classic question is, what happens to the potential when the forager captures a prey item? Actually, many possible things might happen. In our empty/partially full example, the first prey capture represents an enormous jump in potential, but further prey captures have no effect—after the first capture, the potential steadily decreases until all s sites have been visited.

Depletion versus Information

With this framework in mind, one can ask how a prey capture changes the forager's assessment of patch potential. A capture could signal something about patch quality, such as "this is an especially good patch," and this information should increase the potential of the patch. Alternatively, a capture might simply signal that less food is available (i.e., patch depletion), causing a decrease in potential. Crudely speaking, we can think of information and depletion effects as opposing each other. We would expect captures to have high information value (and hence to cause an increase in patch residence time) when the environmental distribution of patch qualities has high variance (i.e., as predicted by the "value of information" calculations developed earlier; see also Valone 1989). If, in contrast, all patches tend to be similar (low variance), then captures will largely be signals of depletion. In addition, it seems reasonable to conclude that prey captures that occur early in a patch visit will usually offer more information about patch potential than later captures.

2.5   Tracking a Changing Environment

So far, we have discussed uncertainty problems that deal with discriminating the properties of a given patch or prey item. This section considers information use at a larger scale, asking how a forager should keep track of changes in its environment. Tracking of environmental changes presents challenging and exciting questions because it has long been thought to be the key evolutionary advantage of learning and memory. As before, I outline a simple model that characterizes the general issues.

Framing the Problem

How should a forager "track" environmental changes? The simplest model imagines an environment in which one resource fluctuates while another is stable (Arnold 1978; Bobisud and Potratz 1976; Stephens 1987). The varying resource, called V, is sometimes in a good state, which yields g units of benefit per unit time, and sometimes in a bad state, which yields b units of benefit per unit time. The mediocre stable resource, called S, always provides s units of benefit per unit time. The states of the varying resource occur in runs, specified by a persistence parameter q, the probability that the state now (in time i) will persist in the next time interval (time $i + 1$). So if $q = 1/2$, the state in the next time interval is just as likely to have changed as to have remained the same, while if q is close to 1, the current state is a good predictor of the state in the next time interval.

We assume that $g > s > b$, so a forager should exploit the varying resource when it's in the good state, but switch to the stable resource as soon as the varying resource "goes bad." A forager might be able to follow this omniscient strategy if some externally visible cue signaled the state of the varying resource, but we will assume that the forager can detect the state of V only via direct experience. In other words, the forager must sample. To keep the problem simple, we assume that experience allows perfect discrimination, so a single sample tells the forager whether the varying resource is in its good or bad state.

Figure 2.5 shows the situation. The varying resource follows the pattern of a square wave that varies between g and b, while the stable resource is a flat line (at s) somewhere between g and b. Now consider what happens when V changes from good (g) to bad (b). The forager detects this immediately and switches to the stable resource, but how long should it stay there? Periodically, the forager needs to check V to see if a transition back to the good state (g) has occurred. An animal that checks too frequently will make many "sampling errors," obtaining b when it could have had s (this error costs $s - b$). On the other hand, an animal that doesn't check frequently enough will make overrun errors, missing the switch back to g and obtaining s when g is available (this error costs $g - s$). We can summarize this logic in a single parameter that we'll call the error ratio, $\varepsilon = (s - b)/(g - s)$ the cost of sampling errors divided by the cost of overrun errors. So, for example, a large error ratio means that sampling errors are relatively expensive, and we expect infrequent sampling. If, instead, the error ratio is small, we would expect frequent sampling to minimize overrun errors. The astute reader may have noticed some familiar elements of signal detection theory in our construction of the error ratio: the consequences g, b, and s neatly fill out a "truth table," as in our development of signal detection (with s filling two cells), and the error ratio itself parallels the ratio of consequences in equation (2.1).

The environmental persistence of a resource, q, also has an important effect on the economics of sampling frequency. One can understand this effect intuitively by considering two special cases. If $q = 1/2$, resource V changes from good to bad at random, and there is, quite literally, nothing to track. So we expect no sampling when $q = 1/2$; the forager should choose either to always exploit S or to always exploit V, whichever provides the higher average gain. On the other hand, if $q = 1$, the current state is a perfect predictor of future states, so we know that if the varying resource V provides g now, it will always provide g. The interesting thing about this "perfect predictor" case is that it makes a single sample extremely valuable—in theory, a single sample can point the forager to a lifetime of correct behavior.

The persistence parameter and error ratio combine to determine the sampling rate (i.e., the time before returning to V to sample its state) that maximizes the long-term rate of resource gain (the optimal sampling rate, σ^*; Figure 2.6). The model predicts sampling in a trumpet-shaped region narrowest where $q = 1/2$ and widening as q approaches 1. A forager should not sample in the region above the trumpet; instead, it should exploit only the stable resource S. Another "don't sample" region lies below the trumpet, in

A. The Environment

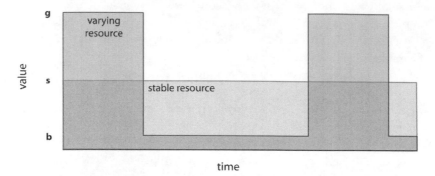

A. High Sampling Rate; many sampling errors(s), few overrun errors(o)

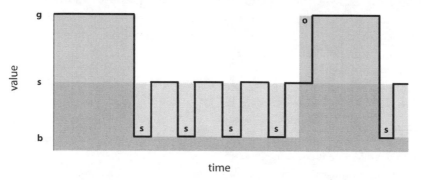

A. Low Sampling Rate; fewer sampling errors(s), more overrun errors(o)

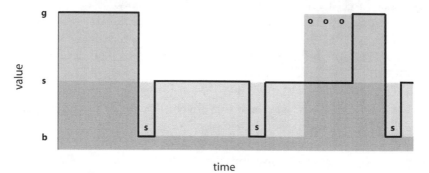

Figure 2.5. Tracking a changing environment. (A) An environment with a varying resource alternating between states *g* and *b* in a square wave pattern and a mediocre stable resource in state *s*. (B, C) The economics of high and low sampling rates. (B) Sampling frequently leads to many sampling errors (s) but few overrun errors (o). (C) Less frequent sampling reduces the number of sampling errors but causes more overrun errors.

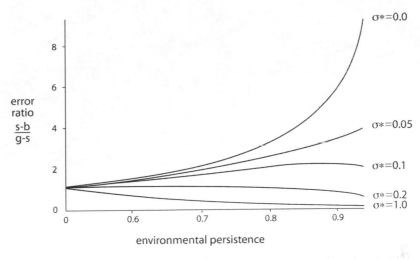

Figure 2.6. The effects of error ratio $(s - b)/(g - s)$ and environmental persistence (q) on the optimal sampling rate (σ^*). The parameter σ^* gives the optimal sampling rate; it is the probably of checking the varying resources during a run of bad luck. Each curve shows combinations of error ratio and environmental persistence that imply a particular optimal sampling rate as shown on the figure. A forager should always exploit the stable resource S in the region above the $\sigma^* = 0.0$ line and should always exploit the varying resource V in the region below the $\sigma^* = 1.0$ line. Sampling, therefore, is predicted only in the trumpet-shaped region bounded by the $\sigma^* = 0.0$ and $\sigma^* = 1.0$ lines.

which the forager should exploit only the varying resource V. As the predictability of the environment (q) increases toward 1, the region in which we predict sampling increases.

While most readers will recognize the logic of this result, it seems surprising if we step back from the particulars and consider the larger context. Animals need to sample because they live in varying environments, yet the conditions that favor sampling steadily broaden as the environment approaches fixity! It seems that sampling is as much about environment regularity as it is about environmental change (see Stephens 1991 for an application of these ideas to learning). The model makes three key predictions:

1. Sampling rates should decrease with s, the value of the stable but mediocre resource, because a decrease in s makes sampling errors more costly while reducing the cost of overrun errors.
2. Sampling rates should increase with g, the value of the varying resource's good state, because an increase in g makes overrun errors more costly.
3. Sampling rates should decrease with q, because q increases the duration of states.

Three separate studies have tested this basic tracking model (Inman 1990 using starlings; Shettleworth et al. 1988 using pigeons; Tamm 1987 using ru-

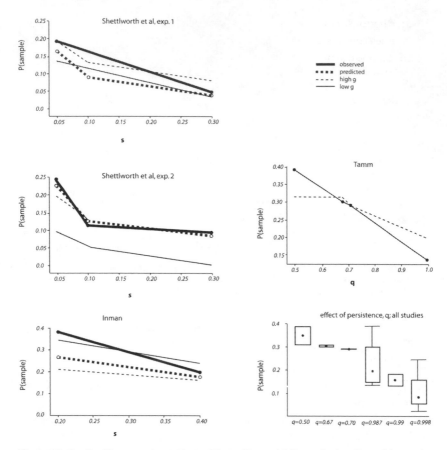

Figure 2.7. Results of three experimental tests of the tracking model. The qualitative effects of the *s* and *q* variables are as predicted, but the effect of *g* seems to contradict the model.

fous hummingbirds). The Shettleworth et al. and Inman studies asked whether the components of ε—especially *g* and *s*—affect sampling behavior as predicted, while Tamm studied the combined effects of ε and *q*. In all three studies, the bad state was "no food," giving $b = 0$ and $ε = s/(g - s)$.

Figure 2.7 presents graphical summaries for these three studies. The figure shows a straightforward pattern: the effects of *s* and *q* agree with the theory. Observed sampling rates decrease with increases in both *s* and *q*. However, the effect of *g* does not agree with the model's predictions. Moreover, the effect of *g* shows no clear pattern: in one case (Shettleworth et al., experiment 1), sampling rates decrease with increasing *g* in direct contradiction of the model; in another (Shettleworth et al., experiment 2), *g* has no effect; in a third (Inman), *g* shifts sampling rates in the predicted direction; in the fourth (Tamm), there is no consistent effect of *g*. The data also suggest several other contradictions. For example, both the Inman and Shettleworth et al. studies had some treatments in which different g and s values predicted the same error ratio (ε); one can do this by changing both g and s by the same factor k (that is,

$$\varepsilon = \frac{s}{g-s} = \frac{ks}{kg-ks}.$$

In both studies, observed sampling rates were lower when k was greater, suggesting possible hunger effects (because when k is large, the subjects obtain more food on average and may be less motivated to feed).

Tracking Prospects

Like so many models in behavioral ecology, our simple tracking model meets the data with mixed success. Some of the economic factors considered in our models influence sampling as predicted, while others do not. The simple model developed here could be improved in several ways. A glaring deficiency is the assumption that foragers can distinguish good states from bad immediately and without error, even though the theory and practice of signal detection tell us that animals make errors even when stimuli seem quite distinct. The process of change applied in this model could also be generalized. The model assumes, for example, that one resource is fixed and the other varies, yet the data suggest that animals "sample" both resources (e.g., checking the stable resource even when the varying resource is in the good state). In short, we could improve the models and experimental studies of tracking in several possible ways. Unfortunately, this important and tractable topic has not received much attention recently.

Tracking and Learning

Tracking foragers learn about the current state of the environment, and it is natural to wonder whether tracking models might provide some insight into the evolutionary significance of learning. The effect of environmental persistence in the tracking model is especially intriguing. The region in which tracking pays off increases as the environment becomes increasingly fixed (high q), yet the conventional wisdom holds that learning exists because it allows animals to adapt to change. Stephens (1991) has modified the tracking model developed here to study this apparent contradiction. The Stephens model asks when a (very simple) learning strategy outperforms a genetically fixed behavior. The model suggests that it is just as reasonable to say that learning is an adaptation to predictability as it is to say that learning is an adaptation to change. Indeed, both statements are naive: learning requires both change *and* environmental regularities that allow today's experience to predict which actions will pay off tomorrow (see Dukas 1998c for an alternative view). The interested reader may want to explore the literature on learning rules (Bush and Mosteller 1955; Harley 1981; Rescorla and Wagner 1972; see also chap. 4). These mathematical models describe the time course and qualitative properties of learning, and an important goal for the future is to reconcile them with models about the evolution of learning.

Theory addressing the evolution of learning has existed for some time, but the difficulties of testing this theory empirically have long frustrated students of learning. However, two emerging research programs have addressed this problem. Mery and Kawecki (2002, 2004) have used *Drosophila* oviposition learning to study the evolution of learning directly. Mery and Kawecki's studies confirm that learning evolves in changing environments when stimuli have predictive power, but they also found that learning evolves when the features of the experimental environment are fixed from one generation to the next. In this case, learning accurately predicted the state of the environment, but a non-learning mechanism would have performed equally well. Another creative research program is exploring the role of learning in the type of naturally occurring behavior that interests behavioral ecologists. One can easily fall into the trap of considering learning as something that happens in laboratories with rats and pigeons, yet learning is a ubiquitous behavioral mechanism that animals use in many contexts. Recent work by Dukas addresses this problem by exploring the role of learning about mates and courtship behavior in *Drosophila* (Dukas 2004b, 2005a, 2005b).

Memory Rules and "Parallel Tracking"

Consider a forager that travels, encounters patches, and exploits them. The forager spends more time in, and extracts more from, each patch when it experiences long travel times between patches. Now suppose that travel times change; say, it experiences long travel times for a few days (demanding long patch exploitation times) and then experiences short travel times (demanding short patch exploitation times) a few days later. It is reasonable, I think most readers will agree, to say that a forager who adjusts to the change in travel time is *tracking* its environment, but this situation differs from the tracking problem outlined above. There the forager had to leave the stable resource to check the state of the varying resource, while in this new situation the forager obtains information about an environmental change in the course of its normal activities. Using a geometric analogy, we will refer to tracking in which the forager has to switch away from current activities *orthogonal*, and tracking in which information can be obtained without a change in behavior as *parallel*.

Parallel and orthogonal tracking problems focus on somewhat different questions. In orthogonal tracking problems, as discussed above, we focus on the allocation of effort to sampling and exploitation. In parallel tracking problems, we ask questions about how foragers use past experience to guide current action. Very early in the development of foraging theory, Richard Cowie (1977) speculated that animals might use a "memory window" to solve the varying travel time problem, using experience from the past to estimate the current travel time. In addition, Cowie speculated that there may be an optimal memory window length: in some situations an animal might do best with a very long memory window, while in others it might be better to devalue the past quite quickly. Since Cowie's early theorizing, it has become traditional to think of parallel tracking problems as problems of memory length and parameter estimation.

Weighting Past and Present

McNamara and Houston (1987a) provide a powerful yet simple way to think about this problem (see also Getty 1985; Hirvonen et al. 1999). Consider an economically important parameter (say, θ, where θ may be the current travel time or the rate of encounter with profitable prey items). At time i, the forager has (1) an estimate of θ—say, μ_i—and (2) a fresh sample—say, X—that provides new information about the value of θ. McNamara and Houston advocate a simple rule for updating the estimate:

$$\mu_{i+1} = \alpha\mu_i + (1 - \alpha)X, \tag{2.6}$$

where α $(1 \geq \alpha \geq 0)$ is the parameter of interest. If α is large, the rule emphasizes the past estimate (μ_i), but if α is small, the past estimate is devalued and the current sample (X) is stressed.

McNamara and Houston point out that, despite its simplicity, this linear updating rule [eq. (2.6)] is quite flexible and general. If, for example, we allow α to depend on i, then many popular rules fit into this framework, including simple averaging and memory windows. McNamara and Houston ask what determines the optimal value of α, and although their mathematical approach is rather advanced, the basic results are straightforward and intuitively appealing. The parameter α should reflect the relative reliability of the past estimate and the current sample. Two things affect this balance: the rate of change in the environment and the extent to which a given sample (X) provides a clean estimate of the current state of the environment. Generally speaking, environmental change decreases the optimal α because it means that past information

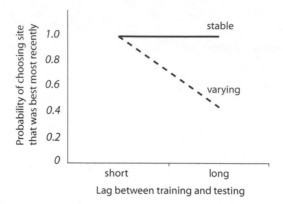

Figure 2.8. Results of Devenport and Devenport's tracking experiment. The data reveal an interaction between environmental change and the interval between training and testing. Animals trained in a stable environment always used their past experience, but animals trained in a variable environment relied on their experience only when tested soon after training.

is less reliable than current information, while sample noisiness (variance in X) makes the current sample less reliable and so should increase the optimal α. Another important variable is the time between samples: when the environment changes, a long lag time between samples should devalue past information (lowering the optimal α).

Devenport and Devenport (1994) performed a simple experiment to test these ideas. They trained ground squirrels to visit a pair of provisioned feeding stations. In the stable treatment condition, the same station always provided the highest feeding rate, while in the varying treatment condition, the two stations alternated. The Devenports then tested the animal's preferences after different delays (1 hour or 48 hours after the end of training). They found an interaction between environmental change and delay (fig. 2.8). In stable environments, the ground squirrels always used their prior experience, but in varying environments, they relied on prior experience only when the delay between training and testing was short.

Psychologists categorize memory into two types: representations of very recent events (working memory, short-term memory) and representations of events archived over longer periods (long-term or reference memory) (see chap. 3). In addition, psychologists usually view the interaction between these two components of memory as a fixed feature of the underlying neural mechanisms. The ideas presented here suggest a relationship that is more dynamic and responsive to economic factors. We do not yet know whether a behavioral ecological approach can contribute to studies of memory, but this approach certainly presents some intriguing possibilities.

Tracking Travel Time

In an important series of studies, Cuthill and his colleagues (Cuthill et al. 1990, 1994; Kacelnik and Todd 1992) manipulated the temporal pattern of travel times and observed the effect on patch exploitation behavior. This experimental paradigm challenges conventional theory because conventional models predict that the long-term rate of patch encounter will control patch exploitation patterns, and that travel time patterns such as "long-short-long-short-long-short . . . " will give the same long-term encounter rate as "long-long-long-short-short-short . . . " Yet these researchers found that observed patch-leaving behavior reflects the most recently experienced travel time, rather than the environmental average (Cuthill et al. 1990), making this one of several lines of evidence against the long-term maximization assumptions of traditional foraging theory.

Cuthill et al.'s (1990) result suggests a small α—experimental animals appear to devalue past experience and emphasize recent experience. In this study, the researchers determined travel times randomly in each patch cycle, with half of all travel times being short and half being long. In a second study (Cuthill et al. 1994), travel times changed much more slowly—on average, only once per day. This study provided evidence of long-term effects because patch exploitation patterns changed gradually after long-to-short (or short-to-long) transitions (the argument being that if only the most recent travel experience was important, then the first short travel time should be sufficient to change observed behavior). Unfortunately, no study has compared different levels of environmental change within a single experiment.

Parallel Tracking and the Behavioral Ecology of Memory

These empirical and theoretical studies suggest how economic variables might influence the way in which animals combine recent and long-term experience, yet behavioral ecologists could do much more. Specifically, no single study has manipulated environmental change and sampling error in a factorial way. In addition, we need more basic theoretical work. We need models of short-term maximization to account for effects like those observed by Cuthill and colleagues, and we need to link these studies with the mechanistic basis of animal memory (see chap. 3).

2.6   Public versus Private Information

On a field edge, a starling hunts for insects in clumps of short grass. As it forages, its success or failure provides it with information about whether a particular clump is rich or poor. But starlings seldom forage alone, and the successes and failures of flockmates also provide clues about resource quality. A growing body of evidence suggests that "neighbors" can provide information about food resources. In most of this chapter, we have assumed that an animal's information comes from its direct experience in the environment—its successes and failures, the food it sees, the cues associated with food, and so on. For many animals, however, "the group" represents a central aspect of the environment, and so it comes as no surprise that an animal may use the actions of its groupmates as a source of information. In addition, behavioral ecologists have long thought that groups can improve feeding rates, and information transfer among group members can account for at least part of this facilitation effect (see, for example, Krebs et al. 1972; Lack 1968; Ward and Zahavi 1973). Recent work (Clark and Mangel 1984, 1986; Templeton and Giraldeau 1995; Valone 1989; Valone and Giraldeau 1993) has sharpened our questions about the distinction between public and private information.

Consider again the orthogonal tracking problem outlined in the previous section, but imagine this time that two individuals exploit our simplified environment of varying and stable resources. If individual A samples according to the model, then individual B can avoid the costs of sampling by watching individual A's behavior. This problem is a game theoretical one (see boxes 1.3 and 1.4). Inman (Inman 1990; Krebs and Inman 1992) has studied this problem theoretically and experimentally. He argues that the only stable equilibria of this game occur when one individual samples at the individual optimum and the other parasitizes the sampler's actions. Intermediate "shared sampling" equilibria are unstable because if one individual increases its sampling rate, the other should decrease its sampling rate, leading inevitably to the stable "sampler-parasite" situation. Inman tested these predictions empirically by testing four pairs of starlings in both "alone" and "paired" conditions. In the paired condition, one individual lowered its sampling rate while the other sampled at the same rate as in the "alone" condition.

The simplest question that one can ask about public information is whether foragers use information from neighbors. This question suggests simple studies in which one manipulates the presence (or absence) of conspecifics. Investigators have conducted several studies of this type with intriguing results. Templeton and Giraldeau (1996) studied foraging starlings in a patch exploitation situation. They paired subjects with a trained stooge, who either gave up quickly or exploited the patch fully. Surprisingly, Templeton and Giraldeau found that the geometry of the experimental patch determined whether the stooge influenced patch-leaving behavior. When the experimental patches were linear arrays (egg-carton-like arrays of food wells) and the subject could exploit them systematically without information from the stooge, the stooge's behavior had no effect on patch leaving. However, when the patches were square arrays of food wells, the stooge's behavior did affect patch leaving, presumably because the subject had more difficultly implementing a simple exploitation rule. In another example, Smith et al. (1999) found that foraging crossbills were (in effect) better "empty patch" detectors when paired with *two* conspecifics, while a single conspecific did not improve their performance.

Results like these suggest that studies of public information must address subtle issues. These issues parallel the methodological problems in the closely allied field of social learning (see, for example, Galef 1988; Shettleworth 1998). For social animals, the presence of conspecifics influences behavior in many ways. A key challenge for students of public information is to disentangle the informational and noninformational effects of sociality. One approach to this problem would combine social and nonsocial treatments with direct manipulations of the value of information. For example, one might create fixed and varying environments and test a focal animal in alone and paired conditions in a factorial way within these environments. If public information influences behavior, we would expect an interaction between the information treatments and the group size treatments. Recent work by Dornhaus and Chittka (2004) on the honeybee dance language provides a masterful example of how we might study the information value of social interactions.

2.7 The Behavioral Ecology of Information and Cognition

Information problems connect behavioral ecology with basic behavioral mechanisms such as learning, memory, and decision making. The mechanisms in question cover a broad swath of animal biology that includes sensory biology, neurobiology, psychology, and cognitive science, which taken together represent an enormous and important research enterprise. Students of foraging information are building connections with these mechanistic research programs in two ways. First, behavioral ecologists can use knowledge of behavioral mechanisms to constrain their models. For example, Kacelnik and his colleagues (Gibbon et al. 1988; Kacelnik et al. 1990) have incorporated the scalar property of animal time estimation (animals remember long intervals less accurately) into foraging models to provide a mechanistic account of risk sensitivity, patch exploitation, and animal preferences for immediacy. According to this view, the scalar property reflects a basic property of the neural timing system (Gibbon et al. 1997) that *constrains* foraging behavior. This approach assumes that some mechanism constrains animals to have less accurate representations of long time intervals and works out the consequences for foraging behavior. The second, and more challenging, type of connection occurs when behavioral ecologists use economic principles to provide novel insights into questions about behavioral mechanisms. An obvious example is signal detection theory, in which an economic model led to the rejection of the mechanistic idea of absolute sensory thresholds. In addition, the growing number of mechanistically based models in cognitive science and neuroscience provide new opportunities for behavioral ecologists. For example, neural network models of learning (Montague et al. 1995, 1996; Sutton and Barto 1981) may provide tools to generalize the simple models of tracking discussed in this chapter.

The models presented here rely on the mathematical machinery of statistics and stochastic processes, but foraging animals do not face information problems that precisely parallel the estimation and testing problems of classic statistics. In introductory statistics courses we learn to estimate quantities and make hypothesis tests. Foraging animals do not need estimates or hypothesis tests; they need to make decisions about how to feed. The development of signal detection theory presented earlier makes this point clearly. A statistician faced with a mixture of tasty and noxious beetles would take a sample and *estimate* the probability that the sample came from the tasty distribution. While this calculation is relevant to a beetle-eating forager, it really isn't the forager's problem. The forager needs to decide what to eat, and as signal detection theory shows, the optimal position of the "eat-don't eat" threshold depends on the tasty and noxious distributions and the costs and benefits associated with eating and avoiding the two types of beetles. The relevant body of statistical theory is statistical decision theory (DeGroot 1970; Lindley 1985; Dall et al. 2005), and not the classic statistics of estimation and hypothesis testing (Getty 1995 provides an elegant example of this difference). Nevertheless, we sometimes find it useful to frame problems as "estimation problems," as we did in our discussion of parallel tracking. This can be a useful modeling strategy in situations in which we don't know, or don't want to specify, how the acquired information will be used.

To say that animals do not *need* to make estimates does not mean that they don't. Animals can solve the same problem in different ways. Cuthill's starlings might have a simple procedural rule, such as "I'm tired so I'll spend a long time in this patch," or they might form some neural estimate of the current travel time between patches and use this to make a more sophisticated decision about patch exploitation. These questions tread in the realm of cognitive science. If starlings maintain some sort of representation or encoding of the current travel time, then a cognitive scientist would describe this as *declarative* knowledge (and according to some views, it would therefore qualify as a truly cognitive process; see Shettleworth 1998). If, instead, the starling uses a simple rule, we describe this as *procedural* knowledge. Studies of these types of questions are difficult, but they can be quite informative, as they have been in studies of navigation (e.g., Dyer 1998). So far as I am aware, there is no general theory about when one would expect a declarative representation to be better than a procedural one, although it should be easier to trick procedural rules by testing them outside of the context where they evolved.

2.8   Summary

Animals obtain information about the state of the environment as they forage. Information is valuable when it can tell an animal something that changes its behavior. The theory of signal detection provides a framework for the analysis of problems in which a decision maker must act in the face of environmental (and neural) noise. The overlap between the signal and noise distributions and the relative costs of false alarms and misses determine the optimal discrimination strategy.

Patch sampling has been an important topic within foraging theory. The distribution of patch types determines how information will affect patch exploitation. Animals must track changing environments, and we recognize two types of tracking problems. In orthogonal tracking, an animal must change its behavior to track a resource that it is not currently exploiting, while in parallel tracking, an animal can observe changes without changing its behavior. In orthogonal tracking problems, one focuses on the sampling rate; environmental change, and the benefits associated with varying and stable resources, influence the optimal sampling rate. In parallel tracking problems, one focuses on how animals should combine past and current information. Two factors, environmental change and sampling error, influence their behavior. When the environment changes rapidly, past information should be devalued. When a sample provides a noisy estimate of the current state, then past information should be emphasized and the current sample should be devalued. Finally, foragers can obtain information from conspecifics and group members. These public information problems should be analyzed using game theory.

2.9   Suggested Readings

The approach of Dall et al.'s (2005) recent review parallels the approach taken in this chapter, but it offers a broader perspective. A recent study by McLinn and Stephens (2006) explores the framework presented here, experimentally focusing on the effects of environmental uncertainty and signal reliability. Giraldeau (1997) provides a review of information in behavioral ecology with a more empirical emphasis. Bradbury and Vehrencamp's (1998) recent book on animal communication covers many of the same issues in a different context. Gescheider (1985) provides an engaging account of psychophysics. Volumes by Egan and Swets (Egan 1975; Swets 1996) provide reviews of signal detection theory and its applications. The volume edited by Dukas (1998a) reviews the relationship between behavioral ecology and cognition. Shettleworth (1998) gives a comprehensive, biologist-friendly treatment of psychological phenomena and practice.

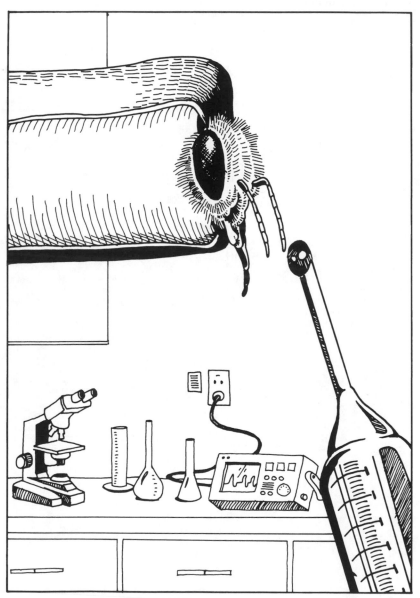

Telander

Neuroethology of Foraging

David F. Sherry and John B. Mitchell

3.1 Prologue

Alive with color, a patch of flowers is also alive with the constant motion of bumblebees, honeybees, syrphid flies, and other pollinators. A
bumblebee lands heavily on a flower, making other insects take flight.
She turns, plunges her head into the corolla, and remains motionless.
After a few seconds, she backs out, rises noisily into the air, and joins the
pollinators shuttling between flowers. Every one of these insects is making decisions about which flowers to visit, how long to remain at each
flower, and how much nectar or pollen to take on board before flying
off. This aerial traffic has a pattern that foraging theorists try to understand with models of energy maximization, efficiency maximization,
and other currencies that they can build into a model and test.

Underneath the rocketing flight from bloom to bloom is another
hubbub invisible to us—the flight of electrical and chemical signals
through the pollinators' nervous systems. Each decision, each choice,
each arrival and departure emanates from unseen neural chatter taking
place on a scale measured in microns and milliseconds. Electrical signals
coursing along neurons carry messages about nectar concentration and
the odor and color of flowers. Chemical signals jump the gap from one
neuron to the next and relay this information to the bumblebee's brain.
Inside neurons, other chemical messengers jot notes on incoming data

while gene transcription records a long-term archive of foraging experience, changing the way the bumblebee's nervous system responds to incoming information. Her next search for a flower worth stopping at will use this information, and her next foraging decision will be based on the neural record of her past experience.

3.2 Introduction

The modeling of foraging behavior has been successful because it makes clear assumptions and explicit predictions about behavior. Part of the appeal of foraging models, and a good deal of their power, is due to their indifference to the cognitive and neural processes underlying foraging choices. This is not to say that researchers working with foraging models are indifferent to causal mechanisms or unaware of the mechanistic questions raised by foraging models. Good foraging models are themselves indifferent to whether a patch departure decision, for example, takes place in the nervous system of an insect, a bird, or a human. Behavioral ecologists can fruitfully construct and test foraging models while remaining uncommitted on the question of how the brain and nervous system arrive at a foraging decision. We expect a foraging model to have broad applicability across taxa and therefore not to depend much on the specifics of mechanism. Increasingly, however, foraging theory has attempted to incorporate information about learning, memory, perception, timing, and spatial ability. One reason for this is that models grounded in accurate information about mechanisms are likely to make better predictions. Another reason is that researchers who are perfectly satisfied with the predictive power of a strictly functional foraging model may eventually ask themselves, "How does it work?"

This chapter explores the relevance of some recent discoveries in the neurosciences to the question of how nervous systems implement foraging decisions. We begin with two caveats: First, our coverage is far from comprehensive. We have selected several recent findings in the neurobiology of animal cognition that seem particularly clear, interesting, and relevant to foraging. Second, there are pitfalls in searching the nervous system for functions that we identify by observing behavior, but which actual nervous systems may not recognize. Research on foraging, like all research on behavior, requires identifying basic conceptual units such as search time, handling time, encounter rate, and intake rate, not to mention memory, variance sensitivity, and state. Most likely, the nervous system does not compartmentalize things in the same way that we conventionally do when observing behavior. This is not to say that the categories of behavior important in foraging models are wrong: they are not. They are categories appropriate to modeling the foraging decisions of animals. We should not be surprised, however, to find that categories useful for observing behavior do not always correspond to how the nervous system actually performs its job of integrating incoming sensory information with prior experience to produce adaptive foraging.

Insect pollinators provide many illustrations of the cognitive processes crucial to foraging. Recent studies reveal how the honeybee brain forms associations at the neuronal and molecular levels among stimuli that are important for successful foraging, such as floral odor and nectar. We begin with a look at the cognitive processes that control honeybee foraging, followed by a more detailed examination of how neurons in the honeybee brain form associations. Similar molecular processes of associative learning turn up in many invertebrates and vertebrates. Next, we look at some more complex aspects of cognition, beyond basic association of stimuli and events. Although associative learning forms an important building block of animal cognition, we can examine many cognitive processes more easily at a level of abstraction once removed from the formation of associations. The hippocampus, a tantalizing and perplexing structure in the vertebrate brain, participates in many cognitive operations relevant to foraging, including spatial memory, episodic memory, declarative memory, and the formation of complex associations. We examine the involvement of the hippocampus in two of these operations, spatial memory and declarative memory. Finally, we discuss the role of the mammalian prefrontal cortex in working memory. Working memory is memory for the ongoing performance of a task and is of central importance in many foraging decisions. The prefrontal cortex and its involvement in working memory illustrate the large-scale integration of neural information processing. We will begin, then, with a description of how foraging animals learn that two stimuli go together, describe some more complex cognitive operations that involve the hippocampus, and end with the role of the prefrontal cortex in keeping track of foraging as it occurs.

3.3   Honeybee Foraging

The Patch Departure Decision

Honeybees leave their hive and travel to nectar sources that may be anywhere from a few meters to 2 km away. A bee visits a series of flowers, draws nectar into its honeycrop, and then begins the journey home, often with only a partially filled crop (Schmid-Hempel et al. 1985). As floral density decreases and travel time to the next flower becomes longer, bees visit fewer flowers before returning home. This correlation between floral density and the number of flowers visited before returning to the hive supports the assumption that honeybees maximize efficiency (net energy gain/energy expenditure) rather than the more conventional currency of net energy gain (Schmid-Hempel et al. 1985; see also section 8.3). In order to respond to the travel time between flowers, foraging honeybees must monitor this variable in some way and then base their decision to cease foraging on their current estimate of travel time, stored in working memory. Memory for travel times between flowers is an important part of honeybee foraging.

Flower Constancy

Honeybees, like other pollinators, can show remarkable constancy within patches of flowers, often specializing on only one of many available species of flowering plants (Chittka et al. 1999). Students of foraging have explained the phenomenon of flower constancy in several ways, including pollinators' limited memory for rewarding flower types, limited memory for flower handling techniques (Gegear and Laverty 1998), and reduced efficiency caused by switching among flower types (Darwin 1876). Chittka and Thomson (1997) found, for example, that bumblebees could learn two flower handling techniques if trained appropriately, but made substantially more errors and wasted more time than bees that learned only a single flower handling technique at a time. The way memory for flowers works in the honeybee brain may make flower constancy advantageous. Memory can have pervasive and unexpected effects on foraging.

Learning Flowers

Honeybees must learn to identify floral nectar sources. Although bees have shape, color, and odor preferences, they do not recognize specific flowers innately and certainly do not know the locations of flowers before they begin foraging. They learn the location, shape, color, and olfactory characteristics of flowers by associating these features with the nectar that a flower provides. As Collett (1996) and others have shown, honeybees learn the locations of nectar sources by remembering a retinotopic representation of the local landmark array around a nectar source. "Retinotopic" means that the bee retains in memory a representation that preserves the relations among objects in the visual world as they impinge on the retina. Bees return to flowers by traveling in a manner that produces a match between their current retinal image of landmarks and their remembered representation of landmarks viewed during the departure flight from the flower. We have known since the work of von Frisch that honeybees learn the color of rewarding food sources (von Frisch

1950). The ways bees learn about the shape and olfactory characteristics of flowers has also been studied extensively (Greggers and Menzel 1993). Learning to recognize sources of food is an essential component of foraging.

3.4   Associative Learning

All of these components of honeybee foraging—whether they deal with travel time, flower handling techniques, or floral features—involve the formation of an association between a food reward and properties of the food source. Whereas nectar in a flower or sucrose solution in a laboratory experiment is the reward, the stimulus properties of the food source are the cues indicating the presence of a reward. The stimulus properties of the food source hold no special significance for the bee until she has experience with the relation between those stimuli and the presence of food and has associated those stimuli with a food reward. The bee's ability to form associations lies at the heart of foraging success.

The simplest way of conceptualizing the formation of associations is classical, or Pavlovian, conditioning. Classical conditioning describes the formation of an association between an unconditioned stimulus (US) that has innate significance for an animal, as nectar does for a honeybee, and a conditioned stimulus (CS) with no such prior significance. As a result of pairing between the CS and US, the CS becomes associated with the US. After repeated pairings, the occurrence of the CS alone produces responses by the animal that the CS did not cause prior to the formation of the association.

Over a century of experimental research has shown how such associations form. Many interesting complications and variations on the simple account of classical conditioning given above have been discovered (Rescorla 1988; Shettleworth 1998). For example, co-occurrence in time of a CS and US is not enough to produce learning. Instead, the US must be contingent upon the occurrence of the CS, or, to put it another way, the CS must be a good predictor of the US. Animals can form associations not only to a CS, but also to the context in which the CS occurs. In addition, animals can form inhibitory associations that reduce the probability of a response to a CS that predicts that the US will not occur.

The fundamental idea underlying the formation of Pavlovian associations, however, is a simple one. Association of a CS with a US causes animals to respond to the CS in ways that they did not prior to learning. Discovering how such associations form in the nervous system has become the Holy Grail of the neurobiology of learning. Somewhere in the nervous system—at a synapse, in the soma of a neuron, or in the combined action of many neurons—there must be a relatively permanent change that *is* the association. Somewhere, neurally encoded information about the CS and the US has to converge. The temporal properties of their co-occurrence must change the nervous system so that subsequent occurrences of the CS have effects that they did not have previously. Not all learning, even in honeybees, consists of the formation of associations, and not all associations are formed in the same way. Nevertheless, much of the neurobiological investigation of learning, as we shall see, has been a search for the mechanisms by which associations form.

Honeybees, like many insects, reflexively extend the proboscis upon stimulation of sucrose receptors on the antennae, mouthparts, or tarsae. Classical conditioning of the proboscis extension response (PER) has been analyzed in detail in honeybees. This unconditioned response is not only of central importance in natural honeybee foraging, but can also be conditioned in restrained honeybees (Takeda 1961). The conditioned response to olfactory and visual cues can be assessed behaviorally by measuring the probability, latency, or duration of proboscis extension, or electrophysiologically by measuring the latency, duration, and frequency of spike potentials in the muscle controlling proboscis extension (Rehder 1989; Smith and Menzel 1989). Olfactory CSs are more readily associated with sucrose than are visual cues (Menzel and Müller 1996), so classical conditioning of olfactory CSs to a sucrose US will be discussed below. The neural pathways responsible for classical conditioning of the PER are well understood and illustrate a general feature of systems that support associative learning: convergence of CS and US inputs at a common neuronal target.

The Mushroom Bodies of the Honeybee Brain

The mushroom bodies of the honeybee brain are bilateral three-lobed structures located in the protocerebrum. Each mushroom body consists of about 170,000 neurons, called Kenyon cells, and their projections. The cell bodies of the Kenyon cells are located around the mushroom body calyces, and the rest of the mushroom body consists of a dense *neuropil* of projections from, and afferent inputs to, the Kenyon cells (see box 3.1 for a glossary of italicized terms). In honeybees, the mushroom bodies receive olfactory afferents from the antennal lobes, visual afferents from the optic lobes, and multimodal input from a variety of other brain areas (Heisenberg 1998; Strausfeld et al. 1998). After examining the firing patterns of individual neurons, Erber et al. (1987) were able to propose several functions for the mushroom bodies, including detection of stimulus combinations, detection of temporal patterns between events, and detection of stimulus sequences. The mushroom bodies are promising candidates as a site for the integration of sensory information, the formation of associations, and the control of honeybee foraging behavior.

BOX 3.1 Glossary

Acetylcholine (Ach) A biogenic amine that acts as a neurotransmitter in vertebrate and invertebrate nervous systems. Neurons using the transmitter acetylcholine are described as *cholinergic*. The *muscarinic* acetylcholine receptor is a membrane protein in the postsynaptic membrane that contains an ion channel activated by the binding of acetylcholine. The action of acetylcholine at this receptor is mimicked by the plant alkaloid muscarine. The *nicotinic* acetylcholine receptor is a G protein-coupled membrane protein with no ion channel. Nicotine mimics the action of acetylcholine at this receptor.

Antagonist A compound that opposes the action of a neurotransmitter, hormone, or drug by acting on its receptor. An *agonist*, in contrast, acts on a receptor with an effect similar to that of a transmitter, drug, or hormone.

Antisense A strand of DNA or RNA that is complementary to a coding sequence. Because it is complementary to the coding sequence, the antisense hybridizes with it and thereby inactivates it. Antisense can be used to precisely target specific proteins and prevent their synthesis.

Biogenic amines Compounds that serve communication functions in both plants and animals. Serotonin (5-hydroxytryptamine), acetylcholine, histamine, octopamine, and the catecholamines adrenaline, noradrenaline, and dopamine are all biogenic amines.

Ca^{2+} The calcium ion. Ca^{2+} acts as a second messenger in neurons. The intracellular Ca^{2+} concentration is maintained at a very low level compared with the extracellular concentration by a calcium pump and a Na^+/Ca^{2+} exchange protein. *Calmodulin* mediates the effect of Ca^{2+} on proteins.

Calmodulin A protein that binds Ca^{2+} and regulates the activation of other proteins, including the Ca^{2+}/calmodulin-dependent (CaM) protein kinases.

CRE (cyclic AMP response element) A highly conserved DNA sequence that acts as a promoter of the transcription of many different target genes. The *cAMP response element binding protein (CREB)* is a transcription factor that is activated by cAMP via the action of protein kinase A (PKA), binds to the CRE promoter site, and initiates transcription of the target gene.

Cyclic AMP (cAMP, 3',5'-cyclic adenosine monophosphate) A cyclic nucleotide that acts as a second messenger in neurons and was the first second messenger discovered. The enzyme *adenylate cyclase* (also called *adenyl cyclase* and *adenylyl cyclase*) converts ATP to cAMP, while the enzyme *cyclic nucleotide phosphodiesterase* rapidly degrades cAMP to 5'-AMP. Activation

(Box 3.1 continued)

of these two enzymes thus regulates the concentration of cAMP within neurons. cAMP activates the cAMP-dependent protein kinase *protein kinase A*.

Glutamate An amino acid that acts as an excitatory neurotransmitter in the mammalian nervous system. There are several different glutamate receptors, named according to the agonist that most effectively mimics the effect of glutamate, including the *NMDA* (N-methyl-D-aspartic acid) receptor and the *AMPA* (α-amino-3-hydroxy-5-methyl-4-isoxazoleproprionate) receptor.

Neuropil (neuropile) A dense feltlike matrix of axons, axon terminals, and the dendrites with which these axons form synapses.

Octopamine A biogenic amine that acts both as a hormone and as a neurotransmitter in invertebrate and vertebrate nervous systems. As a neurotransmitter, it is an adrenergic agonist.

Phosphorylation The transfer of a phosphate group from ATP to a protein. Phosphorylation changes the shape, and hence the activity, of many proteins, including ion channels, second messengers, enzymes, and proteins that regulate gene transcription.

Protein kinase A compound that catalyzes the transfer of phosphate from ATP to a wide variety of proteins, a process called *phosphorylation*. *Protein kinase A* is activated by cAMP, *protein kinase C* is activated by phospholipids and influenced by Ca^{2+}.

The CS Pathway

In honeybees, odors activate chemoreceptors on each antenna, which relay signals to the antennal lobes, where odor characteristics are neurally encoded (Lachnit et al. 2004; Flanagan and Mercer 1989) (fig. 3.1). The projection neurons of the antennal lobe form three main tracts, one of which innervates the calyces of the mushroom bodies. This projection from the antennal lobe to the mushroom bodies serves as the CS pathway for conditioning of the proboscis extension response (PER). Menzel and Müller (1996) suggest that *acetylcholine* is the neurotransmitter in the CS pathway from the antennal lobes to the mushroom bodies because acetylcholine *antagonists* disrupt conditioning of the PER without disrupting olfactory perception (Cano Lozano et al. 1996; Gauthier et al. 1994). This result indicates that acetylcholine antagonists do not impair PER conditioning simply by eliminating the incoming olfactory CS from the antennal lobe, but instead disrupt the CS signal at a later stage of processing.

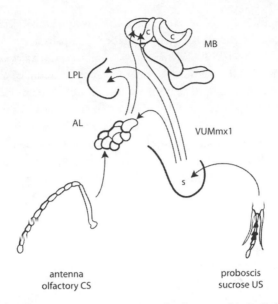

antenna   proboscis
olfactory CS   sucrose US

Figure 3.1. Schematic diagram of the CS and US pathways for olfactory conditioning in the honeybee. The olfactory CS detected by the antenna is relayed to the antennal lobe (AL) and then by acetylcholine-containing projections to the lateral protocerebral lobe (LPL) and the calyx (c) of the mushroom body (MB). The sucrose US detected at the proboscis is relayed to the subesophageal ganglion (s) and then by the octopamine-containing VUMmx1 nerve to the antennal lobe, the lateral protocerebral lobe, and the calyx of the mushroom body. The mushroom body, antennal lobe, and lateral protocerebral lobe are all bilateral structures that occur on both sides of the brain.

Neural signals triggered by activation of chemoreceptors on the antennae thus deliver information about the odor of a nectar source to Kenyon cells of the mushroom bodies via projections from the antennal lobe (Mobbs 1982).

The US Pathway

The unconditioned response of extending the proboscis in response to sucrose begins with sucrose receptors on the proboscis that send projections to the subesophageal ganglion (Rehder 1989). In the subesophageal ganglion, a group of ventral unpaired median (VUM) neurons receive input from the sucrose receptors. One of these neurons, the VUMmx1, responds to sucrose with a long burst of firing that outlasts the actual sucrose US presentation (Hammer 1993). Axons of the VUMmx1 neuron converge with the CS pathway at three different sites: the antennal lobe, the lateral protocerebral lobe, and the lip and basal ring of the mushroom body calyces (see fig. 3.1). There are thus several sites where information about the odor CS and the sucrose US converge.

The VUMmx1 neuron uses the neurotransmitter *octopamine* (Kreissl et al. 1994). Direct injections of octopamine into two of the targets of the VUMmx1 neuron, the mushroom body calyces and the antennal lobe, result in classical conditioning of the PER when the odor CS is paired with octopamine (Hammer and Menzel 1998). When octopamine and other *biogenic amines* are depleted by treatment with the drug reserpine, conditioning of the PER does not occur. Following such depletion, supplements of octopamine can restore conditioning (Menzel et al. 1999). To summarize, the US signal that the honeybee has encountered sucrose is conveyed to the mushroom bodies by the VUMmx1 neuron. Manipulations of the VUMmx1 neurotransmitter, octopamine, confirm this. Depletion of octopamine prevents conditioning, while its application at VUMmx1 terminals is sufficient to produce learning.

The Mushroom Bodies as a Locus for Memory

Although CS and US information converges at both the antennal lobes and the mushroom body calyces, the mushroom bodies appear to be especially important in conditioning, and direct evidence confirms this (Hammer and Menzel 1995). Cooling the calyces of the mushroom bodies produces amnesia similar to that produced by cooling the whole animal (Erber et al. 1980). Mutations resulting in abnormal mushroom body structure cause a loss of conditioning to odors (Heisenberg et al. 1985), and so does destruction of the mushroom bodies (de Belle and Heisenberg 1994).

Associative learning of any kind requires a point of neural convergence between conditioned and unconditioned stimuli. Neurobiological studies of associative learning have begun to describe what occurs at these points of convergence. An important concept introduced by Donald Hebb (1949, 62) serves as a guide for this research: "When an axon of cell A is near enough to excite a cell B and repeatedly or persistently takes part in firing it, some growth process or metabolic change takes place in one or both cells such that A's efficiency, as one of the cells firing B, is increased." In other words, structural changes in the nervous system result from one cell taking part in the firing of another. In the case of the honeybee proboscis extension response, Hebb's postulate leads us to ask what happens to mushroom body neurons when projections from the antennal lobe cause them to fire, and that firing is rapidly followed by further firing of these cells by octopamine release from the VUMmx1 axons. To find the answer to this question, we must now look inside the neurons that are activated in this way.

Cellular Mechanisms

Whereas neurotransmitters are the first line of biochemical messengers carrying signals from one neuron to another, there are also intracellular biochemical signals, known as second messengers. After a neurotransmitter arrives at its target cell and activates its receptor, the next, intracellular step in signaling involves the second messenger system. Numerous second messenger systems have been described in neurons. A complex pattern of interaction occurs among these intracellular second messengers, but several consistent themes emerge concerning the role of second messenger systems in learning and memory.

Within the mushroom bodies, the Kenyon cells are the site of CS and US convergence. Exposing cultured Kenyon cells to acetylcholine (the neurotransmitter conveying the CS signal from the antennal lobes) activates an ion current in these cells that has a high proportion of calcium ions (Ca^{2+}; Menzel and Müller 1996). This means that in the intact animal, olfactory stimulation of the antennal lobes, which causes release of acetylcholine, increases the concentration of Ca^{2+} within Kenyon cells (fig. 3.2A).

Octopamine, the US neurotransmitter, also leads to changes within mushroom body neurons (fig. 3.2B). Octopamine release and the subsequent activation of the octopamine receptor stimulate adenylate cyclase activity within Kenyon cells (Hildebrandt and Müller 1995a; Evans and Robb 1993). The enzyme adenylate cyclase converts ATP into *cyclic AMP* (cAMP); cAMP then has a number of intracellular effects, including activation of *protein kinases*, especially protein kinase A (PKA). In addition to its effect on adenylate cyclase, octopamine, like acetylcholine, can increase intracellular Ca^{2+} levels within mushroom body neurons (Robb et al. 1994).

Thus, the arrival of the CS odor signal and the US sucrose signal at the mushroom bodies activates adenylate cyclase and increases intracellular Ca^{2+} levels. The arrival of both signals produces a greater change within mushroom body neurons than either signal would alone. Olfactory cues alone would lead to a transient increase in Ca^{2+} levels. Stimulation of sucrose receptors would lead to a transient activation of cAMP (through adenylate cyclase activation) and a transient increase in intracellular Ca^{2+} levels. If these two inputs arrive within the appropriate time interval, however, the two effects occur together, and the resulting intracellular change is different, at least quantitatively, from the effect produced by either signal alone.

These CS- and US-induced changes in mushroom body neurons not only have additive effects, but interacting effects as well (fig. 3.2C). Adenylate cyclase activity, and hence the amount of cAMP produced, is potentiated by Ca^{2+} (Abrams et al. 1991; Anholt 1994). The net effect on mushroom body cells is elevated intracellular Ca^{2+} from the CS input, followed by increased adenylate cyclase activity from the US input. The US-induced activation of adenylate cyclase is greater than usual because Ca^{2+} increases adenylate cyclase activity and because the US input arrives at a time when intracellular Ca^{2+} levels are still elevated as a result of the CS signal. The final outcome is a

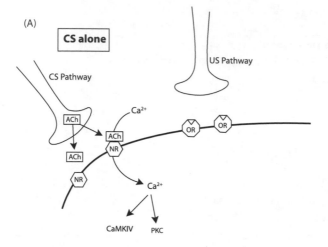

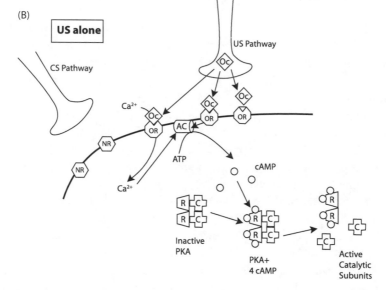

Figure 3.2. Convergence of odor CS and sucrose US signals in Kenyon cells of the honeybee mushroom body. (A) CS alone: CS-induced activity from the antennal lobes arrives in the mushroom bodies, triggering release of the neurotransmitter acetylcholine (ACh). Acetylcholine binds to a receptor (NR) and allows Ca^{2+} to enter the cell. The intracellular Ca^{2+} then activates Ca^{2+}-dependent kinases, such as PKC and CaMKIV. (B) US alone: US-induced activity in the VUMmx1 axon arrives in the mushroom bodies, triggering release of the neurotransmitter octopamine (Oc), which binds to an octopamine receptor (OR). Octopamine has at least two effects on the cell: it activates adenylate cyclase (AC), leading to the conversion of ATP into cAMP, and it increases intracellular Ca^{2+} concentrations. cAMP then activates protein kinase A (PKA) by binding to the regulatory subunits (R), causing them to dissociate from their catalytic subunits (C). Once the catalytic subunits of PKA are dissociated from the regulatory subunits, their active site is exposed, and they can act on various target substrates within the neuron, altering neuronal function.

(C)

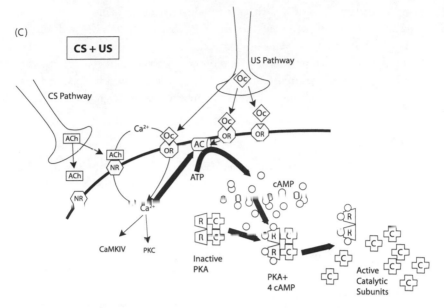

Figure 3.2 *(continued)* (C) CS + US: If the increased intracellular Ca^{2+} from CS stimulation is still present when the US signal arrives, it potentiates the ability of octopamine to activate adenylate cyclase, leading to the production of more cAMP and increasing the number of active catalytic subunits of PKA. For clarity, this illustration omits much of the detail relating to the Ca^{2+}-dependent kinases PKC and CaMKIV. The mechanism of activation of these kinases is analogous to that shown for PKA.

chemical environment within neurons that have received both a CS and US signal that is very different from that in neurons that have received only a CS or US signal alone.

The best-known example of comparable intracellular events in a vertebrate comes from studies of long-term potentiation in the mammalian hippocampus (Bliss and Lomo 1973). Long-term potentiation is a model of synaptic plasticity that may be analogous to the cellular events that occur in learning and memory (Malenka and Nicoll 1999). The excitatory amino acid *glutamate* functions as a neurotransmitter in the hippocampus (and elsewhere). Glutamate activates one type of receptor, the AMPA receptor, as part of normal neurotransmission. A second type of glutamate receptor, the NMDA receptor, is also present in the hippocampus, but it is usually in an inactivated state caused by the presence of the magnesium ion, Mg^{2+}. Because NMDA receptors are blocked in this way by Mg^{2+}, they are not normally involved in neurotransmission within the hippocampus. However, when stimulation produces an action potential and depolarizes a hippocampal neuron, the Mg^{2+} blockade of the NMDA receptor ceases, and glutamate can then activate the NMDA receptor. Such activation leads to an increase in intracellular Ca^{2+}

levels and recruits mechanisms that cause long-term changes in synaptic function (Bliss and Collingridge 1993). Here, too, we can observe the joint effect of the firing of multiple neurons that Hebb envisioned. In neurons of the mammalian hippocampus and in Kenyon cells of the honeybee brain, the arrival of two separate inputs in the correct order and within specific time intervals leads to intracellular changes that neither input can achieve alone.

Lasting Changes in Neurons

The intracellular interactions between CS and US signals are particularly relevant to understanding learning and memory because they can produce lasting changes in neurons when they occur. Research on associative learning has demonstrated the importance of second messenger systems in mediating changes at the synapse (box 3.2). These findings have linked many different second messenger systems and protein kinases to learning and memory across a phylogenetically diverse range of animals (Micheau and Riedel 1999). Studies

---

BOX 3.2   **A Nobel Prize in the Molecular Basis of Memory**

The 2000 Nobel Prize in Physiology or Medicine was awarded jointly to Arvid Carlsson, Paul Greengard, and Eric Kandel for their work on signal transduction in the nervous system. Carlsson received the prize for his discovery that dopamine is a neurotransmitter in the brain and for his research on the function of dopamine in the control of movement. Greengard received the prize for research on how neurotransmitters act on receptors and trigger second messenger cascades that lead to the *phosphorylation* of proteins and modification of ion channels. Kandel's award was for his work on the molecular mechanisms of memory.

Kandel's research on conditioning in the sea slug *Aplysia* revealed many of the basic intracellular processes of memory formation discussed in this chapter. *Aplysia* exhibit a gill withdrawal reflex when the gill is touched, and this reflex can be conditioned to stimulation elsewhere on the sea slug's body. Conditioning results from increases in the levels of second messenger molecules such as cAMP and PKA, leading to protein synthesis and changes in the shapes and properties of synaptic connections between cells. Kandel's recent work has explored comparable mechanisms such as long-term potentiation that may be responsible for memory formation in mammals and has described many similarities to the molecular mechanisms of memory discovered in invertebrates.

of learning in birds, mammals, and the sea slug *Aplysia* implicate protein kinase C (PKC), for example, in changes at the synapse, also known as synaptic plasticity (Micheau and Riedel 1999). Elevation of intracellular Ca^{2+} increases PKC activity. In the honeybee, PKC occurs in both the mushroom bodies and antennal lobes (Grünbaum and Müller 1998; Hammer and Menzel 1995), but its role in conditioning of the proboscis extension response remains unclear. Repeated proboscis extension conditioning trials increase PKC in the antennal lobes, beginning 1 hour after conditioning and continuing for up to 3 days. Blocking PKC activation, however, does not affect initial acquisition of the PER (Grünbaum and Müller 1998). Elevation of intracellular Ca^{2+} may also act through other Ca^{2+}-dependent kinases, such as Ca^{2+}/*calmodulin*-dependent kinase IV (CaMKIV). Activation of this kinase by Ca^{2+} may be an important mechanism underlying long-term memory (see below).

As noted earlier, elevated cAMP levels in the honeybee mushroom bodies activate PKA. There are high levels of PKA in the mushroom bodies (Fiala et al. 1999; Müller 1997), and octopamine is able to activate PKA both in the antennal lobes (Hildebrandt and Müller 1995b) and in cultured Kenyon cells (Müller 1997; but see Menzel and Müller 1996). The activation of PKA by cAMP appears to be a necessary step in the sequence of events that leads to lasting change in mushroom body neurons. The importance of PKA has been tested using *antisense* RNA. Inactivating PKA by injecting antisense RNA complementary to the mRNA sequence of a subunit of PKA impairs long-term memory measured 1 day after training (Fiala et al. 1999). Studies with *Drosophila* have also shown the importance of PKA. A variety of mutations have been identified in fruit flies that produce specific deficits in the flies' ability to form or retain simple associations, and many of these mutations affect the cAMP-PKA pathway (Dubnau and Tully 1998; Waddell and Quinn 2001). The *Drosophila* learning mutant dunce has a mutation of the gene for cAMP phosphodiesterase. Another learning mutant, rutabaga, has a mutation of the gene coding for adenylate cyclase. Both mutants have difficulty learning an association between odor and shock, and what learning they do exhibit decays very rapidly compared with that of wild-type fruit flies.

Converting the Memory Trace to the Engram

Although we do not yet know the full details of how honeybees form associations, we can use results from other species to infer how honeybees convert temporary elevations of cAMP and Ca^{2+} into long-lasting changes in neural pathways. In some animal cells, an increase in cAMP activates the transcription of specific genes. The regulatory region of these genes contains a short DNA sequence called the cyclic AMP response element (*CRE*). This

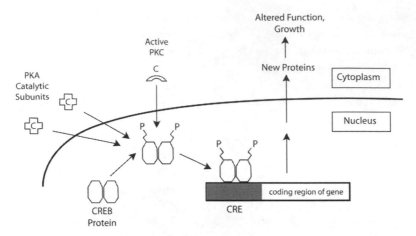

Figure 3.3. The catalytic subunit of PKA, once free of its regulatory subunit, migrates into the cell nucleus, where it phosphorylates proteins that regulate gene expression (phosphorylation is indicated by "P"). One target of PKA is cyclic AMP response element binding protein (CREB). Once activated by PKA, CREB binds to the cyclic AMP response element, CRE, a region of some genes that regulates their transcription. CREB can also be phosphorylated by protein kinases other than PKA, including Ca^{2+}-dependent kinases such as PKC, that would be activated by converging CS-US activity. The activity of genes that contain a CRE sequence is altered by binding with CREB, leading to a change in the production of mRNAs that code for the production of proteins.

CRE sequence is regulated by a specific protein called CRE-binding protein (CREB). CREB is a member of a large family of structurally related proteins that bind to the CRE sequence (fig. 3.3). When CREB is activated by PKA (which is activated by cAMP), it binds to the CRE sequence and regulates gene transcription (Bacskai et al. 1993). Interestingly, other Ca^{2+}-dependent kinases, such as CaMKIV mentioned above, also activate CREB (Ghosh and Greenberg 1995).

Studies of learning in *Drosophila* (Yin et al. 1994), the sea slug *Aplysia* (Bartsch et al. 1995), mice (Bourtchuladze et al. 1994), and rats (Lamprecht et al. 1997) confirm that CREB induces changes in long-term memory that depend on protein synthesis. In the honeybee, inhibition of protein synthesis does not disrupt learning measured 24 hours after training (i.e., learning that does not depend on protein synthesis), but does interfere with long-term changes measured 3 days after training (i.e., learning that does depend on protein synthesis; Wüstenberg et al. 1998).

In summary, high levels of PKA activity in the honeybee mushroom body are caused by an elevated level of cAMP, which results from the convergence of CS odor and US sucrose signals in Kenyon cells. Protein kinase A then activates CREB. CREB, in turn, modulates the activity of particular genes. A Ca^{2+}-dependent mechanism can also increase CREB binding and gene expression. CS- and US-induced activity converge at PKA (because Ca^{2+}

enhances cAMP activation of PKA) and at CREB (because a Ca^{2+}-dependent kinase and PKA each independently activate CREB). These events change the amounts or types of proteins produced in neurons that experience the convergence of the CS and US (see fig. 3.3). Change in gene expression produced by pairings of the CS and US provides a mechanism to translate transient stimulus-induced activation of these genes into lasting change in the nervous system.

Gene Expression

We know relatively little about the gene products that CREB regulates, or about the functions of those proteins. There are, however, several very interesting possibilities. CREB regulates a protein called synapsin I (Montminy and Bilezikjian 1987). Synapsin I anchors neurotransmitter-containing vesicles to the cytoskeletal network, and when phosphorylated by cAMP and Ca^{2+}-dependent kinases, releases synaptic vesicles, allowing them to move to the active zone at the end of the axon terminal for release. In this way, CREB activation can lead to changes in the level of a protein that regulates neurotransmitter release.

Another protein, ubiquitin, may also influence long-term learning (Chain et al. 2000). Ubiquitin acts on the regulatory subunits of PKA, allowing PKA to act on its target substrates. The amount of ubiquitin present in a neuron is regulated by CREB. Ubiquitin thus completes a positive feedback loop that can keep both PKA and CREB levels elevated within a neuron. Enhanced ubiquitin activity leads to greater PKA activity upon subsequent activation of the neuron, and hence greater CREB activity and a continuation of enhanced ubiquitin production (together with sustained change in other CREB-regulated gene products, such as synapsin I). These changes, once induced, can be self-perpetuating if the circuit is periodically activated. In *Aplysia*, an increase in ubiquitin activity occurs along with long-term facilitation (Hegde et al. 1997). Without such a mechanism, we would expect the effects of a change in gene expression to last only as long as the gene product. Most proteins have a life span of a few days (or less). Enhanced ubiquitin activity is one mechanism that may cause these effects to persist and produce long-term change in neurons involved in the formation of associations.

Learning, Memory, and Foraging

There may be considerable redundancy in the mechanisms of learning and memory. Experience-dependent plasticity in the nervous system of the honeybee is unlikely to depend on a single mechanism. Multiple interacting mechanisms are clearly involved in long-term potentiation in the mammalian hippocampus. Both PKA and a Ca^{2+}-dependent kinase can activate CREB, and CREB is only one member of a large family of transcription factors that modulate gene expression (Sassone-Corsi 1995). Similarly, the various protein kinases found in a neuron not only have their own functions, but also have powerful interacting effects on one another (Micheau and Riedel 1999). Other neurotransmitters and neuromodulators, second messenger systems, transcription factors, and gene products are likely to be involved as well. Nonetheless, evidence from a variety of experimental approaches and taxa (both arthropods and vertebrates) indicates that CREB represents a highly conserved mechanism for inducing lasting changes in neuron function.

What does this complex cascade of molecular events in the honeybee nervous system have to do with foraging? For at least one component of foraging—the association of floral odor with the presence of nectar—the causal chain can be followed along axonal projections to synaptic events that activate second messenger systems, initiate gene expression, and alter, both transiently and permanently, the behavior of the foraging bee. Whether the association of nectar with floral color, shape, and location occurs in a similar fashion remains an open question, although the role of second messenger systems in the formation of associations in animals as widely separated phylogenetically as *Aplysia*, *Drosophila*, and laboratory rats follows a broadly similar pattern. It is likely that the estimation of travel time between flowers in a patch, the representation of landmarks, acquisition of flower handling techniques, and many other components of foraging involve similar neurobiological processes. It is likely that foraging decisions and the acquisition of information while foraging, though they may involve many parts of the nervous system and different molecular mechanisms, will ultimately be traceable to comparable processes within neurons.

This section has described the cellular basis of learning and memory. Box 3.3 introduces current thinking about another component of foraging, the neural mechanisms of reward. Foragers not only must learn which events in the world are associated, but also must determine which events are likely to have positive rewarding outcomes. The concept of reward represents an important link between foraging and the neuroscience of behavior.

3.5   The Hippocampus

Many of the cognitive processes involved in foraging, including spatial memory, working memory, episodic and declarative memory, the formation of complex associations, and the integration of experience over time, to name

BOX 3.3   Neural Mechanisms of Reward

BOX 3.3   Neural Mechanisms of Reward
Peter Shizgal

Neuroscientists are striving to identify the neural circuitry that processes rewards and to determine its role in learning, prediction of future consequences, choice between competing options, and control of ongoing actions. The following examples illustrate neuroscientific research on reward mechanisms and its relation to foraging.

Reward Prediction in Monkeys

William Schultz and his co-workers carried out an influential set of studies on the activity of single dopamine-containing neurons during conditioning experiments in macaque monkeys (Schultz 1998, 2000). Midbrain dopamine neurons in monkeys and other mammals make highly divergent connections with widely distributed targets in the brain. These neurons have been linked to many processes important to foraging behavior, including learning about rewards and the control of goal-directed actions.

One of the experimental tasks often employed by Schultz's group is delay conditioning. A typical conditioned stimulus (CS) is a distinctive visual pattern displayed on a computer monitor. After a fixed delay, the CS is turned off, and an unconditioned stimulus (US), such as a drop of flavored syrup, is presented (fig. 3.3.1). An intertrial interval of unpredictable duration (dashed line) then ensues before the CS is presented again.

As shown in figure 3.3.1, dopamine neurons typically respond with a brief increase in their firing rate when the US is first presented (left column, bottom trace). However, after the monkey has learned that the CS predicts the occurrence of the US, the dopamine neurons no longer respond to delivery of the reward (the US). Instead, they produce a burst of firing at the onset of the CS (central column). If a second CS is presented prior to the original one (not shown), the burst of firing transfers to the new CS, which has become the earliest reliable predictor of reward. Omission of the US, after the CS-US relationship has been learned, leads to a brief decrease in the firing rate of the dopamine neurons (right column).

The activity of the dopamine neurons at the time of reward delivery appears to reflect some sort of comparison between the reward that the monkey receives and the reward it had expected. When the monkey encounters the US for the first time, it is not yet expecting a reward; the outcome is thus better than anticipated, and the dopamine neurons increase their firing rate. After training, delivery of the reward merely confirms the monkey's

(Box 3.3 continued)

expectation, and thus the dopamine neurons are quiescent when the anticipated reward is delivered. Omission of the reward constitutes a worse-than-expected outcome, and the firing of the dopamine neurons slows.

Figure 3.3.2 provides a simplified depiction of a model that compares expectations to experience (Montague et al. 1996; Schultz et al. 1997). The moment-to-moment change in the reward prediction is computed by taking the difference between the reward predicted at a given instant

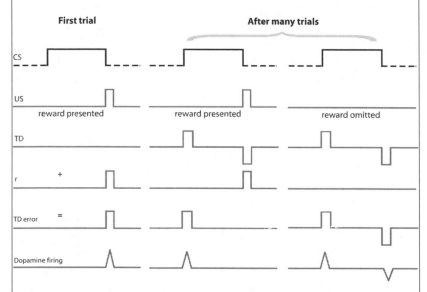

Figure 3.3.1. Responses of midbrain dopamine neurons in monkeys during delay conditioning. Presentations of the conditioned stimulus (CS) are separated by intervals of unpredictable duration (dashed lines). The unconditioned stimulus (US), a drop of juice, is delivered immediately following the offset of the CS. The gray traces represent elements of a model (see Figure 3.3.2) that attributes the changes in dopamine firing to temporal difference (TD) errors. The computation of the temporal difference and the temporal difference error is depicted in Figure 3.3.2. The internal signal that tracks the value of an ongoing reward (the US) is labeled "r."

in time and the reward predicted during the previous instant. Recall that the duration of the intertrial interval is unpredictable. Thus, during the instant prior to the onset of the CS, the monkey does not know exactly when it will receive the next reward. This lack of predictability is resolved in the next instant by the appearance of the CS. The positive "temporal difference" in the reward prediction indicates that the monkey's prospects have just improved.

(*Box 3.3 continued*)

It has been proposed (Montague et al. 1996; Schultz et al. 1997) that the dopamine neurons encode a "temporal difference error." As shown in figure 3.3.2, this error signal is produced when the temporal difference in reward prediction is combined with a signal indicating the value of the delivered reward. Consider the situation of a well-trained subject at CS offset (see fig. 3.3.1, central column). The instant before the CS is turned off, the reward prediction is strong. However, as soon as the CS disappears from the screen, an intertrial interval of unpredictable duration begins. Thus, the occurrence of the next reward has become less predictable, and the sign of the temporal difference is negative (trace labeled "TD"). However, this

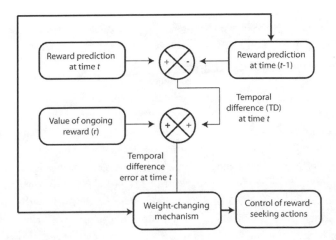

Figure 3.3.2 A simplified depiction of a model that uses temporal difference errors to shape predictions about reward and to control reward-seeking actions.

negative temporal difference coincides with the delivery of the reward. The positive value of the reward ("r") cancels the negative temporal difference. Thus, there is no error signal at the time of reward delivery, and no change in dopamine firing. Omission of the reward (right column) yields a negative temporal difference error and a decrease in dopamine firing. At CS onset in a well-trained subject (central and right columns), the reward prediction has improved. This yields a positive temporal difference error, which is reflected in increased dopamine firing.

In a class of models developed by computer scientists (Sutton and Barto 1998), temporal difference errors are used to form and modify predictions

(Box 3.3 continued)

about future rewards by altering the weights of connections in a neural network. A positive error increases (and a negative error decreases) the influence on reward prediction exerted by stimuli that were present during the previous instant. Thus, the temporal difference error produced in the initial conditioning trial (see fig. 3.3.1, left column) boosts the influence of the final instant of the CS on reward prediction. Over the course of repeated conditioning trials, these weight changes propagate backward through the CS-US interval to the earliest reliable predictor of reward, the onset of the CS.

Independent experiments have demonstrated that brief increases in the release of dopamine can change the sizes of cortical regions that respond to specific sensory inputs (Bao et al. 2001). This finding provides indirect support for the hypothesis that the brief changes in dopamine firing observed by Schultz's group are sufficient to change the strength of connections between neurons that form predictions of future rewards.

The activity of dopamine neurons can be described over multiple time scales (Schultz 2000). Prolonged, slow changes in the average extracellular concentration of dopamine have been observed during events such as the consumption of a tasty meal (Richardson and Gratton 1996). Thus, brief fluctuations in firing rate, such as those observed during conditioning experiments, may be superimposed on a background of slow changes in neurotransmitter release. Given these multiple time scales and the very widespread connections of the midbrain dopamine neurons, it is perhaps not surprising that these neurons have been implicated in many functions in addition to reward prediction, including the exertion of effort and the switching of attention and motor output. Thus, dopamine neurons may make multiple contributions to foraging behavior through several different psychological processes.

Foraging by Model Bees

Forming accurate predictions about future rewards is clearly advantageous to a forager. To reap the benefits of such predictions, the forager must use them to guide its actions. Note that in figure 3.3.2, the temporal difference error not only shapes reward predictions, but also influences reward-seeking actions. A simulation study (Montague et al. 1995) illustrates how temporal difference errors can guide a forager to promising patches.

The core element of the simulation is modeled on the properties of the VUMmx1 neuron of the honeybee, which is described in section 3.4.

(Box 3.3 continued)

This neuron shows some interesting homologies to the midbrain dopamine neurons of mammals. Like the projections of the midbrain dopamine neurons, the projections of the VUMmx1 neuron are highly divergent (see fig. 3.1). The VUMmx1 neuron releases octopamine, a neurotransmitter closely related to dopamine. The VUMmx1 neuron fires in response to certain rewards, and does so more vigorously when the rewarding stimulus is unexpected.

Real VUMmx1 neurons respond to chemosensory inputs (e.g., nectar). The model neuron, which we will call "VUMmxx," responds to visual cues as well and computes a temporal difference error. During encounters with flowers, the model VUMmxx neuron alters weights in a neural network that generates reward predictions. As a result, the model can learn which of several differently colored flower types contains nectar.

The output of the VUMmxx neuron steers the flight of the model bee; weight changes in the model are dependent on contact with flowers, so reward predictions do not change while the bee is flying. The decision rule governing flight is very simple. The stronger the output of the simulated neuron, the larger the likelihood that the bee will continue on its present heading; the weaker the output of the simulated neuron, the larger the likelihood that the bee will reorient randomly.

The distribution of flowers in the artificial field is nonuniform; although the field includes equal numbers of blue and neutral-colored flowers, the random scattering of flower types generates small "clumps" in which one of the colors predominates. Due to the learning that occurred during the model bee's prior contacts with the flowers, the strength of the influence exerted by each flower color on the firing of the simulated VUMmxx neuron varies according to the weights in the network. Let's assume that blue flowers recently yielded nectar and neutral-colored flowers did not.

When the model bee is flying at low altitudes, only a small number of flowers fall within its field of view, and a clump of one color is likely to predominate. If that color is neutral, and the predominance of neutral-colored flowers extends to the center of the field of view, then the firing of the simulated VUMmxx neuron will decrease as the bee descends. The action rule will then cause the bee to reorient, breaking off its approach to the unpromising patch. However, if blue flowers predominate, their prevalence will increase as the bee descends, and the rate of firing of the simulated neuron will tend to increase. This generates a positive temporal difference error, which strengthens the bee's tendency to approach the

(Box 3.3 continued)

blue flowers. Thus, temporal difference errors can guide a forager toward promising patches.

Foraging for Brain Stimulation

Electrical stimulation of the VUMmx1 neuron in the honeybee can serve as the US in a classical conditioning experiment. In the vertebrate brain, there are widely distributed sites where electrical stimulation serves as a most effective reward. Rats will work vigorously to obtain such stimulation by pressing a lever or even leaping over hurdles as they run up a steep incline.

Dopamine neurons play an important role in the rewarding effect of electrical stimulation, but the exact nature of that role has yet to be determined. Altering the synaptic availability of dopamine or blocking the receptors at which it acts changes the strength of the rewarding effect (Wise 1996). What is not yet clear is whether the reward signal is encoded directly by brief pulses of dopamine release or whether the dopamine neurons play a less direct role, for example, by amplifying or suppressing reward signals carried by other neurons.

Under the usual experimental conditions, the activation of dopamine neurons by the rewarding stimulation is mostly indirect, through synaptic input from the neurons that are fired directly by the electrode (Shizgal and Murray 1989). In principle, such an arrangement makes it possible for other inputs (e.g., signals representing reward predictions) to oppose the excitatory drive from the directly activated cells, which could explain why the brief stimulation-induced pulses of dopamine release decline over time (Garris et al. 1999). The input from the directly activated neurons may play the role of a "primary reward signal" ("r" in figures 3.3.1 and 3.3.2), which normally reflects the current value of a goal object, such as a piece of food. Indeed, the rewarding effect of electrical stimulation has been shown to compete with, sum with, and substitute for the rewarding effects of gustatory stimuli (Conover and Shizgal 1994; Green and Rachlin 1991).

It is very difficult to hold the value of a natural reward constant over time because of sensory adaptation and satiety. In contrast, rats and other animals will work for hours on end to obtain rewarding brain stimulation. This property makes brain stimulation a handy tool for studying neural and psychological processes involved in foraging. The strength, duration, and rate of availability of the stimulation are easily controlled, and the experimenter can set up multiple "patches" with different payoffs by offering the subject multiple levers or a maze with multiple goal boxes.

(Box 3.3 continued)

In research modeled on foraging, C. R. Gallistel and his co-workers have studied how the magnitude and rate of reward are combined by self-stimulating rats (Gallistel and Leon 1991; Leon and Gallistel 1998). Two levers are provided, and the rat cannot predict exactly when the stimulation will become available. However, the rat is able, over multiple encounters, to estimate the mean rate of reward at each lever. Faced with two levers that are armed at different rates and that deliver rewarding stimulation of different strengths, the rat tends to shuttle between them. Its allocation of time between these two "patches" matches a simple ratio of the respective "incomes," the products of the perceived rates of reward delivery and the subjective magnitudes of the rewarding effects (Gallistel 1994; Gallistel et al. 2001).

The rats in Gallistel's experiments not only learn about the rates and magnitudes of rewards, but also learn about the stability of the payoffs over time (Gallistel et al. 2001). When the experimenter makes frequent, unsignaled changes in the relative rates of reward, the rats adjust their behavior very quickly so as to invest more heavily in the option that has started to yield the higher payoffs. However, when the experimental conditions have long been constant, the rats' behavior shows much more inertia following a sudden change in the relative rates of reward. Such tendencies would help a forager make use of its past experience in deciding whether a recent decline in returns reflects a bona fide trend toward patch depletion or merely a noisy, but stable, distribution of prey.

Gallistel has interpreted these results within a theoretical framework (Gallistel 1990; Gallistel and Gibbon 2000) very different from the associationist view that changes in connection weights are the basis of learning. In the rate estimation theory proposed by Gallistel and Gibbon (2000), the animal acts like a statistician making decisions on the basis of data on reward rates, time intervals, and reward magnitudes. They argue that these data are stored in representations that cannot be constructed solely from the building blocks posited by associationist theories. In contrast to the division of time into discrete steps in the models in figures 3.3.1 and 3.3.2, time is treated as a continuous variable in rate estimation theory. Decisions such as patch leaving are under the control of internal stochastic processes and need not be driven by transitions in external sensory input.

The debate between proponents of associationist and rate estimation theories concerns the neural and psychological bases of evaluation, decision making, and learning. These processes are fundamental to the ability of

(Box 3.3 continued)

foragers to allocate their behavior profitably. It will be interesting indeed for students of foraging to see how this debate plays out.

Suggested Readings

Dyan and Abbott's (2001) textbook presents temporal difference learning within an overview of computational approaches to many different topics in neuroscience and psychology. Gallistel and Gibbon (2000) challenge associationist accounts of learning and recast both classical and operant conditioning phenomena in terms of rate estimation theory, a decision-theoretical viewpoint based on the learning of time intervals, rates, and reward magnitudes. Schultz and Dickinson (2000) review the role of prediction errors in behavioral selection and learning.

just a few, have at one time or another been attributed to the vertebrate hippocampus. In fact, a number of authors have pointed out the functional similarities between the vertebrate hippocampus and the insect mushroom bodies (Capaldi et al. 1999; Waddell and Quinn 2001). As Waddell and Quinn put it, "Both systems show elegantly regular, only slightly scrutable anatomical organization and appear suited to deal with complex, multimodal assemblies of information" (2001, 1298). In most mammals, the hippocampus, which is part of the limbic system, is an arch-shaped structure deep within the brain. In humans and primates, however, the arch is straightened into an elongated structure that lies entirely within the temporal lobe. In birds, the hippocampus lies at the dorsal surface of the brain along the midline between the hemispheres, the position also occupied by the evolutionarily homologous structure in reptiles, the dorsomedial forebrain. The hippocampus receives input from most sensory modalities via the entorhinal cortex after this sensory information has been processed in other brain areas. The hippocampus sends efferent output to many areas, both within the limbic system and elsewhere in the brain.

Discovering exactly what takes place in the hippocampus, in cognitive terms, has proved elusive. As in the fable of the blind men and the elephant, different research groups concerned with different aspects of behavior have come to very different conclusions about what the hippocampus does. There is good evidence that the hippocampus plays an important role in spatial orientation in birds, mammals, and reptiles. There is also evidence that learning of complex relations among stimuli—spatial or nonspatial—depends on the hippocampus. In humans, damage to the hippocampus disrupts the ability to

form new episodic memories—memories of everyday events or episodes—
but does not seem to impair procedural memory—the ability to learn new
skills and procedures. People with damage to the hippocampus can, for example, learn a new computer skill without any awareness of where or when they
learned it or even any recollection that they now possess this skill. Conflicting
conclusions about the function of the hippocampus are probably the result
of various researchers grasping different parts of what is clearly a complex
beast. In this section, we will discuss two proposed cognitive functions of the
hippocampus, spatial orientation and declarative memory, and describe the
evidence that supports each of these ideas.

The Hippocampus and Space

Animals searching for food need to know where they are. They may need
to know how to head for the next patch of food, how to avoid revisiting a
recently depleted patch, how to return to a reliable food source, or how to
return home at the end of a foraging bout. They may also need to integrate
a great deal of information about the spatial distribution of food or distances
among patches. In some cases, it may be necessary to combine information
about the spatial location of a food source with recent information on its state
of depletion, the time since the most recent visit, or the current estimate of
variability in the occurrence of food.

Three lines of evidence support the idea that the vertebrate hippocampus
serves a central role in spatial orientation: the properties of individual neurons
in the hippocampus, the effects of hippocampal damage on spatial orientation,
and comparative analyses of the hippocampus.

Place Cells

Hippocampal cells called "complex spike cells" have distinctive firing
patterns that respond to where an animal is in its environment. These "place
cells," first described in the rat hippocampus by O'Keefe and Dostrovsky
(1971), produce action potentials when the animal is in a specific place, for
example, in one corner of an enclosure. A place cell is active only in a restricted
region, becoming electrically quiet when the animal moves out of that region.
When the animal returns to the same region, the place cell resumes firing.
Place cells thus have specific physical locations as their receptive fields, and
different hippocampal place cells respond to different locations. O'Keefe and
Burgess (1996) have shown how the geometric properties of the environment
influence the firing patterns of place cells. Place cells respond to the distance
of the animal from an edge and produce a graded firing pattern that varies
with distance from the edge. By moving the walls of an arena while recording

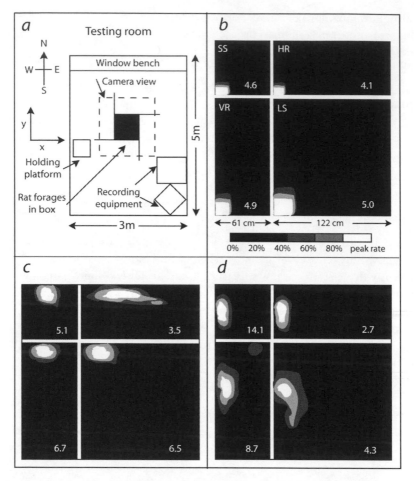

Figure 3.4. Activity of three place cells. Part A is a diagram of the arena in which the activity was recorded. Parts B–D show views from above of regions in the arena where each place cell was electrically active and the effects on each place cell of changes in the size and shape of the arena (SS, small square; HR, horizontally elongated rectangle; VR, vertically elongated rectangle; LS, large square). The gray scale gives the firing rate as a proportion of the peak rate for the cell. Peak rates are shown in Hz in the lower right corner of each plot. Place cell B fires in the southwestern corner no matter what the shape of the arena. Place cell C fires at a fixed distance from the west wall and the north wall, but the distribution of its firing in the east-west direction is sensitive to the shape of the arena. Place cell D fires at a fixed distance from the west wall, but at a fixed proportion of the distance between the north and south walls of the arena. (After O'Keefe and Burgess 1996.)

from cells in the rat hippocampus, O'Keefe and Burgess (1996) found that changing the size and shape of the arena changed the size and shape of the receptive fields of place cells (fig. 3.4). Their results showed that some place cells responded to the animal's absolute distance from an edge, while other place cells responded to the animal's relative position between two edges.

The firing patterns of place cells are influenced by cells of another kind, the "head direction" cells found in the anterior thalamus and elsewhere in the brain (Goodridge and Taube 1995; Taube et al. 1996). These cells are sensitive to the horizontal direction in which the animal's head is pointing. The preferred direction of a head direction cell is remarkably stable despite translational and rotational movement of the animal (Taube and Burton 1995), though, like those of place cells, the firing patterns of head direction cells are affected by the environment in which the animal finds itself. Place cells and head direction cells appear to work together: manipulations that influence the firing patterns of head direction cells produce changes in the receptive fields of place cells (Yoganarasimha and Knierim 2005).

Hippocampal place cells respond to where an animal is in space. The firing pattern of place cells is relatively stable in a stable environment, so assemblies of place cells could act to detect and record the spatial location of a foraging animal. Place cells or groups of place cells acting together could code the locations of food patches, a central place, or routes through the environment.

Hippocampal Lesions

The second line of evidence for a spatial function of the hippocampus is the effect on behavior of lesions of the hippocampus. Researchers frequently use the Morris water maze to assess spatial ability in laboratory rats (Morris 1981). The water maze consists of a circular water-filled pool about 1.5 m in diameter. In the pool, just beneath the surface, is a platform that the rat can climb on to get out of the water. The platform's position is concealed by making the water cloudy or by covering the surface with small floating Styrofoam pellets. In a typical experiment, a rat swims until it encounters the platform and climbs onto it. Over a series of trials, the rat learns that the platform provides a refuge. On test trials, the experimenter places the platform in a novel location, and once the animal has located it, a probe trial is conducted with the platform removed. The amount of time the animal spends swimming over the platform's former location and the number of times it swims directly to this spot measure the rat's memory for the location of the platform. Rats normally solve this problem readily. By varying the positions of prominent objects around the maze, experiments show that rats usually identify the location of the platform by its position relative to these landmarks.

Rats with lesions of the hippocampus perform poorly in the water maze (Morris 1981). They can swim well, but they return to the site of the platform only by chance. Researchers observe similar effects of hippocampal lesions in other tests of spatial memory, such as the radial-arm maze (Jarrard 1993, 1995; Olton et al. 1979). This maze consists of eight or more elevated lanes radiating out symmetrically from a central platform. Laboratory rats or mice learn over repeated trials that a single piece of food is obtainable at the end of each arm. On test trials, the animal starts from the central platform, and an observer records the number of choices required to collect all eight food items. Perfect performance is taken to be entering each arm of the maze only once, with no repetitions, on the assumption that animals minimize the time and energy expended collecting food. As with the water maze, animals identify specific arms of the maze by the configuration of landmarks that are visible around the maze. Hippocampal lesions disrupt performance in the radial-arm maze, producing both spatial disorientation and errors in memory for which arms have already been entered (Jarrard 1993, 1995; Olton et al. 1979).

Comparative Studies of the Hippocampus

A final line of evidence for the spatial function of the hippocampus comes from comparative studies. Passerine birds, breeds of domestic pigeons, and strains of mice all vary in their ability to perform spatial tasks. Among the passerines, most chickadees, tits, crows, jays, and nuthatches hoard food avidly (see chap. 7). These birds create thousands of scattered food caches and retrieve them primarily by remembering where they put their hoarded food (Kamil and Balda 1990; Sherry and Duff 1996; Shettleworth 1995). Lesions of the hippocampus disrupt accurate cache retrieval and selectively impair performance on spatial, but not nonspatial, tasks (Sherry and Vaccarino 1989; Hampton and Shettleworth 1996). Chickadees given an antagonist to the NMDA receptor (see section 3.4) were impaired in their ability to form long-term, but not short-term, spatial memories (Shiflett et al. 2004).

Food-hoarding birds have much larger hippocampi than do nonhoarding birds (Krebs et al. 1989, 1996; Sherry and Duff 1996; Sherry et al. 1989). Even within genera of food-hoarding birds, variation in the intensity of food hoarding correlates with the relative size of the hippocampus (Basil et al. 1996; Hampton et al. 1995; Healy and Krebs 1992; Lucas et al. 2004). Among strains of domestic pigeons, those with homing ability have a larger hippocampus than strains selected for other attributes (Rehkämper et al. 1988). Among strains of laboratory mice, those with the best scores on tests of spatial ability have more and longer neuronal projections running within the hippocampus from the dentate gyrus to the CA3 cell field (Schwegler and Lipp 1995).

In some species, males and females may be subjected to different selection pressures on spatial ability. As brood parasites, female brown-headed cowbirds, but not males, search for host nests in which to lay eggs. Females lay at dawn or earlier, probably in a nest they initially located between one and several days before. After laying, females spend the rest of the morning searching for host nests in which to lay subsequent eggs. They learn the locations of potential host nests and probably retain other information, such as the stage of completion of the nests they find. Female brown-headed cowbirds have a larger hippocampus than males, a sex difference not found in closely related icterid blackbirds that are not brood parasites (Sherry et al. 1993).

South American cowbirds exhibit wide variation in behavior that permits additional comparisons. The bay-winged cowbird is not a brood parasite. It usurps the nests of other birds, but incubates its own eggs and raises its own young. The bay-winged cowbird is, however, the sole host of a specialist parasite, the screaming cowbird. Screaming cowbird males and females search together for bay-winged cowbird nests. Another cowbird, the shiny cowbird, is a generalist brood parasite, and as in the generalist brown-headed cowbird of North America, female shiny cowbirds search for host nests without male aid. Reboreda et al. (1996) found a sex difference favoring females in the relative size of the hippocampus in the shiny cowbird, but not in the other two species, confirming that sex-specific selection can affect the size of the hippocampus in one sex but not the other.

The highly variable mating systems of *Microtus* voles provide a final comparative example of selection for spatial ability and its effects on the hippocampus. Meadow vole males are highly polygynous, and during breeding they occupy home ranges that encompass the home ranges of multiple females (Gaulin and FitzGerald 1986, 1988). These males, in effect, compete spatially for breeding opportunities (Spritzer, Meikle et al. 2005; Spritzer, Solomon et al. 2005). Pine voles, in contrast, are monogamous, and males and females occupy the same home range together. The hippocampus of male meadow voles is larger than that of females, a sex difference not found in monogamous pine voles (Jacobs et al. 1990).

In these examples, variations in spatial ability correlate with differences in hippocampal size. The requirement for augmented spatial ability, and hence the need for a large hippocampus, also appears to vary seasonally in some animals. The hippocampus of food-hoarding birds, for example, varies in size seasonally in step with seasonal variation in food-hoarding activity. The hippocampus of the black-capped chickadee reaches a maximum size in October, at about the onset of seasonal food hoarding in this species. The hippocampus, surprisingly, decreases in size by December, even though food hoarding persists through the winter (Smulders et al. 1995).

Neurogenesis, the birth of new neurons, occurs in the chickadee hippocampus in a seasonal pattern that conforms to the seasonal patterns of change in hippocampal volume and food hoarding (Barnea and Nottebohm 1994, 1996). Incorporation of new neurons into the chickadee hippocampus reaches a peak in October, with relatively low neurogenesis at other times of the year. Why should fall maxima occur in both hippocampal volume and neuronal recruitment in black-capped chickadees? If these changes have anything to do with food hoarding, it would appear that the hippocampus undergoes changes that coincide with the onset of hoarding, but increased hippocampal size and high rates of neuronal recruitment do not persist through the winter, even though hoarding and retrieval of food continues for many months. It is not too surprising that a large hippocampus might not be maintained if it is not needed. Brains are energetically expensive to operate (Laughlin 2001). Smulders and colleagues (Smulders and DeVoogd 2000; Smulders and Dhondt 1997) suggest that the major demands on spatial ability may occur with the initial placement of caches in early fall. Chickadees space their caches widely to safeguard them from systematic pilfering by other animals. Smulders argues that memory for the spatial locations of caches is necessary for this spacing in the fall, but plays a lesser role in cache retrieval later in the winter. There are other possibilities, too. The fall peak in hippocampal size and neuronal recruitment may change the state of the hippocampus, enhancing memory in a way that persists throughout the winter or until the next phase of volume change and neurogenesis. A full understanding of the functional importance of seasonal changes in size and neurogenesis in the avian hippocampus requires a more complete account of how the hippocampus represents and processes information.

Declarative Memory

Neuroscientists have proposed many functions for the hippocampus in addition to spatial memory. One of these is a broader domain of memory, called declarative memory, a memory system that keeps a record of many kinds of experience, only a small part of which is spatial experience (Eichenbaum 2000; Squire 1992). Declarative memory is the directed recall of information. Recalling the provincial capital of Newfoundland, for example, draws on declarative memory. Declarative memory contrasts with procedural memory, in which experience also influences behavior, but which does not involve such directed recall of information. Riding a bicycle, for example, involves procedural but not declarative memory. The improvement that results from practice in bicycle riding is clearly a form of memory, but the effects of this kind of experience are not retrieved in the same directed manner in which declarative memory retrieves the city of St. John's. The effects of hippocampal damage in humans, described earlier, suggest that declarative memory for events and episodes is disrupted in these individuals, but procedural memory for something like a computer skill is spared (Schacter 1996).

Eichenbaum and his colleagues have shown that hippocampal damage in rats disrupts memory in a number of contexts that seem to have little to do with memory for spatial location. In one task, experimenters trained rats in a complex series of odor discriminations. Rats with hippocampal damage acquired these discriminations as successfully as control rats. Control rats, however, were able to combine these discriminations in novel ways, whereas rats with hippocampal damage could not (Bunsey and Eichenbaum 1996). In another experiment, Bunsey and Eichenbaum (1995) showed that hippocampal damage has an effect on the social transmission of food preferences. In a series of experiments, Galef and his colleagues had found that rats develop preferences for food odors that they detect on other rats (Galef 1989; Galef and Stein 1985). In particular, rats prefer food odors associated with the exhaled breath of other rats or with carbon disulfide, a compound found in rat breath (Galef et al. 1988). Rats with hippocampal damage show such food odor preferences initially, but while control rats retain these preferences, rats with hippocampal damage lose them after a few days (Bunsey and Eichenbaum 1995).

These deficits suggest that the hippocampus serves a number of functions in addition to spatial memory. If that is the case, why do comparative analyses consistently show differences in hippocampal size correlated with species and sex differences in the use of space? For many animals, memory for spatial locations may be one particularly important kind of declarative memory. If this is correct, then animals with enhanced spatial abilities represent just one instance in which selection for cognitive processes has affected the hippocampus. There should be correlations between other cognitive abilities and the hippocampus waiting to be discovered by comparative methods.

3.6   Working Memory and the Prefrontal Cortex

Working Memory

One area in which foraging theory has successfully addressed issues of cognitive mechanism is the influence of previous experience with prey on foraging decisions. Many researchers have examined this issue theoretically and empirically (Devenport and Devenport 1994; Hirvonen et al. 1999; Kacelnik and Todd 1992; Shettleworth and Plowright 1992; Stephens 1987). Hirvonen et al. (1999) formulated the problem as follows: Ideally, foraging decisions are made with complete information about the foraging environment. In the classic diet model, the decision to include a particular prey type in the diet depends on encounter rates with more profitable prey types (see chaps. 1 and 5). Foragers are not omniscient, and their only source of information about encounter rates is their memory of previous encounters with prey. Encounter rates may be stable or they may fluctuate, and how much they fluctuate can vary with time and place. How, then, should previous experience with prey be weighted in order to estimate the expected encounter rate with a prey type? Intuitively, in a relatively stable environment, a long memory should give the best estimate of expected encounter rates, while in a fluctuating environment, only the most recent encounters will be informative and previous experience should be devalued. Hirvonen et al.'s model agrees with this intuition. Retrieving information from memory and integrating it with current information about the environment is the domain of "working memory" (see chap. 2), and one structure thought to play a significant role in working memory is the prefrontal cortex.

The Prefrontal Cortex

Working memory is a system for temporarily holding and manipulating information that is currently in use (Baddeley 1986, 1998). Experimental work with primates and rats supports the idea that the prefrontal cortex—the most anterior part of the mammalian brain—serves working memory (e.g., Fuster 1997; Goldman-Rakic 1990). As one might suspect, two structures that neuroscientists implicate in memory, the hippocampus and the prefrontal cortex, have extensive reciprocal anatomical connections (Goldman-Rakic et al. 1984; Swanson 1982).

Memory during Delays

In the 1930s, Jacobsen reported that monkeys with damage to the prefrontal cortex showed deficits in a range of tasks, especially tasks involving a delay between the presentation of a cue and performance of a response (Jacobsen 1936; Jacobsen and Nissen 1937). Monkeys with lesions of the prefrontal cortex performed very poorly in a delayed response task, for example. In the simplest version of the delayed response task, a monkey has two food wells within reach. At the start of the trial, the monkey watches the experimenter place food in one of the wells and then cover the wells with identical blocks of wood. A curtain then descends, preventing the monkey from viewing the food wells. Following a delay, the experimenter raises the curtain, and the monkey can choose one of the two food wells. If it chooses correctly, it gets to retrieve the food. Monkeys with damage to the dorsolateral prefrontal cortex do very poorly at this task. These same animals can successfully perform simple visual discrimination tasks, and can even learn the delayed response task if the food well covers are distinct, demonstrating that the sensory, motor, and motivational components of the delayed response task are within their capabilities (Bachevalier 1986; Fuster 1985, 1997).

The delayed response task seems very simple. Yet it requires a critical capacity that damage to the prefrontal cortex appears to compromise: the ability to remember locations that are briefly out of sight. Lesions of the prefrontal cor-

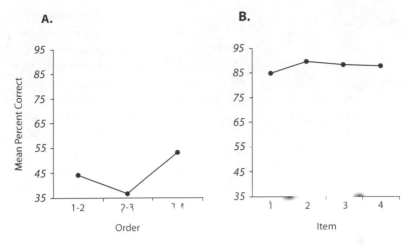

A.

B.

Order

Item

Figure 3.5. Mean percentage of correct responses in tests of (A) order and (B) item memory in rats with lesions of the prefrontal cortex. During the study phase, the animal visited four arms of an eight-arm radial-arm maze. The experimenter ensured that the same four arms were always entered in sequence during the study phase by opening and closing doors at the entrance to each arm. (A) The test for memory of temporal order consisted of opening the doors to the first and second, second and third, or third and fourth arms the animal entered in the study phase. The correct response was to choose the arm that had been entered earliest in the sequence. Animals with prefrontal lesions performed at chance levels; they were not able to discriminate between arms based on the order in which those arms had been visited. Unlesioned rats performed with 75%–85% accuracy (data not shown). (B) The test for item memory consisted of presenting the animal with two open arms, one that had been visited and one that had not been visited. The numbers along the x-axis indicate whether the choice was between the first arm of the visited sequence and a new arm, the second arm of the sequence and a new arm, and so on. The correct choice was to enter the arm that had been visited previously. Performance of the lesioned animals in the item memory test was excellent. That is, rats with prefrontal lesions could discriminate between previously visited arms and arms that had not been visited, but they could not correctly choose between arms based on the order in which those arms had been previously encountered. (After Kesner and DiMattia 1987.)

tex produce a deficit in the ability to hold information in memory temporarily; that is, they disrupt working memory. Effects similar to those described for chronic lesions also occur if neural activity in the prefrontal cortex is disturbed only temporarily. Application of an electric current disrupts neural activity in the prefrontal cortex, and if this current is applied during the delay in a delayed response task, poor performance results (Stamm 1969). Local cooling of the prefrontal cortex has a similar detrimental effect on delayed response tasks and other similar tasks (Bauer and Fuster 1976; Fuster and Bauer 1974).

Since Jacobsen's original demonstration of the effects of prefrontal lesions, neuroscientists have offered many explanations for the cognitive deficits produced by disrupting prefrontal function. Recent theory and data continue to emphasize the role of the prefrontal cortex in working memory and its importance in the active processing of internal representations and their use to guide behavior (Fuster 1997; Goldman-Rakic 1990).

Searching for Food in the Radial-Arm Maze

In rats, lesions of the prefrontal cortex cause deficits in a range of tasks that require preserving information about a cue. The prefrontal cortex plays a selective role in performance in the radial-arm maze, especially under conditions in which rats are allowed to visit some arms of the maze and then a delay is imposed before they can make their next choices (Seamans 1995). As expected, bilateral reversible pharmacological lesions of the hippocampus disrupt an animal's performance in the radial-arm maze with or without a delay between successive choices of maze arms (Seamans 1995). If a delay is imposed between choices, reversible pharmacological lesions of the prefrontal cortex also disrupt performance in the radial-arm maze (Floresco et al. 1997). Previous studies have suggested that insertion of a delay forces the rat to use a "prospective" search strategy. Rather than remembering "retrospectively" which maze arms it has visited, the rat remembers prospectively which maze arms it has yet to visit. That is, rats use information acquired before the delay to predict the remaining locations of food in the maze and then remember just those locations (Cook et al. 1985). If the animal is foraging continuously without an imposed delay, it can perform well by remembering retrospectively which maze arms it has already visited (Cook et al. 1985). Retrospective foraging requires only the hippocampus, but rats need both the prefrontal cortex and the hippocampus for prospective foraging (Floresco et al. 1997).

A further series of experiments by Kesner and co-workers also showed that the hippocampus and the prefrontal cortex perform different memory functions. As figure 3.5 shows, rats with lesions of the prefrontal cortex can correctly discriminate between the arms of an eight-arm radial-arm maze that they have visited and those that they have not. Unlike control rats, however, they cannot correctly discriminate between arms on the basis of the order in which they were visited (Kesner and DiMattia 1987; Kesner and Holbrook 1987). It may be advantageous during foraging not only to remember where food is, but also to retain a record in memory of when and in what order the forager has visited food sources. Many natural food sources are replenished with the passage of time (Bibby and Green 1980; Davies and Houston 1981; Kamil 1978; Prins et al. 1980). Recollection of the relative timing of visits under these conditions should increase foraging success.

Damage to the prefrontal cortex also causes deficits in other test situations, such as delayed alternation and delayed matching-to-sample tasks (Pinto-Hamuy and Linck 1965; Rosenkilde 1979). Delayed alternation tasks simply require the animal to alternate responses on successive trials—to alternate between pressing two different levers, for example—but with a delay imposed between trials. The delayed matching-to-sample task is a very versatile procedure in which the animal views a sample stimulus and then, after a delay, must correctly choose between the sample and another stimulus in order to obtain food. By using a computer touch screen to present these stimuli, an almost limitless variety of abstract shapes, pictures of objects, or real-world scenes can be used as stimuli in delayed matching-to-sample tasks. Damage to the prefrontal cortex impairs performance on both delayed alternation and delayed matching-to-sample tasks.

Rats presented with a novel and a familiar object will direct more exploration toward the novel object (Ennaceur and Delacour 1988; Ennaceur and Meliani 1992). When presented with two familiar objects, they will direct more exploration toward the object they have encountered less recently. Presumably, the passage of time has made them less familiar with this object (Mitchell and Laiacona 1998). Rats with lesions of the prefrontal cortex continue to direct more exploration toward a novel object than toward a familiar object, but they fail to discriminate between familiar objects that differ in how much time has passed since they were last encountered (Mitchell and Laiacona 1998). The ability to discriminate the timing of events appears to be impaired after damage to the prefrontal cortex.

Recording from Neurons in the Prefrontal Cortex

The results discussed above show that when information relevant to the performance of an ongoing task must be kept active, the prefrontal cortex bridges the delay until the response can be performed. The prefrontal cortex, in other words, plays a role in working memory. Another approach to examining the function of the prefrontal cortex is direct recording of the electrical activity of neurons. Fuster (1973) performed the classic electrophysiological study of the prefrontal cortex during a task that required retaining information during a delay. Fuster made recordings from single neurons while monkeys performed a delayed response task. The monkey had to retrieve a piece of apple that it had seen hidden under one of two identical wooden blocks. The majority of neurons recorded in the prefrontal cortex altered their firing frequency as different events occurred during the trial. Some cells increased their firing frequency during exposure to the cues and again during the choice period following the delay. Most strikingly, some neurons increased their firing rate only during the delay (Fuster 1973; fig. 3.6).

In delayed response, delayed alternation, or delayed matching-to-sample tasks, single-cell recordings have identified a large number of neurons in the prefrontal cortex that increase their firing rates during the delay and return

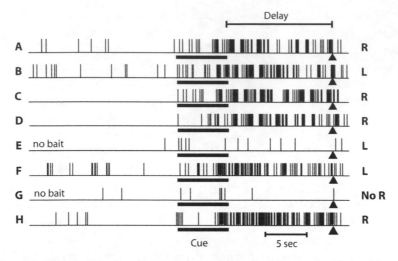

Figure 3.6. Electrical activity of a prefrontal cortex neuron of a monkey (*Macaca mulatta*) during eight delayed response trials (A–H). The heavy horizontal line below each record represents the cue period during which the animal observed the placement and covering of food. The arrows mark the end of the delay and presentation of the choice. The notation to the right of each line shows on which side the monkey responded (R, right; L, left). During trials E and G food was omitted during cue presentation. During trial E, the animal responded to the left, although no food was present. During trial G, the animal did not respond (No R). Note that on trials in which there was no food location to remember, the firing of the neuron was markedly lower than on baited trials. Responses during all other trials were correct. On baited trials, the neuron responded during presentation of the cue and continued to fire at a high rate throughout the delay. Other neurons have been identified that fire during the delay but not during the initial cue presentation. (After Fuster 1973.)

to baseline firing rates after the delay has ended (Fuster and Alexander 1971; Kojima and Goldman-Rakic 1982, 1984). Features such as color, position, or sound that are relevant to the task influence the activity of these neurons, but the pattern of neuronal activity is not specific to any one sensory modality (Bodner et al. 1996). The pattern of activity of these neurons seems to correspond to the activation of working memory, not to sensory modality or the response required for the task. Because the cue-specific differential firing of these prefrontal neurons occurs when the cue is no longer present, these cells take part in the internal representation of the cue. The level of neuronal activity during the delay has a direct and positive relationship to the number of correct responses the animal later makes (Watanabe 1986). Consistent with the results of these electrophysiological studies, Friedman and Goldman-Rakic (1994) have observed increased glucose utilization during spatial working memory tasks in the prefrontal cortex. The activity of individual neurons in the prefrontal cortex may provide a mechanism for keeping a representation active when it is no longer present in perception but is still relevant to future behavior.

Sequential Ordering of Behavior

Together with its working memory function, the prefrontal cortex plays a role in the temporal organization of behavior. In addition to memory in delayed response tasks, the prefrontal cortex may be involved more generally in the temporal organization of behavior, serving the timing functions that are necessary for the sequential organization of behavior (Fuster 1985, 1991; Milner et al. 1985). Lesions of the prefrontal cortex cause impairments in a number of naturally occurring behaviors, such as nest building (Kolb and Whishaw 1983), male social behavior (de Bruin et al. 1983), and food hoarding (Kolb 1974). All of these behaviors involve actions that must be performed in the correct sequence. Rats with prefrontal cortex lesions can produce the individual components of complex behavioral sequences, but they cannot perform the components in the correct order. These findings suggest that the timing and the sequential organization of behavior involve the prefrontal cortex, including behavior that constitutes the complex series of actions animals engage in to forage successfully.

3.7   Conclusions

It is a long journey from the patch departure decision of a bumblebee to the neural localization of working memory in the prefrontal cortex. We have examined the cellular and molecular mechanisms of the formation of associations, two of the numerous proposed cognitive functions of the hippocampus, and the role of the prefrontal cortex in working memory. Much of the research in these areas has used natural components of foraging to investigate how the brain implements cognitive processes. The proboscis extension response of honeybees, the search for food by rats in a maze, and choice by primates among concealed food sites are all components of natural foraging, or very similar to components of natural foraging. It is no accident that feeding and foraging behavior figures prominently in the study of the nervous system; feeding and foraging are things animals do reliably and repeatedly, even under artificial experimental conditions.

Foraging requires a variety of cognitive competencies. We have taken the position that foraging theory makes predictions about what animals should learn and remember if they can. Research in the neurosciences tells us how some of this learning and memory occurs. Foraging theory can make another kind of prediction, however, that neither behavioral ecologists nor neuroscientists typically pursue. Let us return briefly to encounter rates and working memory. Hirvonen et al. (1999) showed that devaluation of previous experience led to better or poorer estimates of the true encounter rate with prey depending on whether encounter rates were stable or variable in the animal's environment. These authors also explored the consequences of making the weighting of previous experience flexible in light of the animal's foraging success. In their approach, they allowed model foragers to adjust the devaluation parameter in a random direction after the first bout of foraging and then make a comparison between success in the first and second foraging bouts. If foraging success improved following the random change in parameter value, animals changed the devaluation parameter further in the same direction. If foraging success deteriorated, they adjusted the devaluation parameter in the opposite direction. As successive foraging bouts continued in this way, foragers converged on the devaluation parameter—that is, the working memory parameter—that best suited the variability of the environment in which they found themselves. Thus, the capacity to change the persistence time of experiences within memory leads to a better match between the current environment and the cognitive process for assessing it.

This conclusion implies that we might expect the neural characteristics of memory and other cognitive processes to vary with foraging conditions in a habitat. Variation in the neural characteristics of memory and cognition could come about either through adaptation to environmental conditions or through selection for modifiable neural mechanisms that are adjusted in light of experience. Although adaptive variation in cognition is the topic of much current research, there have been few concerted attempts to examine the neural implications of adaptive variation in cognition. Foraging is an area of research that shows clearly how functional and causal approaches to the study of behavior can complement each other. Ideas and discoveries about function can lead to new causal questions. Findings about causation can explain how functional outcomes are produced, and can help refine and focus functional questions.

Finally, we would like to draw attention to the remarkably small number of highly conserved cellular and molecular mechanisms that appear to be responsible for a wide variety of cognitive tasks in different brain areas and in different species. Very similar cellular and molecular processes serve conditioning of the gill withdrawal reflex in *Aplysia*, the proboscis extension response in honeybees, and long-term potentiation in pyramidal neurons of the rat hippocampus. The plots are similar, and the parts are often played by the same actors. These neuronal processes are deployed throughout the brain, however, in such a rich variety of contexts that they can provide the fundamental building blocks for the cognitive organization of foraging.

3.8   Summary

Although foraging models usually refrain from any commitment to specific causal mechanisms, it is generally recognized that learning, memory, perception, and other cognitive processes play a crucial role in foraging. One of the simplest forms of animal learning, classical conditioning, enables foragers to learn about environmental cues that predict the presence of food. The neural basis of classical conditioning has been examined in the honeybee. Neural pathways convey information from odor and sucrose receptors to the mushroom bodies of the honeybee brain. Intracellular second messenger systems respond to the co-occurrence of odor and sucrose signals, initiate gene transcription, and cause the long-lasting changes in neurons that are the basis of associative learning. The vertebrate hippocampus has been implicated in many of the cognitive processes that are essential to foraging. Neurophysiological and comparative research has addressed the role of the hippocampus in spatial orientation. Damage to the hippocampus has been shown to disrupt spatial ability and selectively impair some kinds of memory, but not others. The prefrontal cortex plays a role in two cognitive components of foraging, working memory and the sequential organization of behavior. Neural mechanisms from the intracellular to the cortical underlie the cognitive processes of foraging.

3.9   Suggested Readings

The reviews by Menzel and Müller (1996) and Giurfa (2003) analyze learning and cognition in the honeybee at multiple levels, from cellular to behavioral. Micheau and Riedel (1999) discuss the intracellular signaling pathways that are the best understood and most conserved phylogenetically. Their review concentrates on PKA, PKC, and other Ca^{2+}-dependent kinases, and roles are proposed for each of these kinase families in different stages of learning and memory formation. The multiple, complex interactions among kinases are also discussed as a means of fine-tuning memory formation and information processing. Dubnau and Tully (1998) and Waddell and Quinn (2001) describe gene expression and the involvement of gene products in learning-induced changes in neural functioning in mutant and transgenic *Drosophila*. Best et al. (2001) describe recent research on hippocampal place cells, including the effects of experience on place cell activity and computational modeling of the role of these cells in orientation and spatial memory. Written by a major researcher and theorist on the functions of the frontal cortex, the book by Fuster

(1997) is comprehensive, with data on many mammalian species, including humans and nonhuman primates. It presents a theory of the function of the frontal lobe, including the prefrontal cortex, in working memory, timing, attention, motor control, affect, and planning. The book covers comparative anatomy, neurotransmission, neuropsychology, neurophysiology, and recent imaging studies.

Telander

Cognition for Foraging

Melissa M. Adams-Hunt and Lucia F. Jacobs

4.1 Prologue

A hungry blue jay searches for prey along the branch of an oak tree. It scrutinizes the bark closely, ignoring the stream of noise and motion that occur around it. But when it hears a red-tailed hawk cry, it pauses and scans the scene. Seeing no threat, it resumes its search. Prey are difficult to find. Moths have camouflaged wings and orient their bodies to match the patterns of the bark. Dun-colored beetles press themselves into crevices. The jay peers at the bark, but does not immediately see any insects, even though they are within its field of view. Its gaze passes over several moths before it detects one outlined against the brown background. It catches and eats this moth. Renewing its search, the jay soon catches another moth, and then another. As the jay busies itself consuming moths, its gaze passes over many beetles, just as large and tasty, yet it does not detect them. Instead, the jay eats more moths, which it now finds easily, until only a few remain.

4.2 Introduction

An observer might wonder why the jay passes over valuable beetles. Answers to this question can take several forms. According to Tinbergen's classic framework, there are four levels of explanation: phylogeny, ontogeny, survival value, and mechanisms of foraging behavior (Tinbergen 1963). Cognitive scientists focus on mechanisms, the proximate causes of a behavior within the body of an organism. *Cognition* is the set of psychological mechanisms by which organisms obtain, maintain, and act on information about the world. Broadly, these mechanisms include perception, attention, learning, memory, and reasoning. Although humans experience some cognition consciously (but much less than it seems to us; see Kihlstrom 1987), researchers can usually study the information processing aspects of a cognitive process without knowing whether it is conscious. This becomes important when studying nonhumans because we cannot ask them about their conscious cognition. In our prologue, the blue jay's cognitive processing (conscious or not) determines which cryptic prey it will detect, as we will describe in more detail later.

Cognition enables foragers to identify and exploit patterns in the environment, such as by recognizing objects—whether prey, conspecifics, or landmarks—and predicting their future behavior. Evidence suggests that cognitive abilities can affect fitness and evolve (Dukas 2004a). Reasonably, these abilities may have become crucial for survival and reproduction, evolving as their enhancement led to greater fitness. Learning and memory may also have allowed animals to colonize new ecological niches, leading to new selection pressures on their cognitive abilities. Cognition, ecology, and evolutionary processes are intimately connected. This realization has led to a new interest in the role of cognition in understanding species' behavioral ecology and hence to biologists and psychologists collaborating on comparative studies of cognition (Kamil 1994).

Many fields, including ethology, behavioral ecology, comparative psychology, anthropology, neuroethology, cognitive science, and comparative physiology, have informed the study of cognitive processes in nonhuman species. This chapter introduces some of the major phenomena and issues in cognition and foraging research, outlining their diversity and complexity. It discusses four functional problems faced by a forager: perceiving the environment, learning and remembering food types, locating food resources, and extracting food items once found.

4.3   Perceiving the Foraging Environment

Perception begins with sensation: the conversion (*transduction*) of environmental energy into a biological signal (usually neural) that preserves relevant patterns (*information*). When light from the moth and its substratum activates the jay's photoreceptors, the jay senses the moth. The range of sensory abilities among species is impressive, even within taxonomic groups. For example, the auditory sensitivity of placental mammals ranges from the infrasonic vocalizations of elephants to the ultrasonic calls of bats. Diverse sensory modalities exist, including chemo-, electro- and magnetosenses. Animals may also have internal sensations such as proprioception, pain, and hunger. As a consequence of this diversity, the *Umwelt*, or "sensory world" (von Uexküll 1957), of any species is not easily accessible to others—an important realization for humans who study nonhumans. From the available stream of sensory information, an individual must select what is relevant to its current goals. Our jay, for instance, needs to find its prey, the moth.

Feature Integration

To perceive the moth, the jay must separate the moth from the background. This task can involve several cognitive mechanisms. For example, if a mottled white moth rests on a brown oak tree, the jay will immediately perceive the moth by its color, regardless of how closely its texture matches the substratum. Perception researchers call this the *pop-out effect* because under these circumstances items seem to "pop out" from the background. Feature integration theory provides a basic framework for understanding this effect. According to this theory, the visual system treats each perceptual dimension, such as color or line orientation, separately. If a target (the item being searched for) differs from its surroundings in one perceptual dimension, it pops out. When the target lacks a unique feature, pop-out does not occur, and a forager must search more carefully, as when a jay searches for a cryptic moth. In such a *conjunctive search*, the forager must inspect items that share features with the target (*distractors*) one at a time. This necessity decreases search performance linearly. When pop-out occurs, the search, called a *feature search*, proceeds simultaneously on all dimensions. *Attention*—the focusing of limited information processing capacity—is needed in a conjunctive search to *bind* (integrate) separate dimensions, while pop-out occurs without attention (Treisman and Gelade 1980).

Texture segregation experiments with both humans (Treisman and Gelade 1980) and pigeons (Cook 1992) fit this model of feature integration. Displays of small shapes varying in color (e.g., black or white squares and circles), within which a configuration of the small shapes formed a rectangle, were used in one such experiment (fig. 4.1). In the feature search condition, the rectangle contained either all the same shape or all the same color. In the conjunctive search condition, the rectangle contained both shapes, oppositely colored,

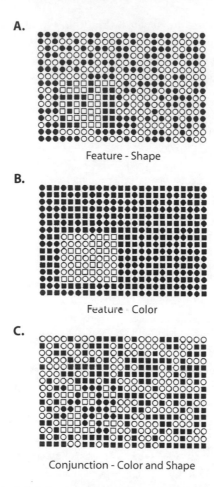

A.

Feature - Shape

B.

Feature - Color

C.

Conjunction - Color and Shape

Figure 4.1. Stimuli used to study texture segregation. Subjects search for a target (the small rectangle) within the display. Displays A and B illustrate targets that differ in a single feature (shape or color) from the background. Note the "pop-out" effect for these single-feature displays. Display C contains a target that differs from the background in a conjunction of features: black circles and white squares in a background of white circles and black squares. Note the difficulty in locating this target. Both pigeons and humans show decrements in performance on such conjunctive searches. (After Cook 1992.)

and the background contained the two remaining combinations. Both humans and pigeons performed poorly in conjunctive searches. Another visual search experiment (Blough 1992) found evidence of serial processing during conjunctive searching in pigeons. Blough used alphanumeric characters as distractors and the letter "B" and a solid heart shape as targets. The number of distractors did not affect search time for the dissimilar heart shape, but increased search time for the cryptic letter "B." Together, these studies suggest that in pigeons and humans, two disparate species that rely on vision, integration of features may require attention. Challenges and extensions to this theory are reviewed in Palmer (1999) and, with additional pigeon experiments, in *Avian Visual Cognition* (see section 4.8 for URL).

Search Image

Luuk Tinbergen (1960) observed great tits in the field delivering insect prey to their young and compared these observations with changing abundances of prey. When a new prey species became available, Tinbergen found that parents collected it at a low rate for a while before the collection rate caught up to its abundance. Tinbergen interpreted this pattern as revealing a cognitive constraint on search: the food-collecting parents behave as if they are temporarily "blind" to the abundance of a newly emerged prey type. He argued that foraging animals form a perceptual template of prey items over time. We now call this phenomenon *search image*.

Laboratory studies have shown that search image effects occur only when prey are cryptic (Langley et al. 1996), suggesting that animals require search images only for conjunctive searching. As reviewed by Shettleworth (1998; see also Bond and Kamil 1999), search image is probably an attentional phenomenon that selectively amplifies certain features relative to others. *Sequential priming* may be the mechanism involved. Every time a predator encounters a feature (e.g., a blue jay encounters the curved line of a moth wing), the perceptual system becomes partially activated (*primed*) for that feature. Priming is a preattentive process that temporarily activates a cognitive representation, often facilitating perception and attracting attention. A classic study by Pietrewicz and Kamil (1979) of blue jays searching projected images for cryptic moths supports the role of sequential priming in search image formation. In these experiments, jays saw photographs of *Catocala relicta* (a light-colored moth) on a light birch background, *C. retecta* (a dark-colored moth) on a dark oak background, and pictures of both types of tree bark with no moth. The apparatus rewarded the jays with a mealworm for pecking at pictures that contained moths. The birds' ability to detect a single moth species improved with consecutive experiences, consistent with sequential priming. Mixing two prey types in a series blocked the improvement.

Bond and Kamil (1998) showed that this search image effect can select for prey polymorphisms because search image formation lags changes in the relative frequency of morphs. The experimental predators, again blue jays in an operant chamber, generated frequency-dependent selection that maintained three prey morphs in a population of digitized images. Jay predation selects for both polymorphisms and crypticity in moths, which may fuel the evolution of the jay's perceptual capacities in turn (Bond and Kamil 2002).

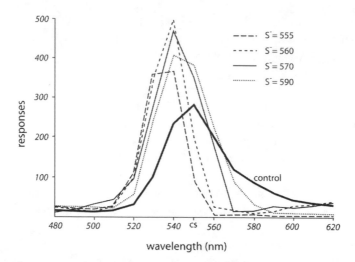

Figure 4.2. Stimulus generalization to a light with a wavelength of 550 nm (the conditioned stimulus, or CS) with no discrimination training and with training to avoid a light of greater wavelength (S⁻). Pigeons trained to respond only to the CS (control) showed a peak response (highest number of pecks) to wavelengths very near the CS. Note the "peak shift" effect caused by discrimination training: the peak response moves away from the negatively trained stimulus. (After Hanson 1959.)

Stimulus Generalization

Because no two moths are identical, the foraging jay must generalize. Stimulus generalization allows a forager to discount minor differences in stimuli. In a classic study, Hanson (1959) trained pigeons to peck at a key that emitted light at 550 nm, a greenish yellow color. When presented with random wavelengths, the trained pigeons also responded to wavelengths close to 550 nm and less strongly to wavelengths farther away (fig. 4.2).

An important characteristic of stimulus generalization is its flexibility. Discrimination training can shift the response peak away from a trained stimulus. When Hanson further trained groups of pigeons to inhibit their response to a second wavelength greater than 550 nm, the pigeons preferred wavelengths less than 550 nm (see fig. 4.2). This *peak shift* effect shows the flexibility of stimulus generalization, which allows animals to group similar stimuli according to behavioral requirements or experience. Peak shift has been shown in animals from goldfish to humans (see Ghirlanda and Enquist 2003 for a review of stimulus generalization).

Categorization

Stimulus generalization may underlie some categorizations. Wasserman and colleagues used a sorting task to investigate visual categorization in pigeons.

First, they trained pigeons to match four classes of objects (cats or people, cars, chairs, and flowers) with the positions of four pecking keys (left or right, upper or lower), where each key corresponded to one object class. Intermittently during training with one set of drawings, the experimenters tested the pigeons with a set of new images from these object classes. This testing demonstrated that the pigeons had not simply memorized the correct response for each image, but were generalizing (Bhatt et al. 1988). In a further demonstration, Wasserman and colleagues required pigeons to sort these same images into "pseudocategories" (classes with an equal number of cats, flowers, cars, and chairs). This greatly impaired the pigeons' performance, suggesting that categorization underlies this behavior (Wasserman et al. 1988). Although this result shows that pigeons can use visual criteria to categorize pictures, because all car drawings resemble one another in many ways, we cannot eliminate an explanation based on stimulus generalization.

To eliminate stimulus generalization, Wasserman and colleagues performed a three-stage experiment. In stage 1, they created superordinate categories of perceptually dissimilar objects. One group of pigeons learned to peck at a key near the upper right corner of a screen if they saw a person or a flower and to peck at a key near the lower left corner if they saw a chair or a car (fig. 4.3). In stage 2, the experimenters changed the response required for each category. The pigeons above saw only people or chairs. When the apparatus showed images of people, the pigeons had to peck the key at the *upper left*. Similarly, when the screen showed images of chairs, the pigeons had to peck the key at the *lower right*. What happened when these pigeons saw flowers again in stage 3? Did they peck at the upper left because that was the correct response for the person-flower category in stage 2, or did they choose between the two new responses randomly? On 72% of stage 3 trials, pigeons in this experiment chose the key corresponding to their category training in stage 2 (e.g., upper left key for flowers and lower right key for cars) (Wasserman et al. 1992). This result demonstrates that pigeons can form a *functional equivalence* between perceptually dissimilar items, a characteristic of true categorization (see Khallad 2004 for review).

Do animals have natural functional categories? Watanabe (1993) trained one set of pigeons to group stimuli into food versus nonfood categories and another set of pigeons to group stimuli into arbitrary categories (with equal numbers of food and nonfood items). Watanabe also trained some individuals with real objects and others with photographs. After training, the experimenter tested subjects on transfer to the opposite condition (real objects to photographs and photographs to real objects). The pigeons trained to distinguish food from nonfood easily transferred their skills from one type of stimulus to the other, but those trained with arbitrary categories did not transfer their skill.

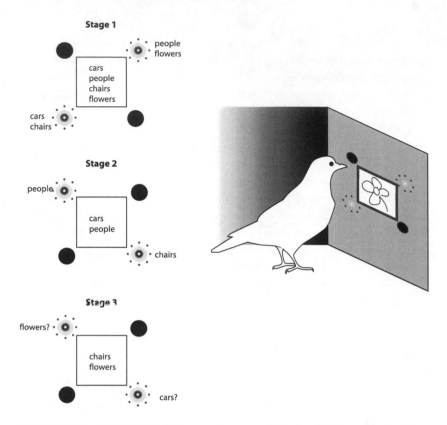

Figure 4.3. Testing for categorization in pigeons using an operant chamber. Subjects pecked at one of two illuminated keys (open circles) in response to a photographic stimulus (listed inside the square) to receive a reward. Correct answers and predicted responses are indicated beside the keys. In stage 1, subjects learned to make a common response to perceptually different pairs of stimuli (cars and chairs or people and flowers). In stage 2, subjects learned a new response for one type of stimulus in each pair. Stage 3 tested whether subjects would generalize this new response to the other stimulus type (cars or flowers). (Experimental design from Wasserman et al. 1992.)

This finding suggests that the subjects in the food/nonfood condition used categories, but those in the arbitrary category condition were making memorized responses to particular stimuli. Moreover, Bovet and Vauclair (1998) found that baboons could categorize both objects and pictures of those objects into food and nonfood groups after only one training trial. Functional categorization is another type of generalization. A forager that can parse its world into groups of related objects can recognize the properties of novel exemplars and predict how they will behave.

Quantity

After determining what objects are around, a forager may need to process information about quantity: How many moths did I encounter in that patch? How many individuals are in my group? An animal might use any of several methods to solve problems about quantity. Detecting *relative numerousness* is simply determining that one set contains more than another. Several species can use relative numerousness to make judgments about quantity, including laboratory rats, pigeons, and monkeys (see discussion in Roberts 1998). In contrast, to discriminate absolute number, the animal must perceive, for example, that four stimuli differ from three and five. Davis and colleagues have demonstrated that laboratory rats can discriminate the absolute number of bursts of white noise, brushes on their whiskers, wooden boxes in an array, and even the number of food items they have eaten (Davis 1996).

How animals accomplish such feats has been the subject of considerable debate. Humans can *subitize*, or perceive the size of small groups of items that are presented for less time than would be needed to count them. Subitizing may be a perceptual process in which certain small numbers are recognized by their typical patterns (or rhythms in the case of nonvisual stimuli). Humans subitize so quickly that the process appears to be preattentive. Animals may subitize, but there is also evidence that they count. Alex, an African gray parrot, could identify the number of objects (wood or chalk pieces, colored orange or purple) by color and/or material on command (Pepperberg 1994). Since selecting the objects to count involves a conjunction of shape and color, Alex may have to count each item serially. Capaldi and Miller (1988) argue that laboratory rats automatically count the number of times they traverse a runway to obtain food because they behave as if they expect reward after a certain number of runs, whether they travel the runway quickly or slowly. This number expectation was transferred when the investigators changed the type of reward, suggesting that rats count using abstract representations rather than specific qualities of the reinforcer. Notwithstanding these impressive numerical feats, some researchers are not ready to conclude that nonhumans meet the strict standard of counting in which each item in a list has a unique tag or identifier (see Roberts 1998 for discussion).

Synopsis

Cognition begins with sensation and perception. Animals possess diverse senses, such as vision, audition, touch, electroception, and proprioception, which provide the information an animal needs to forage effectively. Attention binds complex conjunctions of sensory information. Search image results from these perceptual and attentional processes. Stimulus generalization allows an animal to group stimuli based on sensory similarity. Categorization allows animals to group objects functionally. Finally, numerical competencies allow animals to quantify food items. These processes enable the forager to perceive its environment.

4.4   Learning What to Eat

If a new prey item replaces an old one, a jay that can learn to eat this new prey will be more successful. We will define learning as a change in cognition caused by new information—not by fatigue, hunger, or maturation, which can also cause cognitive changes. Learning has no adaptive value when the environment is completely static or completely random, since learned information cannot be applied (Stephens 1991). In the appropriate environment, learning allows adaptation to occur on an ontogenetic time scale rather than a phylogenetic one. Learning is related to memory: learning is a *change* in information processing, while memory is the *maintenance* of an information state. In practice, students of learning and memory find it difficult to distinguish the two. A forager must, in the end, both learn what to eat and remember what it has learned.

Classical Conditioning

An experienced blue jay may form an association between the shape of a moth and food or between shaking a branch and the appearance of this food item. Known as *associative learning* or *conditioning*, the formation of associations plays an important role in behavior. Classical or Pavlovian conditioning involves passive associations (as in the first case), while instrumental or operant conditioning (which we will discuss later) involves associations between the animal's own behavior and its results. In classical conditioning, the animal learns that something that had been neutral (the conditioned stimulus, or CS; e.g., moth shape) seems to appear predictably with something that it has an innate interest in (the unconditioned stimulus, or US; e.g., food) and to which it will make an innate response (the unconditioned response, or UR; e.g., salivation in the case of Pavlov's original experiments with dogs). Based on this relationship, simply perceiving the conditioned stimulus leads to a response, called the conditioned response (CR), which is often identical to the UR. Common conditioning procedures are described in box 4.1. Modern conditioning researchers generally consider the mechanism underlying the CR to be a cognitive representation of *expectancy*, rather than the Pavlovian "reflex."

These researchers also recognize that all traditional conditioning phenomena may not be explainable by one mechanism, and they acknowledge alternative forms of learning, such as learning by observation, which we will discuss below (see Kirsch et al. 2004 and Rescorla 1988 for excellent discussions).

BOX 4.1   **Learning in the Laboratory**

Researchers studying learning in the laboratory have developed many standard procedures and uncovered numerous replicable phenomena. Here we review some of the best known of these phenomena.

Second-Order Conditioning

A blue jay learns that a rainfall precedes wet leaves, which in turn predict greater abundance of certain invertebrates. Soon, rain by itself will stimulate the jay to look for those prey species. In the laboratory, we first condition a hungry rat to expect food (US) when we switch on a light (CS_1). Then we pair the light with a tone (CS_2), and soon the tone by itself will come to elicit salivation (CR). The conditioning to the tone is second-order conditioning. We have, in effect, chained two conditioned stimuli together.

Conditioned Inhibition

A blue jay that has learned to hunt brown moths on oak trees now learns a new association—that the presence of another blue jay on the same tree is almost always correlated with an absence of moths. This association causes conditioned inhibition of its foraging response. Conditioned inhibition occurs when we pair a CS, such as a tone, with the US (e.g., food) only when the CS appears alone, but not when it appears with a second stimulus, such as a light. This experience inhibits the response to the light-tone combination. Conditioned inhibition allows the forager to learn the circumstances in which a CS (oak tree) does not signal the US (moth).

Sensory Preconditioning

A blue jay encounters an orange butterfly resting on a clump of moss, but sated, it flies away. Later, the blue jay learns that the orange butterfly is toxic. Afterward, the blue jay may show a withdrawal response to the moss, even in the absence of the butterfly. In the laboratory, we present two CSs (such as a light and a tone) together prior to any conditioning procedure. When later, we pair one of these (e.g., the tone) with a US (e.g.,

(Box 4.1 continued)

food) in a conditioning procedure, the second one will also elicit the CR (e.g., salivation) with no direct training. Though this phenomenon seems similar to second-order conditioning, it is actually a form of *latent learning* in which animals gain information (such as an association) in the absence of any apparent immediate benefit for doing so.

Blocking

A blue jay searches for acorns in an oak tree. Every time it finds a branch of a certain diameter, the branch also contains many acorns. It then searches out branches of that diameter. However, on the other side of the tree, branches of this diameter are also covered with lichens. A second blue jay happens to find many acorns on this side, and learns to search for branches of a certain diameter that are covered with lichens. The first blue jay, when it then moves into the lichen area, does not learn that lichens predict acorns. In the laboratory, we condition a subject by pairing a tone with food until the tone reliably produces salivation. After we have completed this conditioning, we present a compound stimulus made up of our old tone and a new light. When we test the subject with the light and tone separately, we find that the tone produces salivation as before, but the light has no effect. We say that the prior conditioning to the tone blocks conditioning to the light. Psychologists view blocking as an important conditioning phenomenon because it demonstrates that correlation with the US is not sufficient for learning to occur; after all, the light has been correlated with food, so one might expect salivation to the light as well, but this is not what we find. Blocking suggests an information model of conditioning: the second CS (the light) adds no new information because the first CS (tone) already perfectly predicts the US (food).

Overshadowing

A blue jay learns that orange wings predict toxicity in butterflies. Black spots also predict toxicity, but the jay has not learned this. In the laboratory, we begin such a conditioning experiment by pairing a compound light-tone stimulus with food until our compound stimulus reliably produces salivation. When we test the light and tone separately, we typically find that one stimulus elicits salivation much more strongly. If we find that the tone and not the light elicits salivation, then we say that the tone overshadows the light. If the light and the tone differ greatly in intensity, size, or saliency (as with a dim light and a loud tone), it is the larger, brighter, louder, or more

(Box 4.1 continued)

critical CS that gains the most strength in eliciting the CR. Studies suggest that subjects learn both CSs, but not equally well. Biological relevance, as found in the Garcia effect (see section 4.4), can be a cause of overshadowing.

Latent Inhibition

A blue jay searching for food never finds any at its nest tree. One morning an infestation of bark beetles takes hold in the tree. The blue jay sees one, but does not stay to forage at the tree. In fact, it takes the jay quite a while to learn that its own tree is now a source of food. In the laboratory, we play a tone to an experimental subject. The subject hears the tone frequently, but it is not correlated with food or other salient events in the subject's environment. If we then try to condition the subject by pairing the tone with food, we find that this prior exposure to an irrelevant tone inhibits conditioning. It is as if what has been learned (that the tone predicts nothing and therefore can be ignored) must be unlearned before the new association can be made. Latent inhibition supports an information model of conditioning and contradicts the expectation that familiarity would facilitate learning.

Extinction

A blue jay foraging for acorns on a particular tree always finds an acorn when it searches in that tree. As the season progresses, the jay is less likely to find an acorn. Eventually, the tree is empty. At the same time, the blue jay becomes less likely to search that tree. In the laboratory, we pair a light with food until a rat reliably presses a lever to get food when the light appears. Now we begin to switch on the light without food. Over subsequent trials, the rat no longer responds to the light. The stimulus that used to provide information about the arrival of food is now useless, and the subject stops responding to it. Like latent inhibition, extinction involves learning not to respond to an unpredictive CS. Psychologists often use the speed of extinction to measure the strength of the original association.

Conditioning Mechanisms

Kamin (1969) first suggested that surprise might cause a new association to form. He proposed that when unexpected events occur, the startle response stimulates an animal to learn. An expected event, in which one stimulus already predicts the occurrence of another, does not facilitate learning, as the blocking phenomenon (see box 4.1) demonstrates. Rescorla and Wagner

(1972) formalized this idea in an elegant model, $\Delta V = \alpha\beta(\lambda - V)$. The term ΔV represents the change in associative value (learning) during a trial. The constants α and β signify the salience of the CS and US, respectively. The difference $(\lambda - V)$ represents the maximum associative strength that the US can support (λ) minus the current associative value of all CSs (V). Behavioral psychologists call the difference $(\lambda - V)$ unexpectedness. Thus, no learning occurs when an animal expects an event [e.g., when $(\lambda - V) = 0$], but learning proceeds quickly when an event is unexpected [$(\lambda - V)$ is large]. This model correctly predicts a negatively accelerated learning curve and also predicts several conditioning phenomena, including the blocking effect. Yet even this influential model cannot explain all standard conditioning phenomena, and theories continue to be developed (see Kraemer and Spear 1993; Miller and Escobar 2001; and other reviews in Zentall 1993).

Ecology and Conditioning

For years, experiments seemed to show that conditioning was equally likely with any arbitrary stimulus—a phenomenon known as "equipotentiality." In 1966, a classic experiment on what became known as "taste aversion" or the "Garcia effect" challenged this dogma. Garcia and Koelling (1966) trained rats to drink saccharine-flavored water while lights flashed and a nearby speaker clicked. This procedure made three neutral stimuli available for conditioning (taste, sound, light). Next, they gave one group mild electric shocks on the feet while they were drinking and made another group nauseated by giving lithium chloride injections or by X-ray exposure several hours later. They then offered each group a choice between flavored water and water near flashing lights and clicking sounds. The shocked and nauseated groups made different choices. Rats from the shocked group avoided the water with lights and noise, but drank the flavored water readily. Rats from the nauseated group avoided the flavored water, but drank the water with lights and noise. This finding demonstrated that the effectiveness of a CS is influenced by its natural relationship to the US. These procedures also violated prevailing wisdom in producing learning after one trial, rather than gradually, and association between events occurring across a long temporal gap (see historical review in Roberts 1998).

Conditioning had also been believed to be the same across species, or universal. Rats are nocturnal foragers that collect and transmit information about what is good to eat via chemical cues, such as a novel odor in the breath of a colony member (Galef 1991). It makes sense that they would associate nausea with a novel flavor, rather than with a food that looked or sounded different. If conditioning effects are adapted to ecological niches, then a visual forager might show the opposite pattern. Exactly this result was found in Japanese quail. Wilcoxon et al. (1971) found that quail could associate the color blue with later nausea.

Aposematic (or warning) coloration trains visual predators more quickly than less intense coloration. First, they see the prey more quickly (the pop-out effect) and learn about them more quickly. In the laboratory, chicks learn to avoid bad-tasting, brightly colored prey more quickly than similar prey that are cryptic (Gittleman and Harvey 1980). But the lessons from cognitive science for the forager do not stop there. These preferences may be transmitted to conspecifics by observation. Day-old chicks (reviewed in Nicol 2004), red-winged blackbirds, and cotton-top tamarins (reviewed in Galef 2004) learn to avoid foods by observing the negative responses of conspecifics. Furthermore, stimulus generalization makes it possible for predators to avoid any species that resembles a poisonous species. This cognitive process underlies the evolution of mimicry, both when the mimic species is palatable (Batesian mimicry) and when it is toxic (Müllerian mimicry, reviewed in Goodenough et al. 1993).

Memory

The blue jay that learns about a new moth species must also remember this information. Memory can be categorized by different characteristics: duration — (long-term vs. short-term), content (episodic, semantic, procedural), use (working memory), or conscious access (declarative memory). Animal cognition researchers commonly recognize three basic types of memory (cf. Roberts 1998 and Shettleworth 1998). *Working memory* is short-term and used within the context of a foraging bout. A blue jay, for example, uses working memory to keep track of which branches it has already searched and to avoid them. *Reference memory* is long-term and is used for other information: where the jay is located in space, where the important resources are, the concept that a moth is food, the rules it has extracted about foraging for moths in that area, and so forth. Finally, there is *procedural memory* of specific skills, such as the movements needed to handle a particular prey species. More fine-grained categories include spatial and serial memory.

Organizing Memories

Animals may organize their memories into *chunks*, smaller lists that are organized categorically, such as places where white moths were found versus places where brown moths were found. Pigeons in an operant chamber learning to peck unique keys in a certain order will learn the task more quickly if the first few keys differ by color (the colored chunk) and the remaining keys differ by pattern (the patterned chunk), or vice versa. When the colored and patterned keys are intermixed, pigeons do not perform as accurately (see reviews in Roberts 1998). The same thing happens with the organization of spatial information: things that are similar are chunked together in memory. For example, rats foraging for three types of food in a twelve-arm radial-arm maze organize their search to retrieve the items in order of preference. If the three types are always found in the same places in the maze, even if these locations are scattered across the maze, the rats become very efficient at increasing their "chunk size," the number of objects of the same type taken in a run. They also learn the twelve arms of the maze more quickly than a second group of rats for which the three food types are placed in random locations in the maze on each trial. The rats therefore seem to categorize the twelve foraging locations (i.e., the ends of the maze arms) by the type of food each contains, and their ability to search proficiently (i.e., one visit to each arm) depends on this ability to organize their memories in this way (Dallal and Meck 1990). Similarly, a blue jay may categorize foraging sites by the prey found there and use this information to organize its foraging routes.

Interference between Memories

If a blue jay first learns about moths on one tree and then about caterpillars on a second tree, the memory of the caterpillars may interfere with the memory of the months. This example illustrates *retroactive* interference, in which a more recent memory interferes with an older one; however, *proactive* interference (in which the moths interfere with the caterpillars) also occurs. Interference occurs at both short and long intervals and thus affects both working and reference memory. For example, pigeons performing delayed matching-to-sample working memory tasks showed both proactive and retroactive interference. In the first task, the experimenter trained pigeons to peck a red key if they saw a red sample stimulus before the delay and a green key if they saw a green sample stimulus. Showing a light of the wrong color before the sample (e.g., green before a red sample) impaired recall in the test phase. Manipulating the interval between the interfering stimulus and the sample changed the degree of proactive interference, demonstrating that competition for encoding does cause proactive interference. Also in a delayed matching-to-sample task, adding distracting stimuli to the interval between sample and test reduced performance and demonstrated retroactive interference (see Roberts 1998).

Maintaining Working Memory

While foraging, the blue jay may need to keep in mind what it is looking for or where it has already looked. This is the role of working memory, which actively filters and prioritizes current data. Active cognitive processes can influence the strength of a memory, increasing it through *rehearsal* or

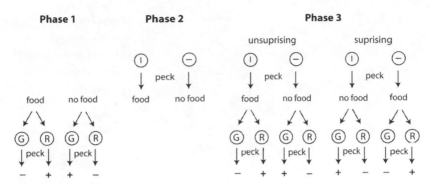

Figure 4.4. Testing for rehearsal in working memory. Pigeons in an operant chamber received the three phases of training diagramed here. Circles represent stimuli on keys: green (G), red (R), vertical line, or horizontal line. + or − indicates reward or no reward. Note that the only difference between the unsurprising and surprising test groups in phase 3 is whether pecking the lined keys resulted in food or no food as expected. (Experimental design from Maki 1979.)

decreasing it through *directed forgetting*. Rehearsal is mentally repeating an event or stimulus (e.g., repeating a phone number), improving memory for that item. Directed forgetting actively decreases or represses working memory for information deemed irrelevant. These two processes may be interrelated.

Studies have demonstrated both rehearsal and directed forgetting in pigeons (see reviews in Roberts 1998). Maki (1979) demonstrated rehearsal using a complicated three-phase delayed symbolic matching-to-sample task (fig. 4.4). In phase 1, the sample stimulus was either the presence or absence of food. In the presence of food, the pigeon had to peck a red key (the "symbolic" match for the food stimulus) to obtain a reinforcement. In the absence of food, a green key resulted in reinforcement. In phase 2, there was no matching, only a contingency. Here pigeons learned that if a vertical line was presented, they would receive food, but if a horizontal line was presented, they would not. Maki divided his phase 3 tests into two types of trials, "surprising" and "unsurprising." During unsurprising trials, the apparatus first showed one of the line stimuli (vertical or horizontal), and then the event the pigeons had come to expect (food or no food, respectively) ensued. Maki then used this event (food or no food) as the sample stimulus for a delayed symbolic matching-to-sample task identical to that in phase 1. In surprising trials, the apparatus showed the line stimuli (vertical or horizontal) as before, but the experimenter switched consequences (no food or food, respectively). As in the unsurprising treatment, Maki then tested the pigeon's memory of the food/no food event using a delayed symbolic matching-to-sample task identical to that in phase 1. "Surprised" pigeons showed better recall. If we assume that surprised pigeons spend more time "mulling over" their surprising observations, then this finding suggests a role for rehearsal in nonhuman memory. Using an entirely

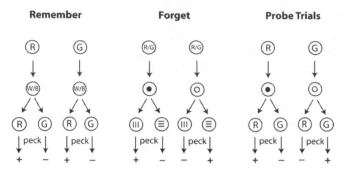

Figure 4.5. Testing for directed forgetting in pigeons. Following a white or blue stimulus (W/B, "remember cues"), pigeons received a memory test for the previous stimulus (G, green, or R, red). Following dot stimuli (solid or open dots, "forget cues"), pigeons received a symbolic matching-to-sample memory test for the dot stimulus: solid dot matches vertical lines, open dot matches horizontal lines. In probe trials, horizontal or vertical lines were replaced with a red/green memory test for the previous stimulus. + or − indicates reward or no reward. (After Roper et al. 1995.)

different design and species (aversive conditioning in laboratory rabbits), Wagner et al. (1973) showed that surprising episodes *after* conditioning trials interfere with learning. Together, these results suggest that surprise does not enhance learning simply by heightening physiological responses, nor do surprising events cause reduced learning due to interference. Instead, a surprising event may draw resources from other cognitive processes.

Animals may also direct working memory resources away from a stimulus. In an experiment that combined a delayed matching-to-sample and a delayed symbolic matching-to-sample procedure, pigeons learned to forget a previously presented sample (fig. 4.5). This procedure presented a pigeon with a red or green sample followed by a white or blue "remember cue." After the remember cue, the subject matched the red or green sample in an ordinary delayed matching-to-sample task. If an open or solid dot (the "forget cue") followed the red or green sample, the experiment tested the pigeon in a symbolic matching-to-sample task using the dot as the sample stimulus and horizontal and vertical lines as the comparison stimuli. Thus, the open or solid dot meant that the pigeon should "forget" the first sample. Periodic probe trials presented "forget cues" followed by red and green comparison stimuli. This manipulation caused a significant decrement in performance on the probes compared with the delayed matching-to-sample tasks, consistent with directed forgetting (Roper et al. 1995).

Maintaining Reference Memory

A foraging jay retrieves information about prey types and locations from reference memory when it returns to foraging after engaging in some other activity. We can compare reference memory to the storage of books in a library. A forager must organize and index memories effectively or they will be lost. Surprisingly, forgetting does not typically erase long-term memories; it just makes them difficult to find. The contextual attributes of a memory at encoding provide the cues needed to locate information from reference memory at retrieval. Memory researchers call this phenomenon *encoding specificity*.

If, for example, a jay learns a new prey type while it is ill or agitated, it will theoretically be better able to retrieve this information when it is again ill or agitated. Memory researchers call this type of encoding specificity *state-dependent memory*. Duplicating external attributes of the learning context (such as being in a meadow or a rainstorm) can reactivate and improve recall. Substantial differences between two learning contexts reduce confusion at recall. Similarly, subjects have better recall when many attributes of the learning context are present because each attribute can potentially reactivate the association (reviewed in Roberts 1998).

Synopsis

A forager can learn what to eat and what to avoid through classical conditioning. As studies of taste aversion show, the biological relevance of the stimuli constrains and facilitates this learning. Learned associations are stored and retrieved in a dynamic and multifaceted memory. Memory retrieval is influenced by events that happen before or after encoding, as in interference. Animals can optimize short-term working memory through rehearsal and directed forgetting, while contextual cues and chunking facilitate retrieval. Learning and memory allow a forager to exploit new biologically relevant patterns in its environment and to recall such information to increase its foraging success.

4.5   Locating Food

A foraging blue jay will have trouble returning to a prime food patch unless it remembers where the patch was and how to get there. Researchers study how foragers orient in space, define locations, and remember locations under the rubric of *spatial cognition*. Because all mobile animals must navigate space, spatial cognition is a central subject in comparative cognitive research. *Scatter-hoarding* species (which make a single deposit to each of many cache sites) represent an extreme case of reliance on spatial cognition: they rely heavily on spatial memory to retrieve their caches. Social animals may exploit a conspecific's spatial knowledge through *social learning* of food locations.

Spatial Orientation

Our hypothetical jay perches on a branch and tries to recall the location of a food cache. It has many cues to the cache location: the sun, distant mountains, the distant odor of the ocean, nearby trees, branches, and leaves. We can divide these external cues into two classes: positional and directional. Positional cues are usually landmarks close to the goal (i.e., *local*), and directional cues are usually distant landmarks, but could also be gradients of concentration, intensity, or size (Jacobs and Schenk 2003). Directional cues provide compasslike information: direction, not distance. Distant landmarks that serve as directional cues are termed *compass marks* (Leonard and McNaughton 1990). A *beacon* is a landmark that coincides with the goal. The forager can, therefore, choose from several *frames of reference*. It can simply approach the beacon, it can triangulate within an array of positional cues, or it can move in the direction of a compass mark or along a gradient.

An object's position within an array of positional cues is its *relative position*, while its position relative to directional cues is its *global* or *absolute position* (Brodbeck 1994). Both positions are relative to some subset of terrestrial cues, but the distinction between them reflects real phenomena. We see the dissociation between relative and absolute frames of reference in rodents and birds, both in the laboratory (Brodbeck 1994) and in the field (Healy and Hurly 1998; Jacobs and Shiflett 1999). For example, rufous hummingbirds searched for an artificial flower in its absolute position if its neighbors were greater than 80 cm apart, but searched at a position relative to an array if the flowers were 10 cm apart (Healy and Hurly 1998).

In many mammalian species, females and males prefer different frames of reference. In several polygamous species, females prefer local cues while males prefer distant or directional cues (reviewed in Jacobs and Schenk 2003). Sexual selection seems to favor this sex difference because males must track the spatial distribution of females (Gaulin and FitzGerald 1989). Tracking females requires long-distance navigation in unfamiliar territory, which cannot rely on familiar local cues.

Gradients

Any forager, regardless of brain size, can orient to a gradient, as in the case of phototaxis. Animals find many gradients in nature, such as polarized light, chemical plumes, and temperature or elevation gradients (Dusenbery 1992). A literal compass is a tool for orientation in a gradient of magnetic polarity (both invertebrate and vertebrate foragers use magnetic polarity to orient; Goodenough et al. 2001). Foragers can use gradients to orient in a one-dimensional map produced by linear changes in a single variable (e.g., temperature or concentration). These maps have the advantage of perceptual simplicity and also allow for extrapolation. A forager following a regular gradient can keep track of its movements, but it can also weather disruptions in continuity by calculating the expected concentration, elevation, or intensity after moving a known distance. This one-dimensional map forms the basis for all spatial orientation and may be necessary for large-scale movements, such as migration (Wiltschko and Wiltschko 1996). Extrapolation to unknown terrain represents the key advantage of this type of orientation, although noise in the signal and the forager's ability to perceive fine gradations limit its accuracy. Animals can, therefore, create only low-resolution maps using gradients (Jacobs and Schenk 2003; Wallraff 1996).

Landmarks

A more complex orienting method requires the ability to perceive and recognize unique objects, such as certain rocks, trees, or mountains. Use of landmarks lets a forager orient within small local arrays of objects. Different species use landmarks in different ways. Some animals encode a "snapshot" of the goal and associated landmarks. Researchers have studied this process in honeybees (Dyer 1996). The foraging bee encodes an image on her retina at the food source. When she returns, she moves such that the incoming visual image matches the stored retinal image. This simple algorithm, *template matching*, returns her accurately to the flower's location. She also uses the earth's magnetic field to encode compass direction. If she learns a retinal image from the south of a flower, for example, when she returns to that flower, she again approaches it from the south to rematch the image (Collett 1996).

We see more complex landmark use in birds and mammals. These foragers can recognize unique features of a specific landmark in three dimensions. In these cases, the forager remembers unique features of the landmarks themselves and the spatial associations among them. With this information, the forager can triangulate to relocate its goal relative to the landmarks. This process, described by different theoretical models (e.g., vector sum model; Cheng 1994), does not require any notion of absolute direction.

These two examples illustrate an important point: different cognitive mechanisms can accomplish the same result. Since the overt behavior is the same (accurate reorientation to a remembered location), we can discover such differences only through experimental manipulation. Collett and colleagues demonstrated such a difference in two classic experiments on spatial memory in honeybees and female Mongolian gerbils (Collett 1996; Collett et al. 1986). Both species accurately recalled a single location that was between two vertical columns. When the experimenters increased the distance between the columns during the forager's absence, the bee and the gerbil responded differently.

The bee matched her retinal image and hence increased her distance from the columns such that their retinal distance from each other matched her stored image. The gerbil, using her mammalian depth perception, searched the correct distance and angle from each of the two columns. Although the gerbil may have encoded more information about the landmarks, the honeybee's simpler solution works just as well under normal foraging circumstances.

Cognitive Maps

Spatial cognition researchers view the cognitive map as the most sophisticated method of spatial orientation. Edward Tolman first proposed that simple stimulus-response mechanisms could not explain the behavior of rats in a maze. He suggested instead that rats store a representation of the maze, a *cognitive map*, independent of immediate contingencies (Tolman 1948). An animal with a cognitive map can demonstrate its capacity by taking novel routes across unknown terrain. For this behavior to be convincing evidence that the animal is following a mental representation of the new route, the animal must create the route without intermediary landmarks or beacons. For example, a squirrel travels 200 meters east to a new foraging area. It then returns to that area using various methods, such as orienting to known landmarks (e.g., arrays of known trees). Later, the squirrel travels 200 meters south to a second novel foraging location. If the squirrel has created a cognitive map, it can then calculate the direction and distance of a vector linking the eastern and southern foraging sites. A squirrel with a cognitive map can navigate between the two sites even without a beacon at the eastern site (e.g., a tall tree, the sound of a waterfall) or a chain of familiar landmarks. The squirrel can recall the cognitive map as often as necessary to create new detours and shortcuts.

Recently, Jacobs and Schenk (2003) proposed a new theory to explain the cognitive map, drawing on Gustav Kramer's map-and-compass hypothesis (Wallraff 1996). Here the cognitive map is composed of two submaps: the *bearing map* (derived from directional cues) and the *sketch map* (derived from positional cues). Two independent neural circuits within the hippocampus subserve these maps. This *parallel map theory* proposes that animals need both hippocampal subfields to create a cognitive map. This may be why cognitive maps are limited to birds and mammals, since other vertebrates have only one subfield enlarged (Jacobs and Schenk 2003). To date, the best evidence indicates that the honeybee does not form a cognitive map (Dyer 1996), but similar experiments have not been conducted using other invertebrates, such as predatory cephalopods, stomatopods, or spiders, which may have greater need for a cognitive map.

Spatial Cognition in Food Hoarders

By storing food and remembering the locations, a forager can even out a food distribution that is clumped in time or space and protect it from competitors. Scatter hoarders use many locations and face special memory demands because they must maintain a large quantity of information over long periods. Scatter hoarding has been found only in birds and mammals (Vander Wall 1990). The study of food-hoarding behavior and how it is related to cognitive specialization is still a new field and has attracted both support and controversy, which has led to several recent reviews of this literature (Hampton et al. 2002; Macphail and Bolhuis 2001; Shettleworth 2003). In general, studies of cognitive specialization in food hoarders have asked how and why such species differ in the ways in which they remember spatial locations and how food hoarding is related to separable, specialized cognitive abilities.

Cue Use and Frames of Reference

The need to encode and forget temporary cache sites may have led to specialization in encoding. Food hoarders might encode spatial information differently from other information, and from nonhoarders, increasing capacity by efficiency. For example, if food hoarders encoded cache sites as unique places on a global map defined by large, distant landmarks (absolute location), this would have several advantages. First, such landmarks are likely to be stable (Biegler and Morris 1993). Second, each site would have unique coordinates, regardless of how similar the closer landmarks (e.g., local vegetation) were between cache sites. Third, unique sites should reduce interference during encoding: the more uniquely a cache is encoded, the less interference among caches. Moreover, if the cache can be encoded not only in terms of a unique place, but also by other characteristics, such as the time of caching or the contents of the cache, all of these features would improve accuracy, based on what we know about memory in general.

When experimenters moved a feeder with a distinctive color and pattern that had been previously baited, scatter-hoarding chickadees searched first at its previous location in the room (absolute location), then at its previous position within an array of feeders (relative location), and finally, after finding no bait, at the feeder that had the correct color and pattern. Nonhoarding juncos, in contrast, searched equally at all locations, suggesting no preference for any available frame of reference (Brodbeck 1994). Clayton and Krebs (1994) found similar results when they compared hoarding and nonhoarding corvids. In the field, free-ranging fox squirrels also preferred to orient first to the absolute location of their goal (Jacobs and Shiflett 1999).

Another method scatter hoarders may use to reduce inference among caches is to distinguish between them by their contents. Sherry (1984) found that black-capped chickadees retrieved preferred seed caches first, suggesting that they chunk items in their memories just as rats chunk baits by type in radial-arm maze studies.

Spatial Memory

Because species vary widely in their reliance on cached food, investigators have devised spatial tasks to examine species and population differences that may correlate with hoarding behavior. For example, within corvids, Clark's nutcrackers rely most heavily on caches, and pinyon jays slightly less. Mexican jays may rely on some caching, but scrub jays do not rely heavily on cached food for survival. The degree of cache reliance paralleled laboratory cache retrieval performance: Clark's nutcrackers outperformed pinyon jays, which in turn outperformed scrub jays (Balda and Kamil 1989). Clark's nutcrackers and pinyon jays also performed more accurately than did Mexican and scrub jays on a radial-arm maze analogue (Kamil et al. 1994). Corvid performance on a *spatial* delayed non-matching-to-sample task was also correlated with reliance on stored food (Olson et al. 1995). Clark's nutcrackers tolerated the longest delay between sample and choice, compared with pinyon, Mexican, and scrub jays. However, when experimenters tested memory for color rather than location, they found a different pattern: pinyon and Mexican jays tolerated a longer delay than nutcrackers or scrub jays. Under certain conditions, Clark's nutcrackers can show accurate cache retrieval over 270 days after caching (Balda and Kamil 1992). In a later study, nutcrackers and pinyon jays once again outperformed Mexican and scrub jays at retrieval intervals up to 60 days (Bednekoff et al. 1997).

The same result was obtained in a working memory task in parids. Biegler et al. (2001) compared the accuracy, capacity, and resolution of spatial memory in coal and great tits using delayed matching-to-sample techniques. Performance decreased for both species with increases in the number of sample locations to be remembered, the delay length, and spatial clumping of the choice objects. Again, the food-hoarding coal tits outperformed the nonhoarding great tits in the delay length they could tolerate—that is, in the persistence of spatial memory.

Scatter hoarding is also found in many mammals, particularly granivores and carnivores (Vander Wall 1990), and similar memory results have been obtained in granivores such as desert rodents and tree squirrels (Jacobs 1995). Scatter-hoarding kangaroo rats are more accurate at cache retrieval than larder-hoarding pocket mice (Rebar 1995). In addition, kangaroo rats can accurately retrieve caches in open spaces without landmarks after a 24-hour delay.

With landmarks, kangaroo rat performance did not change even after a 10-day delay (Barkley and Jacobs 1998).

Such persistent spatial memory might increase proactive interference and degrade performance in some cases. In a simple task in which the correct response varied among a few spatial locations, scatter-hoarding chickadees indeed suffered more interference than nonhoarding juncos (Hampton et al. 1998).

Memory of Caching Events

Perhaps the most advanced organization of spatial memory includes not only a food item's location and contents, but also memory for the unique foraging episode when the item was cached. Recent studies have demonstrated memory for events, or episodic-like memory, previously described only in humans in the scatter-hoarding scrub jay. In these studies, scrub jays learned either that worms spoiled after long storage (5 days) or that they did not. After a long delay between caching and retrieval, the group that had learned that worms spoil searched first for nonperishable peanuts, despite their normal preference for worms. The group without any experience of spoilage expressed their unaltered preference and searched for worms first and peanuts second (Clayton and Dickinson 1998, 1999). Many questions remain about nonhuman episodic memory, yet this experiment demonstrated that a foraging jay could encode a specific event in time and could use this data to optimize subsequent foraging decisions.

Social Learning

Social foragers may initially learn where to find food from other foragers. Social learning can range from guppies locating food by swimming with more knowledgeable conspecifics (Swaney et al. 2001) to the exceptional honeybee dance language (see Shettleworth 1998 for review; Riley et al. 2005 for recent research). Multiple causes can underlie social learning, or the appearance of social learning, so mechanisms must be carefully investigated (see discussions in Galef 2004 and Heyes and Galef 1996). *Local enhancement* (or *stimulus enhancement*) does not require direct contact between individuals. One individual's activity or its effects simply attract the attention of another individual, which then learns on its own. Similarly, in *social facilitation*, the presence of conspecifics may affect the motivation or arousal of the observer and allow it to learn independently. *Imitation* and *emulation*, which we will discuss later, are more complex forms of social learning.

Two recent studies with corvids illustrate *observational learning* of foraging locations. One study showed that free-living Florida scrub jays were able to learn a novel food patch by watching a trained demonstrator forage in the center of a moving ring (Midford et al. 2000). Another found that ravens not only could learn the location of food by observing conspecifics caching, but also cached behind occluders to prevent such observation (Bugnyar and Kotrschal 2002).

Communication is a special type of social learning. The honeybee dance language is one of the best-studied and most sophisticated methods of communicating food location in the animal kingdom. In addition, several social species call in the presence of food, including primates, dolphins, bats, and many species of birds (reviewed in Gros-Louis 2004). Recent evidence suggests that food calls, along with many alarm calls, may be "functionally referential"; that is, the call is given reliably in the presence of the referent, and the receiver of the call behaves consistently whether or not it can detect the referent. Functional referentiality is usually tested using playback experiments. Domestic chickens and tufted capuchin monkeys have both demonstrated responses particular to food calls in playback experiments, indicating that these calls direct individuals specifically to food (reviewed in Gros-Louis 2004).

Synopsis

Foragers rely on a variety of cognitive abilities to locate or store food items. From the simplest phototaxis to a cognitive map, mobile foragers need some form of spatial cognition. Foragers use external cues, such as beacons, gradients, and arrays of landmarks, to orient and to memorize the location of food sources. Different species, and even males and females of a single species, may use different frames of reference for their spatial orientation. Scatter-hoarding species face the additional problem of creating and relocating hundreds or thousands of cache sites, which could explain observed species differences in performance on abstract and naturalistic tasks measuring spatial memory. Social learning can also help a forager locate food by observation or communication.

4.6   Techniques for Obtaining Food

The omnivorous blue jay faces a final cognitive challenge: it must learn to extract food from the environment. It may need to do anything from prying up bark to capture insects underneath to opening a discarded berry container. The jay must learn those food-handling techniques that are not innate.

Instrumental Conditioning

Instrumental or operant conditioning refers to a situation in which an animal learns that its own behavior, in the presence of certain stimuli, is instrumental in causing a particular outcome. The study of instrumental conditioning began with the work of E. L. Thorndike (1874–1949), who conducted the first controlled studies of learning in the laboratory (Thorndike 1911). To compare the "intelligence" of species directly, he developed cages known as *puzzle boxes*, in which a hungry animal had to trigger a release mechanism from inside the box to reach food outside. When first placed in a puzzle box, an animal moved randomly until it accidentally triggered the escape mechanism. In subsequent trials, the animal tended to repeat the behaviors that had occurred just before in escape, whether or not those behaviors opened the apparatus. This process of repeating the behaviors that preceded success produced a gradual, negatively accelerated *learning curve* (as discussed under "conditioning mechanisms" in section 4.4) when Thorndike plotted time to escape against trial number. From this observation, Thorndike formulated the *law of effect*: in a particular context, behavior that is followed by a satisfying event strengthens the association between the context and the behavior, causing the behavior to become more likely should the context recur. This law formed the basis for instrumental learning theory.

Behavioral psychologists use two types of procedures to study instrumental conditioning: discrete-trial and free-operant procedures. In discrete-trial procedures, the subject makes the instrumental response once per trial, such as triggering the escape mechanism of a puzzle box. Likewise, an experiment may require that a rat turn left in a maze to obtain a reward. After the response, the investigator removes the subject from the apparatus. In free-operant procedures, the subject repeats its response freely. The operant chamber is the original and most typical free-operant apparatus and has proved to be a critical tool in the study of instrumental conditioning due to the ease of collecting data.

Both types of procedures rely on the pairing of a behavior with a reinforcing outcome, or *reinforcer*, such as food. One can deliver the reinforcer every time the subject makes the required response (*continuous reinforcement*) or only every so often (*partial reinforcement*). Behavioral psychologists use four basic schedules of partial reinforcement. In an *interval schedule* the subject earns reinforcers for responses after a given time interval. In a *ratio schedule* the subject earns reinforcers after a specified number of responses, such as lever presses or key pecks. The time and number requirements can be fixed (staying the same from trial to trial) or variable (changing from one trial to next), giving four possibilities: variable interval, variable ratio, fixed interval, and fixed ratio schedules. The reinforcement schedule influences the behavior of a subject in predictable ways; for example, subjects in fixed interval schedules begin to respond just before the end of the fixed interval (Roberts and Church 1978; see Domjan 1998 for thorough discussion of instrumental conditioning).

Biology constrains instrumental conditioning, just as it does classical conditioning. Foragers do not have to learn all the behaviors associated with feeding; the corollary of this statement, that some behaviors cannot be unlearned, is *instinctive drift*.

Breland and Breland (1961) first demonstrated instinctive drift in their instrumental conditioning of animals for commercial advertising. For example, they would train a raccoon to drop a coin into a box using the *method of successive approximations*, in which they rewarded the animal for behaviors progressively closer to the desired one. However, the raccoon's behavior proved less malleable than predicted. It would rub the single coin, or later two coins, together, thereby delaying reinforcement. Despite the obvious cost in reinforcements, the raccoon could not suppress its innate foraging movements of rubbing small objects together. These findings have inspired a movement toward a functional perspective in learning theory that emphasizes biological relevance (Domjan 2005).

Imitation

A jay may learn foraging techniques by imitating a conspecific's successful technique. However, as mentioned above, researchers must carefully identify the processes involved. In one famous example, a wild population of English blue tits learned to open milk bottles and drink the cream (reviewed in Shettleworth 1998). Debate ensued over how this skill spread through the population. Sherry and Galef (1990) showed experimentally that the spread of this skill did not require imitation, but could have been accomplished by local enhancement and social facilitation.

Imitation can also be confused with *emulation*. Whereas when an individual imitates, it copies the action of a model, when an individual emulates, it learns that the environment can be manipulated to achieve a particular goal. For instance, an emulator might see a model open a hinge by poking out a pin and learn only that the pin comes out. During replication, an imitator would poke the pin out, whereas an emulator might pull it. Emulation is arguably as cognitively complex as imitation, but may require different mechanisms. The mechanisms involved in both processes are still highly controversial (see reviews in Caldwell and Whiten 2002; Zentall 2004).

In the most definitive test for imitation, the *two-action test*, models demonstrate different solutions to the same problem to different experimental groups. If the subjects use the method they observed, this indicates imitation rather than emulation. For example, demonstrator Japanese quail depressed a treadle with the foot or the beak while one experimental group watched each technique. When tested, the quail generally used the technique they had witnessed. In a further demonstration, observers were more likely to imitate a demonstrator that received food rewards for its actions than one that did not, suggesting that the imitator may also represent the action's purpose—in this case, obtaining food (reviewed in Zentall 2004).

A recent study distinguished between *action* imitation and *cognitive* imitation (Subiaul et al. 2004). In a typical serial learning task, demonstrator rhesus monkeys were taught series of photographs. The monkeys were required to press each photograph on the screen in order, although the location of the photographs was changed in each screen. The observer monkeys were able to gain some information about ordinal position by watching the demonstrators that raised their performance significantly above baseline. This effect was not the result of social facilitation or emulation based on the feedback given by the computer. Therefore, under some circumstances, animals may learn rule-like information from observing conspecifics.

In other cases, animals may learn *not* to imitate one another. Pigeons in a situation in which the actions of a skill demonstrator deliver food to the observer regardless of the observer's behavior do not learn the skill. In contrast, with a small change in the apparatus, the observer is not rewarded during the experience, and under these conditions, observers readily learn to copy the movements of the demonstrator (Giraldeau and Lefebvre 1987). This observation suggests that learning of a particular food-handling technique may depend on whether the subject stands to gain from learning that skill.

Teaching

If animals can learn from others, it stands to reason that behaviors that promote such learning experiences could also evolve. Caro and Hauser (1992) defined teaching functionally as a change in behavior in the presence of a naive individual that is not immediately beneficial to the teacher and helps the naive individual learn. Common chimpanzees may teach their young how to use stone hammers and anvils to open coula nuts (Boesch 1991). Mother chimpanzees in Tai National Park behaved in ways that could facilitate learning, including leaving hammers near anvils when offspring were present, although they usually carried the hammers away (the hammers were used by offspring on 46.2% of 387 such occasions), or bringing nuts or hammers to a young chimpanzee at an anvil (588 occasions, leading to a 20% increase in nuts eaten per minute by offspring). On two occasions, mothers adjusted the orientation of the hammer or the nut, seemingly correcting the infant's use of the technique.

Teaching may be prevalent in species with elaborate predatory behavior, such as birds of prey and carnivores. Among these species, ospreys, domestic cats, and cheetahs demonstrably increase the foraging effort they require from their offspring, from bringing them dead prey to live but wounded prey and finally live prey that are allowed to escape for recapture (reviewed in Caro and Hauser 1992). Some spiders may behave similarly (Wilson 1971). In most of these species, it remains to be demonstrated that this behavior actually facilitates learning. However, a laboratory study with domestic cats found that kittens whose mothers were present and interactive during exposures to live prey learned hunting skills earlier than control kittens whose mothers were not present (reviewed in Caro and Hauser 1992).

As with imitation, cognition researchers want to understand the cognitive processes underlying teaching. It might seem that teachers require a *theory of mind* (a representation another's mental states) to be sensitive to the needs of the pupil. Caro and Hauser maintain that although such a representation would "almost certainly enhance the utility of teaching" and may be present in some species, it is not necessary. To be useful, the teacher must have a mechanism for discriminating which individuals lack skills or knowledge. Distinguishing the actual mechanisms involved will require experimental manipulations. As with other behaviors we have discussed, species differences in the cognitive basis of teaching are likely to emerge.

Insight

Can an animal use existing knowledge to produce a novel foraging technique? One way of doing so might be through *insight*, a novel viewpoint on a situation that can enable undetected relationships to suddenly become apparent. Animals must solve problems without overt trial-and-error learning, innate programmed responses, or observation before insight can be considered. Early experiments by Kohler (1925) are frequently cited as the seminal research on insight in human and nonhuman psychology (reviewed in Ormerod et al. 2002). Working with a group of captive chimpanzees, in one experiment Kohler (1925) hung bananas from a high place and gave the chimpanzees a box. The chimpanzees solved this problem by moving the box so that stepping on it allowed them to reach the bananas. Later they were also able to stack several boxes to solve a similar problem (fig. 4.6). Success tended to come suddenly after a period of no progress, not gradually after many approximations, suggesting insight.

Figure 4.6. Testing for insight in chimpanzees. Captive chimpanzees trying to reach a hanging banana appear to suddenly realize a solution to the problem, suggesting insight. In the drawing at the right, a chimpanzee has stacked three boxes to reach the bananas overhead. In the drawing at the left, another is in the process of stacking four boxes to reach the goal. (After photographs in Kohler 1925.)

Although Kohler's chimpanzees had no previous experience with the exact problem presented to them, an experiment by Epstein et al. (1984) cast doubt on Kohler's results. Pigeons trained separately to push a box toward a randomly placed target and to stand on a box to peck a fake banana put these behaviors together to solve the equivalent problem, reportedly through stimulus-response chaining rather than insight. Pigeons trained to perform only one of the subtasks (e.g., climbing but not pushing) failed to reach the banana. However, why the pigeons pushed the box specifically toward the banana was unclear.

A study of hand-reared ravens controlled more precisely for previous experience (Heinrich 1995). The ravens faced the following problem: how to retrieve food attached to a branch by a long string. A raven had to land on the branch and use its beak and foot to pull up the string in stages. Once the raven obtained the food, it had to suppress its natural tendency to fly away because the food was still connected by the string. Despite the complexity of the motor sequence involved, several ravens performed this task correctly without apparent trial-and-error learning. Although pulling and stepping may be an innate motor pattern in birds (see review in Thorpe 1963), several ravens never completed the task, and the ones that did showed a prolonged delay. Heinrich argued that assembly of the steps into a coherent, novel action, not the origin of the individual steps, is crucial for demonstrating insight. These studies suggest that under appropriate circumstances, animals may create novel foraging techniques without trial and error.

Tool Use

Techniques for obtaining food may include the use of tools. Many animals have been observed using tools, including insects, crabs, rodents, elephants, and many primates (reviewed in Griffin 2001). A *tool* is a material object that an animal manipulates as an extension of its body to achieve an immediate goal. Sea otters, for example, use a rock to crack a prey item's shell; Egyptian vultures and chimpanzees use rocks in a similar way. Many other taxa use a thin stick to extract insects or other food items from crevices; examples include the Darwin's woodpecker finch, common chimpanzee, and New Caledonian crow. Tool use may be acquired by the processes described previously or may be innate. Cognition researchers are particularly interested in whether the tool-using animal understands the relationship between the tool and its use (the *means-ends* or *cause-effect* relationship).

Hauser (1997) demonstrated that cotton-top tamarins can discriminate the functional properties of a tool. Hauser gave tamarins a choice between a functionally intact tool and one that he had modified to make it nonfunctional. For example, the tool might be a cane placed with a piece of candy inside its hook so that the monkey could use it to pull the candy in. A nonfunctional option might be the cane with the candy outside of its hook. In a series of experiments, tamarins chose the functionally intact tool more frequently.

However, capuchin monkeys can successfully use tools without understanding the means-ends relationship. Visalberghi and colleagues (Visalberghi and Limongelli 1996) tested capuchins and chimpanzees using a clear plastic tube with a cuplike depression in the middle, known as the "trap tube" (fig. 4.7), and a reward placed outside the trap at one end of the tube. To extract the food, the animal had to push a stick through the tube, pushing the food out of the tube while avoiding the trap. Previously, three of four monkeys had used sticks to obtain rewards from tubes without traps (Visalberghi and Trinca 1989). With the trap, however, the monkeys needed to push from the correct end. When tested with the trap, three of four monkeys could not extract food more than half the time, even after 140 trials. The fourth monkey learned the task after 90 trials, but apparently learned by rote. She continued to push from the side farthest from the food (as the trap requires), even when the investigators rotated the trap upward (and it no longer acted as a trap). Chimpanzees

Figure 4.7. Testing for means-ends understanding with the trap tube. In this experiment, the subject must use a stick to push a reward out of the tube. If the subject pushes from the wrong direction, the reward will fall into the trap. Here, a capuchin monkey is about to push the reward into the trap. (After a drawing in Shettleworth 1998 of a photograph in Visalberghi and Limongelli 1994.)

showed more signs of means-ends understanding in performing this task. Of five, two solved the original trap tube and transferred this skill to a variant in a way that suggested they understood the intermediate goal of avoiding the trap.

Modification of tools for a particular task also suggests understanding of the means-ends relationship. New Caledonian crows modify their tools into two different shapes (a hook or a jagged tool) as appropriate for removal of insects from different holes, and they shorten the length of a tool when necessary (Hunt 1996). Recent studies have shown that these crows can choose the right length of stick without trial and error (Chappell and Kacelnik 2002), and one individual bent a piece of wire into an appropriate tool (Weir et al. 2002).

Synopsis

Animals can use different cognitive skills to acquire foraging techniques. A forager may learn techniques by trial and error through instrumental conditioning, but within the constraints of innate biases. Imitation may be an efficient way to learn a successful technique from a conspecific. Teaching may also play a role in transferring foraging techniques. Sometimes animals may use insight to produce a correct technique the first time they encounter a problem. Many animals use tools to forage, though they may not always understand why the tool works. The cognitive mechanisms underlying many of these behaviors are still being investigated.

4.7 Summary

Foraging requires a broad range of cognitive skills. Foragers must perceive the environment, learn and remember food types, locate food resources, and learn techniques for extracting food items once found. Students of foraging need an understanding of these processes because they enable and constrain foraging behavior. Theorists can use data on animal cognition to develop more realistic foraging models. Foraging researchers can also pursue cognitive questions that provide potentially relevant information about foraging decisions. The separate traditions of psychology and behavioral ecology have formed a barrier to this interdisciplinary research. Psychologists have focused on process (learning, memory, and so on) using a limited number of species in highly controlled situations (Beach 1950), while behavioral ecologists have focused on functional categories of behavior (foraging, reproduction, etc.) using many species. Investigators are now working to break down these barriers, and foraging is a key point of contact between behavioral ecology and animal psychology. We hope that this chapter will help inspire future interdisciplinary research efforts. New data could bring answers regarding the survival value of cognition and the mechanisms of foraging within our grasp.

4.8 Suggested Readings

There are a number of comprehensive textbooks on animal cognition. *Cognition, Evolution and Behavior* (Shettleworth 1998) provides significant detail suitable for upper-division or graduate students. *Principles of Animal Cognition* (Roberts 1998) offers the most comprehensive discussion of animal memory. Another good introductory text is *Animal Cognition: The Mental Lives of Animals* (Wynne 2001). Conditioning is thoroughly covered in *The Principles of Learning and Behavior* (Domjan 1998), while *Animal Minds: Beyond Cognition to Consciousness* (Griffin 2001) represents the field of cognitive ethology. For a broad sampling of animal cognition, *The Cognitive Animal: Empirical and Theoretical Perspectives on Animal Cognition* (Bekoff et al. 2002) and *Comparative Cognition: Experimental Explorations of Animal Intelligence* (Zentall and Wasserman 2006) are good choices. For in-depth coverage of specific topics, the Comparative Cognition Society (www.comparativecognition.org) publishes free-access online textbooks on animal cognition, including *Avian Visual Cognition* and *Spatial Cognition*.

Part II

Processing, Herbivory, and Storage

Telander

Food Acquisition, Processing, and Digestion

Christopher J. Whelan and Kenneth A. Schmidt

5.1 Prologue

It is a classic moment from a TV nature show: a cheetah pursues a fleeing gazelle. The cheetah's sensory, neural, and muscular systems work at full capacity in support of this unfolding drama. For some predators, however, the real drama begins after consumption. Burmese pythons (*Python molurus bivittatus*) live in a world of feast and famine, going days or even weeks between meals that can be 60% larger than the python's body. As the snake digests these enormous meals, its metabolic rate can increase forty-four-fold. A python digesting quietly on the forest floor has the metabolic rate of a thoroughbred in a dead heat. Metabolic up-regulation is just the beginning: intestinal mucosal mass, total microvillus length, and the mass of the heart and kidney all increase to keep up with the demands of the snake's digestive upheaval. After the snake assimilates its meal, the whole system shifts into reverse. In a prodigious display of physiological flexibility, everything returns to its semi-quiescent between-meal state. Feast-and-famine feeders like the python show the greatest range of gut regulation, but virtually all foragers can regulate their guts to some extent. Typically, the magnitude of this regulation matches the variation in the forager's diet.

5.2   Introduction

The ways in which animals obtain and handle food resources depend on the physiological processes that follow ingestion. Preconsumptive and post-consumptive processes make up an integrated, whole-organism operation (Bautista et al. 1998; Karasov and Diamond 1988; Karasov and Hume 1997; Levey and Martínez del Rio 2001; Penry and Jumars 1986; Whelan et al. 2000). However, while foraging ecologists have made tremendous progress in understanding the ecological factors influencing patch use and prey choice, and while studies of the physiology of digestion have increased our understanding of food processing, theoreticians have made few connections between these two fields. This is unfortunate, because each depends on the other.

Food acquisition and processing are not independent processes. We view foraging as a suite of ecological tools for selecting habitats and diets, which in turn direct foods to the gut that facilitate the gut's processing tools. Foraging and digestion constitute a coordinated and coadapted division of labor. Efforts to secure a resource, or to prepare it for consumption, facilitate efforts to process and assimilate it (Courtney and Sallabanks 1992; Levey 1987). The actions of any one component of the digestive system, right down to the electrophysiological coordination of the two-membrane domains of the absorptive cells of the intestine, the enterocytes (Reuss 2000), facilitate the operation of other components (Caton et al. 2000).

Following Rosenzweig (1981), we recognize a continuum from foraging specialists to foraging generalists. Coadaptation of behavioral, morphological, and physiological traits pertinent to food acquisition and processing shapes the level of a particular animal's specialization. A specialist may need a specialized gut, while a generalist may require a more generalist (jack-of-all-trades) gut (Bjorndal and Bolten 1993; Murphy and Linhart 1999; Sorenson et al. 2004). These scenarios may represent alternative evolutionary strategies of coadaptation of food acquisition and food processing.

Students of feeding will continue to investigate pre- and postconsumptive processes independently, and these separate tracks will often yield important results. Yet, to answer many important questions, we must combine the two fields. Digestive physiologists and foraging ecologists should both "give a hoot" (C. Martínez del Rio, personal communication) about the other field. Foraging ecology matters—omitting or misidentifying the ecological constraints on foraging can render physiological experiments uninterpretable, if not downright meaningless. Likewise, digestive processing matters. Digestive enzymes influence diet choice (Martínez del Rio and Stevens 1989; Martínez del Rio et al. 1992). Internal handling of food in the gut matters—foods compete for processing in the gut, and past consumption influences future consumption (Forbes 2001; Whelan and Brown 2005; and see below).

While this chapter focuses on the interplay between foraging ecology and digestive physiology, we first consider the role of ecology, particularly that of search strategies, in determining diet choice. Without variation in the diet, there is no need for variability in digestive physiology. Next, we briefly review digestive structure and function. Variability in digestive systems reflects variability in foraging ecology. We then describe a variety of approaches to forging tighter links between the two disciplines. We conclude with thoughts on current gaps in our understanding of pre- and postconsumptive processes and their integration, and we offer suggestions for future avenues of research and pertinent readings.

5.3   Physiological Processes

In an Introductory Biology course at the University of Wisconsin, Professor John Neese remarked that we often think of the interior (lumen) of an animal's gut as being inside the animal. In fact, it is actually exterior space that exists as a cavity (as in cnidarians such as jellyfishes) or a tube (as in humans), created by invagination during very early development. When an organism ingests food for processing, what perhaps seems the end of the process to a foraging ecologist is only the beginning of the process to a digestive physiologist: the important work of getting the food *inside* the forager (absorption) has only begun.

The digestive system breaks down macromolecules of carbohydrates, fats, and proteins into sugars, alcohols and fatty acids, and peptides and amino acids. The intestinal wall absorbs these products, transporting them into the circulatory system. But just how this is accomplished, as we will see below, differs considerably among animals with different diets. The last few decades have seen increasing investigations of wild (as opposed to domesticated) animals, revealing an astonishing array of digestive strategies (Hume 1989). Modeling frameworks adopted from optimality and chemical reactor theory have provided new analytic tools.

Preconsumptive Food Handling

The relationship between mouthparts and diet in virtually all taxa clearly reveals the importance of preconsumptive food handling (Labandeira 1997; Lentle et al. 2004; Magnhagen and Heibo 2001; Owen 1980; Schmidt-Nielson 1997; Smith and Skulason 1996). For instance, bill size and shape in birds clearly relate to diet (Benkman 1988; Denbow 2000; Grant 1986; Welty

1975). In their classic investigations of Darwin's finches, Grant and colleagues (e.g., Schluter and Grant 1984; Schluter et al. 1985) elegantly demonstrated the fit of bill size and shape to the prevailing supply of seeds and the remarkably rapid evolution of bill morphology in response to the changing availability of seeds differing in size and hardness. Many other taxa exhibit similar adaptations (Ehlinger 1990; Mittelbach et al. 1999; Smith and Skulason 1996).

Preconsumptive food handling may serve several functions, including preventing escape of the prey organism, preventing injury to the forager by the prey, and preparing the food for ingestion and more efficient postconsumptive processing. Herbivores consume diets of highly fibrous or woody plant parts, and many herbivore species possess grinding mouthparts that fragment cellulose and release cell contents (Owen 1980 and Schmidt-Nielson 1997 provide many examples). This grinding, or mastication, increases the surface area available to digestive enzymes, allowing more efficient chemical breakdown in the intestines. Many birds swallow their food whole and rely on a muscular gizzard (and sometimes, ingested small rocks or pebbles) to physically break down food before it passes into the intestines for digestive processing.

Prinz and Lucas (1997) provide another explanation for mastication in mammals, which combines the physical breaking down of food into small particles with lubrication from saliva. Previous work suggested that initiation of swallowing depended on separate thresholds for food particle size and for particle lubrication. It now appears instead that swallowing is initiated after "it is sensed that a batch of food particles is binding together under viscous forces so as to form a bolus" (Prinz and Lucas 1997, 1715). Bolus formation ensures that swallowed food will successfully pass the pharyngeal region with minimal risk of inhalation of small particles into the respiratory tract, an accident with potentially fatal consequences.

Some spider species chew their prey with their maxillae and then suck out the nutritious body fluids. Other spiders inject hydrolytic enzymes into their immobilized prey and then use their piercing mouthparts to suck out the resulting fluid. Cohen (1995) estimates that 79% of predaceous land-dwelling arthropods use extraoral digestion (EOD). Via extraoral digestion, these small-bodied predators increase their efficiency of nutrient extraction by abbreviating handling time and concentrating nutrients from the consumed foods (Cohen 1995). Venomous snakes inject toxins that not only immobilize prey, but also begin digestion prior to ingestion, when the prey is swallowed whole.

Gut Structure and Function

One can think of the digestive system (gut) as a tubular reactor that extends from the oral opening to the anus. A typical invertebrate's gut has three parts:

the headgut, corresponding to the oral cavity and pharynx; the foregut, corresponding to the esophagus and crop or stomach; and the intestine (Gardiner 1972). The foregut mainly transports food from the oral cavity to the intestine, but in some taxa with an enlarged crop and/or diverticula (blind sacs), the foregut may store food. In some invertebrates (e.g., insects), the intestine consists of the midgut, or ventriculus, and the hindgut, which includes the anterior intestine and the rectum. Most digestion and absorption take place in the midgut. The main role of the hindgut is to transport undigested material away from the midgut for expulsion, but it also is responsible for water, salt, and amino acid absorption, thus playing a role in water and salt balance (Romoser 1973; Stevens and Hume 1995).

Most of the components found in invertebrate digestive systems are also found in vertebrate systems (Stevens and Hume 1995). We divide the vertebrate digestive system into four parts: the headgut (including the oral cavity and pharynx, as well as the gill cavity in fishes and larval amphibians); the foregut (esophagus and stomach); the midgut (often referred to as the small intestine), including the duodenum, jejunum, and ileum as well as the pancreas and biliary system (which secrete enzymes and bile, respectively); and the hindgut (often referred to as the large intestine).

As stated above, food digestion typically begins with the process of mechanical breakdown and lubrication within the oral cavity. Saliva not only lubricates the bolus for transport through the esophagus to the stomach, but in some species, it may also contain the hydrolytic enzyme amylase, which digests carbohydrates (Stevens and Hume 1995). The stomach stores food and secretes HCl and pepsinogen, the precursor of the hydrolytic enzyme pepsin, to physically break down food and initiate protein digestion. After the food has been broken down sufficiently and transformed into a slurry (Karasov and Hume 1997), it moves to the small intestine, the principal site of both digestion and absorption.

The midgut is the primary location of digestion and absorption of digestive products into the circulatory system. The mechanism of absorption (active or passive; more on this below) has been a subject of considerable controversy and interest (Diamond 1991; Lane et al. 1999; Pappenheimer 1993; Pappenheimer et al. 1994; Pappenheimer and Reiss 1987). Within the midgut, mucosal folds and villi increase the surface area available for absorption tremendously (perhaps fractally; Pennycuick 1992). Villi are composed of absorptive cells known as enterocytes, whose own surface area is increased by the microvilli or brush border (Stevens and Hume 1995). The pancreas secretes enzymes that degrade carbohydrates, fats, and proteins. Biliary secretions, which include salts, phospholipids, cholesterol, and hydrophobic apolipoprotein fractions (Karasov and Hume 1997), are emulsifying agents important in fat digestion.

The gastric mucosa secretes pepsin, which digests proteins into polypeptides. A number of additional enzymes (e.g., trypsin, chymotrypsin) break polypeptides into amino acids. Carbohydrases, including amylase (secreted by the pancreas and by the salivary gland in some species), and glycosidases (e.g., sucrase, maltase, lactase) digest carbohydrates. Fats, which are insoluble in water, undergo a two-stage process of emulsification and dispersion, followed by formation of small aggregates of mixed lipids and bile salts suspended within the ingesta, called micelles. Lipase, secreted by the pancreas, attacks the micelles and releases fatty acids, glycerol, and mono- and diglycerides. Other enzymes in the midgut include chitinase (which attacks chitin, a major structural carbohydrate in animals, fungi, and bacteria), found in many vertebrate taxa (Stevens and Hume 1995), and cellulase (which attacks cellulose), found only in microorganisms, some of which are symbionts in some invertebrate and vertebrate guts.

The gut absorbs the products of digestive degradation via passive or carrier-mediated mechanisms. Passive mechanisms include transcellular diffusion, in which particles move through the cells (mainly lipophilic compounds), and paracellular diffusion, in which particles move between the cells (mainly water-soluble compounds, including sugars, amino acids, and some vitamins). Carrier-mediated transport across the (apical) brush border and basolateral membranes of the enterocytes involves carrier proteins. Carrier-mediated transport is either active (involving investment of energy to transport the substance against an electrochemical concentration gradient) or facilitated (in which the substance is transported down an electrochemical gradient). In both cases, saturation of the carrier proteins places an upper bound on transport, following Michaelis-Menten kinetics. Carrier proteins appear to be the primary transport mechanisms for sugars, amino acids, some vitamins, and calcium.

Pappenheimer (Pappenheimer 1993, 2001; Pappenheimer et al. 1994; Pappenheimer and Reiss 1987) proposed an alternative involving passive diffusion of sugars, amino acids, and other small molecules via a mechanism called paracellular solvent drag. Briefly, concentrative sodium-dependent transcellular transport provides an osmotic force that triggers contraction of the cytoskeletal proteins (tight junctions) regulating paracellular permeability, permitting solvent drag between absorptive cells. Pappenheimer (1993) estimated that the paracellular pathway may account for most (60%–80%) absorption of sugars and amino acids.

This controversial proposal has stimulated much research (Afik et al. 1997; Chang et al. 2004; Chediack et al. 2003; Ferraris and Diamond 1997; Karasov and Cork 1994; Lane et al. 1999: Lee et al. 1998; Levey and Cipollini 1996; Weiss et al. 1998). One attractive aspect of this mechanism is an almost instantaneous fine-tuning of the match of absorption to digestive loads because transport is proportional to solute concentration at cell junctions, which is proportional to the rate of hydrolysis (Pappenheimer 1993). A drawback of this mechanism is its nonspecificity, which could lead to the inadvertent uptake of toxins or secondary metabolites (Chediack et al. 2001). Lane et al. (1999) tested paracellular transport of glucose in dogs and concluded that it plays, at most, a minor role (4%–7%) compared with carrier-mediated transport. Some low estimates of the extent of absorption by paracellular transport may be artifactual, however, attributable to inhibition of normal villus microvascular responses to epithelial transport in anaesthetized animals (Pappenheimer and Michel 2003).

In contrast to amino acids, sugars, and vitamins, most products of lipid digestion (free fatty acids and monoglycerides) cross the brush border membrane by simple diffusion. Passive systems transport fatty acids and monoglycerides to the endoplasmic reticulum, where they are transformed into particles called chylomicrons, small milky globules of fat and protein. Chylomicrons enter the lymphatic vessel that penetrates into each villus, and the lymphatic system transports them to the blood.

Digesta are discharged from the midgut into the hindgut. In birds and mammals, a cecum (or paired ceca in birds) at the junction of the midgut and hindgut often serves as a fermentation chamber. The hindgut serves for final storage of digesta, absorption of water (osmoregulation), bacterial fermentation, and feces formation (Laverty and Skadhauge 1999). The extent of these functions differs considerably among taxa in relation to diet. A carnivore's hindgut is a relatively passive structure, while herbivores have greatly enlarged hindguts that are critical fermentation chambers. The hindgut empties into the cloaca (in reptiles, birds, fetal mammals, some adult mammals) or the anus (in most mammals) (Stevens and Hume 1995).

This description of digestive system structure and function is very general, and great variety exists among taxa, as illustrated by the following examples. Ruminants possess a greatly enlarged and compartmentalized stomach (the rumen) and the ability to regurgitate, re-chew, and re-swallow their food. The rumen acts as a fermentation chamber, providing anaerobic conditions, constant temperature and pH, and good mixing (Church 1988). The only known avian foregut fermenter is the hoatzin (*Opisthocomus hoazin*). It has an enlarged muscular crop, containing mixed microflora and protozoans that do the work of fermentation throughout the crop and in the lower esophagus (Grajal 1995; Grajal et al. 1989). In lagomorphs (rabbits, hares, pika), the stomach is simple but elongated. Part of the small intestine has a dilated structure called the sacculus rotundus, and the cecum has a capacity roughly ten times that of stomach (Stevens and Hume 1995).

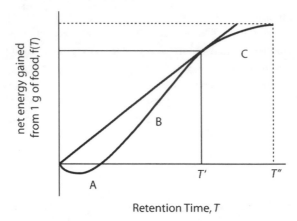

Figure 5.1. Sibly's optimality model relating retention time of food in the gut to net energy gain. Sibly reasoned that following ingestion, energy would at first have to be expended to break the chemical and/or physical defenses of foods against digestion (phase A in the figure). Once the defenses are breached, energy is quickly gained (phase B). As food digestion continues, net energy gain eventually diminishes until all potential energy has been acquired (phase C). T' indicates the length of food retention that maximizes the rate of net energy assimilation. T'' indicates the length of food retention that is associated with complete assimilation of energy. (After Sibly 1981.)

Optimality and Chemical Reactor Models

Sibly (1981) made the first optimality model of the digestive system. He assumed that digestive processes maximized the "rate at which energy is obtained by digestion" (109). He reasoned that, following ingestion, the rate of energy gain at first declines, because energy must be expended on breaking down food defenses before any nutrient absorption can take place. After the digestive system breaches the food's chemical defenses, the rate of energy acquisition rises rapidly at first, then declines as digestion proceeds (fig. 5.1). This scenario is reminiscent of the patch model (Charnov 1976b) because of the strong role of diminishing returns.

Sibly's model identified important relationships between two characteristics of digestive systems, gut volume and retention time. In addition, the model related these gut properties to food characteristics (see also Karasov 1990; Karasov and Diamond 1985). For a given gut volume, higher-quality food should be retained for shorter periods of time than lower-quality food. Letting $E =$ the concentration of enzyme (or equivalent), $C =$ the concentration of substrate, $r =$ reation rate, $T =$ retention time of food in the gut, $k =$ gut volume, and V_0 the flow rate of food through the gut, the model can be summarized by the following relationships (Karasov and Hume 1997):

$$\text{Efficiency of extraction} \propto \frac{(E \times C)}{\pi} \propto T \propto \frac{k}{V_0}. \qquad (5.1)$$

The extent (or efficiency) of the reaction (hydrolysis, absorption, etc.) is thus positively related to the concentration of enzyme and/or substrate and retention time of food in the gut; retention time is itself positively related to gut volume and inversely related to flow of food through the gut.

Chemical reactor theory allows a rigorous examination of the relationships in equation (5.1). Penry and Jumars (1986, 1987) introduced chemical reactor theory to the study of optimal gut design. They recognized the analogy between animal guts and reaction chambers used in industrial applications, and they applied the large body of theory on the physical chemistry of idealized reaction chambers to a variety of gut designs. Penry and Jumars (1986, 1987) analyzed three idealized reactor types: batch reactors, continuousflow, stirred-tank reactors (CSTR), and plug-flow reactors (PFR). These models describe mass transfer between phases (e.g., food reactants and enzyme reagents to products and untransformed reactants) using mass balance equations. Batch reactors are analogues for the gastrovascular cavities found in some invertebrates, including hydras and coelenterates; plug-flow reactors are analogues for the tubular guts found in most multicellular invertebrates and vertebrates; and continuous-flow, stirred-tank reactors are analogues for the large chambers found in foregut and hindgut fermenters. Models of actual animal guts often allow different idealized reaction chambers to be connected serially. For instance, a ruminant may be modeled as a large continuous-flow, stirred-tank reactor serially followed by a plug-flow reactor and then a small continuous-flow, stirred-tank reactor (Alexander 1994).

Chemical reactor models of guts have been heuristically useful by helping investigators diagnose the configurations of digestive systems and digesta flow within them; by specifying how the interplay of processing costs, reactant volumes, and reaction kinetics affects digestive system performance; and by spawning empirical tests of the predictions of specific models (Alexander 1991, 1993; Dade et al. 1990; Hume 1989; Jumars 2000a, 2000b; Jumars and Martínez del Rio 1999; Levey and Martínez del Rio 1999; Martínez del Rio and Karasov 1990; Martínez del Rio et al. 1994). Early models were general and permitted broad comparisons among widely different digestive systems. These early models indicated, for instance, that plug-flow reactors outperform both batch and continuous-flow, stirred-tank reactors for a given reactor volume and when reactions are catalytic, but continuous-flow, stirred-tank reactors outperform plug-flow reactors when reactions are autocatalytic. They also showed that a digestive system consisting of a continuous-flow, stirred-tank reactor/plug-flow reactor series was superior in performance on the low-quality foods eaten by foregut fermenters (Alexander 1991; Penry and Jumars 1987).

Later models, aimed at capturing the digestive systems of particular animals, incorporated specific physiological and/or ecological traits of the foragers under investigation (herbivorous fishes—Horn and Messer 1992; frugivorous and nectarivorous birds—Karasov and Cork 1996; Levey and Martínez del Rio 1999; Martínez del Rio and Karasov 1990; herbivorous insects—Yang and Joern 1994b; Woods and Kingsolver 1999). Some of these more specific models did not produce predictions that were upheld by empirical tests. For instance, Karasov and Cork (1996) and López-Calleja et al. (1997) tested a model proposed by Martínez del Rio and Karasov (1990). In their work with the rainbow lorikeet (*Trichoglossus haematodus*), a species that absorbs sugars passively, Karasov and Cork expected that increased sugar concentrations in the diet would result in decreased retention times and extraction efficiencies. Neither prediction was upheld. Karasov and Cork (1996) suggested that the response of the lorikeet in their experiments was better interpreted as being consistent with the goal of time minimization and extraction efficiency.

López-Calleja et al. (1997) found that captive green-backed firecrowns (*Sephanoides sephanoides*), which absorb glucose actively (by carrier-mediated transport), exhibited close to complete assimilation of sugars and increased both food retention and inter-meal interval times with increasing sugar concentrations, as predicted by Martínez del Rio and Karasov's (1990) model. In contrast, they did not observe the predicted correlation between sugar concentration and daily energy intake. López-Calleja et al. (1997) concluded that one objective function of the original model, energy maximization, was inappropriate for birds that were not growing, storing fat, or reproducing, and that a more appropriate objective under these conditions might be "satisficing" (Ward 1992).

The chemical reactor paradigm has proved useful as an organizing framework for constructing models and tests of gut structure and function. Jumars and Martínez del Rio (1999) and Levey and Martínez del Rio (2001) provide excellent discussions of chemical reactor models, including several explanations for why they sometimes fail: inaccurate estimation of the physiological parameters (processing [foraging] costs, gut volumes, reaction kinetics) or incorrect specification of the objective function (optimization criterion) itself. Section 5.4 (below) considers the challenges of measuring foraging costs. In addition, some important assumptions of the approach may not hold; for example, real guts may seldom be at a steady state (Penry and Jumars 1986, 1987).

Diet Composition and Modulation of Gut Structure and Function

Foraging ecologists often consider gut morphology, digestion and absorption biochemistry, and the flow rate of food through the gut as constraints on foraging behavior (Stephens and Krebs 1986). But digestive physiologists have long known that diet composition influences gut structure and gut function in a flexible way (Afik et al. 1995; Karasov 1996; Karasov and Hume 1997; Starck 2003). The interplay between gut function and diet composition gives the forager some leeway, allowing it to bend the rules (Foley and Cork 1992). In the following discussion, we use the term "modulation" to include acclimatization and regulation of gut structure and function in response to changes in diet composition.

The most dramatic example of gut modulation yet investigated involves foragers that undergo extreme bouts of feast and famine: sit-and-wait-foraging snakes that feed at infrequent intervals, but consume 25%–160% of their body mass when they do. Examples include the boa constrictor (*Boa constrictor*), the Burmese python (*Python molurus*), and the sidewinder rattlesnake (*Crotalus cerastes*) (Secor and Diamond 1995, 2000; Secor 2003; but see Starck and Beese 2002; Starck 2003; Starck et al. 2004). In these snakes, the gut responds to extreme variation in contents: it is empty most of the time and only occasionally full. Changes in the structure and function of the gut at meal ingestion are among the highest recorded (Secor and Diamond 2000; Secor 2003; see also Hopkins et al. 2004). Less extreme variation in diet composition, such as seasonal switches between fruits and insects in passerine bird species (Levey and Karasov 1989, 1992), leads to more modest, but nonetheless significant, changes in gut function (Karasov 1996; Whelan et al. 2000).

Why do animals modulate their guts so dramatically? Why aren't they geared up for efficient food processing whenever the chance presents itself? Intuitively, it seems that active guts must be costly to maintain (Karasov and Diamond 1983; Karasov 1992, 1996), as the dramatic "up-regulation" in gut morphology and function after feeding in snakes suggests. Stevens and Hume (1995) summarize a number of studies showing that the contribution of the digestive system to total (whole-animal) oxygen consumption ranges from 12% in rats to 25% in pigs. They also document that protein synthesis is particularly high in actively proliferating or secreting tissues. In ruminants, for example, the gut wall constitutes a mere 6% of body protein, but accounts for a whopping 28%–46% of whole-animal protein synthesis.

When they fed fasting snakes, Secor and Diamond (2000) found a "10- to 17-fold increase in aerobic metabolism, 90%–180% increase in small intestinal mass, 37%–98% increases in masses of other organs active in nutrient processing, three- to 16-fold increases in intestinal nutrient transport rates, and five- to 30-fold increases in intestinal uptake capacities [integrated over the entire intestine]" within a single day. Following digestion, the digestive organs quickly atrophied to preconsumptive levels. Starck and Beese (2001, 2002) found that the mass of the snake's small intestine increases without cell proliferation because the mucosal epithelium, a transitional epithelium, can reversibly undergo enormous size changes. The cost of gut modulation in snakes may therefore owe more to changes in gut function (specific dynamic action, gastric processes involving digestion, protein synthesis, action of associated organs) than to changes in gut structure (Overgaard et al. 2002; Secor 2003; Starck 2003).

American robins (*Turdus migratorius*) change their diets seasonally. Robins consume arthropods during the breeding season, but eat mostly fruit during the rest of the year (Levey and Karasov 1989, 1992; Martin et al. 1951; Wheelwright 1986, 1988; Whelan et al. 2000). In contrast to the dramatic short-term changes in snake guts, American robins do not increase absorption rates of sugars and amino acids when they switch to their fruit diet, nor do they compensate via changes in gut length, surface area, or volume. Instead, fruit-eating robins pass food more quickly than insect-eating robins. Short retention time is the key adaptation to frugivory in this (and other bird) species (Karasov 1996; Levey and Karasov 1989, 1992).

In the face of infrequent feedings, it is not surprising that the gut should atrophy (Piersma and Lindström 1997; Karasov et al. 2004). What is perhaps more surprising (and impressive) is how quickly the gut structure and function can be reconstituted. The robin-snake comparison tells us that the degree of modulation reflects the degree of diet change: from feast to famine in the python; from one food type (insect) to a second (fruit) in the robin. Digestive physiologists have observed gut modulation in many taxa (Starck 2003). This modulation can include changes in digestive enzymes, nutrient absorbers, gut structure, or gut retention time. Digestive modulation increases digestive efficiency (Karasov 1996; Whelan et al. 2000) and helps foragers meet their metabolic demands in the face of a shifting and sometimes unpredictable resource base.

5.4   Integrating Ecological and Physiological Processes

This section examines a number of ways to integrate digestive physiology and foraging ecology. To begin, we compare the disparate cost accounting practices of foraging ecologists and digestive physiologists. We argue that better integration of these costs will increase our understanding of both ecological and physiological processes.

Costs of Foraging

Foraging ecologists and digestive physiologists focus on different aspects of the costs of foraging. These differences reflect distinct perspectives on the intrinsic and extrinsic factors that influence foraging. To a foraging ecologist, intrinsic factors include the forager's search and attack strategies, habitat preferences, and susceptibilities to predation. Extrinsic factors are properties of the environment, such as the abundance and distribution of resources and predators, together with properties of the resource, such as ease of detection and capture. In contrast, to a digestive physiologist, intrinsic factors include the structure and function of the gut, including gut capacity, the suite of digestive enzymes, and transport mechanisms (active and passive) for moving nutrients from the gut lumen into the forager's bloodstream. Extrinsic factors include properties of the resource, such as the proportion of digestible versus refractory components, nitrogen content, and energetic value (see Karasov 1990 for extensive review and discussion).

Both perspectives offer valid insights, but they emphasize different costs. Improper accounting of either ecological or physiological costs can lead to errors in both ecological and physiological models, and thus to experimental manipulations that do not test the predictions of the models (see Jumars and Martínez del Rio 1999). Thoughtful integration of ecological and physiological approaches can help avoid errors.

From a physiological perspective, constraints on gut emptying impose frequent bouts of inactivity as a hummingbird waits for its crop to clear before it can resume foraging. However, foraging hummingbirds may experience high predation risk (Lima 1991; Martínez del Rio 1992). From a foraging ecology perspective, we suggest that because hummingbirds are highly vulnerable while foraging, they have evolved a foraging strategy and an accompanying gut processing system that allows them to minimize their exposure to predation while maintaining a high rate of energy gain. Relyea and Auld (2004) present a related scenario involving tadpoles.

A difficulty arises because the physiological costs of foraging are quantifiable in joules expended, but not all ecological costs are. Physiological costs include the metabolic cost of foraging, the fixed cost of maintaining the digestive system, the variable cost of moving food through the digestive system, and the cost of specific dynamic action (also referred to as the thermogenic cost of foraging, which includes the enzymatic costs of food processing and the costs of chemosynthesis). Ecological costs not directly quantifiable in joules expended include the costs of predation risk and missed opportunities.

Foraging theory has solved the problem of costs measured in different currencies (see chap. 1). The fitness costs of predation danger or lost opportunities can be translated into a common currency by using experimental manipulations (Abrahams and Dill 1989; Nonacs and Dill 1990; Todd and Cowie 1990; Brown 1988) or the economic concept of marginal rates of substitution (Brown 1988; Brown, Kotler, and Valone 1994; Mitchell et al. 1990). The most powerful and flexible approach is that of dynamic state variable models, described in chapters 1 and 7.

Linking Ecological and Physiological Processes

Ecological Consequences of Physiological Modulation

The harvest rate of a consumer in relation to resource abundance is known as the functional response. A widely used functional response model, Holling's disc equation [similar to equation (5.1.1)], includes variables representing conversion of food biomass to consumer biomass (e) and time needed to handle food (h). Whelan et al. (2000) developed models of gut function in which they assumed that these terms of the functional response implicitly incorporate physiological parameters, nutrient absorption, and gut handling of food (box 5.1). These models allow e and h to vary (independently or jointly) in response to changing diet composition in a manner that simulates physiological modulation. Through such modulation, two digestive modes emerge, each of which is more efficient at processing a particular diet. Modulation thus promotes diet switching and specialization. The models also indicate, as suggested by physiological investigations (Levey and Karasov 1992), that modulation incurs an initial cost, though it ultimately increases efficiency.

BOX 5.1   Modeling Digestive Modulation in an Ecological Framework

Christopher J. Whelan

Consider two perfectly substitutable resources denoted as 1 and 2 (Whelan et al. 2000). Let the forager's per capita growth rate be a monotonically increasing function of its feeding rate, f. Let Holling's disc equation describe the feeding rate for an opportunistic forager seeking two co-occurring foods:

$$f = \frac{(e_1 a_1 R_1 + e_2 a_2 R_2)}{(1 + a_1 h_1 R_1 + a_2 h_2 R_2)}, \quad (5.1.1)$$

where e_i is net assimilated energy from consuming a food item i, a_i is the encounter rate for a resource, h_i is the handling time for a resource, and R_i represents the density of a resource (see Royama 1971 for a derivation).

We define a consumption isocline as all of the combinations of abundances, R_1 and R_2, such that a forager has the same feeding rate, k (Holt 1983; Brown and Mitchell 1989). To solve for the consumption isocline, we set equation (5.1.1) equal to a constant feeding rate k and solve for R_2 in terms of R_1:

$$R_2 = \frac{k}{a_2(e_2 - h_2 k)} - \left[\frac{a_1(e_1 - h_1 k)}{a_2(e_2 - h_2 k)} \right] R_1. \quad (5.1.2)$$

(Box 5.1 continued)

In the state space of resource abundances 1 and 2, this equation describes a straight line that has a negative slope when $e_i/h_i > k$. Combinations of R_1 and R_2 that lie outside this isocline yield harvest rates greater than k; combinations inside the isocline yield feeding rates less than k. When k represents the subsistence level of resource consumption by the forager, the corresponding consumption isocline is the zero net growth isocline, ZNGI, at which the forager's per capita growth rate is zero (Vincent et al. 1996).

Gut modulation may take the form of adjustments in the rate of nutrient assimilation, which we model by allowing e_1 and e_2 to increase or decrease in relation to a changing diet composition. Similarly, gut modulation may take the form of variation in the rate of food transport through the gut (gut retention time), modeled by assuming that h_1 and h_2 implicitly include both pre- and postconsumptive handling of food, and thus change in response to changing diet. We will restrict our development here to the case of active nutrient transport involving the e terms; the h modulation case is very similar (Whelan et al. 2000).

This model allows the e terms to vary between two gut modulation modes, which we will designate A and B, respectively. Each mode has its own consumption isocline (see below). In some circumstances, the isoclines will intersect so that one mode is more efficient at certain resource abundances, while the other is more efficient at other resource abundances. We assume that the modulation mode is chosen to maximize the forager's fitness, written as $G = \max\{f_A, f_B\}$. This objective function applies for a family of fitness functions (Brown 1992).

Assume that gut modulation strategy A increases the rate of assimilation of resource 1 via an increase of active 1 transporters, e_1, coupled with a decreased rate of assimilation of resource 2 via a decrease of active 2 transporters, e_2. Let the opposite be true for gut modulation strategy B. The variables e_{1A} and e_{1B} represent the assimilation rates for resource 1 under modulation strategies A and B, respectively. For a given constant feeding rate k, we now have two consumption isoclines, one for each gut modulation strategy:

$$R_2 = \frac{k}{a_2(e_{2A} - h_2 k)} - \left[\frac{a_1(e_{1A} - h_1 k)}{a_2(e_{2A} - h_2 k)}\right] R_1, \quad (5.1.3A)$$

$$R_2 = \frac{k}{a_2(e_{2B} - h_2 k)} - \left[\frac{a_1(e_{1B} - h_1 k)}{a_2(e_{2B} - h_2 k)}\right] R_1. \quad (5.1.3B)$$

(Box 5.1 continued)

Each equation is a straight line with negative slope. When assimilation of resource 1 is greater for modulation mode A than it is for modulation mode B ($e_{2A} > e_{2B}$) and assimilation of resource 2 is less for modulation mode A than it is for modulation mode B ($e_{2A} < e_{2B}$), then the two lines must cross at positive values for resource abundances. This indicates that each digestive strategy yields a higher feeding rate (in terms of assimilated energy per unit time) at some combinations of resource abundances. At the point of intersection, both gut modulation strategies yield the same feeding rate.

Resource abundance combinations for which the two gut modulation strategies yield the same feeding rate define the modulation isoleg (*sensu* Rosenzweig 1981). Some algebra shows that this is

$$R_2 = \left[\frac{a_1}{a_2}\right] \left[\frac{(e_{1A} - e_{1B})}{(e_{2B} - e_{2B})}\right] R_1, \quad (5.1.4)$$

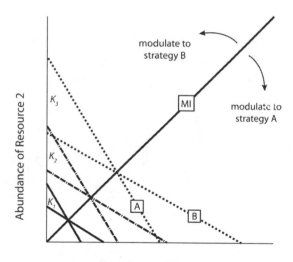

Figure 5.1.1. Graphical representation of the effect of modulation of digestive processing on consumption isoclines. Families of paired equal consumption rate isoclines for three levels of harvest rate or fitness, *k*, when the forager modulates between strategies *A* and *B* (labeled for harvest level k_3). Each isocline represents the relative combinations of resources 1 and 2 that result in a constant harvest rate. Note that the modulation isoleg (indicated by MI), that combination of resources 1 and 2 that results in an equal harvest rate for both modulation strategies, cuts through the intersection of each pair of consumption isoclines.

(Box 5.1 continued)

a straight line with positive slope. Points on the isoleg further from the
origin represent higher feeding rates, k (fig. 5.1.1). Above the isoleg, gut
modulation strategy B $(f_A < f_B)$ yields a higher feeding rate, and below the
isoleg, modulation strategy A $(f_A < f_B)$ yields a higher feeding rate. When
resource abundances lie above the isoleg, the forager should modulate
nutrient transport to become more efficient on resource 2. Similarly, below
the isoleg, the species should modulate nutrient transport to become more
efficient on resource 1. The net effect of modulation results in an "effective"
consumption isocline that is piece-wise linear and is composed of the part of
each component isocline [equations (5.1.3A) and (5.1.3B)] that lies within
that of the other. This effective consumption isocline approximates that
for antagonistic resources (see Tilman 1980, 1982), despite the fact that the
model specifically treats resources as perfectly substitutable (fig. 5.1.2).

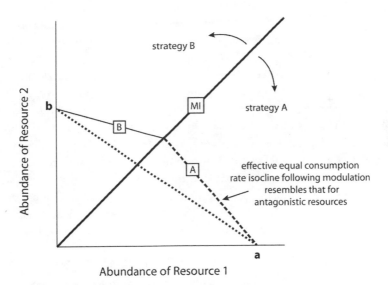

Figure 5.1.2. Graphical representation of the effect of modulation of digestive processing on
consumption isoclines. Following completion of modulation to digestive strategy A and B, respectively, the piece-wise linear "effective" consumption isocline approximates that for antagonistic
resources. Note that this "effective" consumption isocline bows out from the simple line connecting the intercepts of the abscissa (*a*) and ordinate (*b*), which would be the expected consumption
isocline for two perfectly substitutable resources.

Whelan et al. (2000) analyzed the consequences of their functional response equations in consumer-resource models that allowed analysis of gut modulation modes and diet selection under three ecological scenarios. First, when the consumer does not deplete its resources, the resource standing crop determines the optimal modulation strategy. Second, when a consumer population of a fixed size depletes its resources, and the standing crop of resources results from a dynamic equilibrium between resource renewal and resource consumption, the equilibrium between renewal and consumption determines the optimal gut modulation strategy. Finally, when resource renewal, depletion, and consumer population size all equilibrate, the intersection of the consumer's depletion trajectory with the modulation isoleg at the consumer's zero net growth isoclines (ZNGIs) determines the optimal gut modulation strategy (Whelan et al. 2000).

These analyses show that we cannot fully understand the consequences of modulating gut physiology independently of an organism's ecological circumstances. They also hint at reasons why some foragers modulate digestive processes while others do not. Foragers that exploit nondepletable resources should show rapid modulation in response to changes in the standing crop of food. The situation is more complex and nonintuitive for foragers that exploit depletable resources. To illustrate, consider a scenario in which resource renewal, depletion, and consumer population size equilibrate. In this circumstance, the relation of the carrying capacity of the resources and the depletion vector (the trajectory of resource consumption) that intercepts the "elbow" of the modulation isoclines (fig. 5.2) determines the optimal modulation mode. When the resource supply points lie above this special depletion vector, the forager should modulate its physiology appropriately for resource 2 in figure 5.2, even though it may consume mostly resource 1 (a surprising result!). When the resource supply points lie below this special depletion vector, the forager should modulate its physiology appropriately for resource 1, the resource it is consuming predominantly (a much more intuitive result).

Nutrient Transfer Functions

Raubenheimer and Simpson (1998) present a graphical framework that views the digestive process as nutrient transfer between serially connected processing compartments. The nutrient transfer functions that apply at each junction are key points of integration between the behavioral and physiological components of input regulation. Raubenheimer and Simpson's framework focuses on two nutritional variables, the rate ("power") and efficiency of nutrient processing, and the transfer from one processing compartment to the next.

Raubenheimer and Simpson (1998) plot the processing time for a given quantity of food at stage S_i (where $i = 1, 2, \ldots, n$ serial stages of processing,

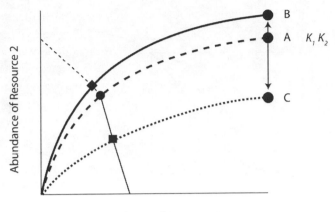

Figure 5.3. Graphical representation of consumer-resource model when resource renewal, depletion, and consumer population sizes equilibrate. K_1 and K_2 represent the carrying capacity of resources 1 and 2, respectively, for three resource supply points, A, B, and C. In this case, the optimal digestive physiology modulation mode is determined by the depletion vector connecting the resource supply point to the intersection of the ZNGI (resource supply point A). When below this depletion vector, the consumer should always modulate to digestive mode A (resource supply point C), and when above, the consumer should always modulate to digestive mode B (resource supply point B).

and may represent foraging, ingestion, digestion, absorption, etc.) against the cumulative release (or transfer) of the product of processing at stage S_i to the following stage (stage S_{i+1}). Following Sibly (1981), they assume that a sigmoidal curve represents this nutrient transfer relationship (fig. 5.3). Given this sigmoidal relationship between time of processing in compartment S_i and transfer of the product to the next serial compartment, S_{i+1}, the model finds the maximal rate of transfer using tangent construction techniques, as in graphical solutions of the marginal value theorem. If natural selection maximizes efficiency, rather than rate, then processing in compartment S_i should proceed until the transfer curve reaches its asymptote (Raubenheimer 1995; Raubenheimer and Simpson 1994, 1995, 1997; Simpson and Raubenheimer 1993b, 1995, 2001).

A potential flaw in Raubenheimer and Simpson's graphical approach may be that sigmoidal enzyme reaction kinetics pertain to allosteric enzymes, but many digestive enzymes and carrier-mediated (saturable) transport mechanisms follow Michaelis-Menten kinetics, which are monotonically increasing with decelerating slope [as in the type II functional response of equation (5.1.1)]. The logic of the marginal value theorem may still apply, however. For instance, if one considers a nutrient's "travel time" (say, from oral cavity to reaction chamber), the marginal value theorem approach can still be applied in the manner of Raubenheimer and Simpson (1998; see, for instance, fig. 3b in Penry and Jumars 1986; see also Cochran 1987).

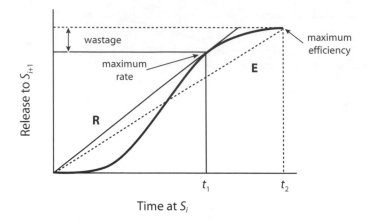

Figure 5.3. An example of nutrient transfer relationships hypothesized by Raubenheimer and Simpson. Note the general similarity to the graphical model of Sibly (fig. 5.1). The x-axis represents the amount of time that digesta (or nutrient) is processed at one stage in a serial nutritional pathway, and the y-axis represents the cumulative release to the next stage in the pathway. The slope of the smooth linear line (labeled R) connecting the origin to the nutrient transfer function (heavy solid line) represents the maximal rate of transfer from stage i to stage i + 1. Dropping a vertical line from this point on the transfer curve to the x-axis indicates the associated processing time in compartment S_i. The slope of the dashed linear line (labeled E) represents the rate of transfer from stage i to stage i + 1 when processing at stage i is allowed to proceed to completion (maximum efficiency). Maximum rate of transfer from stage i to stage i + 1 is accomplished at time t_1. Maximum efficiency is accomplished at time t_2. (After Raubenheimer and Simpson 1998.)

Incorporating Digestive Processing into the Functional Response

Recently, a number of investigators have considered the influence of digestion and food quality on the functional response (Verlinden and Wiley 1989, 1997; Hirakawa 1997a, 1997b; Farnsworth and Illius 1998). We refer to their closely related models as digestive rate models (DRM), after Verlinden and Wiley (1989). In these works, digestive capacity is modeled as an on/off inequality constraint. These studies suggest that under some circumstances (e.g., high food abundance, low food quality), digestive quality (energy gain per throughput time) determines diet selection when digestive rate is limiting. Under these circumstances, the diet is composed of a smaller number of food types of higher quality, partial preference is expected for one food type, and all other food types are either always accepted or always rejected (the zero-one rule; Hirakawa 1997a). A critical conclusion of the digestive rate model is that the digestive properties of foods, which we refer to as their bulk properties, can play a major role in diet selection.

In a recent review of the functional response, Jeschke et al. (2002) suggested that most predators (in the broad sense, including carnivores, herbivores, parasites, and parasitoids) are, in fact, digestion limited. They proposed the steady-state satiation model, which incorporates both the handling and digestion of prey. Digestion influences a predator's hunger level, and this in turn determines the likelihood that the predator will search for prey. While digestion is a background process, gut fullness influences feeding rate as a sliding motivational state that can result in the forager choosing to cease foraging—it does not forage when satiated. This model has similarities to the "digestive pause" of Holling (1965). Because Jeschke et al. (2002) consider foraging on only a single food type, their model does not suggest how this digestive pause should influence diet selection.

Whelan and Brown (2005) developed an extension of Holling's (1965, 1966) disc equation that incorporates the passage rate of food through the gut (referred to as postconsumptive handling) as an integral component of total food handling time. In a manner similar to Jeschke et al. (2002), they modeled the extent of the "digestive pause" on a sliding scale, but one that reflects gut fullness (rather than satiation). In contrast to Jeschke et al. (2002), they developed their model for a forager that consumes two or more food types, and thus their model considers the effect of digestive processing on both harvest rates and diet selection. Postconsumptive handling time may be partially exclusive of time spent searching for and handling additional food items (preconsumptive activities). In contrast to the DRM, in which the effect of internal gut passage on harvest rate is a step function (operable or inoperable), it is continuous in Whelan and Brown's model. However, in a manner similar to the DRM, the bulk properties of foods, via their effects on postconsumptive handling, can also have strong effects on harvest rates and diet selection.

Whelan and Brown (2005) begin with a modification of the type II functional response (Holling 1965, 1966), in which they include terms for external (preconsumptive) and internal (postconsumptive) handling of food:

$$H = \frac{(aR)}{\{1 + aR[h + gm(B)]\}}. \tag{5.2}$$

External handling, h, is identical to that in the original disc equation. Internal handling consists of two variables. The first, g, represents the actual processing of food within the gut; the second, $m(B)$, represents the proportion of gut handling time that is exclusive of alternative foraging activities, and can take any functional form with a monotonically positive slope. External handling, h, and internal food processing, g, have units of (time/item). Internal food processing, g, is determined by the quotient of food bulk per item, b (ml/item), and the volumetric flow rate of food through the gut, V_0 (ml/time): $g = b/V_0$. But V_0 = gut capacity, k (ml), divided by retention or throughput time, T (time) (see Jumars and Martínez del Rio 1999; McWhorter and Martínez del Rio 2000). Thus, passage time per item is given by $g = (bT)/k$. For simplicity, let $m(B) = B$ (a linear function), the proportion of gut volume occupied by food. Gut fullness, B, is given by the bulk rate of intake (bulk of the resource, b, multiplied by its ingestion or harvest rate, H) and the retention time of food in the gut (the quotient of throughput time, T, and gut volume, k): $B = (bHT)/k$. This definition of $m(B)$ allows the exclusivity of internal handling to be a continuous sliding scale that reflects the extent to which gut volume is filled from food consumption. Substituting g and B into equation (5.2) and simplifying yields

$$H = \frac{(aR)}{\{1 + aR[h + b^2 H(T^2/k^2)]\}}. \tag{5.3}$$

This model can be solved explicitly for H (see Whelan and Brown 2005), but this explicit expression obscures the way in which external and internal food handling influence the forager's consumption rate. Equation (5.3) has three interesting consequences, which we explore graphically in figure 5.4. First, we now see the intimate connection between harvest rate and gut processing: we need the harvest rate to specify the gut processing rate, and we need the gut processing rate to specify the harvest rate. Harvest rate and gut processing rate mutually feed into each other. Second, equation (5.3) shows transparently that pre- and postconsumptive food handling jointly limit harvest rate. Third, we see that external handling and internal handling are qualitatively different phenomena. External handling, h, has a fixed cost per item consumed that is paid in time—it operates qualitatively like a batch reactor (Martínez del Rio et al. 1994) that is full (on) or empty (off). Internal handling, $gB (= g(bHT)/k)$, in contrast, has a variable cost paid in time because one component, harvest rate, H, is continuous (see also Jumars and Martínez del Rio 1999). In other words, internal handling operates like a continuous reactor, such as a plug-flow reactor (Martínez del Rio et al. 1994).

An analogous expression can be written for consumption of two (or more) food types:

$$H_T = \frac{(a_1 e_1 R_1 + a_2 e_2 R_2)}{\{1 + a_1 R_1[h_1 + b_1(T^2/k^2)(b_1 H_1 + b_2 H_2)] + a_2 R_2[h_2 + b_2(T^2/k^2)(b_1 H_1 + b_2 H_2)]\}}. \tag{5.4}$$

The behavior of equation (5.4) is qualitatively similar to that of equation (5.3) and is illustrated by plotting H_T as a function of R_1 and R_2 (fig. 5.5). In all cases, increasing R_i increases H_i and decreases H_j, where $i \neq j$. This occurs because resource 1 (or resource 2) reduces the forager's consumption of resource 2 (or resource 1) through both external and internal handling times. By handling an item of resource 1, the forager spends less time looking for food. The external handling time is independent of the forager's overall harvest rates on resources 1 and 2. However, the internal handling time increases with harvest rates and

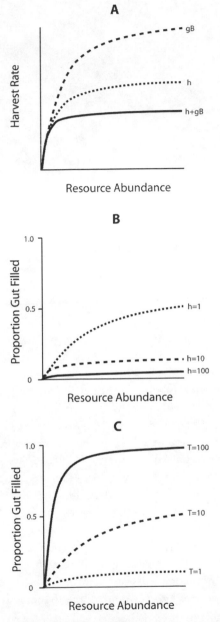

Figure 5.4. Graphical results of Whelan and Brown's (2005) foraging model incorporating both external (preconsumptive) and internal (postconsumptive) handling of food. (A) Harvest rate as a function of resource abundance when both external and internal food handling operate ($h = gB$); when only external food handling operates (h); and when only internal food handling operates (gB). (B) Proportion of gut filled as a function of increasing resource abundance as external handling time (h) increases from 1 to 100. The proportion of the gut filled rises monotonically with decreasing slope. Note that the proportion of gut filling declines sharply with longer *external* handling times. (C) Proportion of gut filled as a function of increasing resource abundance as internal handling time (T) increases from 1 to 100. The proportion of the gut filled rises monotonically with decreasing slope. Note that the proportion of gut filling declines sharply with shorter *internal* handling times.

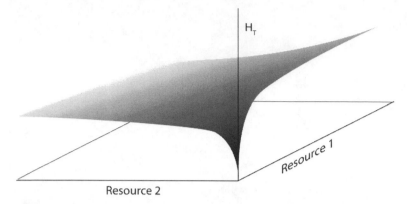

Figure 5.5. Harvest rate surface for two resources, 1 and 2, as a function of their abundance. Resource 2 has half the energetic value and 10 times the bulk volume of resource 1. Note that because the forager has greater external handling efficiency on resource 1, and because resource 1 is richer than resource 2, the total harvest rate is maximal when the forager consumes only resource 1.

gut fullness. These different properties of external and internal handling time result directly from their batch reactor–like and continuous flow reactor–like properties, respectively.

Referring to the food type with the greater profitability e/h as resource 1, there can be two relationships between resources 1 and 2 in terms of bulk properties. First, both profitability (external handling efficiency) and food richness (ratio of energy to bulk) are greater for resource 1 (e_1/h_1 e_2/h_2 and $e_1/b_1 > e_2/b_2$). Second, profitability is greater for resource 1, but food richness is greater for resource 2 ($e_1/h_1 > e_2/h_2$ and $e_2/b_2 > e_1/b_1$). These relationships between the food types lead to different expectations regarding diet selectivity. Under the first relationship ($e_1/h_1 > e_2/h_2$ and $e_1/b_1 > e_2/b_2$), the forager will exhibit either complete selectivity for resource 1, partial selectivity for resource 2, or complete opportunism, depending on the relative abundances of the two foods (fig. 5.6). Interestingly, the relative food abundances that result in partial preferences depend on the ratio of the richness of resource 1 to that of resource 2: the larger the ratio, the greater the range of partial preferences. Under the second relationship ($e_1/h_1 > e_2/h_2$ and $e_2/b_2 > e_1/b_1$), the forager will either show complete selectivity for resource 1 or take both foods opportunistically, depending again on resource abundances.

Food Preference Reconsidered

When gut capacity is not limiting, food preference is determined by a descending ranking of external handling efficiency, e_i/h_i. Whether a food item is included in the diet is determined by the position of the (vertical) preference isoleg (see fig. 5.6). Once gut capacity limitations enter the equation, however,

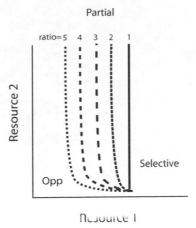

Figure 5.6 Selectivity isolegs in state space of abundance of resources 1 and 2. In the region of the state space to the right of the solid vertical isoleg, the forager is selective on resource 1. In the region of the state space between the vertical isoleg and the curved isolegs, the forager is selective on resource 1, but takes only some of encountered resource 2. The degree of spread between the two isolegs is determined by the ratio of the richness (e/b) of resource 1 to that of resource 2. As this ratio increases, the spread between the isolegs increases. To the left of the curved isoleg, the forager takes both resources opportunistically. For this figure, $a_1 = 0.1$; $e_1 = e_2 = h_1 = b_1 = b_2 = 1$; $b_2 = 1$ to 5; $T = 1$; $k = 2$; $R_1 = 0.01$; $R_2 = 0, 1, \ldots, 50$.

the world of food preference gets more complicated and interesting. Now preference is determined by the relative ranking of $\{e_i / [h_i + b_i(T/k)B]\}$, where $B = (\Sigma b_j H_j)(T/k)$, the extent to which the gut is "bulked up" from previous consumption. Now let

$$\rho = \frac{e_i}{[h_i + b_i(T/k)B]}, \tag{5.5}$$

where ρ, which has units of (energy $\cong$ time^{-1}), represents an equal acceptability threshold for a food item to be included in the diet. Noting the similarity of equation (5.5) to the Michaelis-Menten equation for enzyme kinetics, Whelan and Brown (2005) used reciprocals of both sides to find a solution:

$$\left(\frac{b_i}{e_i}\right) = \frac{k}{\rho(TB)} - \left(\frac{k}{TB}\right)\left(\frac{h_i}{e_i}\right). \tag{5.6}$$

Equation (5.6), a straight line with negative slope within the state space of external handling time:energy (h_i/e_i) and food bulk:energy (b_i/e_i), represents an equal preference isocline such that all food items lying on it are of equal acceptability. The region above the line represents higher ratios of bulk:energy and external handling time:energy. Any food item lying in this outer region is therefore less preferred. The region lying below it represents lower

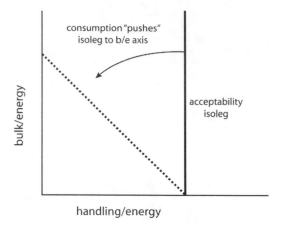

Figure 5.7. Preference isoleg in state space of bulk:energy and external handling time:energy. When internal handling is fast relative to external handling, the isoleg is a vertical line at the value of $1/\rho$, and any food on or to the left of it is preferred. As the forager takes on bulk through resource consumption, the isoleg takes on a negative slope (k/TB), and is "pushed" toward the y-axis. Consumption thus decreases the universe of preferred food items.

ratios of bulk:energy and external handling time:energy, and items in this region are preferred. When the line is a threshold of acceptability, then it is an isoleg separating acceptable (below) from unacceptable (above) food items. We can see now that preference is influenced by previous consumption, so that as the gut takes on bulk, B, fewer items will be included in the diet. As B increases due to consumption, both the y-intercept and the slope of equation (5.6) may decline (fig. 5.7): food consumption narrows the acceptable diet. Box 5.2 discusses how learning and taste affect foraging.

BOX 5.2   **More Than a Matter of Taste**

Frederick D. Provenza

Palatability is considered to be a matter of taste, and all popular definitions focus on either a food's flavor or its physical or chemical characteristics. Yet, if palatability is merely a matter of taste, why do herbivores supplemented with polyethylene glycol increase their intake of unpalatable plants high in tannins? Why do goats eat wood rat houses? Understanding these odd behaviors requires a fuller understanding of palatability.

(Box 5.2 continued)

Flavor-Feedback Interrelationships

Palatability is more than a matter of taste alone because a food's flavor is affected by its postingestive effects (Provenza 1995a; Provenza and Villalba 2006). Flavor is the combination of odor, taste, and texture. Postingestive effects are due to feedback from the cells and organs mediated by nutrients and toxins. Feedback is positive—increases palatability—when foods meet nutritional needs. Feedback is negative—decreases palatability—when foods are inadequate or excessive relative to nutritional needs or if foods are toxic. Thus, flavor-feedback interactions are influenced by the nutrient and toxin content of the food and the nutritional needs of the animal. The senses—smell, taste, vision—enable animals to select among foods and provide pleasant or unpleasant sensations associated with eating. Feedback modulates palatability as a function of utility.

Polyethylene Glycol

Tannins reduce the digestibility of protein and energy in foods, and some tannins are toxic. Polyethylene glycol binds with tannins, preventing their adverse effects (Silanikove et al. 2001). Animals supplemented with polyethylene glycol eat much more of foods high in tannins because the tannins no longer produce negative postingestive effects. Thus, the aversive postingestive effects of tannins, not their flavor, render high-tannin plants unpalatable, and the positive postingestive effects of nutrients make otherwise nutritious, high-tannin foods palatable.

Goats and Wood Rat Houses

The shrub blackbrush is deficient in macronutrients—energy and protein—during winter. During a winter grazing study, we placed ten goats on each of six blackbrush pastures (Provenza et al. 1983). As the study progressed, the goats became increasingly averse to blackbrush. In one pasture, the goats began to eat wood rat houses. Goats acquired a preference for wood rat houses because the houses contained urine-soaked (nitrogen-rich) vegetation that helped the goats rectify their deficiency. By the end of the 90-day study, goats that ate wood rat houses lost 12% of their body weight, whereas goats that did not lost 20%. Animals deficient in nutrients seek out new foods, and animals are likely to form a preference for a food, no matter how odd, if the food corrects a nutritional deficit or imbalance (Provenza and Villalba 2006).

(Box 5.2 continued)

Flavor-, Nutrient-, and Toxin-Specific Satiety

So, palatability is the interrelationship between flavor and feedback, but if that's all there is to palatability, then why is the grass always greener on the other side of the fence? For example, why do sheep prefer to eat clover in the morning and grass in the afternoon, even though clover is more digestible and higher in protein than grass (see chap. 6)? Why do animals perform better when offered choices of different foods?

Interactions between the senses and the body help to explain why palatability changes within meals and from meal to meal (Provenza 1996; Provenza, Villalba, Dziba et al. 2003). Flavor-, nutrient-, and toxin-specific satiety refer to the decrease in preference for the flavor of a food during and after eating because of interactions involving a food's flavor and postingestive feedback from cells and organs in response to nutrients and toxins. Flavor receptors respond to taste (sweet, salt, sour, bitter), smell (a diversity of odors), and touch (astringency, pain, temperature). Flavor receptors interact with receptors in the body (liver, gut, central nervous system, and elsewhere) that respond to nutrients and toxins (chemoreceptors), concentrations of salts (osmoreceptors), and gut distension (mechanoreceptors).

Preference for the flavor of a food declines automatically as that food is eaten because of interactions between the senses and the body (Provenza 1996; Provenza, Villalba, Dziba et al. 2003). These interactions cause transient decreases in the preference for foods just eaten. The decrease in preference, which is influenced by an animal's nutritional needs relative to a food's chemical makeup, is more persistent when a food has either too many or too few nutrients or when the food contains excess toxins. Aversions also occur when foods are deficient in nutrients. They even occur when animals eat nutritionally adequate foods, particularly if those foods are eaten too often or in too great an amount. Thus, eating any food to satiety causes a transient aversion to the flavor of that food.

These interrelationships also help to explain why the efficacy of plant defenses may vary with the mixture of plants in a community, and why the chemical attributes of a single plant must be considered within the context of the entire plant community (Provenza, Villalba, and Bryant 2003). Herbivores satiate on nutrients and toxins, and that limits their intake of particular foods and combinations of foods. Thus, flavor-nutrient-toxin interactions set the asymptote of a functional response curve that defines the relationship between plants and herbivores. These dynamics

(Box 5.2 continued)

are influenced by plant abundance because the chemical defenses of a species will satiate the detoxification capabilities of herbivores at a critical threshold of plant abundance. Above this threshold, herbivory will favor domination by a plant chemotype. Below it, local extinction is more likely as a plant species becomes increasingly less abundant. An animal's preference for a mix of plants and habitats may range from strongly aversive to strongly positive, depending on complementarities of nutrients and toxins in different forages. Learning how nutrients and toxins interact will clarify how biochemical diversity influences herbivore abundance and plant diversity.

Sheep and Clover

Sheep satiate on clover in the morning and switch to grass in the afternoon (Newman et al. 1992). In the morning, hungry sheep initially prefer clover because it is highly digestible compared with grass. As they continue to eat clover, however, sheep satiate—acquire a mild aversion—from the effects of nutrients such as soluble carbohydrates and proteins, from the effects of toxic cyanide compounds, and from eating the same flavor. The mild aversion causes them to switch to grass in the afternoon. During the afternoon and evening, the sheep recuperate from eating clover, and the aversion subsides. By morning, they are ready for more clover. The combination of clover and grass enables sheep to eat more each day than if only one species were available. The satiety hypothesis helps to explain why the growth of insects, fishes, birds, and mammals improves with varied diets (reviewed in Provenza, Villalba, Dziba et al. 2003). Because satiation varies with the kinds and amounts of nutrients and toxins in foods and the physiological state of the animal, animals that eat a variety of foods can ingest diets that are biochemically complementary.

Finally, researchers and managers typically consider foraging only in terms of how the physical and chemical characteristics of plants influence an animal's ability to achieve high rates of nutrient intake. However, social learning, especially from mothers, helps young herbivores learn about kinds and locations of nutritious and toxic foods. Learning from mother about foods begins early in life as the flavors of foods mother eats are transferred to her offspring in utero and in her milk. Lambs given a choice of palatable shrubs such as mountain mahogany or serviceberry—one of which their mother has been trained to avoid—show a preference for the shrub they ate

(Box 5.2 continued)

with mother. Through her actions, mother models appropriate foraging behaviors for her offspring, who learn what to eat and where to forage. The strong and central role of learning in this process means that it is likely, but not inevitable, that individuals will learn to select the best foods and habitats within an area (Provenza 1995b; Provenza, Villalba, Dziba et al. 2003). It also helps to explain why both wild and domesticated animals introduced into unfamiliar environments often suffer from malnutrition, overingestion of toxic plants, and predation. With increasing awareness of this problem, conservation biologists are beginning to advocate food and habitat training before introducing wild animals into unfamiliar habitats.

What Have We Learned, and Where Should We Go?

Knowledge of digestive physiology enriches our understanding of foraging ecology, yet many challenges remain before we can fully integrate these two research traditions. It may be especially difficult to incorporate ecological costs into the laboratory environment of digestive physiology. Studies attempting such integration could consider the following questions:

1. A variety of studies suggest that birds primarily absorb glucose via passive paracellular transport, rather than by active, carrier-mediated absorption. Do avian intestines differ from those of other vertebrates, or are the results of these studies due to technical or methodological problems (Ferraris and Diamond 1997)? If glucose (hexose) transport in birds is indeed passive, how does this relate to diet switching and modulation of digestive function?

2. Reserve capacity is still a matter of controversy. Whelan and Brown's model (2005) suggests that reserve capacity is a design feature of gut structure and function. How does spare capacity relate to metabolic demand and regulation (e.g., torpor), resource availability and foraging, and digestion (C. Martínez del Rio, personal communication)? Ferraris and Diamond (1997) report that the reserve capacity of the brush border glucose transporter is typically about 2, but why it has this value is not clear.

3. Can knowledge of how enzymes and transporters are distributed along a plug-flow reactor, in relation to harvest rates and food characteristics, improve chemical reactor models of guts?

4. Can we relate resource availability, resource depletion, and digestive traits such as modulation of gut function and structure? Does the capacity of an animal to modulate its gut physiology result from diet plasticity per se, or to

the effect (depletion or nondepletion) of the forager on its resources? Will relationships among those factors vary during the annual cycle? For instance, in spring, migratory birds require a diet of arthropods to prepare for breeding. Many studies show that birds deplete arthropod resources (Marquis and Whelan 1994, 1998). In contrast, in fall, when many fruiting species reach peak standing crops (Willson and Whelan 1993), birds can switch among different resource types in relation to local availability. Do these seasonal differences in food availability lead to differences in gut modulation seasonally?

5. How do the ecological costs of foraging influence digestive processing? For example, does predation risk shift diets toward generalization, and hence toward less specialized guts? In other words, is digestive flexibility yet another adaptation to predator avoidance?

6. Can experimentalists manipulate the profitability (e/h) and richness (e/b) of foods? These manipulations could be used to test Whelan and Brown's model.

5.5 Summary

Knowledge of the physiological mechanisms governing food processing has advanced rapidly, particularly in the growing body of information on the digestive physiology of nondomesticated animals. These studies have revealed physiological adaptations to specialized diets and the ability to adjust or modulate gut function. Alternating between feast and famine can cause gut modulation, but more subtle changes in diet composition also produce gut modulation.

The digestive systems of animals share many basic properties and characteristics, but each species appears to possess unique characteristics related to the resources constituting its own diet. Optimality modeling based on chemical reactor theory provides one approach to understanding the diversity of digestive strategies. This approach models the gut as a chemical reactor and uses reaction kinetics to derive predictions about gut processing, such as efficiency of absorption and changes in retention time.

Application of chemical reactor theory to digestion allows investigators to diagnose the configurations of digestive systems and digesta flow within them; to specify how the interplay of processing costs, reactant volumes, and reaction kinetics affects digestive system performance; and to design empirical tests of the predictions of specific models. However, improper accounting of foraging costs can lead our analyses to error. Similarly, omission of digestive costs may affect estimates of metabolic costs of foraging in ecological models. The failure

to fully account for ecological and physiological costs will limit future progress.

Mathematical models incorporating both ecological and physiological parameters illustrate that both kinds of processes influence harvest rate, diet selection, and the extent of gut filling. Although physiological processing capacity
clearly limits foraging, theory suggests that guts are seldom full, suggesting
further that spare capacity is an intrinsic feature of gut design. More importantly, this approach demonstrates the coadaptedness of ecological and physiological processes. We encourage and challenge others to further integrate
ecological and physiological approaches.

5.6 Suggested Readings

Additional information on digestive structure and function is available for
vertebrates (Karasov and Hume 1997), mammals (Chivers and Langer 1994),
and invertebrates (Wright and Ahearn 1997). Warner (1981) provides a classic
review of techniques for characterizing and measuring the passage of food
through the digestive systems of birds and mammals. A discussion of ecological
versus digestive constraints, which contrasts somewhat from the view of this
chapter, and a comparison of chemical reactor and compartmental models
of digestive systems can be found in Penry (1993). Starck (2003) provides
a fascinating review of the cellular mechanisms underlying modulation of
gut structure and function in mammals, birds, and snakes. All those with an
interest in the physiological aspects of foraging will want to consult Starck
and Wang (2005).

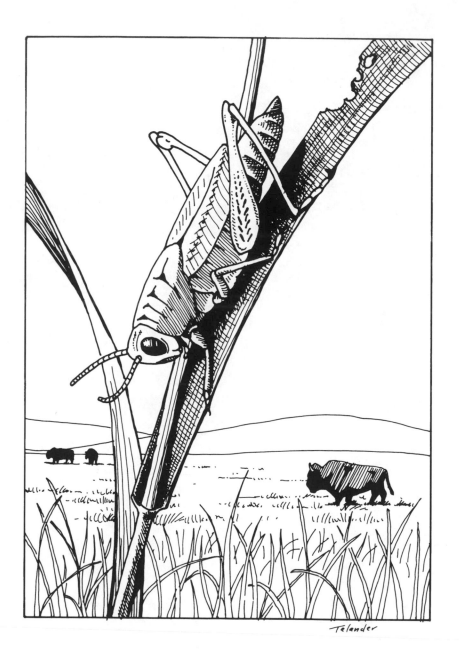

Herbivory

Jonathan Newman

6.1 Prologue

It is 4:00 a.m. on a cold, wet midsummer's day in Southwest England. The 500 kg dairy cows have been grazing for 30 minutes. A network of eighteen video cameras in weatherproof cases stands ready to record events across the study site. By 8.30 p.m. the cows have grazed for 9 hours and spent another 7 hours ruminating (regurgitating and chewing). A bite recorder (fig. 6.1) has logged every jaw movement (more than 72,000 of them). Each cow has ingested more than 6 kg of food while roaming across the 11-hectare field. Meanwhile, in a nearby greenhouse, an experimenter places individual peach aphids onto small melon plants growing in 12 cm pots. Each 2×10^{-6} kg aphid wanders across the plant for 10–15 minutes, occasionally stopping to probe the plant, then inserts its stylet into the leaf phloem and remains motionless for the next 2 hours, sucking in sap and expelling honeydew. It repeats this process, continuously, day and night.

What could possibly be interesting about these two foraging situations? Who cares, and why? Milk production depends critically on crude protein ingestion. Are the cows selecting a diet that maximizes their protein intake? Can we manipulate their natural behaviors to increase milk production? How can we maintain the pasture species composition and density in the face of the cows' foraging behavior? The

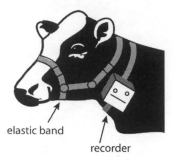

elastic band

recorder

Figure 6.1. Cow wearing the Penning bite recorder. The recorder works by recording the stretching of the elastic band under the jaw. Jaw movements of different types stretch the elastic in characteristic ways. A computer program then converts these data into jaw movements of different types based on their characteristic shapes. See www.ultrasoundadvice.co.uk for more information.

aphids' population dynamics are intimately linked to their diet, mainly to amino acid concentrations. Aphids can go through a generation in about 10 days, doubling their population size every 3 days under ideal conditions. Even at low densities, aphids can significantly reduce crop yields, and aphids are the most important vectors of plant viruses. Virus acquisition and transmission depends on aphid feeding behavior and movement on and between plants. Winged aphids facilitate the long-distance dispersal of viruses. Winged morph production increases with increasing aphid density or decreasing plant quality. Both of these problems have major financial implications, and a complete understanding of foraging behavior will inform our responses.

6.2   Introduction

A videorecording of herbivores feeding is not the sort of footage that leads to many *Trials of Life*-type, glossy documentaries, narrated by important natural historians with English accents. Predation, parasitism, and other animal-animal interactions dominate these documentaries. Yet, when it comes to foraging, herbivory is vastly more common. Insect herbivores make up 25% of the extant macroscopic organisms on earth, and every green plant (another 25%) has insect herbivores (Bernays and Chapman 1994, 1). Most nonaquatic vertebrate herbivores can be found in four orders of eutherian mammals: Lagomorpha (ca. 60 spp.), Proboscidea (2 spp.), Perissodactyla (ca. 18 spp.), and Artiodactyla (ca. 174 spp.); in addition, many of the Rodentia (ca. 1,700 spp.) are at least sometimes herbivorous. Herbivores are also by far the most common vertebrate animals housed by humans—from laboratory rodents (tens of millions) to farmed cattle, sheep, and goats (hundreds of millions each) to horses, asses, and camels (tens of millions each). Whether you look at numbers of species, numbers of individuals, total biomass, or rates of flow of mass and energy, there is no denying the practical significance, ecological dominance, and evolutionary importance of herbivory.

Elephants (ca. 6,000 kg) and grasshoppers (ca. 0.001 kg) differ in body mass by more than six orders of magnitude, yet they face essentially the same foraging problems: where to eat, what to eat, how fast to eat, and how long to spend eating. I ignore taxonomic boundaries for most of this chapter and focus on how herbivores answer these questions. I will use two important ideas as my framework: first, that the answer to each of these four questions lies in the animal's objectives and constraints; second, that the answer to any one question depends, at least in part, on the answers to the others. Herbivory is a compromise or trade-off between these four related questions. Finally, I will consider the dynamic nature of the herbivore plant interaction. Herbivory and plant growth are tightly coupled. Short-term studies of individual foraging behavior provide important glimpses of the herbivore's behavioral repertoire, but rarely provide a complete picture of its interaction with its food plants. Plant and animal respond dynamically to each other, and ultimately we must understand this dynamic to solve important applied problems such as ecosystem management, agricultural production, and the conservation of rare plants and animals.

Herbivory is the concern of ecologists, entomologists, agricultural scientists, range scientists, animal welfare scientists, conservation biologists, and marine scientists; even plant biologists get into the act. As one might imagine, there is relatively little communication across these disciplines. The literature on herbivory is very extensive, and the amount that any scientist can read is necessarily limited. Moreover, it is unevenly distributed among fields. For example, there are many more publications on the grazing behavior of sheep and cattle than on that of all 70 species of African ungulates combined. Can we learn much about the behavior of wild animals from the investigation of domesticated animals, or vice versa? I believe that a cross-disciplinary approach is beneficial and offer the following personal experience to support this view. In the early 1990s, I proposed to some colleagues that we should look at how sheep respond to predation pressure. They were, of course, incredulous, because there are no predators on sheep in Southwest England. Of course, they were correct—but sheep have lived on farms for only a small fraction of their evolutionary history, and there was no a priori reason to suppose that their antipredator behaviors had been lost. Indeed, predator avoidance was probably so heavily selected that there might be little genetic variance left in this suite of traits! Sure enough, sheep responded behaviorally to increases in feeding aggregation size in much the same way that wild animals do, by increasing their feeding time and decreasing their vigilance behavior (Penning et al. 1993). The evidence was not merely correlational, as it would probably have had to be if the subjects were antelope on the Serengeti. The data came from an experiment in which we randomly assigned individuals to different group sizes—something impossible on the African plains. My colleagues doubted the role of predation partially because their training as agricultural scientists did not prepare them for this possibility, even though predator effects seem basic to someone trained as an ecologist.

I believe that we can gain insight into the behavior of domesticated herbivores by studying their wild relatives, and vice versa. However, we must also remember that agricultural animals often result from unnatural husbandry practices (e.g., abnormally early weaning ages, small enclosure sizes, etc.) that can cause lifelong behavioral abnormalities. Such abnormalities can influence the outcome of any foraging experiment, sometimes subtly, sometimes overtly. Furthermore, those interested in applied problems may have to consider these abnormalities when implementing management strategies (see box 6.2 below).

Synthesizing the vast and disparate literature on herbivore foraging behavior across disciplines, taxonomy, and body size in one book chapter is a tall order for anyone. So let's start by limiting the scope just a bit. I will focus on terrestrial herbivores, specifically generalist insect herbivores and vertebrates that are always or predominantly herbivorous. I will ignore seed eaters and root feeders, sticking mainly with animals that remove photosynthetically active material (although I will occasionally mention sap-sucking insects). With these obvious limitations in mind, let's start by looking broadly at foraging behavior along traditional taxonomic lines.

6.3 Herbivory: A Traditional Taxonomic Viewpoint

Entomologists categorize insect herbivores along a continuum from strictly monophagous (feeding from a single plant genus or species) to oligophagous (feeding on several genera within the same plant family) to polyphagous (feeding on plants from different families). Although examples of each type occur in all major insect taxa, the Orthoptera (grasshoppers and katydids) are the most polyphagous. Proven cases of monophagy are rare in this order. In other insect orders, 70% or more of the species are mono- or oligophagous (Bernays and Chapman 1994). Among the more specialized insect herbivores, some use more or less the entire plant, but more commonly species tend to be associated with particular plant parts. Specialization is the norm among holometabolous larvae (flies, beetles, and Lepidoptera), and in particular among the leaf miners (Bernays and Chapman 1994). Another good example of specialization is the approximately 3,000 species of aphids that feed almost entirely on sap from the phloem of a single species of host plant.

These observations about herbivorous insects lead to two remarks about the literature. First, much of the literature on their foraging behavior (in particular, on diet choice) consists of work on grasshoppers (over 2,500 papers in the last 25 years, more than 300 of which were on feeding behavior; CAB Agricultural Abstracts). Second, because many herbivorous insects are monophagous, students of insect herbivory see diet choice (host plant selection) as uninteresting. However, as Bernays and Chapman (1994) point out, females do not always select the most appropriate host, and some do not even lay eggs on the host plant, but rather nearby. Even when on the proper host species, larvae often need to move as the quality of the present host individual declines, so it is probably safe to say that the majority of insect herbivores show some form of host plant choice. When entomologists have studied host plant selection, they have typically focused on chemical cues in the form of attractants, repellents, phagostimulants, and deterrents. A quick survey of this literature will give the impression that we know a great deal about the mechanisms of host plant selection, but this impression would be wrong, since we've studied only a small fraction of the total number of phytophagous insects.

Vertebrate herbivores are less numerous and less diverse than insect herbivores, but their sheer size means that they have large effects on plant communities. For this reason, they have attracted the attention of ecologists. Pastoral agriculture occupies some 20% of the global land surface and is the focus of agricultural and range scientists. It is obviously economically important, and as a predominant form of land use in some of the more fragile areas of the world, it is of considerable interest to conservation biologists (Hodgson and Illius 1996, ix).

In comparison with animal tissue, plant material is low in nitrogen and high in fiber, and animals can digest it only slowly. While animals can easily digest the contents of plant cells, they cannot digest the cellulose and hemicellulose that constitute plant cell walls, in most cases because they lack cellulase enzymes. Many vertebrate herbivores solve this problem using fermentation in the gut, where symbiotic bacteria digest the cell walls. The rate of clearance of the indigestible plant components from the gastrointestinal tract limits the ability of most vertebrate herbivores to process large quantities of food. David Raubenheimer considers this topic further in box 6.1.

BOX 6.1 **Herbivory versus Carnivory: Different Means for Similar Ends**

David Raubenheimer

When the nineteenth-century American psychologist William James (James 1890) wrote that living organisms are characterized by attaining "consistent ends using variable means," he was referring to the fact that an animal's homeostatic responses (e.g., alterations in the rate of food intake) counteract environmental variations (e.g., in the nutrient density of foods), thus maintaining a constant outcome (e.g., satisfying its nutrient requirements). He could just as well have been referring to the nutritional responses of animals at the longer, evolutionary time scale. There is, for instance, no evidence that groups as trophically divergent as herbivores and carnivores differ substantially in their tissue-level *requirements* for nutrients, but there are major differences in their *means of satisfying* those requirements.

The means of satisfying tissue-level nutrient requirements can, broadly speaking, be separated into two processes: the acquisition of foods from the environment (foraging) and the acquisition of nutrients from foods (food processing). Broadly speaking, the nutritional challenge for carnivores is to find, capture, and subdue scarce or behaviorally sophisticated packages of high-quality food, while herbivores target abundant but nutritionally inferior foods. Not surprisingly, therefore, the conspicuous nutritional adaptations of carnivores are concerned with acquiring food from the environment, and those of herbivores with extracting nutrients from foods. Here I will briefly outline some of the behavior-related adaptations involved in food acquisition by carnivores before turning to the food-processing adaptations of herbivores.

Food Acquisition

As a consequence of the relative scarcity of their food, carnivores typically maintain larger home ranges than do herbivores (McNab 1963; Schoener 1968; for an exception, see Garland et al. 1993). Their body size, too, tends to be larger than that of their quarry (Carbone et al. 1999). While this helps in subduing prey, it also has disadvantages, such as reduced maneuverability (Harvey and Gittleman 1992) and a reduction in nutritional gain per prey captured. Not surprisingly, therefore, there are predators that have adapted to eating prey larger than themselves; among the most spectacular examples are some snakes that eat animals up to 160% of their body weight (Secor and

(Box 6.1 continued)

Diamond 1998). Some mammalian predators use cooperative hunting as a means of capturing prey larger than themselves (Caro and Fitzgibbon 1992).

Carnivores typically have morphological and sensory features in common. These features include forward-facing eye sockets, which help in judging distances (Westheimer 1994) and also enhance visual sensitivity at low light levels (Lythgoe 1979). The eye sockets of prey species, by comparison, tend to be laterally placed, increasing the overall angle of vision in which predators can be perceived (Hughes 1971). The retinas of predators typically have specialized areas of high-resolution vision called foveae and areae. These are particularly well developed in birds of prey (Meyer 1977), but are also found in mammals (Dowling and Dubin 1984), and analogous structures occur in the compound eyes of insect predators (Land 1985). Predatory fishes, too, have specialized visual adaptations. Game fishes often feed in twilight, since they have a visual advantage over their prey at low light intensities. This advantage is achieved by having unusually large, and hence more sensitive, photoreceptors compared with those of their prey (Munz and McFarland 1977).

The challenges of a predatory lifestyle are also reflected in brain structure (Striedter 2005). Among small mammals, for instance, those that prey on insects tend to have larger relative brain sizes than do herbivores (Mace et al. 1981). However, Bennett and Harvey (1985) failed to find an overall correlation between diet and relative brain size in birds. This might be because it is not the size of the brain as a whole that is selected in relation to the animal's lifestyle, but rather the relative sizes of a number of functionally distinct subsystems (Barton and Harvey 2000). For example, the relative size of the tectospinal tract, a pathway involved in movements associated with the pursuit and capture of prey, increases with the proportion of prey in the diets of different mammalian species (Barton and Dean 1993). Interpreting such differences as evolutionary adaptations for predation should, however, be done with caution, since brain size and structure are notably susceptible to activity-dependent developmental influences (Elman et al. 1996). Thus, London taxi drivers have an enlarged posterior hippocampus (involved in spatial memory) (Maguire et al. 2000); I doubt whether even the most ardent adaptationist would attribute this to differential survival in the urban jungle!

(Box 6.1 continued)

Nutrient Acquisition

Compared with animal prey, plant tissue is generally more abundant and more easily captured and subdued, but once ingested, it is nutritionally less compliant. The contents of plant cells are enclosed in fibrous cell walls consisting predominately of compounds such as lignin and cellulose that are difficult to degrade enzymatically. These structural compounds both impede access to the nutrients contained in the cytoplasm (Abe and Higashi 1991) and lower the concentration of nutrients such as protein and digestible carbohydrate (Robbins 1993). Plant tissue is also highly variable in its ratios of component nutrients (Dearing and Schall 1992) and often contains deterrents and toxins (Rosenthal and Berenbaum 1992).

Foragers can ameliorate these problems to some extent via food selection, as suggested by the observation that many mammalian herbivores favor foliage with a relatively high nitrogen and low fiber content (Cork and Foley 1991). Since the fiber that produces leaf toughness is likely to be tasteless, it has been suggested that this selectivity might be achieved through perceiving toughness directly (Choong et al. 1992; Lucas 1994); it is, however, also possible that taste perception of low levels of nutrients is involved (Simpson and Raubenheimer 1996). The avoidance of plant fiber might be particularly important for small endothermic animals, which have a high relative metabolic rate and hence high energy requirements. Evidence from mammals supports this prediction: the proportion of species eating fibrous plant tissues declines, and the proportion selecting low-fiber plant and animal tissues increases, with decreasing body size (Cork 1994). This might explain the scarcity of herbivorous species among birds (Lopez-Calleja and Bozinovic 2000).

Rather than avoiding plant fiber, many herbivores have structures that are adapted for degrading it mechanically, releasing the cell contents for digestion and absorption. These structures include specially adapted teeth and jaws in mammals (Lucas 1994), mandibles in insects (Bernays 1991), and teeth, jaws, and post-oral pharyngeal mills in fishes (Clements and Raubenheimer 2005). An alternative, or complement, to mechanical breakdown is the enzymatic degradation of plant fiber. In mammals, which do not produce cellulytic enzymes, fiber digestion is achieved with the aid of symbiotic microorganisms, usually bacteria or protozoans and occasionally fungi (Langer 1994). Some herbivorous fishes (Clements and Choat 1995), birds (Grajal 1995), and insects likewise have microbe-mediated

(Box 6.1 continued)

fermentation, while some insects and other arthropods can synthesize endogenous cellulases (Martin 1991; Slaytor 1992; Scrivener and Slaytor 1994; Watanabe and Tokuda 2001). Enzymatic degradation of structural carbohydrates has the added advantage of making the energetic breakdown products available to the herbivore, and where microbes are involved, microbial proteins and B-complex vitamins are further useful by-products (Stevens and Hume 1995).

Despite (and in many instances because of) these mechanisms for cellulose digestion, the guts of many herbivores have structural specializations for subsisting on plant tissue. Gut size is known to increase with decreasing nutrient content of foods (both within and between species) in a wide range of animals, including mammals (Martin et al. 1985; Cork 1994), birds (Sibly 1981), fishes (Horn 1989; Kramer 1995), reptiles (Stevens and Hume 1995), insects (Yang and Joern 1994a), and polychaete annelids (Penry and Jumars 1990). Larger guts allow a greater rate of nutrient uptake and, in some cases, greater efficiency of digestion (Sibly 1981).

Not only the size, but also the shape of the gut is modified in many herbivores. All else being equal, digestion is thought to occur most rapidly where there is a continuous flow of food through a slender tubular gut, with little opportunity for the mixing of foods ingested at different times (Alexander 1994). Such "plug-flow reactors" (Penry and Jumars 1986, 1987) are often found in carnivores (Penry and Jumars 1990; Alexander 1991). They are less suitable for herbivores that rely on microbial symbioses for cellulose degradation, because in such a system the microbes would be swept away in the flow of food through the gut (Alexander 1994). A population of microbes can, however, be maintained indefinitely in a digestive chamber wide enough to ensure continuous mixing of its contents (a "continuous-flow, stirred-tank reactor"), and indeed, such chambers are a conspicuous feature of the guts of herbivores. Many, including ruminants such as cows, have developed fermentation chambers in the foregut, while others (e.g., horses) have an enlarged hindgut (caecum and/or colon). Foregut and hindgut fermentation are very different strategies for dealing with low-quality foods; the former is associated with long digestion times and particularly poor-quality foods, and the latter with differentially retaining the more rapidly fermented component and egesting the rest (Alexander 1993; Björnhag 1994). Not surprisingly, therefore, mammalian herbivores tend to be either foregut or hindgut fermenters, but not both

(Box 6.1 continued)

(Martin et al. 1985). It is generally only large herbivores, with low mass-specific metabolic rates, that can afford the slow passage times associated with foregut fermentation of high-fiber foods (Cork 1994). Interestingly, some herbivorous mammals (Hume and Sakaguchi 1991) and fishes (Mountfort et al. 2002) have significant levels of microbial fermentation without appreciably specialized gut morphology.

An important but relatively neglected problem associated with herbivorous diets is nutritional balance (Raubenheimer and Simpson 1997; Simpson and Raubenheimer 2000). Compared with animal-derived foods, plants are believed to be more variable in the ratios of nutrients they contain (Dearing and Schall 1992), and they are generally poor in nutrients, such that "most single plant foods are inadequate for the growth of juvenile animals and their development to sexual maturity" (Moir 1994). This observation leads to the expectation that herbivores should be significantly more adept than carnivores at independently regulating the levels of different nutrients acquired (i.e., at balancing their nutrient intake). Some insect herbivores do, indeed, have a remarkable ability to compose a balanced diet by switching among nutritionally imbalanced but complementary foods (Chambers et al. 1995; Raubenheimer and Jones 2006). Such responses are mediated largely by the taste receptors, which "monitor" simultaneously the levels of proteins and sugars in the food and in the hemolymph, and also involve longer-term feedbacks due to learning (Simpson and Raubenheimer 1993a; Raubenheimer and Simpson 1997). Mechanisms for nutrient balancing might also exist at the level of nutrient absorption (Raubenheimer and Simpson 1998).

It remains uncertain, however, whether nutrient balancing is in general better developed in herbivores because some carnivores, too, have been shown to perform better on mixed diets (Krebs and Avery 1984; Uetz et al. 1992) and to select a nutritionally balanced diet (Mayntz et al. 2005). One possibility, suggested by physiological data, is that both groups are adept nutrient balancers, but with respect to different nutrients. For example, domestic cats (which are obligate carnivores) apparently lack taste receptors for sugars and have low sensitivity to sodium chloride (neither of which are important components of meat), but have impressive sensitivity for distinguishing among amino acids (Bradshaw et al. 1996). Similarly, unlike some omnivores and herbivores, cats are unable to regulate the density of carbohydrate absorption sites in the gut in response to nutritionally

(Box 6.1 continued)

imbalanced diets, but do regulate the activities of amino acid transporters (Buddington et al. 1991).

Why should carnivores have evolved mechanisms for nutrient balancing? Perhaps the nutritional variability of their food has been underestimated. Alternatively, the answer might be found not on the nutritional supply side, but on the demand side. If variation in tissue-level demand for, say, different amino acids by a predator is high (e.g., with different activity levels, diurnal cycles, reproductive state, etc.), then no single food composition will be balanced, and the animal will require specific adaptations to differentially regulate acquisition of the various amino acids. Although little is known about such variation in the nutrient needs of either carnivores or herbivores, if it turns out to be significant, then William James's dictum might need revising: animals are characterized by attaining *"variable* ends using variable means."

The nutritional limitations of plant material have important consequences for body size. Comparative work shows that the metabolic requirements of mammals increase with body mass$^{0.75}$, but the capacity of the gastrointestinal tract increases with body mass$^{1.0}$ (Iason and Van Wieren 1999). Therefore, smaller animals have higher mass-specific energy requirements, but lack proportionally large gut capacities, and therefore require more nutritious forage (sometimes known as the Bell-Jarman principle, after Bell 1970 and Jarman 1974). These allometric considerations suggest that the smallest ruminant mammal should be at least 15 kg and the smallest nonruminant mammal at least 1 kg (see, e.g., Iason and Van Wieren 1999 for more discussion).

For much of the remainder of this chapter, I will ignore taxonomy and attempt an organization of the current state of the field around what I call *the big questions*. Herbivores can differ in many ways, but they all must answer the four questions listed above.

Four Big Questions

In the study of herbivore foraging behavior, four big questions interest us.

Where Will the Animal Eat?

Although I pose this as a single question, the problem exists at several spatial scales. At large scales, the question is one of habitat selection. Should the animal forage in the uplands or the lowlands? Should it graze near the forest edge or by the river? Within a habitat, at a smaller spatial scale, the question is one of patch selection. Should the animal graze from patches of tall vegetation, sacrificing plant quality for a faster intake rate, or should she graze from shorter patches where the plant quality is higher, but the intake rate is lower? At even finer spatial scales, some animals choose among parts of a single plant. For example, aphids prefer to feed at the base of grass plants, rather than out on the leaves.

This question may concern patch exploitation (see chap. 1 in Stephens and Krebs 1986), but it is important to consider both the "attack" decision (which patches to use) and the "exploitation" decision (how long to use a patch). An example may help to clarify this distinction: consider the cows from the prologue (section 6.1). What may seem to be a homogeneous pasture is likely to comprise patches that differ in plant species composition, vegetation height and density, presence of parasites, plant quality (often due to previous grazing and dung and urine deposition), and so on. These patterns may follow an environmental gradient (e.g., the slope), or they may arise through the previous grazing patterns of the cows or other animals. How does a cow choose among these patches? The patch exploitation model in its simplest form is ill equipped to deal with such heterogeneity.

What Will the Animal Eat?

What to eat is, principally, a question of diet selection. It is the kind of question addressed by the classic diet model, but often complicated by the continuous nature of some vegetation and the postingestive consequences of food choice (see chap. 5 in this volume). When the animal is faced with an array of potential food sources, which should be included in the diet and which ignored? This question applies not only to different plant species, but also to plants of the same species that differ in growth or regrowth states. Should a grasshopper eat young ryegrass leaves but avoid older leaves? Should a sheep graze patches of tall fescue when they are 5 to 7 days from their last defoliation, but not sooner (because the bite mass is too low) or later (because the plant quality is too low)?

At some finer spatial scales, we may ask what parts of the plant the animal feeds from, but I don't view this as *the big question*. I include host plant selection by invertebrate herbivores here, rather than in the previous question, although this may be just an issue of semantics.

How Fast Will the Animal Eat?

How fast to eat is the question of intake rate. The animal's environment and morphology sometimes constrain its intake rate, but often intake rate is a behavioral choice. I will elaborate on this distinction in section 6.5. There are digestive consequences that accompany the choice of intake rate. An animal can increase ingestion by chewing less thoroughly, but this can slow passage rate and reduce digestion.

How Long Will the Animal Spend Eating?

While how long to eat might be a question of bout length, for most herbivores *the big question* is total foraging time. Time spent foraging incurs opportunity costs because it is time not spent avoiding predators, engaging in social interactions, reproducing, ruminating, and so on. There are environmental and physical/morphological constraints on foraging time, and I will elaborate on these in section 6.6, but often foraging time is a behavioral choice.

Reductionism and the Big Questions

Animals rarely answer the big questions piecemeal. Available diet choice and intake rate can determine habitat choice. Intake rate can determine diet choice, and vice versa. Diet choice can determine grazing time, and vice versa. Ultimately, herbivory is the *integration* of these four questions. The study of one question in isolation may help us to determine how these questions are integrated, but it will rarely yield the total picture. In the next four sections, I will consider what we know about herbivore behavior in light of each of these questions, but remain mindful of the interactions among the questions.

Different animals have different constraints and objectives, and so they come to different compromises between the answers to these questions. There is no "grand unified theory of herbivory," but an understanding of these trade-offs and accommodations will help to provide a coherent framework for studying herbivory. I will discuss the experimental treatment of interactions among the questions in section 6.8.

6.4   Diet Selection

To understand herbivore diet selection, we need to think about how the animal's goals and constraints operate. Students of herbivory have expressed considerable interest in classic "optimal" foraging theory, but the literature contains a variety of misconceptions, owing perhaps to the fact that many researchers who study herbivory come from a background not in behavioral ecology, but in agriculture, range science, or entomology. As Ydenberg et al. make clear in chapter 1, foraging theory *is not* synonymous with intake rate maximization (cf. Dumont 1995). Foraging theory is about maximizing an objective function. In early studies of "optimal foraging," the objective function of interest was the rate of energy intake, as it was thought that, in some cases at least, the rate of energy intake might be a good surrogate for evolutionary fitness. A smaller family of models considered minimizing the foraging time required to meet a fixed intake requirement. It should be obvious that these two objective functions are similar, although not identical.

Authors have used the term "rate maximization" rather cavalierly in recent publications on herbivory, particularly in the agricultural literature. The original foraging theory models clearly hypothesized that the objective being maximized was *net* energy intake rate, not *gross* energy intake rate. Tactics that maximize gross intake rate also maximize net intake rate *if* there are no differences in digestibility or foraging costs. Vegetation varies dramatically in gross energy content, digestibility, passage rates, concentrations of secondary metabolites, and so on; thus, maximization of gross energy intake is rarely an appropriate objective.

Both intake rate maximization and time minimization remain popular objective functions in models of herbivory and as alternative hypotheses in experiments (e.g., Distel et al. 1995; Focardi and Marcellini 1995; Forchhammer and Boomsma 1995; Farnsworth and Illius 1996, 1998; Van Wieren 1996; Torres and Bozinovic 1997b; Ferguson et al. 1999; Illius et al. 1999; Fortin 2001), but there are other objective functions that one should consider. For some animals, a particular nutrient acts as a limiting resource (for example, crude protein); in these cases, energy is clearly the wrong surrogate for fitness (see Berteaux et al. 1998). Researchers have considered several currencies other than rate maximization, including optimization of growth rate (Smith et al. 2001), ruminal conditions (Cooper et al. 1996), oxygen use efficiency (Ketelaars and Tolkamp 1991, 1992; Emmans and Kyriazakis 1995; Nolet 2002), and survival maximization (Newman et al. 1995). I will come back to the question of objective functions when I consider intake behavior. For now, I will simply state that the appropriate objective function surely differs among herbivores of differing body sizes, guilds, and digestive physiologies.

Empiricists often use simple foraging models a straw man. These models may fail because, as a mathematical convenience and as a first level of simplification (one goal of a model, after all), they ignore important constraints. Foraging theory says that animals should maximize their fitness (or some appropriate surrogate) *subject to their constraints*. Indeed, the goal of an optimality research strategy is to identify the objective function and important constraints—not to test whether animals are optimal per se (Mitchell and Valone 1990). Let's consider the potential constraints, which I will refer to broadly as environmental and physiological/morphological. In many cases, the constraints are those that the animal has evolved to work within or around, but in other instances (e.g., intensive farming), they are not.

Herbivores face many of the same constraints as nonherbivores. For example, many large vertebrate herbivores are social animals, and social context often plays a role in determining their diet choice. Dumont and Boissy (1999, 2000) have shown that sheep may forgo an opportunity to graze more selectively if this means they must leave their social group, even temporarily (though Sevi et al. [1999] failed to find this effect). Rearing conditions also may alter diet selection (Sutherland et al. 2000; box 6.2). We'll revisit the issue of gregariousness when we look at intake rate decisions.

BOX 6.2   Animal Farm: Food Provisioning and Abnormal Oral Behaviors in Captive Herbivores

Georgia Mason

Drooling, the stalled cow rhythmically twirls her tongue in circles. She does this for hours a day, as do many of her barnmates. Next door, the stabled horse repeatedly bites his manger, pulling on the wood with his teeth (fig. 6.2.1). He has done this for years—all his adult life. A foraging biologist should find such bizarre activities interesting because they raise new questions about the control of herbivore feeding. They also highlight a real need for more fundamental research—one made urgent by the welfare problems that these behaviors probably indicate. Here I will describe these abnormal behaviors before discussing their possible causes and the research questions they raise.

Figure 6.2.1. Stabled horses may perform a number of abnormal oral behaviors, including crib biting. (After a photo by C. J. Nicol.)

(Box 6.2 continued)

Strange, apparently functionless oral behaviors are common in ungulates on farms and in zoos. Some, like the tongue twirling and crib biting described above, resemble the pacing of caged tigers and other "stereotypies" (Mason 1991) in having an unvarying, rhythmic quality and no obvious goal or function (e.g., Redbo 1992; Sato et al. 1992; McGreevy et al. 1995; Nicol 2000). Others, like wool eating by farmed sheep or wood chewing by stabled horses (e.g., Sambraus 1985; McGreevy et al. 1995), involve more variable motor patterns and an apparent goal, but still puzzle us by seeming functionless and different from anything seen in the wild. These activities can be time-consuming—stall-housed sows may spend over 4 hours a day in sham chewing, bar biting, and similar behaviors—and common—for example, shown by over 40% of the cattle in a barn (reviewed in Bergeron et al. 2006). Dietary regime seems to be the main influence, with abnormal behaviors most evident in populations fed only processed foodstuffs (e.g., milled, highly concentrated pellets; reviewed in Bergeron et al. 2006). Sometimes it is unclear what elicits individual bouts, but often it is eating, with the behaviors being displayed soon after the animal has consumed its food (e.g., Terlouw et al. 1991; Gillham et al. 1994).

In form and timing, this pattern differs from the typical picture for captive carnivores, which pace, and do so before feeding, even when they are fed highly processed food (e.g., Clubb and Vickery 2006). But are these differences caused by underlying biological traits or merely by differences in husbandry (Mason and Mendl 1997)? Would captive carnivores bar-bite and tongue-roll if taken from their mothers before natural weaning (as happens to most pigs and cattle), underfed (the case for many pigs), or kept in narrow, physically restrictive stalls? A survey controlling for these factors (Mason et al. 2006) showed that ungulates are inherently prone to abnormal oral behaviors (fig. 6.2.2), with wall-licking giraffes (Bashaw et al. 2001), tongue-rolling okapis, and dirt-eating Przewalski's horses (e.g., Hintz et al. 1976; Ganslosser and Brunner 1997) just some of the cases adding to the agricultural data. These observations do not provide sufficient phylogenetically independent contrasts to link abnormal oral behaviors with herbivory per se, but their form, timing, and links with feeding regimes strongly implicate foraging. How could ungulates' specializations for herbivory lead to these behaviors? Three hypotheses have been advanced, each essentially untested.

(Box 6.2 continued)

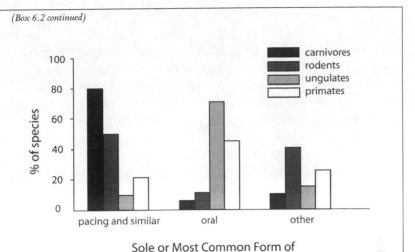

Figure 6.2.1. Taxonomic distribution of abnormal behaviors across four mammalian orders (carnivores, 61 species; rodents, 15 species; ungulates, 26 species; primates, 19 species). (From Mason et al. 2006.)

1. Ungulates cannot completely abandon foraging, even when it is redundant.

On farms and in zoos, ungulates are typically fed in a way that requires minimal foraging: homogeneous hay, browse, or artificial food—milled, low-fiber mash or pellets—is placed in a manger under their noses. It thus does not need to be searched for, it neither demands nor allows diet selection, and it often needs little chewing. Consequently, captive ungulates eat their daily rations in a fraction of the time it would take naturally. For instance, horses on pasture may graze for 16 hours a day, yet in stables, horses commonly consume all their food within 2 hours (Kiley-Worthington 1983); similar contrasts apply to all provisioned ungulates (reviewed in Bergeron et al. 2006). Several authors have therefore hypothesized that abnormal oral behaviors represent foraging behaviors that ungulates are unable or unwilling to abandon, despite their now being unnecessary for ingestion (e.g., reviewed Rushen et al. 1993). Evidence consistent with this hypothesis includes the observation that stalled pigs bar-chew for lengths of time similar to those they would naturally spend in grass chewing, rooting, and stone chewing if kept outside (Dailey and McGlone 1997). If correct, this idea raises new questions about what

(Box 6.3 continued)

ungulates are defending (a minimum time spent in foraging behavior? a minimum number of bites per day?) and functional questions as to why. It could simply be that selection has not favored complete flexibility in foraging time. As this chapter shows, foraging time does generally decrease if intake rate goes up, but investigators obtained these findings in naturalistic conditions that may not extrapolate to the extreme intake rates that occur in captive situations. Alternatively, defending a certain minimum level of daily foraging could bring functional benefits independent of nutrient gain, such as information gain, preventing excessive tooth growth, or maintaining gut flora and other aspects of digestive function.

2. Oral movements help maintain gut health.

As this chapter shows, ungulate foraging involves thousands of daily bites that do more than break down food: they stimulate saliva production (100 or more liters per day in cattle), which helps buffer gastrointestinal acidity. Processed diets, however, take less chewing per unit time (Abijaoude et al. 2000), much less total foraging time per day, and overall, involve far fewer mouth movements. Could these reductions impair gut health by reducing salivation? Processed, low-fiber diets certainly cause gastrointestinal acidity—and even ulceration—in cattle, horses, and pigs (Blood and Radostits 1989; Hibbard et al. 1995; Sauvant et al. 1999; Nicol 2000). The second hypothesized explanation for abnormal oral behaviors is thus that they are attempts to generate saliva to buffer gut acidity. Thus, horses' crib biting can be reduced by antacids and by antibiotics that control the gut's lactate-producing bacteria (Johnson et al. 1998; Nicol et al. 2001). Some oral behaviors are linked with gut health: tooth grinding and crib biting are associated with gastritis and ulcers in horses (Rebhun et al. 1982; Nicol et al. 2001), but tongue rolling and similar behaviors in calves correlate negatively with stomach lesions (Wiepkema et al. 1987; Canali et al. 2001). This idea raises several unanswered questions: How do ungulates monitor the pH of their digestive tracts, and does this vary with foraging niche? Do some or all ungulates monitor saliva production levels? Do abnormal oral behaviors effectively generate saliva, and does this help alleviate abnormal gut pH? If so, are these learned or innate responses—or does this vary with dietary niche?

(Box 6.2 continued)

3. Captive ungulates are deficient in nutrients and so stay motivated to forage.

Naturally, diet selection is the principal means of modulating gastrointestinal acidity; for example, ruminants respond to acidosis with increased fiber intake (Keunen et al. 2002). Herbivores also have excellent abilities to detect specific nutrient deficits and respond to them behaviorally (see section 6.4 and box 6.1). Yet, in captivity, humans constrain the quantities ungulates eat and the diets they can select. The last explanation for abnormal oral behaviors is therefore that they represent state-dependent foraging attempts driven by dietary deficiency. For example, simple energy deficits play a major role in pigs' oral stereotypies (e.g., Appleby and Lawrence 1987; Terlouw et al. 1991), while deficits of copper, manganese, or cobalt can induce tongue rolling in cattle (Sambraus 1985). It is unclear at the mechanistic level why such behaviors are then sustained, but evolutionarily, it may be that it is adaptive to search for food until successful. In some instances, however, the abnormal behavior is a "pica" (the ingestion of nonfood items) that may actually redress deficits, as has been argued for dirt eating by free-living horses (Blood and Radostits 1989; McGreevy et al. 2001). Thus, in captive ungulates, horses' wood chewing may be an adaptive response to a lack of dietary fiber (Redbo et al. 1998), and the chewing of urine-soaked wood slats by sheep a way of gaining nitrogenous urea when deficient in protein (e.g., Whybrow et al. 1995). Protein deficiency could also explain wool chewing by sheep, since the soiled wool from other animals' rear ends is preferred (Sambraus 1985). In these instances, we do not know whether foragers identify the required nutrients via specific taste receptors, or the extent to which associative learning about physiological consequences reinforces the behavior.

Overall, these three interlinked hypotheses ask fundamental research questions about which aspects of herbivore foraging are inherently "hardwired" and difficult to modify, which respond facultatively to state and circumstance, and how these design features relate to dietary niche. We can also see that abnormal oral behaviors reflect deficiencies. These may be nutritional deficiencies or a mismatch between the feeding methods imposed in the captive situation and the foraging mode that the free-living animal prefers. Some abnormal oral behaviors almost certainly indicate gastrointestinal discomfort, even pain. Addressing the questions they raise is thus ethically important as well as scientifically interesting.

The opposite occurs in some insects, in which the presence of conspecifics may lower host plant attractiveness. Feeding by conspecifics may reduce plant quality or induce plant defenses (e.g., see Raupp and Sadof 1991). Insects seldom gain the antipredator benefits of group foraging (except in cases of predator satiation), but they do pay the costs of intraspecific competition and perhaps increased conspicuousness.

Like other animals, herbivores may alter their diets in the presence of predators or parasites. For example, Cosgrove and Niezen (2000) have shown that sheep infected with gastrointestinal parasites shift toward diets that contain higher proportions of protein than uninfected animals. Even the *risk* of predation or parasitism can cause such dietary shifts. Hutchings et al. (1998, 1999, 2001; Hutchings, Gordon et al. 2000; Hutchings, Kyriazakis et al. 2000) have shown that sheep may forage less selectively in response to differences in intake rate if more selective foraging also means a higher exposure to parasitic worm larvae. Abrams and Schmitz (1999) modeled the results of Rothley et al. (1997), who showed that the presence of a spider caused grasshoppers to shift their foraging effort from high-quality grasses to low-quality forbs. Smith et al. (2001) showed a similar result for herbivorous crane flies. Kie (1999) provides an excellent review of this trade-off in ungulates.

Herbivores face many other trade-offs. For example, Torres and Bozinovic (1997a) demonstrated a diet selection–thermoregulation trade-off in the degu (*Octodon degus*), a generalist herbivorous rodent from central Chile. Degus preferred low-fiber diets to high-fiber diets at 20°C, but were indifferent at 38°C, preferring to minimize their thermoregulatory risk rather than maximize their digestible energy intake.

Herbage quality may change during the day, creating another environmental constraint. The relative qualities of two plant species may change from dawn, when water-soluble carbohydrate concentrations are low, to dusk, when they are higher after a day of photosynthesis (e.g., Ciavarella et al. 2000). Orr et al. (1997) have shown that the dry matter, water-soluble carbohydrate, and starch content of grass and clover increase *differentially* over the course of the day (0730–1930), and that sheep bite rate and chewing rate decline while bite mass increases, apparently in response to the changes in the plants. Plant quality may vary over longer time scales as well. There are strong seasonal variations in both herbage quality and, of course, quantity; for example, Luo and Fox (1994) have nicely demonstrated seasonal shifts in the diet of the eastern chestnut mouse (*Pseudomys gracilicaudatus*). Many plant secondary metabolite concentrations vary seasonally, requiring animals to track these changes (e.g., Dearing 1996). Provenza (1995b; see box 5.2) reviewed the use of individual memory of the postingestive consequences of nutrients and toxins to track temporal variation in plant secondary metabolite concentrations generally, and Duncan and Gordon (1999) re-viewed the effects of these conflicting demands of intake rate maximization and toxin intake minimization on diet choice in larger herbivores.

Spatial distribution of the vegetation clearly influences diet selection. Indeed, many workers believe that this is the key difference between "diet preference" (diet choice when unconstrained by the environment) and "diet selection" (diet choice under environmental constraints; for more discussion, see Newman et al. 1992; Parsons, Newman et al. 1994). Here we are thinking not only about differences in encounter rates with each plant species (these are adequately considered in even the simplest diet choice models), but also about differences in the total, vertical, and horizontal abundance and distribution of herbage mass. To a grazing mammal, what does it mean to "take a bite of perennial ryegrass"? Ryegrass may be finely interspersed with other plant species, it may occur higher or lower in the grazed horizon, it may be younger or older than other available bites, it may include reproductive stems, and so on. Many researchers have addressed these issues. Harvey et al. (2000) showed that sheep traded off diet preference and pasture height in a complex manner (fig. 6.2). Edwards et al. (1996a) used an artificial pellet system to test the influence of spatial variation on sheep diets while keeping total food availability constant (see also Dumont et al. 2000). They found that the proportion of the preferred cereal pellet in the diet declined when its horizontal distribution (equivalent to fractional cover) declined, but only when the vertical abundance of cereal was low. They concluded that diet selection experiments that ignore how the food alternatives are distributed horizontally and vertically could be misunderstood. Although my examples here have been of large vertebrates, invertebrates also show responses to the spatial distribution of host plants that simple encounter rate considerations cannot explain.

The environment presents constraints enough, but, as discussed in chapter 5, herbivores must also deal with an array of physiological and morphological constraints. The classic physiological constraint is nutritional, as exemplified by the sodium constraint for browsing moose (see also Forchhammer and Boomsma 1995); protein provides another example (e.g., Tolkamp and Kyriazakis 1997; Berteaux et al. 1998). Belovsky's (1978) now classic paper spawned a cottage industry of linear programming models of herbivore behavior (e.g., Nolet et al. 1995; Randolph and Cameron 2001). Linear programming is a mathematical technique for solving an optimization problem subject to linear constraints. While this approach remains popular today, it has not been without controversy in the study of herbivory (e.g., Hobbs 1990; Owen-Smith 1993, 1996, 1997). Hirakawa (1997a) has modeled digestive constraints using a more sophisticated nonlinear programming approach. Hirakawa shows that when foraging time is long or food is abundant, the digestive

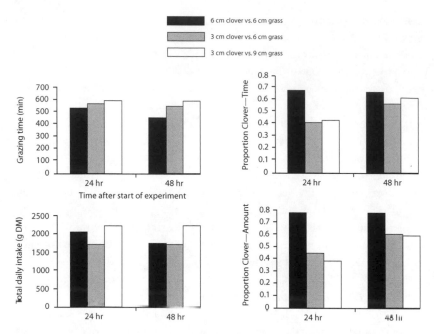

Figure 6.2. Results of a grazing study examining how sheep trade off diet preference against intake rate. In this experiment, replicate flocks of sheep were stocked on replicate paddocks in which one-half of the paddock contained white clover and the other half contained perennial ryegrass. Different paddocks were managed to achieve different contrasts in sward surface height (SSH): 6 cm clover vs. 6 cm grass, 3 cm clover vs. 6 cm grass, or 3 cm clover vs. 9 cm grass. The investigators estimated species-specific intake rates for these sward surface heights to be 3 cm clover = 3.58 ± 0.4 g dry matter/min; 6 cm clover = 4.66 ± 0.8 g dry matter/min; 6 cm grass = 2.49 ± 0.4 g dry matter/min; 9 cm grass = 3.99 ± 0.4 g dry matter/min. The nature of the results is complex. Animals could easily have achieved a monospecific diet. Their expressed diet preference is neither based entirely on intake rate nor on plant species, but on some combination of the two. To complicate matters, in addition to changing their diet preference, the animals also altered their grazing time and hence their total daily intake. (After Harvey et al. 2000.)

constraint intensifies, and animals should concentrate on the digestive process, choosing fewer diet items of higher digestibility. However, when time is short or food is less abundant, animals should concentrate on the ingestion process, choosing more food types that have faster handling rates. This requirement for flexible diet selection nicely illustrates why the prior ranking of food types (as in the diet model) may be irrelevant when digestion constrains foraging.

In arid and semiarid environments, water constrains diet selection. For example, Manser and Brotherton (1995) demonstrate this constraint on the diet selection of dwarf antelopes during the dry season. They show that in order to meet minimum daily water requirements, dik-diks (*Madoqua kirkii*) fed on plant species they normally avoided during the wet season. Given a choice between foods with differing water contents, grasshoppers' diet choice depends on their state of dehydration (Roessingh et al. 1985); they choose higher water content over energy content when dehydrated. Digestive constraints operate for many animal species, whether it's too much sugar in the phloem sap ingested by the aphid seeking nitrogen or too much lignin in the grass eaten by the goat seeking digestible organic matter. David Raubenheimer contrasts the nutritional challenges faced by herbivores with those faced by carnivores in more detail in box 6.1.

Foraging theorists often think of digestion as a constraint, but students of herbivory have considered the adaptive design of digestive processes. For example, Mathison et al. (1995) have suggested that ruminants have some control over gut retention time, which they can adjust to optimize assimilation rates. Many disagree, noting that the weight of evidence suggests that mechanistic factors such as particle size determine passage rate (for more discussion, see, e.g., Illius et al. 2000). Ultimately, the animal controls mastication and rumination, which in turn control particle size, so clearly, ruminants do have some degree of control over this process.

Many plants produce secondary metabolites that either make the plant less nutritious to some animals (e.g., tannins) or make the plant toxic in sufficient quantities (e.g., alkaloids). Guglielmo et al. (1996) demonstrated that the presence of coniferyl benzoate in aspen leaves strongly influenced ruffed grouse (*Bonasa umbellus*) diet selection. Dearing (1996) found similar results for the North American pika (*Ochotona princeps*), and Tibbets and Faeth (1999) demonstrated that the presence of alkaloid-producing endophytic fungi altered leaf-cutting ants' choice of grass leaves. Bernays and Chapman (1994, chap. 2) and Launchbaugh (1996) give general introductions to the role of plant secondary metabolites in herbivory.

Plant secondary metabolites may also influence diet selection among parts of the same plant. Boer (1999) showed that pyrrolizidine alkaloid concentrations were higher in the youngest (and most nutritious) leaves of *Scenecio jacobaea* plants, so that cotton leafworms (*Spodoptera exiguq*) and a noctuid moth (*Mamestra brassicae*) both preferred the older leaves. More generally, Hirakawa (1995) noted that when the classic diet model is modified to consider toxins, partial preference may occur for one diet item while all others follow a zeroone rule (see chap. 5 in this volume). Hirakawa also showed that the prey selection criterion changes with the intensity of the toxin constraint, making it impossible to rank diet items a priori. These results are qualitatively different from those reported by Stephens and Krebs (1986).

An animal's state can strongly influence nutritional, digestive, and some secondary metabolite constraints. One approach to the study of current physiological state has been to alter an animal's state through fasting. Experiments routinely use fasting to motivate animals to feed, but fasting should be used with caution because it can alter both diet preference and diet selection (Newman,

Penning et al. 1994; Edwards et al. 1994). States other than hunger per se can be important as well. My colleagues and I demonstrated that sheep that had previously grazed grass had a stronger preference for clover when given a choice between the two, and that sheep that had previously grazed clover had the reverse preference (Newman et al. 1992). Parsons, Newman et al. (1994) demonstrated that such effects can influence diet preference over a period of several days. While the desire to compensate for some imbalance in the previous diet *might* explain these results, the missing component has yet to be identified. Bernays et al. (1997) suggest that "novelty" per se is the mechanism for incorporating even unpalatable food items into the diet and provide experimental evidence to support this hypothesis in a grasshopper (*Schistocerca americana*).

Previous state sometimes appears in experiments in the form of hidden variables. Many large mammalian herbivores are maintained on high-energy, low-bulk pelleted foods when not taking part in experiments. These diets can cause gastrointestinal acidity and even ulcers, and subsequent diet selection may be greatly influenced by these pathologies. For example, acidosis leads cattle to self-select more fiber in their diet (see box 6.2).

Raubenheimer and Simpson (1993) have introduced a useful framework for examining the effects of physiological state on diet choice as well as total intake (or feeding time). I describe their framework in figure 6.3, showing how animals may use complementary plants to reach some target intake. More interestingly, their framework gives some insights into foraging behavior when the animal's diet is nutritionally deficient. This basic framework has proved powerful in a variety of situations with a variety of species. Here is but one recent example. Behmer et al. (2001) provided locusts (*Locusta migratoria*) with pairs of synthetic food sources that differed in their protein and digestible carbohydrate content (7% P:35% C and 31% P:11% C). Neither food source alone was optimal (for growth), but together they were complementary. The locusts were able, over the course of 4 days, to respond to their physiological state by adjusting their intake of the two complementary diet items to satisfy their target intake of 19% P:23% C. However, when fed each of these diet items singly, locusts attempted to defend both their protein and carbohydrate goals, as in figure 6.3E. In addition to levels of specific macronutrients (or even micronutrients), digestion rate itself may be a physiological state variable that influences diet selection. For example, degus selected food plants based on plant quality (water content and nitrogen:fiber ratio) and on mean gut retention time (Bozinovic and Torres 1998).

So far, the physiological and morphological constraints we have considered affect the processing of food—in other words, the "postingestive" consequences of diet choice. However, many constraints act *before* herbivores

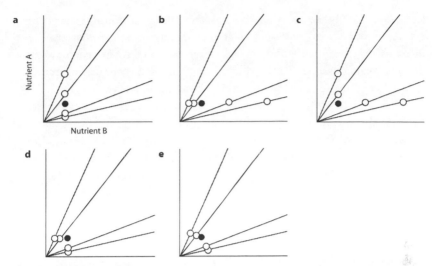

Figure 6.3. Graphs of nutrient space, with nutrient A on the *y*-axis and nutrient B on the *x*-axis, both measured in grams. The target intake of each nutrient is shown as a solid circle. Any given food item has a fixed ratio of the two nutrients, and we can represent that food item as a line from the origin. Raubenheimer and Simpson (1993) call these lines "rails." If two complementary foods are available (one rail on each side of the intake target), then the animal can achieve its target by selecting a mixed diet. This system is particularly powerful for investigating dietary priorities. This can be done by feeding animals on a variety of single food items, one at a time, and examining their intake. In each graph, there are several hypothetical food items, each available one at a time. The open circles represent hypothetical intake of each item. (A) If we saw this intake behavior, it would tell us that the animal is more concerned about its intake of nutrient B than of nutrient A and always seeks to satisfy this requirement (although sometimes gut constraints might prevent this, particularly for food items that are quite different from the target ratio). (B) Similarly, this intake behavior would demonstrate a desire to always satisfy the nutrient A requirement. (C) Intake behavior that always seeks to satisfy both nutrient requirements, even if this means exceeding the total intake target (sum of the *x* and *y* coordinates). (D) Intake behavior that seeks to meet one nutrient requirement while maintaining total intake at or below the target. (E) A forager that seeks the optimal compromise between its two nutrient requirements. This forager eats until a point on the rail that is geometrically closest to the target intake. Raubenheimer and Simpson demonstrated that locusts tend to behave as in part E with respect to carbohydrate and protein. (After Raubenheimer and Simpson 1993.)

ingest their food. A herbivore's spatial memory for locations of different foods or their qualities may limit its diet selection.

Memory constraints are perhaps less important in large vertebrates than our intuition might suggest. Edwards et al. (1996b), Laca (1998), and Dumont and Petit (1998) have demonstrated that some large grazing mammals possess excellent spatial memory and can use it to improve the quality of their diets. For example, sheep with 6 days' experience were able to visit exclusively four patches containing food among thirty-two patches in an 800 m² grid, using spatial memory alone (Edwards et al. 1996b). Provenza and others have demonstrated that these same animals have very good temporal memories about toxins (e.g., Provenza 1995a, 1995b, 1996). Of course, there may be significant fitness costs to forgetting that a plant contains a toxin, but in cases that do not involve toxins, the penalty of forgetting the postingestive consequences may be small. Consider an animal grazing two species of grass that differ in protein and carbohydrate. A herbivore may take a mixed diet simply because it cannot remember the nutritional consequences of the less preferred species, and must resample that grass to refresh its memory.

In addition to memory constraints, animals may face pre- or postingestive perceptual constraints. For example, can a grazer recognize the difference between two species of grass without eating them? Edwards et al. (1997) showed that sheep could distinguish grass and clover (a common pasture mixture) without eating them. Howery et al. (2000) showed that cattle aided by visual cues associated with preferred and non-preferred foods were more efficient at achieving their preferred diets than uncued animals. This difference was particularly evident when the food items were not located in fixed positions (and hence the cattle could not use spatial memory). Other researchers have demonstrated that several large herbivore species can tell the difference between preferred and non-preferred pelleted foods without eating them (for a review, see Baumont 1996).

Odor can play an important role in phytophagous insect diet choice. Omura et al. (2000) show that oak sap odor stimulates feeding and influences host choice behavior in two butterflies (*Kaniska canace* and *Vanessa indica*). Chapman and Ascoli-Christensen (1999) discuss the physiological mechanisms by which sucrose acts as a phagostimulant and nicotine hydrogen tartrate acts as a feeding deterrent in grasshoppers. Of course, odor is not the only cue used by phytophagous insects. Fereres et al. (1999) show that some aphids use color cues to select host plants. Leaf surface chemicals may also be important. Lin et al. (1998a, 1998b) show that alpha-tocopherylquinone acts as a feeding stimulant and forms the basis for host plant choice in cottonwood leaf beetles (*Chrysomela scripta*) feeding on poplar trees (*Populus deltoides*). For a thorough and thoughtful discussion of the role of chemical cues in host plant selection by phytophagous insects, see Bernays and Chapman (1994, chap. 4).

Perceptual constraints and cues are obviously important in phytophagous insects, and entomologists have studied them intensively. We do not know how important such perceptual constraints are for larger vertebrate herbivores. Postingestive perceptual constraints may also include the ability to match postingestive consequences with some external cue (Provenza et al. 1996). For example, grass with higher nitrogen content (hence more crude protein) may also be greener. Villalba and Provenza (2000) easily conditioned sheep to use strong flavors to distinguish between forages with different postingestive consequences. Whether they use such cues in nature awaits further investigation.

6.5   Intake Rate

The product (encounter rate) × (handling time) × (bite mass) specifies a forager's intake rate. Classic foraging models such as the diet model consider
handling time and bite mass to be constants, even though these parameters
vary considerably. For discrete food items, this is a reasonable simplification
because there is likely to be little variation in item size, implying that variation
in bite mass and handling time might be small and conveniently ignored. In
some cases, however, researchers have found that what we treat as a constraint should be treated as a decision (e.g., Newman et al. 1988). In the case
of herbivores, although there are physical limits to bite mass and handling
time, studies have often demonstrated that foragers can *voluntarily* adjust their
handling time. Before we consider this possibility, let's first consider bite mass
and handling time as constraints.

Short-Term versus Long-Term Intake Rates

We can roughly divide studies of intake rates into two types: (1) studies of
short-term intake rates as a *basis* for diet selection and (2) studies of long-term
intake rates as a *consequence* of diet selection. Neither provides us with a complete picture. Students of herbivory have debated the relevance of short-term
intake rates to large grazing mammals (e.g., see Newman et al. 1992; Illius
et al. 1999). Short-term rate studies may not be informative because they look
at what happens over only a few hundred bites, while large grazing mammals
may take tens of thousands of bites in a day. Looking at behavior over a
few minutes in isolation ignores the importance of total grazing time. That
said, however, diet selection by grazing mammals sometimes correlates with
achievable short-term intake rates from the plant species on offer (e.g., Illius
et al. 1999).

Short-Term Intake Rate

As before, I divide constraints into environmental and physiological or
morphological. On the environmental side, the major constraint is vegetation
structure. To ingest vegetation, animals must first sever (prehend) the herbage
and then, perhaps, chew (masticate) it. Allden and Whitaker (1970) point out
that intake rate in grazers is prehension bite rate multiplied by bite mass.
This simple observation has led to extensive (indeed, obsessive) consideration
of the determinants of bite mass. Most bite mass studies focus on the role
of vegetation structure. Plant height and density influence the maximum
bite mass achievable from vegetation of a given species. Researchers have

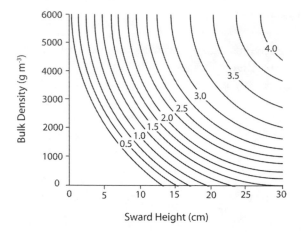

Figure 6.4. Laca and colleagues used hand-constructed sward boards to investigate the relationship between sward bulk density (g/m³) and sward height (cm) in determining bite mass (g dry matter/m²; shown as contours) for cattle grazing alfalfa and dallisgrass. (After Laca et al. 1992.)

investigated these relationships repeatedly, often using hand-constructed swards or turves; studies by Black and Kenney (1984; Kenney and Black 1984) and Laca et al. (1992; fig. 6.4) provide classic examples. Other approaches have also been tried; a particularly amusing example is that of Burlison et al. (1991; fig. 6.5). Ungar (1996) provides a nice review of this area of research.

Physiological and morphological constraints on intake rate (as opposed to total daily intake, which we will consider shortly) largely focus on jaw morphology. The time it takes to sever the vegetation may be a physical limitation of the jaw muscles (see Newman, Parsons et al. 1994 for discussion), but it may also depend on the tensile strength of the vegetation (see, e.g., Prache and Peyraud 1997). Illius and Gordon (1987) demonstrate that incisor arcade breadth (the distance between the right and left fourth incisors), more closely than body mass, predicts variation in bite mass.

One difficulty with this area of work is that all measurements of animals' short-term intake rates are measurements of what animals *do*, not what they *can* do. We know from numerous studies that animals can voluntarily increase their short-term intake rate with no change at all in vegetation structure (e.g., Greenwood and Demment 1988; Dougherty et al. 1989; Newman, Penning et al. 1994). My colleagues and I used a simple mechanistic model to demonstrate the flexibility that grazers have to use behavior to increase intake rate. We showed that within a forage species, grazers have little latitude to alter their handling times, but some flexibility to alter their bite mass (Newman, Penning et al. 1994). I will consider behavioral decisions regarding intake rate in more detail in the next section.

Researchers in the field now appreciate the mechanistic aspects of bite mass and hence intake (e.g., see review by Baumont et al. 2000), and these mechanisms form the basis of several widely used models of grazing behavior (e.g., Spalinger and Hobbs 1992; Newman, Parsons et al. 1994; Parsons, Thornley et al. 1994; Thornley et al. 1994; Pastor et al. 1999; Illius 2006).

Long-Term Intake Rate

Long-term intake rates are not easily measured in the field. It is difficult or impossible to see the size of each bite. Peter Penning has developed an excellent device for recording details of intake behavior in larger herbivores at pasture; figure 6.1 shows the device on a cow. By weighing the animal before it goes on the pasture, collecting its dung and urine, weighing it again after a period of time, and correcting for insensible weight loss, intake rates can be estimated over longer time periods. Penning et al. (1991), for example, used the bite recorder to investigate the relationship between pasture surface height, tiller density (which tends to be negatively correlated with pasture surface height), and bite mass. They found that grazing time and prehension bite rate declined with increased pasture surface height, while rumination time and mastication bite rate increased with pasture surface height.

Clearly, the constraints discussed for short-term intake rates sometimes determine long-term intake rates as well, but there are also times when animals behaviorally adjust their short-term intake rates to manipulate their long-term intake rates. A number of studies that have examined the effects

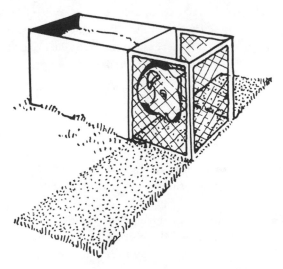

Figure 6.5. Modified metabolism crates allowed sheep access to a 0.56 × 0.46 m area of pasture. (After Burlison et al. 1991.)

of different physiological states for animals grazing the same pastures clearly demonstrate this. For example, Greenwood and Demment (1988), Dougherty et al. (1989), and Newman, Penning et al. (1994) report that fasting in sheep and cows can cause a *voluntary increase* in intake rate by 27% to 72% with no changes in sward structure. Similar results have been reported during lactation in these animals (e.g., Penning et al. 1995; Gibb et al. 1997; Prache 1997; Patterson et al. 1998). Moreover, Iason et al. (1999) have shown that when grazing time is limited, sheep may voluntarily increase their intake rates; if food is abundant enough, this behavior can compensate for the time limitation (see also Ydenberg and Hurd 1998).

So if animals can voluntarily raise their intake rates, why don't they always eat this quickly? Increased intake rates may reduce vigilance behavior, which imposes a cost (real or perceived) of increased predation in the longer term (Underwood 1982; Illius and FitzGibbon 1994). The choice of intake rate additionally implies a corresponding digestion rate as well as rumination requirements, both of which may have fitness (opportunity cost) consequences (Greenwood and Demment 1988). While physical, morphological, and environmental constraints on intake rate have received extensive attention, researchers have largely ignored the potentially important contribution of behavioral decisions. This area certainly requires further research.

Intake rate in social animals should represent a balance between the need for vigilance and intraspecific competition (both scramble and interference). Rind and Phillips (1999) nicely demonstrated this with cows. They found that prehension bite rates were lower in both small and large foraging aggregations. However, the effects of social constraints on intake rates have not been extensively studied, largely due to the technical difficulty of estimating intake rates of animals at pasture. It is more common to consider the effects of social constraints on grazing time rather than intake, something I do in the next section.

6.6 Grazing Time

Intake rate multiplied by grazing time determines total daily intake. Many theoretical studies take grazing time to be a constraint (see, e.g., Verlinden and Wiley 1989). In many cases, this assumption is appropriate, as a number of environmental factors can constrain grazing time. Again, the social context may be important. Penning et al. (1993) showed that grazing time is a negatively accelerating function of aggregation size for sheep grazing a monoculture (hence with no diet selection; fig. 6.6), and Sevi et al. (1999) and Rind and Phillips (1999) found similar results in cows and sheep. In addition, Rook

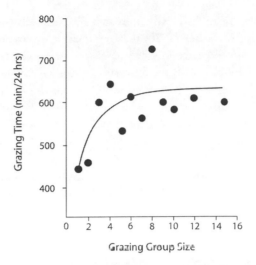

Figure 6.6. Penning et al. replicated flock sizes and used bite recorders (see figure 6.1) to record the grazing behavior of sheep. The sheep were maintained on a monoculture of perennial ryegrass. The data clearly show that individual animals and small groups (≤ 5 animals) behave differently from larger grazing groups. The best fit curve has the equation [grazing time (min/24 hr)] = 629 − 311 × exp(−0.46 × group size). (After Penning et al. 1993.)

and Penning (1991) and Rook and Huckle (1995) present strong evidence for synchronization in grazing behavior in sheep and cows. Such synchronization may well be a general phenomenon in large social animals. In addition, day length itself can constrain grazing time. On some pastures, lactating ewes may need to graze nearly all the daylight hours to meet their daily energy requirements. The requirements for other fitness-enhancing behaviors, such as vigilance, may also constrain the time available for foraging (Underwood 1982; Illius and FitzGibbon 1994).

Gut passage time constrains the behavior of some herbivores. Forage quality can be so poor that animals can starve to death, even on ad libitum food. Plant quality characteristics such as lignin and cell wall content affect gut passage time (for review, see Iason and Van Wieren 1999). The size of food particles entering the gut, which depends in part on the animal's intake behavior, also affects passage time (Gidenne 1992; Kennedy 1995; Wilson and Kennedy 1996; Schettini et al. 1999). Sheep, for instance, have a relatively constant number of jaw movements per minute when grazing (ca. 150; Penning et al. 1991), and these jaw movements must be partitioned between prehension and chewing bites. To increase the prehension bite rate, a herbivore will have to reduce the amount of chewing it does, and this can lead to slower passage rates through the gut (or increase rumination requirements).

Gut size itself, which scales with body size allometrically (see section 6.2), constrains passage time. Bell (1970) and Demment and Van Soest (1985) describe how food use relates to body size (see also Illius and Gordon 1990; Iason and Van Wieren 1999). Size-based differences in forage use operate both between and within species (e.g., sexual dimorphism), and forage use may change as an individual develops. Bernays and Chapman (1970b, 1970a) showed that changes in the mandible of the grasshopper *Chorthippus parallelus* from the first to the fourth instar correlate with a shift in diet from thinner to thicker-leaved grasses.

The animal's ability to detoxify or excrete plant secondary compounds also constrains grazing time. Lauriault et al. (1990; Dougherty et al. 1991) have shown that grazing time in cattle declines in the presence of alkaloids from endophyte-infected tall fescue. Pfister et al. (1997) have shown that cattle grazing tall larkspur (*Delphinium barbeyi*), which contains a potentially toxic alkaloid, can regulate their intake to remain below the toxic threshold.

When energy requirements increase (e.g., due to lactation) or food availability declines, animals often voluntarily increase their grazing time in response (Arnold 1975; Clutton-Brock et al. 1982; Penning 1986; Penning et al. 1991). As with the question of intake rate behavior, we might ask why animals don't spend more time grazing when they can. Ultimately, the likely evolutionary reason is that grazing takes time away from other fitness-enhancing activities, such as vigilance, rumination, and social interactions (for a theoretical consideration of this issue, see Thornley et al. 1994; Newman et al. 1995). However, it may be possible to gain some mechanistic insight into the flexibility of the behavioral repertoire by looking at what happens when grazing time constraints are no longer applicable—a situation that happens in captivity when animals are provided with high-quality concentrated feed. Ungulates evolved some 40 million years ago, but we have only housed them in ways that severely curtail their foraging behavior for a few decades. Such studies have led, for example, to the realization that jaw movements serve a function beyond their mechanical effect on food: they promote salivation, which, in ruminants at least, has a vital buffering effect on fermentation in the rumen. It is entirely possible that such a mechanism could, in some pastures, lead to grazing times *longer* than necessary to satisfy energy demands. Georgia Mason considers this issue in more detail in box 6.2.

6.7   Return to Question One: Where to Eat?

In many cases, habitat or patch choice reflects both diet selection and intake rate considerations. For example, Wallis de Vries and Daleboudt (1994) found that cattle may select patches based on long-term rate maximization (however, see Distel et al. 1995), though additional considerations such as nutrient content (e.g., phosphorus) may also be important (Wallis de Vries and Schippers 1994). More generally, the presence of preferred forage species (e.g., Crane et al. 1997; Watson and Owen-Smith 2000) and differences in forage quantity and nutritional quality (e.g., Wallis de Vries et al. 1999; Van der Wal et al. 2000) seem to be important in these decisions. In some cases, abiotic factors, such as the time since the last fire event (e.g., Irwin 1975; Coppedge and Shaw 1998), affect plant quality in patches. In others, the behavior of the animals themselves alters patch quality; for example, through dung and urine deposition (e.g., Keogh 1975; Day and Detling 1990; Lutge et al. 1995). There is evidence that animals shift their patch preferences in response to both kinds of considerations.

Stokke (1999) found a sex-based difference among elephants in habitat use. Sex-based differences often reflect body size differences, although Perez-Barberia and Gordon (1999) observed sex-based differences in the patch choices of Soay sheep even when the body size differences where removed. Nevertheless, body size differences can mean that "patch quality" is a relative quality. A model by Illius and Gordon (1987) showed that allometric relationships between bite size, metabolic requirements, and body size can explain differences in habitat choice within species, especially between males and females of dimorphic species. Body size difference may be the mechanism that determines the outcome of interspecific competition, as in the example of cattle and mule deer (Loft et al. 1991).

Mysterud et al. (1999) point out that factors other than food availability can determine habitat choice, but may reflect trade-offs between, for example, food and protective cover. They demonstrated that roe deer (*Capreolus capreolus*) habitat use did not correlate with the availability of herbs, but did correlate with the availability of canopy cover. Similarly, Ginnett and Demment (1999) found that when female Masai giraffes (*Giraffe camelopardalis tippelskirchi*) were caring for offspring, they selected habitats without cover for predators. Parasites, too, can influence patch and habitat choice. Cooper et al. (2000) demonstrated a trade-off between patch quality and the presence of sheep dung (infected with *Ostertagia circumcincta* larvae), and Duncan and Cowtan (1980) showed that horses may choose foraging habitats based on the densities of blood-sucking flies.

Just as they influence other foraging decisions, social interactions can influence habitat and patch choice. A model by Beecham and Farnsworth (1998) demonstrated that a species-specific spacing preference can constrain patch choice and resource utilization, resulting in a short-term reduction in intake rate and an increase in the variability of resource utilization. Bailey (1995) showed that initial patch selection by groups of steers was often determined by the behavior of one or two individuals. Social learning in early life accounts for sexual differences in roe deer habitat selection that earlier authors attributed to competition (Conradt 2000).

Patch use leads to intake rate depression. The work of Laca et al. (1994) nicely demonstrates this for cattle. They showed that bite mass decreased more in tall, sparse patches than in short, dense patches. While bite mass declined, time per bite did not change, resulting in intake rate depression as exploitation time increased. Patch depletion naturally leads to marginal value-like considerations for patch-leaving rules. Baharav and Rosenzweig (1985) found behavior consistent with Charnov's model (Charnov 1976b) in Dorcas gazelles (*Gazella dorcas*). According to the marginal value theorem, patch exploitation depends on travel time between patches, and the spatial distribution of patches determines the travel time. Dumont et al. (1998) showed that sheep exploit patches more intensively when they must travel greater distances between patches. More generally, Wallis de Vries (1996) modeled the interacting effects of group size, inter-patch distance, and resource distribution pattern (degree of aggregation of patches) on the spatial distribution of foraging time for an ungulate. He showed that travel costs can be very important, even when small.

Nevertheless, support for the patch model has not been unanimous. For example, Jiang and Hudson (1993) preferred a mechanistic explanation for patch leaving in wapiti based on their lateral neck angle and biokinetic considerations. Lundberg and Danell (1990) argued that marginal value theorem explanations are less useful than optimization of handling time for each ramet for moose browsing birch stands.

6.8 Decision Making with Multiple Objectives

I have been developing a view that herbivore foraging should be understood as a function of the animal's constraints and objectives. For an animal attempting to answer each of the four big questions (where to eat, what to eat, how fast to eat, how long to eat), there are often multiple objectives. For example, sheep choosing a diet from grass/clover pastures would like to graze with conspecifics; avoid areas of previous defecation to guard against parasite infection; eat a 65% clover, 35% grass diet; take bites of 53 mg clover and 30 mg grass at a rate of 83 bites per minute from each, with 17 chews per gram of clover and 27 chews per gram of grass; and of 660 available minutes, spend 334 grazing clover, 166 grazing grass, and 160 not grazing (all weights measured as dry mass; Dumont and Boissy 1999, 2000; Hutchings et al. 1998; Parsons, Newman et al. 1994; Newman, Penning et al. 1994). To some extent, the animal can control each of these objectives. For example, we know that hungry sheep can increase their grazing time by 80–185 minutes, increase their bite masses 54%–290%, decrease their bite rates 5%–41%, and decrease their mastication by 9%–20% per gram (Newman, Penning et al. 1994). We know that sheep whose previous diets comprised a monoculture of either grass or clover choose more or less clover, respectively, than sheep that were recently grazing a mixture of grass and clover (Parsons, Newman et al. 1994). The constraints are the same for the fasted and non-fasted sheep. The animals were all tested on the same pasture; the changes were entirely voluntary.

The reductionist approach to herbivore foraging behavior will not always work, however. In the previous example, we know a good deal about all of these objectives and others, but very little about their relative or absolute importance. When sheep find it impossible to meet all of these objectives simultaneously, how do they respond? Are there some objectives that sheep defend vigorously and others that they sacrifice for higher-order objectives? Which of these multiple objectives takes priority, and in which circumstances?

Foraging location, diet choice, intake rate, and grazing time all have fitness consequences, and how herbivores trade off the multiple objectives within and between these broad categories of behavior differs by species, body size, ecosystem, time of year, age, and many other factors. There is not a one-to-one mapping of any of these four dimensions onto fitness. An animal may compensate for a low long-term intake rate by becoming less selective, thus increasing its encounter rate; or more selective, thus improving its diet quality; or it may increase its grazing time, or some combination of these tactics. Studying just one of the big questions in isolation may not yield tremendously clear results.

Marc Mangel and others (e.g., Mangel and Clark 1988; Houston and McNamara 1999; Clark and Mangel 2000) have advocated the use of dynamic state variable models (see chap. 1 in this volume). I used this technique to consider how animals select diets and grazing time in the face of intake and passage rate constraints and predation danger (Newman et al. 1995). That work provided a behavioral explanation for total grazing time, which earlier investigators had always treated as a simple constraint. Useful as we felt that model was, it still considered only two of the big questions simultaneously, and I do not believe that the dynamic programming approach will be useful in addressing the integration of the myriad of objectives herbivores routinely face. The problem is computational. Beyond a few state variables and a few decisions, the computational demands of dynamic programming quickly make it intractable. The technique can provide useful analyses of some of the questions in isolation or in combination, but not all of them.

Computational limitations aside, to apply dynamic programming to understand how herbivores balance their multiple objectives, we ultimately need to know how each objective *and their combinations* map onto evolutionary fitness. It is unlikely that we will ever reach that goal. How should we proceed? In my opinion, the problem with our ongoing research programs is that they are too mired in "traditional" experimental design. We rely heavily or exclusively on the univariate analysis of variance or multiple regression approaches, but decisions with multiple objectives are, necessarily, multivariate problems. To address the questions posed at the start of this section, I think we need to borrow some techniques from economists. Microeconomic theory embraces the notion that preferences are based not on single attributes, but jointly on several attributes. Economists get people to reveal the utility that the attributes of goods or services have for them by examining the trade-offs that they make between those attributes in the process of making choice decisions.

Economists have developed extensive theory and techniques for addressing decisions with multiple objectives. I find this approach most intuitive when I think about how people value "the environment." All other things being equal, we would like to have clean air, clean water, high biodiversity, charismatic species, unspoiled natural landscapes, rainforests, and so on. However, we would also like to eat; care for our children, the sick, and elderly; improve our education and social welfare systems; control our agricultural pests; and so on. All other things are rarely equal, and we have to make choices and trade-offs among our multiple objectives. The field of research called "economic valuation" is aimed at understanding the choices we make, finding out how much we value particular states of nature and how combinations of these states map onto our utility. In my opinion, this is exactly what we must do for herbivore foraging decisions. Substitute "eat highly nutritious forage" for "have clean air"; "avoid predators" for "control our agricultural pests"; and so on, and the usefulness of economic valuation becomes apparent.

Economists use two broad categories of methods to understand how humans value their multiple environmental objectives: revealed preference techniques and expressed preference techniques. In the remainder of this section, I will briefly introduce some of these techniques and point out how I think they can be used to study herbivore foraging behavior. It is not my intention to teach the full background theory or discuss all the caveats and problems with each technique. Rather, my goal is to stimulate researchers to find out more about these techniques and then apply them where appropriate.

Revealed Preference Techniques

With free-roaming animals, replicated controlled experiments are difficult or impossible. In these cases, we can make use of revealed preference techniques. Economists sometimes call these "indirect techniques" because the researcher does not directly ask people questions, but rather studies what people do and then deduces their preferences from their observed behavior. In a sense, students of foraging behavior do this already, but in relatively unsophisticated ways. Revealed preference techniques offer us a chance to gain deeper insight into how herbivores sort out their multiple objectives. I will quickly review two such techniques, hedonic pricing and travel cost methods. In my opinion, travel cost methods show more promise.

Hedonic Pricing

The most common use of hedonic pricing deals with housing prices. This technique relies on the assumption that an individual's utility for a house is based on the attributes it possess, such as size, location, school district, air quality, general level of environmental quality, and so on. In certain circumstances, one can use multi-market data to derive society's "willingness to pay" for attributes such as clean air.

It may be possible to use either total energy costs or total time costs as a substitute for housing price and the foraging habitat as a substitute for the house. Then, by comparing the choices of many individual animals for nearby foraging habitats that differ in many foraging attributes (e.g., forage species available, intake rates achievable, social context, predator/cover attributes, etc.), it may be possible to derive explicit values and trade-offs (substitutions) for each of the foraging attributes. One hedonic pricing study showed that a particular group of people was willing to pay $5,500 for a marginal improvement in nitrogen oxide levels. By knowing the price they were willing to pay, we can see how they might trade off air quality for, say, water quality, whose value a similar study estimated as $41,000. Applying this method in a herbivore study might tell us, say, that gazelles are willing to pay 3.7 MJ of energy for a marginal improvement in predation danger and 2 MJ for a marginal improvement in grazing time. It might then be possible to estimate the trade-off between predation avoidance and grazing time. I reiterate that many caveats would go along with such a study, and the investigator should fully appreciate these problems before undertaking it.

Travel Cost Methods

Travel cost methods use the price people are willing to pay to travel to a non-priced recreation site as a means of inferring the value of environmental attributes that a person experiences at that site. These methods consistently show that as the price of travel (distance) increases, the rate of visits to the site falls. For example, one travel cost study found that a group of people was willing to pay an extra $7 per use to improve water quality from "boatable" to "fishable" and a further $14 per use to improve the quality from "fishable" to "swimmable."

Much as with the hedonic pricing method, we can examine the travel costs (time or energy) that individual animals are willing to pay for a change (improvement) in the dimensions of foraging quality as they move from one foraging habitat to another. The animals reveal their preferences, and their willingness to make trade-offs among preferences, through their willingness to pay the cost of travel. For example, we know from Parsons, Newman et al.'s (1994) work that sheep prefer about 70% clover in the diet when grass and clover monocultures occupy adjacent sides of a pasture. How important is this particular mixture to them? How willing would they be to trade off this mixture for, say, grazing time? One way to see this is to increase the distance the animal has to travel between grass and clover. As we do so, the animal has to pay increasingly higher travel costs to defend the mixture, and we may see how important the 70% objective is versus whatever grazing time objective it has. The sheep example could obviously be studied experimentally, but similar kinds of studies could be conducted in nonexperimental situations with free-roaming animals. We could use travel cost methodology by examining the choices of individuals that must travel different distances for access to each of the choices.

Two comments are warranted here. First, revealed preference techniques will be most useful if we do our best to measure *all* of the relevant behaviors and objectives. The point is to see how the animal combines or trades off these multiple objectives. This cannot be done if we ignore one or more of the major behavioral decisions the animal must make. Second, we need to recognize that these approaches identify correlations, not mechanisms (for this we need expressed preference approaches). If we are to ultimately understand, and so predict, herbivore behavior, we must understand the relevant mechanisms.

Expressed Preferences

With invertebrates and captive (often agricultural) vertebrates, replicated controlled experiments are possible. In these cases, we don't need to rely on animals revealing their preferences to us; we can ask them directly. Economists use two main expressed preference methods: contingent valuation and conjoint analysis. Again, I will briefly review these two methods with an eye toward how they might be applied to foraging studies. There are extensive

Table 6.1   Measurements of the value of resources to farmed mink

Resource	Elasticity of demand	Consumer surplus (kg)		Reservation price (kg)	Total expenditure (kg)
		Travel cost	Aggregate		
Water pool	$0.26 \pm 0.04^{A\S}$	$81.41 \pm 9.97^{A\S}$	24.00	$1.25 \pm 0.00^{A\S}$	$134.33 \pm 11.18^{A\S}$
Alternative nest site	$0.41 \pm 0.08^{AB\S}$	$60.72 \pm 5.67^{AB\S}$	22.75	1.17 ± 0.05^{A}	$114.83 \pm 13.27^{AB\S}$
Novel objects	0.58 ± 0.08^{BC}	$54.58 \pm 5.02^{AB\S}$	22.50	1.16 ± 0.04^{A}	$83.62 \pm 9.93^{D\S}$
Raised platform	0.57 ± 0.07^{BC}	$50.78 \pm 7.65^{B\S}$	22.25	1.14 ± 0.06^{A}	82.17 ± 16.11^{B}
Toys	0.62 ± 0.05^{BC}	24.30 ± 3.25^{C}	21.00	1.06 ± 0.07^{A}	$34.11 \pm 6.39^{C\S}$
Tunnel	$0.73 \pm 0.07^{C\S}$	$21.61 \pm 1.72^{C\S}$	20.75	1.06 ± 0.06^{A}	$262.33 \pm 3.66^{C\S}$
Empty cage	$0.77 \pm 0.06^{C\S}$	$9.19 \pm 0.90^{D\S}$	17.00	$0.84 \pm 0.07^{B\S}$	$8.79 \pm 1.34^{D\S}$

Source: Mason et al. 2001.

Note: The price elasticity of demand was calculated from the slope of the log-log plot of visit price versus number of visits for each resource. Consumer surplus was calculated by estimating the area under two types of demand curves: a plot of visit price versus visit number (analogous to the travel cost method discussed earlier) and an aggregate plot of price versus the number of subjects willing to pay each price. Reservation price is similar to the "break point" used by experimental psychologists and was calculated as the maximum price paid for each resource. Total expenditure per unit time is a measure most behavioral ecologists would use. Letters denote resources whose values differ significantly at the $P < 0.01$ level (Tukey's t test). § denotes resources that contribute significantly ($P < 0.01$) to the general linear model value = sex + mink(sex) + resource + sex × resource. Low elasticities denote little reduction in visit number as visit cost increases.

literatures on each method; consequently, the limitations of such studies are rather well understood.

Contingent Valuation Methods

This group of methods asks an animal directly what it is "willing to pay" to secure a preference. Mason et al. (2001) published a nice example of this approach (table 6.1), albeit not with a herbivore. Nevertheless, their example is instructive. Mason et al. offered caged mink, in a closed economy, the option to use seven different compartments that each contained a different resource. The animals had to push a weighted one-way door to gain access to the resource. By examining each animal's "willingness to pay" for access, Mason et al. were able to assess how much the animals valued the resource. Mason et al. did not examine resource trade-offs, but they could have. For our purposes, the desirable characteristic of this study is that we get a good sense of how important particular preferences are—by increasing the costs of expressing a preference, we can see how much the animal is willing to defend that preference.

Conjoint Analysis

Conjoint analysis is a group of techniques used to estimate utilities based on subject responses to combinations of decision attributes. In essence, these techniques ask subjects to choose among (sometimes to rank) various packages that differ from one another in the degree to which they satisfy the consumer's multiple objectives. By carefully constructing these packages and using the appropriate analysis, the researcher can estimate the utility that any particular attribute gives to the consumer and *how these attributes trade off against one another*.

To my knowledge, conjoint analysis techniques have not been used in behavioral ecology, but I think this is the area that holds the most promise for future herbivore foraging research. Unfortunately, it is also the most theoretically complex, and so is beyond the scope of this essay. In section 6.11, I provide a few references to get the interested researcher started.

6.9 Spatially and Temporally Dynamic Interactions between Herbivores and Plants

A herbivore may choose roughly the same diet for a long time. More often, however, foraging behavior changes dynamically, because plants change dynamically. Sheep may choose 6 cm ryegrass over 3 cm clover now, but in the future, the clover will grow and the ryegrass will become shorter, and their preferences may change.

Early students of foraging behavior were interested in foraging behavior because they thought it would yield a deeper understanding of predator-prey dynamics. They thought we could use an understanding of predator behavior to predict the population dynamics of predator and prey. The now classic paper by Noy-Meir (1975) provides a good example of this approach. Noy-Meir showed that the interaction between grazing animals and pasture growth could result in "dual stability," in which there are two stable equilibrium plant densities for the same intake rate of the herbivores (fig. 6.7A). This occurs because the relationship between plant growth rate and leaf area index (LAI, m^2 leaf/m^2 ground) is roughly parabolic, and the herbivore functional response crosses the growth curve (producing an equilibrium) more than once. Herbivory can have profound effects on both the stability and the dynamics of plant growth. Figure 6.7B shows what can happen as we change the herbivore density when we have the type of dynamics show in figure 6.7A. At densities between 3 and 5 animals per hectare, there are two steady states, with LAI of 1 and 5.5. For these densities, if the pasture starts out at low values of LAI, then only the lower equilibria are reached. If the pasture starts out at a high LAI, then the upper equilibria are reached. However, the system can easily flip from the higher stable state to the lower one through changes in,

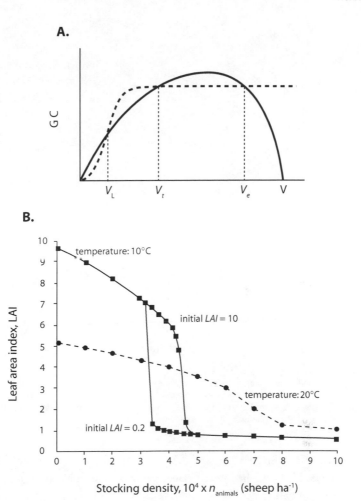

Figure 6.7. (A) This graph shows how combining the vegetation growth curve (GC) with a type II or type III (in this case) functional response can result in two alternative stable states. V denotes vegetation mass. V_L is the lower stable equilibrium, V_t is a transient unstable equilibrium, and V_e is the high stable equilibrium. (B) This graph shows how dual stability can arise through changes in foraging pressure. (After Noy-Meir 1975 and Thornley 1998.)

for example, management. This dual stability is temperature dependent. As seen in figure 6.7B, the bifurcation disappears as the temperature increases.

Noy-Meir assumed that grazers defoliated plants in a deterministic and continuous manner. Schwinning and Parsons (1999) relaxed these assumptions and considered the grazing process as discrete bites in a spatially heterogeneous environment, with selection behavior by the animal, and used a more realistic plant growth function. Most importantly, they also considered how the herbivore's behavior alters and responds to the spatial heterogeneity in the pasture. Figure 6.8 shows that the animals' foraging behavior can indeed have

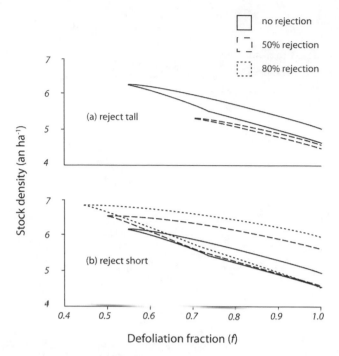

Figure 6.8. Dual stability domains for different patch selection decisions. Patch encounter occurs randomly, and there is no extra cost associated with patch rejection. The defoliation fraction is the fraction of the standing biomass that a grazer removes before moving on to the next patch or feeding station. (A) Foragers probabilistically reject patches that are > 0.2 kg/m². (B) Foragers probabilistically reject patches that are < 0.12 kg/m². (After Schwinning and Parsons 1999.)

a strong effect on the dynamics of the herbivore-pasture interaction. Dual stability is less likely when animals reject patches of high biomass, but more likely when animals reject patches of low biomass. Schwinning and Parsons (1999) point out that there are several detailed and spatial analyses of foraging (e.g., Farnsworth and Beecham 1999; Grunbaum 1998) that take into account the costs of different foraging strategies, but fail to realistically incorporate resource regeneration (but see Hutchings and Gordon 2001).

Schwinning and Parsons (1999) suggest that progress now depends on the merging of spatially explicit approaches that consider the dynamics of the interaction between herbivores and plants (for more on this subject, see Parsons et al. 2001) with developments in foraging theory. I would agree. In order to adopt a truly dynamic view of herbivore foraging behavior, we must understand the dynamics of plant growth. We need to become experts, not just on animal behavior, but also on plant growth and metabolism (or find someone who is and work very closely with them!). To understand foraging behavior in herbivores, we must understand the entirety of the plant-animal interaction.

6.10   Summary

In this chapter, I present a view of herbivory based on four interacting behaviors: habitat choice, diet choice, intake rate, and foraging time. We can view each of these behaviors as resulting from objectives and constraints, both environmental and physiological. Sometimes it is useful to think of one or more of these behaviors as a constraint, but at other times our analysis will be better served if we frame these behaviors as decisions. Most importantly, this complex of interacting behaviors must be seen as inducing change in the plants and responding to these changes in a dynamic fashion.

Although there are clearly many gaps in our understanding of the answers to each of the four big questions, I want to reiterate two areas that need further attention. One important frontier is the integration of the four basic behaviors. Under which circumstances does one action take priority over another, and can we develop a sufficient understanding to predict this? We need to develop mathematical models that incorporate all four big questions, and we need to use these models to generate and experimentally test predictions about how the four questions interact in various environmental circumstances. This is the area of decision making with multiple objectives discussed in section 6.8. The second frontier, which I discussed in section 6.9, is the dynamic aspect of the plant-animal interaction. We need models that integrate population dynamics with foraging behavior, and we need experimental tests of these models. A worthy, but as yet unfulfilled, goal is to develop a dynamic understanding of even a simple herbivore-plant system.

My final comment is a plea for openness and integration. I have attempted, perhaps too superficially, to integrate literature from a variety of fields because I think that trading ideas and perspectives leads to unexpected gains. While there are many well-written reviews on aspects of this chapter's material, they are invariably limited to particular taxonomic groups, and in many cases even more limited than that. A true integration of vertebrate and invertebrate foraging behavior, across applied and basic science disciplines, would yield great benefits. Hmmm, that sounds like a topic for another book . . .

6.11   Suggested Readings

E. A. Bernays and R. F. Chapman's *Host-Plant Selection by Phytophagous Insects* (1994), chapters 4 and 5, provide an excellent introduction to herbivorous insect foraging behavior. This book is a little dated now, but serves as a good starting point for the field. In *The Ecology and Management of Grazing Systems*, edited by J. Hodgson and A. W. Illius (1996), chapters 5 (Laca and Demment),

7 (Ungar), and 9 (Murray and Illius) provide a good starting point for larger vertebrates. For a much broader view of herbivory, see *Herbivores: Between Plants and Predators*, edited by H. Olff, V. K. Brown, and R. H. Drent (1999). Parsons and Chapman (2000) offer an excellent discussion of the dynamic nature of the plant-animal interaction. O'Connor and Spash (1999) is a good starting point for the field of economic valuation, and Gustafsson et al. (2003) and Louviere (1988) provide details on conjoint analysis.

Telander

Energy Storage and Expenditure

Anders Brodin and Colin W. Clark

7.1 Prologue

The snow creaks under our winter boots as we walk along the snow
scooter track to our study site. The cold is overwhelming, and though
we have been walking for an hour, we do not feel warm. The air is
perfectly still, and the heavy snow on the branches of the surrounding
conifers absorbs all sounds. When we arrive at the bait station, we spill
some seeds onto the feeding tray and retire to the nearby trees. The
seeds soon attract the attention of some willow tits. It is astonishing
that these 10 g animals with their high-speed metabolism can survive in
an environment where the temperature can remain below freezing for
months. We know they need to eat three or four food items per minute
throughout the short winter day to survive the long night. Surprisingly, the willow tits do not consume the seeds. Instead, they begin ferrying
seeds from the tray to hiding places nearby. They conceal them carefully
under flakes of bark, in broken branches, and in tufts of lichen. Evidently, willow tits can exploit the temporary abundance of seeds most effectively by hoarding them, deferring their consumption until later. sophisticated energy management makes their survival in these extreme
conditions possible. Their daily regimen combines use and maintenance
of external (thousands of individually stored items) and internal (several

grams of fat) energy supplies, augmented when necessary with tactics such as hypothermia.

7.2 Introduction

Organisms need energy to sustain their growth and metabolism. Most animals do not forage continuously and must store energy for periods when foraging is not possible. They also need to perform other activities that may not be compatible with foraging. Periods when energy expenditure exceeds energy intake may be short; for example, between two meals or overnight. They may also be long, lasting through the winter or throughout extended periods of drought. Energy can be stored in the body as fat, carbohydrates, or sometimes as proteins, or in the environment as hoarded supplies.

Many forms of energy storage are well known. Bears become very fat in autumn before they go into hibernation. Honeybees store large supplies of honey in the hive to be used as food during the winter. Many avian and mammalian species hoard thousands of seeds and nuts in autumn and depend on these foods during the winter. Energy storage is also common in organisms such as plants and fungi. Many of our most common root vegetables, such as potatoes, rutabagas, and carrots, are good examples of plants that store energy for future growth and reproduction.

Animals must actively regulate their energy expenditure. During hibernation, most animals reduce expenditure by lowering their body temperature and thereby their metabolism. Many humans try to decrease their body fat energy stores and get slimmer; for example, by reducing food intake. Others instead try to increase their energy stores. Before a race, cross-country and marathon runners may actively deplete the glycogen reserves in the liver and muscles. The evening before the race, they gorge on carbohydrates, attempting to enlarge those reserves and so increase their endurance (e.g., Åstrand and Rodahl 1970). For animals that live in seasonally fluctuating environments, finely tuned management of the energy supply may be crucial for survival and reproduction. Indeed, without such adaptations, these organisms could not inhabit these environments.

We begin this chapter by presenting examples of how animals store and regulate energy. Next, we adopt an economic perspective that focuses on the costs and benefits of energy storage. This leads to a brief overview of how behavioral ecologists have modeled energy storage. We devote the second half of the chapter to dynamic state variable modeling (Houston and McNamara 1999; Clark and Mangel 2000). From the simplest possible model, we proceed through models of increasing complexity to illustrate the key factors controlling energy storage. The text considers the problems of small passerine birds in a cold winter climate as a convenient model for problems of energy storage and regulation. We focus on evolutionary aspects of energy regulation. Box 7.1 introduces neural and endocrine mechanisms of energy regulation.

BOX 7.1   Neuroendocrine Mechanisms of Energy Regulation in Mammals

Stephen C. Woods and Thomas W. Castonguay

Myriad approaches have been applied to the study of how animals meet their energy requirements. A century ago, the predominant view was that events such as gastric distension and contractions determine food intake, with signals from the stomach relayed to the brain over sensory circuits such as the vagus nerve. One of the most influential theories of energy balance, the "glucostatic hypothesis" posited over 50 years ago by Jean Mayer (1955), proposed that individuals eat so as to maintain a privileged level of immediately available and usable glucose. When this commodity decreased, either due to enhanced energy expenditure or to depleted energy stores, hunger occurred and eating was initiated; as a meal progressed, newly available glucose was able to reduce the hunger signal. While theories such as this were highly influential, subsequent research has found them to be simplistic and limited, and it is now recognized that an intricate and highly complex control system integrates signals related to metabolism, energy expenditure, body fat, and environmental factors to control food intake.

Most contemporary research has concentrated on the question "How much do we eat in a given meal, or in a given period of time?" Over 50 years ago, Adolph (1947) pointed out that when we eat energetically diluted foods, a greater bulk of food is consumed. Conversely, we eat smaller meals when food is energetically rich. This simple observation implies that we eat to obtain a predetermined number of calories of food energy. In fact, we humans adjust our caloric intake with remarkable precision, with our intake under free feeding conditions matching our energy expenditure with an error of less than 1% over long intervals (Woods et al. 2000).

The Control of Meals

Energy is derived from three macronutrients: proteins, fat, and carbohydrates. The carbohydrate glucose and various fatty acids provide energy to most tissues. The brain is unique, requiring a steady stream of glucose from

(Box 7.1 continued)

the blood in order to function. This reliance of the brain on glucose formed the basis of the glucostatic theory, and other theories over the years have focused on available fat or protein as being key to energy regulation. The premise underlying all of these hypotheses is that the level of some important commodity (glucose, fatty acids, total available energy to the brain or some other organ) waxes and wanes during the day. When the value gets low, indicating that some supply has become depleted, a signal is generated to eat; when the value is restored (repleted), a signal is generated to stop eating (Langhans 1996). While the logic of these "depletion-repletion" theories has considerable appeal, the bulk of evidence suggests that energy flux into the brain and other tissues is remarkably constant and that small fluctuations cannot account for the onset or offset of meals.

What, then, determines when a meal will begin, especially when an individual could, in theory, eat whenever it chooses? The best evidence, at least for omnivores such as humans and rats, suggests that eating occurs at times that are convenient given other constraints in the environment, or at times that have resulted in successful eating in the past. We eat at particular times because of established patterns, or because someone has prepared food for us, or because we have a break in our busy schedules (Woods et al. 1998). If depletion of some critical supply of energy provided an impetus dictating that we put other behaviors on hold until the supply is replenished, daily activity patterns would be much different. Instead, animals enjoy the luxury of eating when it is convenient, and they regulate their energy needs via controls over how much is eaten once a meal is initiated.

Signals that Influence Intake

Armed with the tools of contemporary genetics, molecular biology, and neuroscience, scientists have discovered literally dozens of signals over the past 20 years that either stimulate or inhibit food intake (Schwartz et al. 2000; Woods et al. 1998). As depicted in figure 7.1.1, these signals fit into three broad categories. The first are signals generated during meals as the ingested food interacts with receptors in the mouth, the stomach, and the intestines. Most of these signals are relayed to the brain via peripheral nerves (especially the vagus nerve) and provide information as to the quality and quantity of what is being consumed. These are collectively called "satiety" signals because as their effect accumulates during a meal, they ultimately lead to the sensation of fullness or satiety in humans, and their

(Box 7.1 continued)

administration reduces meal size in animals including humans. As an example, mechanoreceptors in the stomach respond to distension, and this information is integrated with chemical signals generated in response to the content of the meal. The best-known satiety signal is the intestinal peptide cholecystokinin (CCK). CCK is secreted in proportion to ingested fat and carbohydrates, and it elicits secretions from the pancreas and liver to facilitate digestion. CCK also stimulates receptors on vagus nerve fibers.

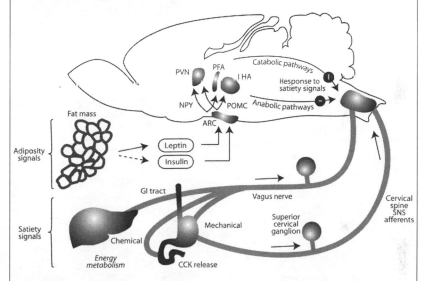

Figure 7.1.1. Schematic diagram of the signals that control caloric homeostasis. Satiety signals arising in the periphery, such as gastric distension and CCK, are relayed to the nucleus of the solitary tract (NTS) in the hindbrain. Leptin and insulin, the two circulating adiposity signals, enter the brain and interact with receptors in the arcuate nucleus (ARC) of the hypothalamus and other brain areas. These adiposity signals inhibit ARC neurons that synthesize NPY and AgRP (NPY cells in the diagram) and stimulate neurons that synthesize proopiomelanocortin (POMC), the precursor of α-MSH. These ARC neurons in turn project to other hypothalamic areas, including the paraventricular nuclei (PVN) and the lateral hypothalamic area (LHA). Catabolic signals from the PVN and anabolic signals from the LHA are thought to interact with the satiety signals in the hindbrain to determine when meals will end. (From Schwartz et al. 2000.)

If individuals are administered an antagonist to CCK receptors prior to eating, they eat a larger meal, implying that endogenous CCK normally helps to limit meal size. Analogously, if CCK is administered prior to a meal, less food is eaten (Smith and Gibbs 1998). CCK is but one example of peptides secreted by the stomach and intestine during meals that act as satiety signals (table 7.1.1).

(Box 7.1 continued)

Table 7.1.1   A partial list of signals known to influence food intake

Signals arising from peripheral organs

Catabolic (satiety signals)	Anabolic
Leptin	Ghrelin
Insulin	
Amylin	
Cholecystokinin (CCK)	
Bombesin family (gastrin-releasing peptide or GRP, neuromedin B, bombesin)	
Glucagon	
Enterostatin	
Apolipoprotein AIV	
Somatostatin	
Peptide YY (PYY)	
Glucagon-like peptide 1 (GLP-1)	

Signals that act within the hypothalamus

Catabolic	Anabolic
Leptin	Neuropeptide Y (NPY)
Insulin	Galanin
Amylin	Corticosterone
Corticotropin-releasing hormone (CRH)	Cortisol
Urocortin	Dopamine
Urocortin II	Melanocyte-concentrating hormone (MCH)
Neurotensin	Orexins
Oxytocin	Ghrelin
Serotonin	Agouti-related peptide (AgRP)
Histamine	Beacon
Glucagon-like peptide 1 (GLP-1)	Cannabinoids
Glucagon-like peptide 2 (GLP-2)	β-Endorphin
Tumor necrosing factor-α (TNF-α)	Dynorphin
Interleukin-6 (IL-6)	Norepinephrine
Interleukin-1 (IL-1)	Amino acids
Peptide YY (PYY)	
α-Melanocyte-stimulating hormone (α-MSH)	
Cocaine-amphetamine related transcript (CART)	
Prolactin-releasing hormone (PRL-RL)	

(Box 7.1 continued)

At least one stomach-produced signal has the opposite effect. Ghrelin is a hormone secreted from gastric cells just prior to the onset of an anticipated meal, and its levels fall precipitously once eating is initiated. Exogenously administered ghrelin stimulates eating, even in individuals that have recently eaten (Cummings et al. 2001). Hence, ghrelin is unique among the signals that have been described that arise in the gastrointestinal tract and influence food intake, since all of the others act to reduce meal size (see table 7.1.1). An important and as yet unanswered question concerns the signals that elicit ghrelin secretion from the stomach. It is probable that the brain ultimately initiates ghrelin secretion from the stomach at times when eating is anticipated.

The second group of signals controlling food intake is related to the amount of stored energy in the body. The best known of these "adiposity" signals are the pancreatic hormone insulin and the fat cell hormone leptin. As depicted in figure 7.1.1, each is secreted into the blood in direct proportion to body fat, each enters the brain from the blood, and receptors for each are located in the arcuate nucleus of the hypothalamus in the brain. When either leptin or insulin is administered directly into the brain near the arcuate nucleus, individuals eat less food and lose weight in a dose-dependent manner. Likewise, if the activity of either leptin or insulin is reduced locally within the brain, individuals eat more and become quite obese (Schwartz et al. 2000; Woods et al. 1998). Hence, both leptin and insulin could hypothetically be used to treat human obesity, but only if they could be administered directly into the brain, since their systemic administration has proved relatively ineffective and elicits unwanted side effects.

The third category of signals controlling energy homeostasis includes neurotransmitters and other factors arising within the brain. These signals are generally partitioned into those with a net anabolic action and those with a net catabolic action. When their activity is stimulated in the brain, anabolic signals increase food intake, decrease energy expenditure, and increase body weight. In contrast, when the activity of catabolic signals is enhanced in the brain, anorexia and weight loss occur (fig. 7.1.2). While numerous neuropeptides and other neurotransmitters have been reported to alter food intake (see table 7.1.1), a few will serve as examples. Neuropeptide Y (NPY) is synthesized in neurons throughout the brain and peripheral nervous system. One of the more important sites of synthesis with regard to energy homeostasis is the arcuate nucleus of the hypothalamus, where NPY-synthesizing cells contain receptors for both leptin and insulin (see

(Box 7.1 continued)

figs. 7.1.1 and 7.1.2). These NPY neurons in turn project to other regions of the hypothalamus, where they stimulate food intake and reduce energy expenditure; administering exogenous NPY near the hypothalamus results in robust eating (Schwartz et al. 2000; Woods et al. 1998).

A separate and distinct group of neurons in the arcuate nucleus also has receptors for both leptin and insulin, but these neurons synthesize a peptide called proopiomelanocorticotropin (POMC). POMC, in turn, can be processed to form any of a large number of active compounds. POMC neurons in the arcuate nucleus process the molecule into α-melanocyte-stimulating hormone (α-MSH), a potent catabolic signal (see fig. 7.1.2). Like NPY,

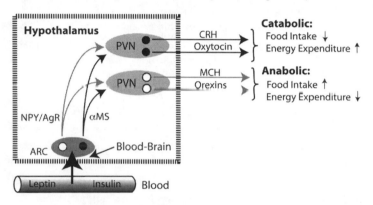

Figure 7.1.2. Hypothalamic circuits that influence caloric homeostasis. The adiposity hormones, leptin and insulin, are transported through the blood-brain barrier and influence neurons in the arcuate nucleus (ARC). ARC neurons that synthesize and release NPY and AgRP are inhibited by the adiposity signals, whereas ARC neurons that synthesize and release α-MSH are stimulated by the adiposity signals. NPY/AgRP neurons are inhibitory to the PVN and stimulatory to the LHA, whereas α-MSH neurons are stimulatory to the PVN and inhibitory to the LHA. The PVN, in turn, has a net catabolic action, whereas the LHA has a net anabolic action.

α-MSH is released in other hypothalamic areas, where it elicits reduced food intake, increased energy expenditure, and loss of body weight. An important feature of this network is that α-MSH causes its catabolic actions by stimulating melanocortin (MC) receptors (specifically, MC3 and MC4 receptors). Activity of these same receptors can be reduced by a different neurotransmitter called agouti-related peptide (AgRP), which is also made in the arcuate nucleus; specifically, within the same neurons that synthesize NPY. Thus, arcuate POMC neurons, when stimulated by increased leptin and insulin (as occurs if one gains a little extra weight), release α-MSH at MC3 and MC4 receptors to reduce food intake and

(Box 7.1 continued)

body weight. At thesame time, elevated leptin and insulin inhibit arcuate NPY/AgRP neurons. If insulin and leptin levels decrease (as occurs during fasting and weight loss), the POMC neurons are inhibited and the NPY/AgRP neurons are activated. The NPY stimulates food intake while the AgRP inhibits activity at the MC3 and MC4 receptors. This complex system therefore helps to keep body weight relatively constant over time, and the transmitters involved (NPY, AgRP, and α-MSH) are but three of a long list of transmitters that influence the system (Schwartz et al. 2000; Woods et al. 1998).

Integration of the Different Categories of Signals

An area of considerable research activity at present is determining how the various types of signals interact to control energy balance. The picture that is emerging is that most regulation occurs at the level of meal size. That is, there is flexibility with regard to when meals begin, since most evidence suggests that idiosyncratic factors based on convenience, environmental constraints, and experience are more influential than energy stores in determining meal onset (Woods 1991). However, once a meal starts and food enters the body, satiety signals are secreted, and as they accumulate, they eventually create a sufficient signal to terminate the meal (Smith and Gibbs 1998). Evidence suggests that the sensitivity of the brain to satiety signals is in turn regulated by adiposity signals. That is, when leptin and insulin are relatively elevated (as occurs if one has recently gained weight), the response to signals such as CCK is enhanced. In this situation, meals are terminated sooner and less total food is consumed, leading to a loss of weight over time. Conversely, when leptin and insulin are decreased (as occurs if one has lost weight), there is reduced sensitivity to satiety signals, and meals tend to be larger. Many other factors, of course, interact with this system. For example, seeing (or anticipating) a particularly palatable dessert can easily override the signals so that an even larger meal can be consumed.

It is important to remember that the biological controls summarized in this short review must be integrated with all other aspects of an individual's environment and lifestyle. Because of other constraints, the actual effect of satiety and adiposity signals is not always apparent when food intake is assessed on a meal-to-meal basis. Rather, energy balance (the equation of intake and expenditure in order to maintain a stable body weight) becomes evident in humans only when assessments are made over several-day intervals (de Castro 1988).

(Box 7.1 continued)

Although most of the research on the signals that control food intake has used humans, rats, or mice as subjects, sufficient analogous experiments have been performed on diverse groups of mammals as well as on several species of birds and fish, and the results are quite consistent with the conclusions above. Another important point that has recently come to light is that the same intercellular as well as intracellular signals that control energy homeostasis in mammals have been found to have comparable functions in many invertebrates, including insects and roundworms, as well as in yeasts (see review in Porte et al. 2005). What differ are the sources of energy used by different organisms and the foraging methods used to obtain them.

7.3 Forms of Energy Storage and Regulation

Food Stored in the Gut

The digestible contents of the gut will eventually become available as energy and can be considered an energy store. The supply varies depending on how much and how recently an animal has eaten. During winter, food in the crop of the willow ptarmigan (or red grouse), *Lagopus lagopus*, weighs on average 15% of body mass, enough to sustain the grouse for 24 hours (Irving et al. 1967). Yellowhammers (*Emberiza citrinella*) fill their crops with wheat before going to roost in early winter (Evans 1969). The arctic redpoll (*Carduelis hornemanni*) has a larger crop than similar species of southern latitudes, presumably because extra stores are more important in a cold climate (White and West 1977). However, in most species of small birds, food stored in the crop is a minor energy reserve.

Fat and Carbohydrates

Animals cannot store food in the digestive tract for very long. Even a large animal will digest the contents of its crop or stomach relatively quickly, and its blood glucose level will soon fall unless the animal consumes more food. Glycogen lasts longer, but animals can store only limited amounts. In order to build up larger or longer-lasting energy supplies, animals must either gain body fat or hoard food outside the body.

Animals commonly store lipids as fat and carbohydrates as glycogen, while plants normally store lipids as oils and carbohydrates as starches. Some marine organisms store waxes (Pond 1981). In most animals, carbohydrates primarily

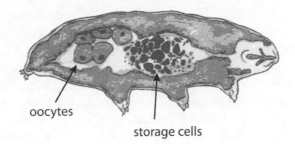

oocytes

storage cells

Figure 7.1. A tardigrade with the body cavity around the gut and the gonad (here with five oocytes) filled with a large number of circular storage cells that contain fat and carbohydrates. These cells represent the system for both energy storage and circulation in tardigrades. The storage cells show a distinct pattern of buildup and utilization of energy stores (reflected by variation in the sizes of the cells) strongly connected with the cycle of egg maturation. (After a photo by K. I. Jönsson.)

serve as fuel for short-term, high-intensity work, since they generate more energy per oxygen molecule than does fat. Fat, on the other hand, is better for long-term storage in the body. Being hydrophobic, it contains twice as much energy per unit weight as the hydrophilic carbohydrates (Weis-Fogh 1967). Animals can also metabolize proteins to produce energy, although these mainly serve other functions.

Many examples of energy storage come from studies on birds and mammals, but invertebrates also store energy. Tardigrades have special cells for storing fat and glycogen (fig. 7.1). These small animals use the energy in these cells for reproduction. The storage cells vary in both size and contents. When the tardigrade reproduces, the cells shrink or disappear and growing eggs take their place (Jönsson and Rebecchi 2002). Vetch aphids (*Megoura viciae*) store lipids in special fat cells and use this energy for reproductive investment (Brough and Dixon 1989). Benthic amphipods of several species (*Pontoporeia* spp. and some close relatives) accumulate lipids during the spring diatom bloom (Hill et al. 1992). Some amphipod species may store lipids in their bodies for as long as a year. Amphipods use these stores during periods of food scarcity, reproduction, and metamorphosis.

Insects that normally fly long distances use fat deposits as fuel, while those that normally only make flights of short duration use carbohydrates (Yuval et al. 1994). In the mosquito *Anopheles freeborni*, male mating success depends on swarming ability (Yuval et al. 1994). Swarming occurs after sunset, and the males feed on nectar after swarming. Since the next swarming flight will not occur until the following evening, the mosquito must store energy, primarily in the form of glycogen, for the rest of the night and the following day. The mosquitoes also have body lipid stores, but they use these for resting metabolism and not for flight.

Animals also use carbohydrates as short-term fuel and fat as long-term fuel in many contexts other than flight. Wood frogs (*Rana sylvatica*), for example, breed explosively during a mating period that lasts only 3–5 days (Wells and Bevier 1997), fueled by large glycogen reserves in muscle tissue. The males do not feed during the breeding period, being preoccupied with calling and searching for females. Spring peepers (*Pseudacris crucifer*), on the other hand, have a prolonged mating period that may last up to 2 months. During this period, males call at extremely high rates—3,000 to 4,000 notes per hour. Males draw 90% of the energy used for calling from fat and only 10% from glycogen (Wells and Bevier 1997). Most hibernating animals rely on fat for their winter metabolism, though carbohydrates can also be important in this respect. In the common frog (*Rana temporaria*), glycogen forms 40%–50% of the energy stores at the onset of hibernation and supplies 20%–30% of the energy metabolized during the winter (Pasanen and Koskela 1974).

Two forms of avian fat regulation have attracted special interest from researchers: migratory fattening and fat regulation in wintering songbirds. Box 7.2 deals with migratory fattening, and we develop some specific models of winter fat regulation in this chapter. Some bird species require large fat reserves for reproduction. Northern populations of geese build up larger fat deposits for breeding than southern populations (Mainguy and Thomas 1985). In harsher northern environments, geese must rely on fat for both yolk production and the female's own metabolism. At more southerly latitudes, the earlier growth of vegetation can support the female's metabolism during incubation, but females must still rely on fat for yolk production.

BOX 7.2   Energy Stores in Migrating Birds

Åke Lindström

Humans imagine migrating birds as free and unfettered in long and spectacular flights, but the truth is a little more prosaic: most of a migrant's time is spent on the ground. As much as 90% of its total time, and 66% of its total energy, is spent on foraging and resting ("stopovers") before and between migratory flights (Hedenström and Alerstam 1997). Migration can therefore be seen largely as a foraging enterprise, now and then interrupted by flight.

The long flights of migrating birds would not be possible without the deposition of extensive fuel stores. Even swallows, masters of feeding

(*Box 7.2 continued*)

while in flight, put on substantial fuel stores during migration (Pilastro and Magnani 1997), presumably because they and other migrants often cross large ecological barriers where foraging is not possible at all, such as oceans and deserts. Migrants on stopovers must work hard and consume much more food than usual to deposit the necessary fuel. Accordingly, foraging capacity and conditions during stopovers are crucial for successful migration. The constitution of avian fuel stores, the amount and rate of fuel deposition, and the rate of foraging and energy acquisition during fuel deposition are therefore of particular interest to researchers trying to understand bird migration.

What Kind of Fuel?

It has long been thought that birds use only fat as their fuel for migration. This makes sense, since fat is by far the most energy-dense fuel available. Although fat catabolism is indeed responsible for about 95% of the energy used for flight, some protein is also metabolized during flight. Therefore, it is appropriate to speak of "fuel" rather than "fat" deposition.

About 30% of the total mass loss during a flight (and subsequent mass increase during a stopover) may be due to protein catabolism (Jenni and Jenni-Eiermann 1998). The protein fuel is "stored" as active tissue, mainly in muscles, liver, gut, and heart. Some level of protein catabolism may be physiologically necessary for the active animal, but the rapid cyclic metabolism of organs may mainly reflect adaptive rebuilding of the bird's body (Piersma and Lindström 1997). During flight, the birds have a large "flying machine" (muscles and heart), whereas digestive organs are small to avoid extra flight costs. During stopovers, the birds have a large "eating machine" (gut, intestines, liver), whereas heart and flight muscles are relatively small.

How Much Fuel?

The size of migratory fuel stores varies enormously between individuals and species, from very small (5%–10% above lean body mass) to huge (> 100% above lean body mass; Alerstam and Lindström 1990). That is, some birds more than double their mass before they take off for a migratory flight. Fuel stores for migration are regularly much larger than stores for winter survival, which rarely exceed 50% (Biebach 1996). Obviously, many birds do not store as much fuel in winter as they are physically capable of.

(Box 7.2 continued)

Numerous factors influence the amount of fuel stored by a migratory bird. The minimum is obviously set by the distance that needs to be covered, especially when migrants must cross ecological barriers (Alerstam and Lindström 1990). Stores may also be larger than the minimum set by distance, as a safety measure against potentially unfavorable arrival conditions (Gudmundsson et al. 1991). Other strategic decisions that influence the size of fuel stores relate to how much (or rather, how little) time and energy ideally should be spent on migration (Alerstam and Lindström 1990). If birds try to minimize time spent on migration, maximizing the speed of migration to reach the destination as soon as possible, then they should put on more fuel at a given site the faster the rate of fuel deposition (Lindström and Alerstam 1992). If minimizing energy expenditure is more important, they should put on relatively small stores, independently of fuel deposition rate (Dänhardt and Lindström 2001). The risk of predation may also be an important factor to take into account. One way to minimize predation risk is to keep fuel stores small, reducing the negative effects of weight on maneuverability and takeoff ability (Kullberg et al. 1996).

The upper limit to the size of fuel stores is set by the capacity for takeoff and flight (Hedenström and Alerstam 1992). Some migrants have been reported as being so heavy that they could barely take off from the ground (Thompson 1974). At the other end of the spectrum, poor feeding conditions may almost preclude fuel deposition. The smallest fuel stores reported (10%) are found in irruptive species ("invasion species") such as tits, woodpeckers, and crossbills (Alerstam and Lindström 1990). This is not surprising, however, since these birds are on the move because of food shortage in the first place.

Rate of Fueling?

When time is short, which it may be for migrants that need to cover great distances during a short migration period, the fueling rate is crucial. The fueling rates reported for migratory birds are normally 0%–3% per gram of lean body mass per day. For example, a 100 g lean bird adding 3 grams per day has a fueling rate of 3%. For this bird, it takes 20 days to put on 60% fuel. The highest fueling rates known in wild birds are 10%–15% (Lindström 2003).

The maximum fueling rate is achieved when the food intake rate is maximized and the energy expenditure rate is minimized (the minimum possible energy expenditure rate is the basal metabolic rate, BMR). Maximum

(Box 7.2 continued)

fueling rates are negatively correlated with body mass, being 10%–15% in small birds (less than 50 g) and 1%–2% in large birds such as geese (more than 1 kg). The explanation for this important relationship is as follows. The maximum energy intake rates of animals are about 5–6 times BMR, independently of body mass (Kirkwood 1983). BMR scales allometrically (the energy turnover rate per gram decreases with increasing body mass), so fueling rates are lower in larger birds. As a result, a small songbird with a fueling rate of 10% can reach a given proportional fuel load—for example, 50%—in 5 days, whereas a large goose with a rate of 2% will need 25 days to reach the same fuel load. On average, the relative amount of fuel needed to cover a given distance is independent of body size (for example, a 40% fuel load is needed to cover 2,000 km). Therefore, fueling rates largely determine the speed of migration. Large birds may thus be limited in how far they have time to migrate within a given migration season.

The actual rate of fueling in a migrant is most often determined by food abundance. However, some migrants experience unlimited food supplies, such as spilled seeds on fields and invertebrate eggs and larvae on beaches. In these birds, it is mainly the capacity of the digestive system that limits fueling rates (Lindström 2003). In addition, the amount of time per day that feeding is possible is important (Kvist and Lindström 2000). For diurnal feeders, it is therefore advantageous to migrate when days are long (for example, at high latitudes in summer).

Migratory birds in captivity display many traits that they would in the wild; for example, they consume large amounts of food whenever possible. Such studies have shown that migratory birds have among the highest energy intake rate capacities measured in any homeothermic animal (Kvist and Lindström 2003). Intake rates of up to 10 times BMR have been measured. A contributing factor is certainly the capacity to rapidly enlarge the digestive organs during fueling. Natural selection has obviously favored traits that make large energy turnover rates possible during migration.

Some female pinnipeds fast during lactation so that they can remain with their pups. Female gray seals (*Halichoerus grypus*) lactate for 16 days. Their milk contains 60% fat, and the pups gain an average of 2.8 kg per day, most of it as body fat (Boness and Bowen 1996). This weight gain allows the pup to stay on the ice until it has molted and is ready to go to sea. During her fast, the mother uses fat in the blubber layer and loses almost 40% of her body mass (Iverson et al. 1993).

Food Hoarding

Some species accumulate external food reserves, typically called hoards or caches, that they can use as substitutes for or supplements to energy reserves stored in the body. In honeybees (*Apis mellifera*), queen and workers survive the winter by eating honey that they stored in autumn. To make sure that there is enough food for the hive, workers usually kill the drones, which the hive no longer needs, but if honey stores are large, the workers may allow the drones to live (Ohtani and Fukuda 1977). Under cold conditions, the bees form a cluster so that a dense mantle of workers insulates the brood (Michener 1974; Seeley 1985). To save energy, the bees actively reduce the oxygen level in the hive, thereby reducing their metabolic rate. In cold weather, the hive may be nearly dormant, with an oxygen level of only 7.5% in the core (van Nerum and Buelens 1997). The bees can also increase the hive's temperature by active heat production, such as movements of the flight muscles (Michener 1974; Seeley 1985).

European moles (*Talpa europeae*) store earthworms in underground "fortresses" (Funmilayo 1979). The mole decapitates the worm and pushes its front end into the earth wall. Without a front end, the worm cannot move, and it stays alive and fresh until it is eaten, often after several months. A single mole may store over a kilogram of worms in this way, which serves as an important energy reserve (Skoczen 1961).

Beavers (*Castor fiber* and *C. canadensis*) stay in their lodges most of the winter. During this time, they exploit caches of preferred foods, such as twigs and branches of aspen (*Populus* spp.), birch (*Betula* spp.), and hazel (*Corylus* spp.). They stick the branches vertically into the bottom mud or stock them under floating rafts that they construct of less palatable trees (Doucet et al. 1994). The rafts and the upper ends of the vertical branches will freeze into the ice, and the palatable underwater parts will then become a safe underwater supply of winter food (Vander Wall 1990).

Male northwestern crows (*Corvus caurinus*) store mussels found at low tide. The stores ensure that the crows can eat mussels even when the high tide makes them unavailable. Males feed incubating females stored mussels, which makes it possible for females to stay on their eggs (James and Verbeek 1984). The South Island robin (*Petroica a. australis*) stores earthworms during the early morning when they are most available. Robins eat the stored worms later the same day (Powlesland 1980).

Regulation of Energy Expenditure

An alternative to increasing the amount of stored energy is to reduce energy expenditure. Since energy stores will last longer if an animal reduces its metabolic rate, strategies such as hibernation, torpor, and hypothermia are closely connected to energy storage. We will discuss such strategies that mainly aim to reduce energy expenditure in this chapter. Aestivation, or summer torpor, is a functionally equivalent way to escape drought or high temperatures.

In temperate and boreal regions, ectotherms and many small endotherms hibernate by entering a state of torpor. Their body temperatures may be close to zero and their heart rates reduced to only a few strokes per minute. Endotherms that hibernate are typically small, insectivorous mammals, such as bats and hedgehogs. Some birds, such as hummingbirds and nightjars, also use torpor to save energy. Large mammals such as bears and badgers "hibernate" with body temperatures only a few degrees below normal (Hissa 1997). The basis for this difference between small and large mammals is largely allometric. Larger animals have more heat-producing mass in relation to cooling surface, and hence can have lower metabolic rates, than small ones. Hibernation at a high body temperature requires large energy reserves, but has other benefits. A hibernating bear can flee or defend itself almost immediately if startled. In addition, pregnant females can give birth and lactate in the protected den, which would be impossible under torpor.

7.4   The Economy of Energy Reserves

Benefits of Energy Reserves

The previous section gave a sampling of the forms of energy storage. Energy storage allows animals to perform activities, such as sleeping or breeding, that are not compatible with foraging, to inhabit areas with temporarily harsh conditions, to survive periods of food shortage, and so on. Though the most obvious benefit of storing fat in the body is the energy that becomes available when it is metabolized, there are other possible benefits, such as insulation, protection, support, and social and sexual signals (Witter and Cuthill 1993). Furthermore, energy stores can provide an insurance benefit, even if the animal rarely has to metabolize them (Brodin and Clark 1997).

Long-term food hoarding provides a good example of how active energy regulation allows animals to inhabit temporarily harsh environments. Nutcrackers (*Nucifraga* spp.) spend most of the autumn hoarding food (Swanberg 1951; Tomback 1977; Vander Wall 1988), and they depend on this stored food during the winter. Hoarding makes their regular food source—pine seeds or hazelnuts—available during a predictable time of food shortage—the winter. When pine or hazelnut crops fail, nutcrackers turn up in large numbers in areas far from their breeding grounds (Vander Wall 1990). These massive emigrations illustrate the nutcrackers' dependence on stored food.

Family groups of acorn woodpeckers (*Melanerpes formicivorus*) maintain granaries of acorns consisting of specially excavated holes in tree trunks or telephone poles. They use the stored acorns during brief periods of food shortage, but not as a regular winter food source (Koenig and Mumme 1987). So, for acorn woodpeckers, food hoarding seems to be a hedge against unpredictable periods of low food availability. In contrast, nutcrackers need stored food to survive the predictable onslaught of winter.

These two benefits of food hoarding frequently act at the same time. The willow tit (*Parus montanus*) is a small boreal parid. Like nutcrackers, they store a large proportion of their winter food during autumn. An individual may store 40,000 to 70,000 items in one autumn (Haftorn 1959; Pravosudov 1985; Brodin 1994c). Other, less well-known parid species may store even more (Pravosudov 1985). Willow tits probably do not remember the specific locations of all these caches (Brodin and Kunz 1997). Instead, they place their caches in locations where they will forage during the winter (Brodin 1994b). These stores increase the hoarder's general winter food level (Brodin and Clark 1997) and, as in nutcrackers, they constitute a regular source of winter food (Haftorn 1956; Nakamura and Wako 1988; Brodin 1994c).

Besides this massive hoarding in autumn, willow tits also store smaller numbers of seeds if there is surplus food during the winter (Haftorn 1956; Pravosudov 1985; Brodin 1994c). Over shorter time periods, tits can remember the precise locations of seeds they have stored (e.g., Sherry et al. 1981). Tits can retrieve these remembered seeds more quickly than the larger store of unremembered seeds. They are too few to be a substantial energy source, but provide insurance against unpredictable conditions. Such small caches that are retained in memory may allow willow tits to maintain lower fat reserves than nonhoarding species, avoiding fat levels that would be costly to carry (Brodin 2000).

The importance of energy storage as a bet-hedging strategy increases as the environment becomes less predictable. Avian ecologists assume that ground foragers experience more variation in winter than tree-foraging species. Rogers (1987) compared fat reserves in species of similar size and physiology foraging in different habitats. He found that tree foragers carried smaller fat reserves than similar-sized species foraging on the ground.

Small birds in boreal regions are fatter in winter than the rest of the year (Lehikoinen 1987; Haftorn 1992). They also have a larger daily amplitude of mass gain and loss in winter, which depends on the fact that winter nights

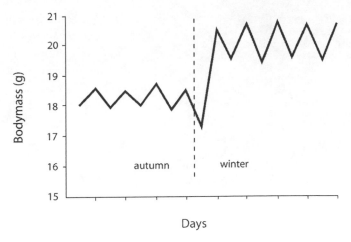

Figure 7.2. Winter fattening in small birds. The figure shows a hypothetical example with a sudden onset of winter (dashed vertical line) when temperatures fall below zero and the environment becomes covered with snow. Since nights in winter are longer and colder than in autumn, the amplitude of the daily weight gain and loss is larger, but minimum reserves are larger as well. This phenomenon was labeled winter fattening by Lehikoinen (1987).

are longer and colder than summer nights (fig. 7.2). Their reserves at dawn are higher in winter than in summer, meaning that the birds maintain a larger buffer against poor feeding conditions in winter, a phenomenon called winter fattening (Lehikoinen 1987). Winter fattening occurs both in the field (Rogers and Rogers 1990) and in the laboratory. Great tits (*Parus major*) increased their fat reserves in response to stochastic variation (Bednekoff et al. 1994; Bednekoff and Krebs 1995). Thus, stored energy serves both as a regular energy source and as a bet-hedging strategy.

Costs of Storing Energy

Acquiring and maintaining energy stores can be costly in several ways. In humans and domestic animals, excessive fat deposits can increase mortality, mainly through increased strain on the heart and vascular system (Pond 1981). An energy-storing animal spends time and energy foraging that it could have allocated to other behaviors. Furthermore, foraging may entail exposure to predators that the animal would not otherwise have experienced (see chap. 13).

Behavioral ecologists have extensively studied the costs of storing body fat in birds, both theoretically and empirically. Pravosudov and Grubb (1997) have reviewed energy regulation in wintering birds. Witter and Cuthill (1993) have reviewed the costs of carrying fat in birds, noting especially that mass-dependent costs may be important. Small birds should carry the smallest

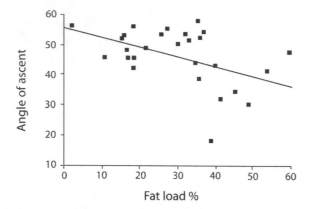

Figure 7.3. Angle of ascent in relation to fat load (as a percentage of fat-free body mass) in a warbler, the blackcap. To make these measurements, birds foraging in a cage were startled by an attacking artificial predator. (After Kullberg et al. 1996.)

reserves possible to escape an attacking predator, but they should carry the largest reserves possible to avoid starvation. This means that they face a trade-off between starvation and predation that may not be evident in nonflying organisms. In section 7.6 we explore this trade-off in detail.

Behavioral ecologists have focused on both mass-dependent predation risk and mass-dependent metabolic expenditure. Houston and McNamara (1993) have also suggested that body mass may reduce foraging ability, especially for birds that forage on the wing. Mass-dependent predation risk seems obvious; physical laws tell us that increasing fat loads must affect a bird's acceleration and takeoff angle. Kullberg et al. (1996) have shown this empirically using blackcaps (*Sylvia atricapilla*) (fig. 7.3). They measured takeoff angles and velocity during premigratory fattening, when fat loads were as large as 30%–60% of the lean body mass. It is less clear, however, whether smaller fat loads also affect takeoff ability. In boreal regions, wintering passerines gain about 10% of lean body mass in the course of every winter day and metabolize this store during the night when they cannot forage. Empirical evidence suggests that body mass fluctuations of this magnitude have little or no effect (Kullberg 1998a; Kullberg et al. 1998; Veasy et al. 1998; van der Veen and Lindström 2000; but see Metcalfe and Ure 1995). Either we cannot detect the effects of these small increases, or birds somehow compensate for the extra mass. Although we have no firm evidence, birds might compensate by increasing flight muscle tissue, and hence flight power, in parallel with fat. Lindström et al. (2000) have demonstrated a rapid buildup of wing muscles in parallel with fat reserves in migrating knots (*Calidris canutus*), so wintering passerines might do this as well. Small birds may be able to compensate for small or moderate fat loads, but probably not for large fat loads (fig. 7.4).

Changing environmental conditions may require that animals make major adjustments to their energy reserves. In autumn, migrating or hibernating animals require large fat reserves. Animals that spend the winter at northern latitudes build up larger minimum reserves in winter than they carry in summer and autumn. Houston et al. (1997) and Cuthill and Houston (1997) labeled the costs of such seasonal transitions "acquisition costs," whereas they called costs emanating from the daily regulation of reserves "maintenance costs." If we consider the daily fluctuations in figure 7.2, it is clear that fat is acquired and lost on a daily as well as a seasonal basis. This means that "maintenance costs" may also result from the acquisition of fat. The main difference is that acquisition costs result from increasing the average level of reserves, rather than just compensating for daily fluctuations.

Hoarding food externally also incurs costs. Hoarding will be wasted effort if precipitation, temperature, or microorganisms cause stores to spoil. Honeybees invest considerable time and work in converting stored nectar into a more durable form, honey. They produce an enzyme that converts simple sugars into more concentrated forms that have antibacterial effects (e.g., Vander Wall 1990).

An important ecological consideration is that competitors can steal hoarded supplies. To reduce theft, hoarders can defend larders or scatter caches widely. Typical larder hoarders are small burrowing mammals such as various rodents (Rodentia), pygmy possums (*Burramys parvus*), shrews (Soricidae), and pikas (Ochontidae) (Vander Wall 1990). Larder hoarders can easily retrieve stored

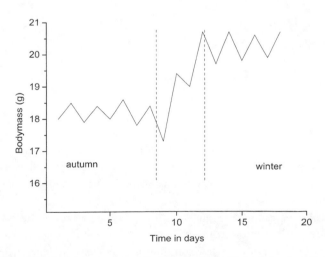

Figure 7.4. The effect of body fat mass on predation risk as suggested by Brodin in a theoretical model. The x-axis shows fat as a percentage of lean body mass. At low levels of fat, a bird can compensate for the extra mass carried by increasing its flight muscle tissue. (After Brodin 2001.)

items, while scatter hoarders face a more challenging retrieval problem. But the consolidation that makes retrieval from a larder so simple also means that the whole supply can be lost if a larger competitor finds the larder. In eastern chipmunks (*Tamias striatus*), only individuals that can defend a burrow store food in larders. Newly emerged juveniles scatter-hoard until they become older and stronger (Clarke and Kramer 1994).

Scatter hoarders do not risk losing all their stored items if a competitor discovers a cache, but they need some mechanism for retrieval of their concealed and scattered caches, which can also be costly. The most accurate way to retrieve cached items is probably to remember their exact locations. However, if thousands of caches are stored for several months, this may require special adaptations of spatial memory. Implementation of memories may require repair of neurons and synapses, redundancy or backup in the form of extra brain tissue, and so on. Dukas (1999) discusses the potential costs of memory.

As mentioned earlier, animals can reduce energy expenditures instead of building up energy stores, but this strategy also incurs costs. In winter, small birds at northern latitudes frequently use nocturnal hypothermia to save energy (e.g., Haftorn 1972; Reinertsen 1996). Small passerines use their high metabolic rate to achieve body temperatures of up to 42°C. A 10°C reduction in nighttime body temperature can save a considerable amount of energy. Hypothermia, however, might also be risky. At dawn, it may take 15 minutes to regain a normal body temperature, and the bird might be vulnerable to predation during this warm-up period. We know little, however, about the possible costs of nocturnal hypothermia (see section 7.7).

7.5   Modeling Energy Storage

Optimization models can help us understand the selective forces that have shaped energy storage and expenditure strategies. Such models have become standard in evolutionary and behavioral ecology (Stephens and Krebs 1986; Mangel and Clark 1988; Bulmer 1994; Houston and McNamara 1999) and range from simple analytic to complex computer models. While analytic models may be appropriate for studying foraging efficiency, they seldom provide sufficient detail for studies of the acquisition, storage, and use of energy supplies.

As a rule, we cannot measure the fitness consequences of stored energy directly. Instead, we must use some measurable currency that, we assume, is ultimately linked to fitness. Foraging models typically use currencies based on averages, such as the average net rate of energy gain (rate maximization) or the average time required to obtain the necessary daily food intake (time minimization). Models of energy storage have used the net rate of energy gain (e.g., Lucas and Walter 1991; Tamura et al. 1999), the ratio of energy gained to energy spent (Wolf and Schmid-Hempel 1990; Waite and Ydenberg 1994a, 1994b), or survival rate (Lucas and Walter 1991). We will use the probability of survival to the end of winter as the fitness currency in the dynamic models in this chapter. In cases in which winter mortality is high, it is reasonable to assume that winter survival is directly related to Darwinian fitness. In other cases, ending the winter with adequate reserves for future activities may also be important; for example, in models that include breeding events.

As section 7.4 shows, collecting food to store is costly. We can model these costs in various ways, depending on the currency and the aim of the model. Sometimes it may be convenient to see these costs as a probability of death, while at other times it may be more convenient to see them as energy losses. We will give two specific examples here.

In a model that aimed to investigate the potential effects of dominance rank on optimal food hoarding effort, Brodin et al. (2001) assumed that the cost of food hoarding consisted of an increase in predation risk while foraging. In this model, hoarding in autumn increased winter survival by making more winter food available at the same time as it reduced present survival in autumn by a probability of death, P_D. If predation risk is proportional to the amount of food stored, h (or more generally, foraging effort), this probability can be expressed as

$$P_D = 1 - e^{-kh} \tag{7.1}$$

(modified from Schoener 1971). Here k is a scaling constant. The probability of survival then becomes

$$1 - P_D = e^{-kh}, \tag{7.2}$$

which can be multiplied by some fitness measure.

In some cases, it might be better to model costs as energy losses. In a field experiment on hoarding gray jays (*Perisoreus canadensis*), Waite and Ydenberg (1994b) used the time and energy spent hoarding as costs. The net rate of storing, γ_H, is

$$\gamma_H = \frac{\overline{g}_H p_R - C_e - C_T}{t_H + t_W}, \tag{7.3}$$

where $\overline{g}_H$ is the average energetic gain from one cache, p_R the probability of recovering it, c_e the energetic cost of transporting and storing it, c_T the energetic cost of waiting for food at the feeder (a time cost controlled by the

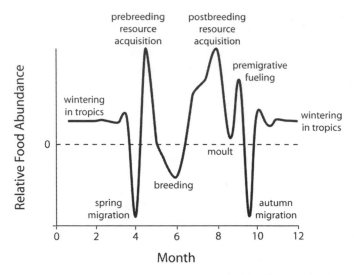

Figure 7.5. A hypothetical graph of a migratory bird's daily food availability (solid curve) in relation to its average energy requirements (dashed line) over a year. During some periods food availability exceeds energy requirements, while food availability falls below energy requirements on other occasions.

experimenters), t_H the time needed to store one cache, and t_W the manipulated waiting time.

A Graphical Paradigm

A graph (fig. 7.5) of an animal's daily food availability and energy requirements over a year shows periods of positive energy balance (food availability exceeds energy requirements) interspersed with periods of negative balance (food availability falls below energy requirements). Prolonged periods of positive energy balance might coincide with breeding episodes, whereas periods of negative energy balance would place emphasis on survival. This chapter focuses on periods of potential negative energy balance. Such periods must follow periods of positive energy balance because animals need to build energy reserves for use during subsequent periods of negative energy balance.

This graphical paradigm oversimplifies the problems of energy storage and retrieval in several respects. For example, a simple graph of the type in figure 7.5 cannot indicate uncertainty. In reality, the supply of and demand for energy resources may fluctuate randomly (though with predictable, seasonally dependent patterns) on both long-term and short-term time scales. Exceptionally high food availability during a period of positive energy balance may result in above average reproductive success. Conversely, low food availability during the normally productive season may limit reproduction and lead to increased risk of mortality. Under such circumstances, parents may sacrifice current reproduction to enhance survival.

ESS Models

In an influential paper, Andersson and Krebs (1978) showed the necessity of a recovery advantage for hoarders over nonhoarding conspecifics for hoarding to constitute an evolutionary stable strategy (ESS). In a group of foragers of size n, it is necessary that

$$F_H(n_H) > F_{NH}(n_H). \tag{7.4}$$

F_H is the fitness of a hoarder in a group with n_H hoarders, and F_{NH} is the fitness of nonhoarders in the same group. For hoarding to be an ESS, the probability that the hoarder will find its own cache, p_H, must exceed the probability that a scrounger will find the cache, p_S, by

$$\frac{p_H}{p_S} > \frac{C}{G}(n-1)+1, \tag{7.5}$$

where C is the cost of hoarding one item and G the gain from eating it in the future. In addition, p_H must exceed the probability that an unstored item will remain available until retrieval:

$$p_H > \frac{C}{G} + p_N m, \tag{7.6}$$

where m is a deterioration factor (e.g., decay) and p_N is the probability that a food item that was not stored will remain available.

If a hoarder stores food in a location where any member of the group is equally likely to find it, the hoarder will be at a disadvantage. If stored supplies are communal property, the individuals that refrain from assuming the costs of hoarding will gain the same benefit from the stored supplies as others. Even if population size is decreasing due to a lack of stored food, a hoarder will always do worse than a nonhoarder will.

The probability that a hoarder will retrieve its own cache, p_H, can be divided into two probabilities: the probability that the cache will be found, p_f, and the probability that a stored item will remain until retrieval, p_r (Moreno et al. 1981):

$$p_H = p_f p_r. \tag{7.7}$$

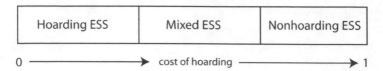

Figure 7.6. In a small group of foragers, as the cost of hoarding increases from 0 to 1 (= death), the ESS will change. If the cost of hoarding is low, hoarding may be the ESS even if there is no recovery advantage for hoarders. (After Smulders 1998.)

With substitution into equation (7.6), this gives

$$p_f > \frac{p_N}{p_r}m + \frac{C}{Gp_r}. \tag{7.8}$$

A low probability that food that is not stored will remain where it is found, p_N, will decrease the right-hand term and may make hoarding worthwhile even if retrieval probability is low.

In a game theoretical model of the "playing against the field" type (Maynard Smith 1982), Smulders (1998) showed that under certain conditions, hoarding may evolve in species that forage in small groups even when all members of a foraging group share caches equally. If hoarding provides a net benefit, a rare hoarding mutant can always invade a pure population of nonhoarders. On the other hand, nonhoarders will have the highest fitness in groups that contain both strategies because they do not pay the costs of hoarding. Even if nonhoarders have higher fitness within the group, the global fitness of nonhoarders will be low if other groups do not contain hoarders. The reason is that nonhoarders must pilfer from hoarders to achieve their higher fitness. The result is a mixed ESS (fig. 7.6).

Modeling hoarding effort as a function of dominance rank, Brodin et al. (2001) showed that differences in hoarding behavior in a population may depend on rank. Dominants can steal caches more easily from subordinates than vice versa, so dominants should be less willing to pay the costs of hoarding. Thus, under the right conditions, subordinates should hoard more than dominants. According to this model, dominant group members should not hoard at all unless they have a recovery advantage. Without a recovery advantage, dominants should specialize in pilfering. In addition, compared with that of subordinates, the dominants' investment in hoarding should increase as the environment gets harsher (fig. 7.7).

7.6   Dynamic State Variable Models

Neither the graphical paradigm nor analytic models consider how an organism might alter the two quantities, resource supply and energy expenditure, over the year. In our modeling approach, we consider strategies that either alter the natural resource supply by means of energy storage, or alter energy expenditure by methods that reduce energy use, or both. Our approach allows animals to store energy reserves either internally—for example, as body fat—or externally—as food caches or hoards. Methods of reducing energy expenditure include wintering in the tropics, physiological changes, the use of hypothermia, and hibernation.

Such problems require dynamic state variable models (Houston and McNamara 1999; Clark and Mangel 2000), which explicitly involve temporal dynamics, including seasonal and stochastic effects, and also consider dynamic state variables, including the current levels of energy reserves, both internal and external. One can include other important aspects of energy storage and use, such as predation risk and social interaction, in these models. Behavioral ecologists have applied the dynamic state variable approach to many aspects of winter survival strategies for small birds in northern climates; we review this literature below.

State variables are essential components of these models. They characterize the organism's internal state and, at the same time, model the effects of short-term decisions on fitness. In models of energy storage typically, state variables may be fat reserves, gut contents, or cache sizes. These models determine optimal (i.e., fitness-maximizing) time- and state-dependent behavioral strategies. In the simplest cases, such models treat individual fitness

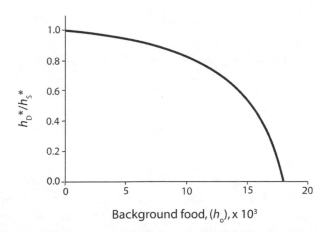

Figure 7.7. In a model by Brodin et al., animals of different rank experienced a period of food surplus (e.g., autumn) followed by a period of food scarcity (e.g., winter). The figure shows the ratio between optimal hoarding by a dominant animal, h_D^*, and optimal hoarding by a subordinate animal, h_S^*, as a function of unstored food available during the period of food scarcity, h_0. Decreasing background food (h_0) will have the same effect as making the environment colder, since both changes increase energy expenditure. (After Brodin et al. 2001.)

maximization, but game theoretical models may be required where fitness is frequency-dependent.

Dynamic game theoretical models can be quite complicated (e.g., Houston and McNamara 1999), so various simplifying assumptions are often employed. For example, the dynamic aspect may be suppressed, or a dominant-subordinate social structure may be assumed (e.g., Clark and Ekman 1995; Brodin et al. 2001).

Management of Avian Fat Reserves

In a series of papers, McNamara and Houston and their co-workers have used dynamic state variable models to study various aspects of avian fat regulation during winter (McNamara and Houston 1990; Houston and McNamara 1993; Bednekoff and Houston 1994a, 1994b; McNamara et al. 1994; Houston et al. 1997). These models have considered (1) optimal winter fat regulation strategies; (2) the sensitivity of overwinter survival, and of daily fat levels, to ecological parameters, such as seasonal changes in day length and ambient temperature; and (3) the relative risks of starvation and predation in bird populations during winter. As always in modeling, the modeler seeks an improved understanding of complex natural phenomena. The construction, analysis, and testing of models adds intellectual rigor to what is otherwise often little more than speculation.

The framework of McNamara and Houston's models includes the following basic assumptions:

1. Birds store energy reserves as body fat.
2. Birds die of starvation if their energy reserves fall to zero at any time.
3. Daily food intake and overnight metabolic costs vary stochastically.
4. Metabolic costs during daytime activities increase with body mass.
5. Predation risk while foraging also increases with body mass.
6. Natural selection has favored fat regulation strategies that maximize the probability of overwinter survival.

The starvation-predation trade-off discussed in section 7.4 is essential in these models. As Houston and McNamara (1993) point out, the fact that many birds maintain lower levels of fat reserves in winter than they could probably reflects the importance of this trade-off, rather than the occurrence of resource shortages. Lima (1986) made this point previously.

A Basic Dynamic State Variable Model

To begin, we will describe a basic dynamic state variable model of optimal energy reserves for a generic "small bird in winter" (McNamara and Houston 1990; Bednekoff and Houston 1994b). This basic model can be extended in many ways.

Time periods $d(d = 1, 2, \ldots, D)$ correspond to the days that make up the winter, with D being the last day. The state variable X denotes a bird's energy reserves in kJ and has a present value of x. $X_0(d)$, then, is the bird's energy reserves at dawn on day d. The daily foraging effort, $\varepsilon(d)$, represents the fraction of daylight hours spent actively foraging ($0 \leq \varepsilon(d) \leq 1$). This is also the decision variable; the foraging bird can "decide" how much time it should spend foraging. The rest of the time, $1 - \varepsilon$, is then spent resting. Food intake on day d is $\varepsilon(d)f(d)$, where $f(d)$ is a random variable (measured in kJ of usable energy). Thus, food intake depends on both foraging effort and environmental stochasticity.

Total daytime metabolic costs are

$$c(x, e) = c_1(x)\varepsilon + c_2(x)(1 - \varepsilon), \ x = X_0(d), \tag{7.9}$$

where $c_1(x), c_2(x)$ are activity and resting costs, respectively. Since c is a function of x, these costs depend on body mass; for example, fatter birds may have higher metabolic rates.

Let $X_1(d)$ be the bird's reserves at dusk, after a day's foraging. The expression

$$X_1(d) = X_0(d) + \varepsilon(d)f(d) - c[X_0(d), \varepsilon(d)] \tag{7.10}$$

relates reserves at dusk, $X_1(d)$, to reserves earlier that morning, $X_0(d)$. Fat reserves cannot be negative or immensely large. The inequality

$$0 \leq X_1(d) \leq X_{max} \tag{7.11}$$

constrains both X_0 and X_1, where X_{max} is the maximum capacity for body reserves. If reserves at dusk in equation (7.10) fall to zero or less, then starvation kills the bird.

Overnight metabolic costs are $c_n(d)$; for simplicity, these costs are random and serially uncorrelated, with stationary distribution. In a more realistic (and complex) model, we could instead assume that an unusually cold night was likely to be followed by another cold night. Before the next day $(d + 1)$, the bird will metabolize fat reserves. This is described by the overnight state dynamics,

$$X_0(d + 1) = X_1(d) - c_n(d),$$ (7.12)

subject to the constraints of equation (7.11); as before, the bird dies of starvation if reserves the next morning, $X_0(d + 1)$, fall to zero.

While foraging, the bird faces a predation risk, $\mu(x)$, that depends on body mass; the daily survival probability S [cf. equation (7.2)] then depends on both body mass and foraging effort:

$$S(x, \varepsilon) = \exp[-\mu(x)\varepsilon].$$ (7.13)

Hence, we have two sources of mortality, predation and starvation.

The fitness currency is the probability of survival to the end of winter at $d = D$. We define the function

$$F(x, d) = \max \Pr(\text{bird survives from day } d \text{ to } D), \text{ given } X_0(d) = x,$$ (7.14)

where the maximization refers to the choice of how much of the day the bird should spend foraging, $\varepsilon = \varepsilon(x, d)$. Since our model allows foraging decisions to depend on state, the bird can control its reserve level, for example, by increasing foraging activities when $X_0(d)$ is low and vice versa. As in all dynamic state variable models, we work backward from the last time period; in our case, the end of winter. We can find the probability of survival on the last day of winter (day D) from the assumption that the bird dies of starvation if $X_0(d) = x = 0$. This gives us the terminal condition,

$$F(x, D) = \begin{cases} 1 & \text{if } x > 0 \\ 0 & \text{if } x = 0 \end{cases}.$$ (7.15)

That is, the bird survives the winter if it has more than zero reserves on day D, and dies otherwise.

The dynamic programming equation for the day before the end of winter (for $d < D$) follows from the definition of $F(x, d)$ and the model specifications (cf. Clark and Mangel 2000):

$$F(x, d) = \max_{0 \le \varepsilon \le 1} S(x, \varepsilon) E[F(x', d + 1)].$$ (7.16A)

We now have an equation that can get us through the whole winter, since the function at day d connects to itself at day $d + 1$. The first step in our backward iteration would then be to calculate fitness for the penultimate day of winter, $D - 1$, when equation (7.16A) would become

$$F(x, D - 1) = \max_{0 \leq \varepsilon \leq 1} S(x, \varepsilon) E[F(x', D)]. \tag{7.16B}$$

On the right-hand side we have $F(x', D)$, which is either 1 or 0 [eq. (7.15)], depending on the value of x'. We can calculate x' from equation (7.10), which gives the change in fat reserves over the day, and equation (7.12), which gives the overnight energy loss. If we write these together, we get

$$x' = x + \varepsilon f(d) - c(x, \varepsilon) - c_n(d). \tag{7.17}$$

The expectation operator, E, in equation (7.16) may look complicated, but makes it more readable than if we explicitly expressed the probabilities for the two random variables $f(d)$, $c_n(d)$ that it encompasses. For simplicity reasons, modelers frequently use two complementary probabilities rather than some continuous distribution in this type of model. For example, the $c_n(d)$ variable could be a small energy loss during mild nights, $c_{n(GOOD)}$, with probability p_{GOOD}, and a larger loss during cold nights, $c_{n(BAD)}$, with probability $1 - p_{GOOD}$. The $\max_{0 \leq \varepsilon \leq 1}$ expression tells us that only the optimal foraging effort needs to be considered. The daily survival probability $S(x, \varepsilon)$ from equation (7.13) is a mass-dependent variant of equation (7.2). Fat reserves can never have negative values. Thus x' in equation (7.16) is replaced by 0 if the right-hand expression is negative. In addition, we have

$$F(0, d) = 0, \tag{7.18}$$

reflecting the assumption that the bird has died if $x = X(d) = 0$.

Equation (7.16) shows that the trade-off between starvation and predation risk that we discussed previously has two linked components. First, greater foraging effort ε increases reserves at dusk, $X_1(d)$, and this increases the probability of overnight survival. But greater foraging effort also increases predation risk, $\mu(x)\varepsilon$. Second, increased foraging activity also increases the expected level of future reserves, $X_0(d')$, where d' is any day between d and D. Increased reserves lower the risk of future starvation (for example, during a prolonged period of unusually cold nights), but they increase mass-dependent metabolic costs and predation risks.

If we imagine how natural selection optimizes foraging behavior with respect to this trade-off, intuition might suggest an optimum at which equally many birds die from starvation and predation. Field observations of northern passerines do not support this prediction: most birds that die in winter

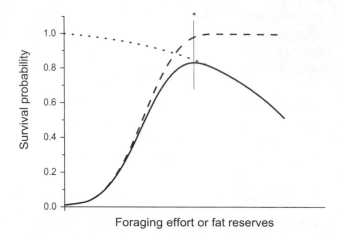

Figure 7.8. The effect of foraging effort (or fat reserves) on survival probability in wintering birds. Total survival (solid curve) depends on the combined effects of starvation risk (dashed curve) and predation risk (dotted curve) on survival. The optimal foraging effort (or optimal level of fat reserves) will be where the marginal values of these risks are equal.

die of predation, not starvation (Jansson et al. 1981). Only unusually extreme conditions result in starvation, but predation risk is always present and unavoidable. McNamara and Houston (1987b, 1990) provided a theoretical explanation for this observation. Total mortality, $\mu_T(x)$, at the level of fat reserves x is the sum of mortality from both starvation (S) and predation (P):

$$\mu_T(x) = \mu_s(x) + \mu_p(x). \tag{7.19}$$

The minimum of this equation is where the derivative equals 0, or

$$-\mu_s'(x) = \mu_p'(x). \tag{7.20}$$

Thus, the optimal foraging effort is where the rate of decrease in starvation equals the rate of increase in predation (fig. 7.8), or alternatively, where the marginal changes of these components are equal. However, even this may be vague, since it is not always clear what variable marginality should refer to, foraging effort or average level of reserves. Most models predict that the probability of starvation is much smaller than the probability of predation for individuals using the optimal strategy (see Bednekoff and Houston 1994b and Clark and Mangel 2000, chap. 5).

In this type of model, the predicted reserves will exceed the level needed to meet average nocturnal costs. The optimal strategy pays a "premium" to guard against nights with unusually high energetic demands. On nights with normal energetic costs, the model bird's body mass is higher than needed, and hence increases predation risk in the morning. This is a weak form of "bet hedging" (Seger and Brockmann 1987). Even though the model suggests that predation presents a greater risk than starvation, an extended period of unusually cold nights can cause the starvation of most members of the population.

7.7 More Complex Dynamic State Variable Models

Optimizing the trade-off between starvation and predation involves more than regulating fat reserves. Modelers have used the dynamic state variable approach to study several other aspects of the starvation-predation trade-off, including seasonal variations in the environment, daily foraging routines, food hoarding, nocturnal hypothermia, and social interactions within flocks. We will discuss each of these topics briefly, indicating appropriate changes to the basic model. Box 7.3 discusses the benefits and limitations of dynamic state variable models in more detail.

BOX 7.3   What Current Models Can and Cannot Tell Us about Adaptive Energy Storage

Alasdair Houston and John McNamara

Many of the models of avian body mass regulation described in the text have a number of common elements. These models are concerned with the behavioral strategy that maximizes the probability that an animal will survive an extended period of many days (winter). At night an animal rests. At each decision epoch during the day, it must choose one action from a set of available actions. One option is to rest. A resting animal gets no food and is safe from predators. Other actions involve active feeding. If an animal takes one of these actions, it may obtain food, but may be killed by a predator. These feeding options incur a mass-dependent cost; either metabolism or predation risk increase with body mass and hence with energy reserves. The set of available actions is the same at each decision epoch in winter. The consequences of taking a given action—that is, the probability distribution of food found and the predation risk—are the same at a given time of day on every day in winter. They may depend on time of day, but in a predictable manner. For example, food availability may depend on time of day in the models developed by Houston and McNamara

(Box 7.3 continued)

(1993) and McNamara et al. (1994), and predation risk depends on time of day in Pravosudov and Lucas's model (2001a).

In the above models, the optimal strategy specifies a critical threshold at each time of day. If reserves are above that threshold, the animal rests. If reserves are below that threshold, the animal feeds. The lower an animal's reserves, the greater the energy gain and predation risk under the feeding option that it should choose.

Things that Follow from the Models but Aren't Widely Known

A bird should accept free food.

When reserves exceed the critical threshold, it is optimal to rest rather than feed. This does not mean that it would be disadvantageous to the bird to have even higher reserves. It is just that all feeding actions incur a predation risk, and this risk outweighs the advantage of obtaining extra food. If the bird has the opportunity to raise its energy reserves without risk, then it should do so. A bird that raised its reserves well above the critical threshold could then rest (safely) until reserves fell to the threshold. The more food that it obtains without risk, the longer it can adopt this completely safe option. These arguments can be put more formally by considering how reproductive value depends on energy reserves. When resting provides a safe option, reproductive value always increases as reserves increase. The bird should accept any predation risk up to a critical value in order to get the extra food (Houston and McNamara 1999). Only if resting has a sufficiently strong mass-dependent predation risk, so that reproductive value is no longer increasing, should a bird refuse the offer of extra food at no risk.

Models can be used to predict the effects of manipulations.

Dynamic state-dependent models predict how behavior depends on state (and time). They can thus be used to predict the effect of experimental manipulations on subsequent behavior. For example, suppose that an animal is given an extra item of food. The immediate effect of the extra food is to raise the animal's reserves. Thus, the effect of supplementary food on subsequent behavior can be investigated by using the model to compare the predicted behavior of a control animal with that of one that initially has higher reserves. In comparing behaviors, it is assumed that both control and experimental animals continue to use the optimal strategy after the time at which the manipulation is carried out. Any differences in behavior thus result from differences in state. Houston and

(Box 7.3 continued)

McNamara (1999) illustrate the effect of supplementary feeding in a model that incorporates both an animal's level of immunocompetence and its energy reserves as state variables. Their predictions concern the effect of the timing of the manipulation on the frequency and timing of subsequent breeding. Supplementary food in midwinter is predicted to have its main effect on the probability that a bird will breed. Supplementary feeding just prior to the breeding season has its main effect on the timing of breeding.

In the above approach, it is assumed that the only effect of supplying extra food is to change energy reserves. However, more food now may also act as a cue indicating more food in the future. In modeling the effects of supplementary feeding, there is a continuum of cases. At one extreme, the animal may ignore the informational aspect of the cue and continue to use the original optimal strategy, as envisaged above. At the other extreme, it may behave as if the improved food supply will continue into the future (McNamara and Houston 1994). Similar possibilities apply in the context of optimal fuel loads for a migrating bird (see box 7.2). Lindström and Alerstam (1992) provided extra food for migrating bluethroats at a stopover site in Sweden. Predictions based on time minimization are not accurate if it is assumed that future stopover sites are unchanged, but reasonably accurate predictions are obtained if it is assumed that the extra food indicates improved conditions at other sites (Houston 1998; Weber et al. 1999).

In many instances, models can make predictions only when the model environment is richer than in the standard models.

If a supplementary food source is not included in a model, it is not possible to use that model to predict where on the continuum of cases outlined above behavior will lie. To make a prediction, the model must be extended as follows: The feeding options now include one that delivers the supplementary food, which we can refer to as the free food option. Not all options are available at every time period. Instead, the model must specify the frequency with which combinations of options occur and the time correlations between them (cf. McNamara 1996). For example, it is important to know how often the free food option occurs, and whether one occurrence indicates that the option is likely to persist for some time in the future. Once these ingredients are specified, it is possible to determine how optimal choice depends on current options and past experience. Thus, it is possible to predict how the animal will respond to supplementary feeding.

(Box 7.3 continued)

The above example, in which an animal has to deal with a complex and changing environment, illustrates a limitation on what optimization models can achieve. We do not expect animals to be exactly optimal. Rather, we expect them to have behavioral mechanisms that perform well on average in the animal's natural environment. Such mechanisms may not, however, perform well in specific circumstances, especially if those circumstances rarely occur. So, for example, if the free food experienced during supplementary feeding hardly ever occurs in the wild, an animal's response to the food may well not be optimal. There is then probably little that optimization models can do to predict the response to the food.

Things that Don't Follow from the Models but are Assumed To
Effect of Season

It is tempting to take the long-term behavior predicted by a model under given conditions to be the behavior that should be observed if those conditions occur during part of a winter. This view ignores the fact that conditions change over the winter. There are few models that include changes within a season. Brodin and Clark (1997) analyzed a model in which there are two seasons, autumn and winter, but within a season, conditions do not depend on time of year. Bednekoff and Houston (1994b) investigated a very simple model of behavior over winter in which the only seasonal change during winter was in the daily energy requirement. This requirement was low at the start and end of winter and high in midwinter. The dependence on day was symmetric about the middle of winter, so that the same energy requirement could occur on two different days. Bednekoff and Houston found that the optimal behavior on a given day depends not just on the energy requirement for that day, but on whether it is early or late in winter. (Analogous effects occur in the context of daily changes; e.g., Pravosudov and Lucas 2001a. For a model of seasonal changes applied to fishes, see Reinhardt 2002.)

Effect of the Level of Predation Risk

Standard models predict that increasing the predation risk of feeding options decreases the critical threshold of reserves at which birds rest. In other words, birds are predicted to be lighter under high predation risk. Consider an experimental procedure in which a bird is occasionally shown a predator (cf. the procedures of Lilliendahl 1997; Pravosudov and

(Box 7.3 continued)

Grubb 1998; van der Veen 1999). Do we expect the bird to be lighter than if it had not been subjected to this treatment? The standard models cannot make any prediction here. These models assume that predation risk does not vary with time. In the experiment, however, there are sudden, intense increases in predation risk. At one extreme, we might model these sudden increases as an interruption (see Lilliendahl 1998; Pravosudov and Grubb 1998; Rands and Cuthill 2001; van der Veen and Sivars 2000 for discussion), with no increase in predation risk when the predator is not visible. Models incorporating interruptions (e.g., Lima 1986; Houston and McNamara 1993; McNamara et al. 1991, Clark and Dukas 2000; Clark and Mangel 2000) then predict that the treatment should increase fat levels. Alternatively, the occasional appearance of the predator might be modeled as indicating a higher risk between appearances as well as when the predator can be seen. In general, a model would need to specify how often the predator appears, how appearances alter the probability of subsequent appearances, and how appearances alter the predation risk for subsequent options when there is no predator to be seen. Using this approach, McNamara et al. (2005) showed that when a small bird is exposed to a predator, its optimal response can be either to increase or decrease its reserves, depending on the information about future levels of predation that is provided by seeing the predator.

Things that Aren't Considered in the Models
Other State Variables and the Time Scales at Which They Change

It may be advantageous to have high energy reserves at the end of winter. The criterion that overwinter survival probability is maximized ignores any such advantages and would usually not be appropriate for an animal that is about to breed at the end of winter. Whether pure survival models are appropriate for analyzing behavior well before the end of winter depends on the time scale at which state variables fluctuate. The fat reserves of small birds fluctuate on a time scale of a few days. For such organisms, the value of reserves (i.e., the terminal reward) at the end of winter can be ignored when predicting reserves a long way from its end (McNamara and Houston 1982, 1986; Houston et al. 1997), and the strategy that maximizes expected reproductive value at the end maximizes the survival rate well before the end (McNamara 1990).

The above remarks apply to models that use energy reserves of body fat as the only state variable. If there are other state variables that change

(Box 7.3 continued)

on a longer time scale, then it may not be appropriate to ignore the terminal reward, even when considering behavior in midwinter or before. Damage or injury to the organism is an example of a state change whose effect is long-lived. Thus, if foraging options differ in their probability of incurring injury, both the immediate trade-off of energy gained versus injury sustained (see Houston and McNamara 1999) and the long-term consequences of injury need to be considered. Possession of a territory is another state variable that has potentially long-lived effects.

Life history models usually assume that there is a trade-off between current reproductive effort and future reproductive success. This could be because effort leads to immediate mortality through predation. However, there is evidence that some effects of reproductive effort are not immediate, but can affect the organism months or years later (Daan et al. 1996; Gustafsson et al. 1994). When this occurs, it must be because some state variable has been affected by the reproductive effort, the effects of the state variable are long-lived, and the value of the state variable is an important determinant of reproductive value. As we have indicated, the energy reserves of a small bird change on a rapid time scale, so it is unlikely that long-term trade-offs can be mediated by energy reserves alone. Other variables that could be important are feather quality (e.g., Chai and Dudley 1999; Hedenstrom and Sunada 1999) and the state of the immune system (e.g., Norris and Evans 2000; Sheldon and Verhulst 1996). It seems reasonable that if high reproductive effort has a long-term deleterious effect, then so might high foraging effort during winter. Thus, it could be important to include state variables other than energy reserves in models. Houston and McNamara (1999) illustrate how this can be done when the state variables are energy reserves and level of immunocompetence. For an example of the application of this type of model to optimal migration routines, see McNamara et al. (1998). Barta et al. (2006) include the condition of a bird's inner and outer primary feathers as part of its state and investigate optimal annual routines of molting.

Geometric Mean Fitness

Even if there are no significant carryover effects at the end of the season, maximization of survival probability may not be the appropriate optimization criterion. That criterion is appropriate if fitness can be taken to be the expected total number of surviving offspring produced over a lifetime. But when there are environmental fluctuations affecting all

(Box 7.3 continued)

population members, this measure of fitness is not appropriate. Instead, some fitness measure based on geometric mean fitness needs to be used. Small birds in winter are likely to be significantly affected by unpredictable changes in weather conditions. In modeling these organisms, should we be using the geometric mean probability of survival as the currency that is maximized?

To define this currency, suppose that weather conditions $w_1, w_2, \ldots,$ w_n occur with probabilities $p_1, p_2, \ldots, p_n$, respectively. Let $S(w_i)$ be the probability of overwinter survival when the weather is w_i. Then

$$A = p_1 S(w_1) + p_2 S(w_2) + \cdots + p_n S(w_n)$$

is the arithmetic mean survival probability. All of the models using survival probability as a criterion implicitly use this quantity as the currency being maximized. In contrast, the geometric mean probability of survival is given by

$$G = S(w_1)^{p_1} S(w_2)^{p_2} \cdots S(w_n)^{p_n}.$$

When birds are long-lived and weather conditions affect only juvenile survival, the two currencies will produce similar predictions. However, if adults and juveniles are equally affected, then the currencies may make significantly different predictions, even for long-lived species. There may also be large differences for short-lived species, even when only juveniles are affected. Differences will be most pronounced when there are large, unpredictable fluctuations in conditions and it is important not to be caught out with low energy reserves when the weather suddenly turns bad.

The main difference between the predictions of the two currencies is that under maximization of G, an animal should put more emphasis on bad conditions than under maximization of A. This emphasis can be seen from various perspectives. From one, an animal that maximizes G should maximize its survival probability relative to other individuals (Grafen 2000). Alternatively, maximization of G is equivalent to maximization of a modified arithmetic mean fitness,

$$A' = p_1' S(w_1) + p_2' S(w_2) + \cdots + p_n' S(w_n),$$

where the probabilities $p_1', p_2' \cdots, p_n'$ are distorted from their true values to make poor environmental conditions more likely and good conditions less likely (McNamara 1995; Houston and McNamara 1999).

Seasonal Changes

Bednekoff and Houston (1994b) modified the basic model by allowing metabolic costs to vary seasonally. They assumed that costs increase gradually to
a midwinter maximum and then decline. These nocturnal costs, $c_n(d)$ in equation (7.12), now have a mean and variance that depend on d. The predicted optimal dusk reserve level, $X_1^*(d)$, closely tracks this contour of daily costs, with
additional reserves to hedge against unpredictable daily variation. This modification adds the realism of changing day length to the scenario in figure 7.2.

Two aspects of this model could help explain winter fattening (see fig.
7.2). First, optimal dusk reserves exceed the level required on average nights,
and this safety margin increases with variation in nocturnal costs. Thus, dawn
reserves are positive after most nights, more so when nocturnal variation is
high. Second, it may be optimal to deliberately increase dawn reserves as a
hedge against poor foraging success during the day. If midwinter conditions
imply increased uncertainty in nocturnal costs or diurnal foraging success,
or both, the model will predict greater average dawn reserves in midwinter,
corresponding to "true winter fattening."

One can also model other aspects of seasonality, although no one has published a complete analysis. For example, Clark and Dukas (2000) included
the possibility of sudden, unpredictable, prolonged "cold snaps." Such cold
snaps, which may affect food availability as well as thermoregulation costs,
can be particularly dangerous.

An important feature of dynamic state variable models is that predicted
optimal strategies generally anticipate environmental changes. For example,
if the probability of cold snaps increases in midwinter, the optimal reserve
level should increase in anticipation of such events. Similarly, as noted by
Bednekoff and Houston (1994b), fattening strategies may change toward the
end of winter, with a prebreeding (or premigratory) increase in fat reserves.

Daily Foraging Routines

To model daily foraging routines, we alter the time scale in the basic model.
Time periods $t = 1, 2, \ldots, T$ correspond to T time intervals (typically a few
minutes' duration) in a single day. Index $d = 1, 2, \ldots, D$ again corresponds
to days over the winter. The reserve state $X(t, d)$ equals body reserves at the
beginning of period t on day d. The decision variable is simplified to $i = 1$ or
2, denoting rest or forage, respectively, in the current period t. The equation
for state dynamics becomes

$$X(t + 1, d) = X(t, d) + f_i(t) - c_i[X(t)], \tag{7.21}$$

where $f_i(t)$ is energy intake for the period (i.e., zero if $i = 1$ and a random variable if $i = 2$) and $c_i[X(t)]$ is the metabolic cost of decision i. Overnight changes to X are

$$X(1, d + 1) = X(T, d) - c_n(d), \qquad (7.22)$$

provided this number is positive. Otherwise, the bird dies overnight.

Corresponding to equation (7.14), we define the function $F(x, t, d)$ as the maximum probability of surviving from period t on day d to period T on day D, given that $X(t, d) = x$. The dynamic programming equation is analogous to equation (7.16).

Besides predicting optimal dusk reserves $Y(T, d)$, the present model also predicts the optimal timing of foraging activities over the day. A commonly observed pattern is for birds to forage intensively at dawn, to decrease their activity at midday, and then to increase their activity again as dusk approaches. Students of natural history have usually attributed this dawn and dusk feeding pattern to daily variations in food availability, but, as explained below, the model suggests other reasons for this feeding pattern.

Maintaining low body mass for most of the day benefits the bird because body mass increases predation risk. This would explain the burst of foraging toward dusk, but not the dawn burst. However, if foraging success is uncertain, a burst of dawn activity may act as a hedge against later failure, especially if morning fat reserves are low. Food-hoarding species might show this effect most strongly, since their hoarded food lets them maintain lower fat levels (Brodin 2000). If it is successful, the bird ceases foraging; otherwise, it persists (McNamara et al. 1994). The two-burst feeding pattern of birds can thus be interpreted as an adaptation to uncertainty and mass-dependent predation risk, rather than to daily variation in resource availability.

Digestive constraints (rate of digestion and stomach capacity) can also influence daily foraging routines. To model digestive constraints, we add a state variable $Y(t, d)$ for stomach contents (Bednekoff and Houston 1994a). This broader model predicts the most intense foraging at dawn; body mass increases rapidly as the stomach is filled, but fat reserves grow at a constant rate over the day.

Food Hoarding

The models discussed so far consider only reserves that the animal retains within its body. Although theoreticians have published several dynamic models of external food hoarding, the complexity of this topic remains poorly understood. Here we will describe some existing models and outline other aspects in need of investigation.

Hoarding as a Substitute for Body Reserves

Hitchcock and Houston (1994) developed a model (based on acorn woodpecker data) in which hoarding substitutes for high levels of body fat. This model specifically considers the bet-hedging effect of stored food. Food acquisition can, on average, cover metabolic costs, but variation in foraging success means that dangerous deficits can accumulate. To hedge against this possibility, a bird must either maintain large body reserves, use its hoard, or both. The model predicts that moderate initial hoards can substantially increase fitness (i.e., survival). For example, a hoard equal to approximately 20% of total (100-day) energy requirements can, under realistic assumptions for the model parameters, increase the probability of survival from under 20% to 60%. The model also makes other predictions. First, predation, not starvation, will be the main cause of mortality. Second, birds will compensate for smaller hoards by maintaining higher body reserves. Thus, hoarding, under the model assumptions, mainly hedges against variation, resulting in lower predation risk. Hoarded supplies are not essential as an energy resource (unless the variation in food availability is very high). Hitchcock and Houston's model does not consider the cost of establishing a hoard, which one would have to consider in any complete analysis of hoarding.

In more northern habitats than those inhabited by acorn woodpeckers, short-term variation in metabolic (thermoregulation) costs is likely to be at least as important as variation in food availability. Clark and Mangel (2000, section 5.4) discuss a model incorporating variable nocturnal costs [see equation (7.12)] and including hoard use. Its predictions are similar to those of Hitchcock and Houston (1994): relatively small hoards can increase fitness substantially in a fluctuating environment.

In a model with variable nighttime costs, Brodin (2000) showed that stored supplies hedged against variation, permitting hoarders to start each new day with small fat reserves. This model related to small birds such as parids in cold habitats and, unlike Hitchcock and Houston's model, incorporated hoarding. Throughout the winter, the model birds continued to store small amounts of food as insurance against unexpected variations (see section 7.4).

However, these and most other models ignore one important aspect of environmental stochasticity, the availability and use of information. For example, if the day provides no information about the demands of the coming night, the bird must store adequate body reserves at dusk to meet worst-case night costs. The bird will not use all of its reserves unless the worst-case conditions actually prevail. The phenomenon of winter fattening (see fig. 7.2) suggests that animals do not posses perfect information about conditions in the nearest future. It is more realistic to assume that the current day's weather provides some (but not perfect) information about energy demands for the following night. In other words, current environmental "cues" influence the probability distribution of nightly costs, $c_n(t)$. Taking advantage of such cues might have strong fitness consequences, particularly for animals facing mass-dependent costs. Models of this situation could lead to novel, experimentally testable predictions.

McNamara et al. (1990) developed a dynamic model of optimal hoarding and fat regulation for the case of ephemeral hoards. Specifically, they assumed that a hoard can be built up and utilized later the same day, and that unused hoards disappear overnight. In spite of conservative assumptions (e.g., lack of mass-dependent predation risk and brief persistence of hoards), the optimal strategy involves the buildup and use of daily hoards. The main advantage of hoarding relative to immediate consumption is cost saving: consumed food increases metabolic costs, which daytime food storage can prevent. Secondarily, a hoard (built up in the morning) can hedge against lack of foraging success later in the day.

The model also predicted differences in foraging patterns between hoarders and nonhoarders. Since hoarders can rely on stored food, they can delay fat gain until late afternoon. Nonhoarders, on the other hand, face a more unpredictable access to food and must therefore start gaining fat earlier. So far, there is no empirical evidence for this prediction; on the contrary, hoarding species may even gain fat at a higher rate in the morning than nonhoarders (Lilliendahl 1997).

Changing three assumptions of McNamara and co-workers' model produces the pattern that Lilliendahl observed in the field (Brodin 2000; Pravosudov and Lucas 2001b). First, these revised models allow birds to leave caches overnight and retrieve them in the morning. Second, small fat reserves have little or no effect on predation risk (see section 7.4). Finally, the bird's scattered caches may be time-consuming to retrieve, decreasing the profitability of retrieval. Under these assumptions, hoarders can carry small overnight fat reserves. If the morning weather is worse than expected, hoarders can retrieve caches, whereas similarly lean nonhoarders may starve. Since hoarders start the day with small reserves, they need to secure sufficient fat in the morning to hedge against uncertainty later in the day (fig. 7.9). After reaching a more secure level of reserves, they can afford to spend time on hoarding and other activities. This more secure level of reserves increases predation only slightly because it is relatively small (see fig. 7.4).

Lucas and Walter (1991) modeled hoarding strategies in Carolina chickadees (*Poecile carolinensis*). They listed four advantages of hoarding that earlier

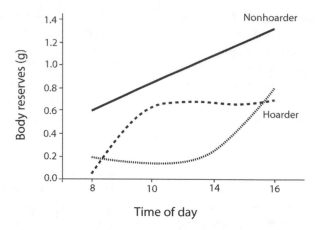

Figure 7.9. Hypothetical daily mass gain of hoarding and nonhoarding birds in winter. The nonhoarder has less predictable food availability and must hedge with larger body fat reserves. There are two cases for hoarders, baseline assumptions (dotted curve) and the three added assumptions described in the text (dashed curve). The most important difference is that under baseline assumptions, predation risk is mass-dependent at low levels of fat, whereas under the three added assumptions, predation risk is mass-dependent only above some limit.

authors had suggested: (1) hoarding takes advantage of ephemeral food supplies; (2) hoarding provides additional food during shortages; (3) hoarding provides an alternative to body fat reserves; (4) hoarding birds can exploit low-cost foraging opportunities—for example, when predators or competitors are scarce. Note that this list does not explicitly mention hoarding as a strategy for counteracting uncertainty in future supplies and costs.

Lucas and Walter discussed two models of optimal hoarding, a harvest rate maximization model and a survival rate maximization model; only the latter was a dynamic state variable model. They assumed that hoards are short-lived (≤3 days), as observations of Carolina chickadees in the wild suggest. Lucas and Walter tested their predictions experimentally; the results of the experiments supported the survival rate maximization model. In contrast to simpler rate maximization models, dynamic state variable modeling can predict complex state-dependent decisions; as Lucas and Walter pointed out, "caching behavior is an example of a behavior pattern that is strongly affected by the state of the animal."

Long-Term Hoarding

In contrast to Carolina chickadees, some passerines from more northern climates use both short-term and long-term hoards (Brodin 1992, 1994a; Brodin and Ekman 1994). Brodin and Clark (1997) modeled this situation. Our model uses three state variables, all measured in kJ: $X(t) =$ body reserves,

$Y(t) =$ stored seeds retained in memory, and $Z(t) =$ forgotten long-term stores. Newly hoarded seeds first enter the remembered store, $Y(t)$, but when the bird no longer remembers their location, the model transfers these seeds to long-term stores, $Z(t)$. Both stores are subject to loss (decay, pilferage) at a constant rate. The bird can retrieve remembered seeds quickly when needed, but long-term stores merely enrich the natural, background supply.

The model considers fall and winter as two sets of environmental conditions that occur in sequence. Otherwise, the details are fairly similar to those of the models described above: fitness given by survival, mass-dependent costs, stochastic environment. The model bird can decide whether to rest, forage and eat, forage and hoard, or retrieve food from remembered hoards. We used parameter values derived from observations of willow tits in central Sweden.

The model provided several intuitively reasonable predictions. Long-term stores held at the start of winter strongly influence overwinter survival. So the optimal fall strategy builds long-term stores as much as possible. Short-term stores at the start of winter have a smaller influence on overwinter survival. Nevertheless, the optimal strategy in winter includes continual hoarding, because remembered seeds provide a hedge against bad weather, without the need to maintain high levels of body fat. Increasing the capacity for remembered stores, Y_{max}, had minimal effects on overall fitness, suggesting relatively weak selection for additional memory capacity in willow tits.

The Effect of Pilferage

Hoarders often lose a portion of their cache to pilferage, and it would seem that this pilferage imposes a cost on hoarding. Lucas et al. (2001) used a dynamic state variable model to investigate the effects of pilferage on hoarding rates. Intuitively, one might think that high rates of pilferage would make it optimal to store fat instead of caching food. Lucas et al. showed that this is not always the case. Hoarders can compensate for increased pilferage by hoarding at a higher rate as long as the marginal value of caching exceeds that of resting.

Hypothermia

Animals can use nocturnal hypothermia to manage their energy budgets. Small birds can reduce their nighttime body temperature to 30°C–38°C (see the end of section 7.4). Willow tits and other small passerines at northern latitudes use hypothermia during cold winter nights. Although one can easily understand how hypothermia saves energy, its costs, if any, remain mysterious. Perhaps the hypothermic state increases predation risk. Alternatively, hypothermia may have long-term physiological consequences that reduce subsequent performance.

Clark and Dukas (2000), Pravosudov and Lucas (2000), and Welton et al. (2002) have developed dynamic models of optimal hypothermia, assuming that hypothermia increases predation risk. Nocturnal hypothermia may be an adaptation to a temporary deficit of energy reserves, but birds can use hypothermia in other situations as well. As explained below, in extreme conditions, a small bird may use hypothermia to prevent or control the gradual depletion of body reserves over a period of several days. While Pravosudov and Lucas considered hypothermia to be a single drop in temperature, Clark and Dukas considered the depth of hypothermia to be variable. In these two models, increased predation risk at night or in the morning is the cost of hypothermia. Welton et al. included an additional warming-up cost in their model.

Clark and Dukas's model had two decision variables, foraging effort ($0 \leq e(d) \leq 1$) and θ, the energy saved overnight by using hypothermia ($0 \leq \theta \leq \theta_{max}$). The model was an elaboration of the basic model we present in equations (7.9–7.18). We will not give all the details of this model here, but only explain how it included hypothermia. In a manner similar to equation (7.12), dawn reserves are given by

$$X_0(d+1) = X_n(d) - c_0 - c_{th}(d) + \theta. \tag{7.23}$$

Minimal hypothermia, $\theta_{minimal}$, is defined as the nocturnal energy savings needed to prevent overnight starvation:

$$\theta_{minimal} = \begin{cases} c_0 + c_{th}(d) - X_n(d) & \text{if this is positive} \\ 0 & \text{otherwise} \end{cases}; \tag{7.24}$$

i.e., $\theta_{minimal}$ makes up for any overnight deficit in reserves. Hypothermia is facultative, in the sense that the bird knows the current thermoregulation cost, $c_{th}(d)$, when θ is chosen. The options on a given night are then to use minimal hypothermia, $\theta_{minimal}$ (possibly zero), avoiding starvation and starting the next day with zero body reserves, or to use $\theta > \theta_{minimal}$, starting the next day with positive reserves. The model determines which of these options is optimal in terms of overall survival.

Clark and Dukas modeled the cost of hypothermia through its effect on overnight survival,

$$S_{night}(\theta) = \exp(-\mu_n \theta), \tag{7.25}$$

where μ_n is the hypothermic mortality coefficient. How large this cost is depends on the bird's energy reserve strategy as expressed in its dusk reserves,

$X_n(d)$. Exactly how the starvation-predation trade-off involving both reserves and hypothermia is realized depends critically on the distribution of nocturnal costs, $c_{th}(d)$.

Clark and Dukas (2000) modeled possible spells of bad weather by assuming that such spells were initiated randomly and, once initiated, persisted for a random number of days. This assumption required a second state variable, $N(d)$, the number of bad days in the current spell. This way of looking at winter weather patterns seems realistic and is much simpler computationally than more traditional approaches.

We can think of hypothermia as an emergency measure that birds use only when nocturnal conditions are unexpectedly severe. Even if relatively costly when used, hypothermia may be an adaptive alternative to large, seldom needed energy reserves. On the other hand, birds may need hypothermia to survive in cases in which the maximum reserve capacity is too low to meet nocturnal requirements. A third possibility is that birds use hypothermia only in long spells of bad weather that affect both the energy supply, $f(d)$, and the nocturnal costs of thermoregulation, $c_{th}(d)$. Hummingbirds are known to combine high metabolism with small body mass. At least in regions with cold nights, as in higher regions of the Andes, hummingbirds seem to use nocturnal hypothermia on a regular basis (Hainsworth 1981; Carpenter and Hixon 1988).

As Clark and Dukas varied the food supply during bad weather, they found (1) that hypothermia did not improve fitness (i.e., survival) when the food supply was large; (2) that for intermediate food supplies, the optimal level of hypothermia was minimal ($\theta_{minimal}$), but had a large effect on fitness; and (3) that when the food supply was small, the optimal level of hypothermia exceeded $\theta_{minimal}$, with a large effect on fitness. In addition, optimal reserves, $X_1(d)$, increased at lower food supply values and were also higher under the no-hypothermia assumption than for optimal hypothermia. These findings suggest that hypothermia serves mainly as an alternative to large body reserves in intermediate environments. As in previous models, the main source of mortality is predation rather than starvation, expect in very harsh environments. Clark and Dukas (2000) did not consider the interrelation of hoarding and hypothermia, a topic that might be worth pursuing.

Social Interactions

Ekman and Lilliendahl (1993) discovered that the dominant bird in a pair of willow tits has lower fat reserves than the subordinate. They suggested that this occurs because the dominant monopolizes the best foraging sites, especially in bad weather; the subordinate has a less secure and more variable food supply. Thus, the subordinate needs to carry extra reserves as a hedge against interruptions in its supply.

Clark and Ekman (1995) modeled dominant-subordinate fattening strategies as a one-sided dynamic game. The dominant maximizes its fitness independently of the subordinate. The subordinate then maximizes its fitness, subject to the condition that the dominant always excludes it from foraging in the dominant's current foraging habitat. Clark and Ekman considered three habitats: H_0 was a refuge with no food and no predators, H_1 provided an inferior intake rate but was relatively safe (e.g., interior branches of trees), and H_2 provided a high intake rate, but was risky (e.g., outer branches).

Clark and Ekman assumed that food in the mediocre but safer H_1 habitat was sufficient to cover metabolic costs on normal days, but not on cold days (which occurred with a certain probability p). Thus, both dominant and subordinate had to use the more dangerous H_2 habitat at least part of the time. The model predicted that the dominant would use the mediocre H_1 habitat most of the time, thereby excluding the subordinate from this preferred habitat. Consequently, the subordinate experienced higher risk of predation than the dominant. In addition, the dominant excluded the subordinate from the rich high-risk habitat (H_2) on cold days. The model thus predicted that the subordinate would carry greater fat reserves than the dominant as a hedge against inadequate food on cold days.

The model formalized the hypothesis of Ekman and Lilliendahl (1993) and explained how the observed differential between dominant and subordinate survival rates could arise. Clark and Ekman also considered how changes in the food supply in the mediocre H_2 habitat affected their predictions. At low levels of food supply, both birds suffered increased mortality rates. In addition, the dominant switched to carrying the same level of reserves as the subordinate. Farther north, where winter conditions are harsher than in Ekman and Lilliendahl's field area, dominants carry as much or even more fat than subordinates (Koivula et al. 1995; Verhulst and Hogstad 1996).

7.8 Summary

Organisms must store energy because they will experience periods when energy expenditure exceeds energy intake. Animals can store energy either internally, as fat or carbohydrates, or externally, as hoarded food. Alternatively, animals can reduce energy expenditure during periods when energy intake is not possible. Hibernation and temporary hypothermia provide examples of this strategy.

Energy storage incurs both costs and benefits, and the trade-off between these makes storing behaviors well suited for optimization modeling. Stored energy not only permits energy use when expenditure exceeds intake, but it may also hedge against unpredictable variation in intake rates. For example, hoarded food may provide insurance against rare food shortages even if the hoarder never consumes it. In order to build energy stores, animals must find more food than they need for immediate consumption. This added effort may increase energy expenditure and predation risk. Large fat reserves may increase mortality, both because fat stresses the heart and vascular system and because it decreases the animal's ability to escape from an attacking predator (especially for birds). Hoarded supplies may incur costs if hoarders must defend them, or if hoarders forget their locations.

Behavioral ecologists have modeled the regulation of energy reserves in several ways, including game theoretical models, rate maximization models, and other analytic models. However, most models on this topic have been dynamic state variable models, which are better suited for the complexity of energy storage problems. Simple analytic models cannot simultaneously incorporate phenomena such as temporal dynamics, stochastic effects, nonlinear fitness effects, and predation effects. Small birds in winter offer an especially appealing modeling problem, since they face a delicate trade-off between predation and starvation that nonflying animals do not face. At the same time, their high metabolic rates further complicate their energy storage problems.

7.9 Suggested Readings

Vander Wall's book (1990) offers the most comprehensive summary of food hoarding. It is now over sixteen years old, but still contains a valuable summary and an impressive list of references. Most reviews of energy storage focus on birds, but Vander Wall covers food hoarding in all animals, including insects. Källander and Smith (1990), who reviewed avian food hoarding in the same year, concentrated on the evolutionary aspects of hoarding. Witter and Cuthill (1993) and, more recently, Pravosudov and Grubb (1997) cover fat storage by birds. Cuthill and Houston (1997) provide a more general perspective on energy acquisition and storage. Blem (1990) gives a physiological perspective on fat storage in birds. Bulmer (1994) gives an overview of the theory of evolutionary ecology. Houston and McNamara (1999) and Clark and Mangel (2000) explain the modeling techniques and concepts that we have used in this chapter.

Part III

Modern Foraging Theory

8

Provisioning

Ronald C. Ydenberg

8.1 Prologue

A honeybee (*Apis mellifera*) colony contains thousands of foragers that collect large amounts of nectar, pollen, propolis, and water and deliver them to the hive. The colony's activities and, ultimately, reproduction depend on these resources. Millions of years of honeybee evolution and thousands of years of domestication have selected for proficient resource provisioning.

Bees divide the labor of resource acquisition and provisioning. Scout bees specialize in finding ephemeral resources and recruiting foragers to good locations. Foragers fuel up on the communal honey supply and leave the colony knowing where to go and what to expect. En route, they regulate their flight speed, micromanage their body temperature, and carefully collect a load for transport back to the hive. In the hive, a system of feedbacks involving behaviors, odors, and pheromones regulates the quantity and quality of future resource deliveries. Using this system, the colony can quickly refocus its activities on the commodities it needs most.

Many predators, including bears, honey badgers, honeyguides, honey buzzards, and hornets, covet the contents of a hive, and the bees must defend it. Outside the hive, bee wolves and other predatory insects, as well as a suite of birds such as bee-eaters, make a forager's life hazardous. If she eludes all these dangers, she faces a routine of grueling work: after only 20 days or so, her wings are tattered and her body pile worn. Soon all internal systems fail. When she dies, her comrades unceremoniously dump her body onto a pile of other spent workers outside the hive. Selection has not built workers to last, but to provision.

8.2   Introduction

This chapter considers provisioning: the collection and delivery of materials such as food, nesting material, or water. The quintessential feature of provisioning is that provisioning animals deliver material to a site where they either feed it to others or store it for later use (Ydenberg 1998). The earliest provisioning studies considered a parent bird working as hard as possible to deliver prey items to its altricial offspring. This chapter will show that provisioning raises important questions and issues that go far beyond the problems of a parent bird feeding its young.

The "parent bird" paradigm focuses attention on only a few key features of animal provisioning. Many other animal taxa provision, using a wide range of behaviors (table 8.1), and the realm of interesting provisioning phenomena includes aspects other than foraging theory's classic problems of prey choice and patch exploitation (Stephens and Krebs 1986; see chap. 1 in this volume). The extensive literature on diverse provisioning systems makes it clear that we must consider selective benefits beyond simple energy acquisition to understand the diversity of provisioning behavior.

Like the rest of behavioral ecology, provisioning models emphasize costs and benefits, and they ask how costs and benefits select for certain types of behavior. Fundamentally, these models assume, often implicitly, that selection (natural, sexual, or artificial) has acted on the structure and function of "decision mechanisms" (Ydenberg 1998). Physiological processes and morphological structures inside the provisioner control these decision mechanisms. The models do not require cognitive functions such as memory, consciousness, or forethought, but they do not preclude them either. Decision mechanisms integrate information from sensory organs and internal indicators of state, such as hunger or weariness, to produce behavior. For example, seabirds such as thin-billed prions (see table 8.1) decide whether their next provisioning trip will be a short outing of a few days to the edge of the continental shelf or a long pelagic excursion (apparently evaluating their nestling's condition, their own condition, their recent provisioning history, and the availability of prey). Chapters 3, 4, and 5 consider some of the mechanisms animals use to integrate this information (see also Dukas 1998a).

For selection to act on provisioning behavior, decision mechanisms must affect the provisioner's reproduction or survival. This could happen in a variety

Table 8.1   Selected examples of provisioning tactics documented in free-living animals

Selection of prey for self-feeding
  Sonerud (1989) describes how a kestrel (*Falco tinnunculus*) and a shrike (*Lanius excubitor*) direct small, medium, and large prey to self-feeding versus delivery.

Foraging destination
  In thin-billed prions (*Pachyptila belcheri*), parents deliver undigested meals after short trips, and lose condition themselves, evidently because they power the excursion from body reserves. After long trips, parent prions deliver prey (partially) concentrated into energy-rich stomach oil; parental condition improves (Duriez et al. 2000).

Foraging mode
  Bumblebees may fly or walk between flowers.

Travel speed
  On the outbound flight from hive to feeder, honeybees fly faster when the sucrose concentration at their destination is greater (von Frisch and Lindauer 1955). They fly more slowly on the return trip, and speed does not vary with concentration. Load size increases with concentration.

Body temperature
  Honeybee workers have higher body temperatures and cool more slowly after landing on higher-concentration sucrose solutions (Schmaranzer and Stabenheiner 1988).

Prey processing into parts
  Rands et al. (2000) describe models and observations of prey dismemberment for transport by a provisioning merlin (*Falco columbarius*).

Prey processing into partially digested material, or nutritious secretions
  Carnivores may carry whole prey to the den or regurgitate partially digested prey. Of course, female mammals also lactate (Holekamp and Smale 1990).

Time devoted to provisioning
  Spotted hyenas (*Crocuta crocuta*) vary attendance times at the den depending on prey availability (Hofer and East 1993).

Adjusting brood location
  Lapland longspurs (*Calcarius lapponicus*) divide broods evenly into two units after nest departure, each tended by one parent (McLaughlin and Montgomerie 1989).

Body weight or constitution alteration and metabolic rate adjustment
  The wet body mass of a worker honeybee drops 40%, and maximal thorax-specific oxygen consumption increases 10%, during the transition from hive bee to forager (Harrison 1986).

Adjustment of participation in brood rearing or helping
  Adult pied kingfishers (*Ceryle rudis*) facing high demand recruit helpers (Reyer and Westerterp 1985).

Egg size, brooding, or delivery
  Birds can supply materials to the nest in the egg itself, by brooding offspring, and by provisioning nestlings. These alternatives have different costs and benefits, and birds can adjust them accordingly (Hipfner et al. 2001).

Table 8.1   *(continued)*

Workload

Honeybees in large colonies work harder than those in small colonies (Wolf and Schmid-Hempel 1990; Eckert et al. 1994).

Offspring gender ratio

In many mass-provisioning hymenopterans, provisioners adjust the gender ratio of the brood (Rosenheim et al. 1996) between small males and large females. Ovipositing females can also adjust the sequence and position of offspring sexes within the nest.

Trophic eggs

Many animal taxa provision young with trophic eggs (eggs used as food). In the poison arrow frog *Dendrobates pumilio*, mothers deliver trophic eggs to tadpoles secreted in phytotelmata (tiny pools of water up in trees; Brust 1993). In other poison arrow frog species, parents supply water to these pools to prevent them from drying.

---

of ways. Most often, investigators have considered direct effects of the amount of food provisioned on the quantity or quality of offspring. However, sexual selection could also act on decision mechanisms through their effects on the number or quality of mates attracted. For example, stickleback (*Gasterosteus aculeatus*) nests are built by males from material delivered to the assembly point and may advertise a male's qualities (Barber et al. 2001; see also Soler et al. 1996). Provisions placed in hoards can enhance survival when resources are scarce (see chap. 7), and in some species the size or quality of structures built from delivered material affects reproductive success (e.g., stone ramparts; Leader and Yom-Tov 1998).

In addition to evaluating benefits, researchers must carefully characterize the fitness costs of provisioning. Provisioning always involves work because provisioners must expend time and energy to collect and deliver materials. Whether provisioners deliver food, water, stones, or mud, the provisioner's metabolism generates the necessary power, and the provisioner must feed itself to provide the fuel for provisioning. Provisioning models pay careful attention to the relationships between self-feeding, metabolism, and delivery capacity, but they must also recognize the importance of factors other than energetics. Collecting materials or the extra food needed to fuel their delivery may expose the provisioner to danger or distract it from important tasks such as offspring care or the management of stored food.

This chapter outlines the structure of provisioning models and their relationship to traditional foraging models, investigates the rate of work and its relation to metabolism, considers how provisioners should respond to demand, and discusses provisioning in a life history context. It focuses throughout on the underlying ecological selective factors that shape the morphology, physiology, and life history of provisioning behavior.

8.3   Basic Models of Provisioning Behavior

This section outlines the history of provisioning models, focusing on how foraging models provided a framework for ideas about provisioning. It develops the most basic provisioning model and touches on the issues that connect provisioning and foraging models.

Central Place Foraging

Great tits (*Parus major*) are small songbirds living in European woodlands. Each spring a pair raises a brood of about eight nestlings in a tree cavity or nest box. While provisioning the nestlings, each parent spends almost all of its time searching through the trees in its territory for insect prey, especially caterpillars, which are fed to the brood. Each parent makes hundreds of back-and-forth trips each day, delivering prey to fuel the growth of the nestlings. Better-fed broods grow faster and survive better.

Orians and Pearson (1979) invented the term "central place foraging" to describe this and similar situations in which animals make repeated foraging excursions from a central location. Their model introduced the basic concepts of central place foraging, developed the idea of "loading" prey, and distinguished "single-prey" and "multiple-prey" loaders, appreciating the different nature of the decisions that these foragers face. The simplicity, novelty, and applicability of this model inspired many field and experimental studies. Its simple framework can be applied to a variety of situations: box 8.1 considers as an example the effect of social interactions on central place foraging.

Central place foraging models consider the amount or type of prey that foragers should deliver to their central place. "Single-prey loaders" deliver a single prey item from a capture site on each trip, and the decision they face is the minimum size of prey acceptable for delivery. This decision implies a trade-off, because low selectivity (capture any prey) means that the forager may spend too much time in transit with small prey, while high selectivity (capture only large prey) means that the forager may spend too much time at the capture site searching for suitable items. The selectivity giving the highest rate of energy gain depends on the size (energy content) distribution of prey, prey density, and travel time. Single-prey loader models predict that foragers should set a higher minimum prey size when prey are more abundant and when they must travel greater distances to capture sites.

Krebs and Avery's (1985) studies of bee-eaters (*Meriops apiaster*) provisioning their broods provide a field example. Bee-eater parents captured both small (mostly bees and wasps) and large (mostly dragonflies) prey, but delivered a lower percentage of small prey when returning from more distant capture sites.

Social interactions at resource collection points often affect the tactics that provisioners use. Eastern chipmunks (*Tamias striatus*) defend territories and usually avoid one another. They compete aggressively at rich resource points, and even the mere proximity of a conspecific can reduce the rate at which they load seeds into their cheek pouches. Ydenberg et al. (1986) incorporated this interference effect into a central place foraging framework to explain Giraldeau and Kramer's (1982) observation that chipmunks collected smaller loads and spent more time exploiting experimentally provided seed piles as interference increased over repeated visits to the experimental patches (fig. 8.1.1; see Lair et al. 1994).

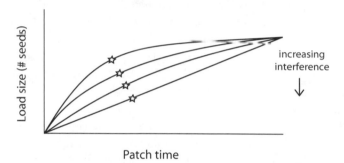

Figure 8.1.1. Interference among chipmunks slows loading, and so reduces load size, but increases patch residence time. The star indicates the predicted rate-maximizing load size at each level of interference.

Other creatures cooperate rather than compete in resource collection. Leaf-cutter ants, for example, travel along trails to particular bushes and trees, where they cut semi-discs from leaves, often stripping entire branches in the process. Trails to collection sites bustle with two-way traffic as ants transport leaf fragments to their large underground colonies, where they are processed into mulch. The ants grow fungus on the mulch and feed this fungus to the brood.

Foraging ants cut semi-discs from the leaf margin. Larger pieces are more profitable because cutting time increases linearly with the radius, while mass rises as the square of the radius. However, it takes more time to cut large pieces, so workers looking for cutting sites along the leaf margin may have to wait in a queue for the next available cutting site. So,

(Box 8.1 continued)

while cutting large leaf fragments may increase the delivery rate for an individual, it can reduce the overall delivery rate. Students of social insects must frequently address similar conflicts between benefits at the individual and colony levels (e.g., Ydenberg and Schmid-Hempel 1994).

Burd et al. (2002) analyzed how this conflict affects delivery in the leaf-cutter ant *Atta cephalotes*. For an individual worker, the expression load mass/(outbound time + queuing time + cutting time + inbound time) gives the rate of delivery of leaf material. The size of the leaf fragment influences every term of this expression except outbound time. (Ant size influences all of the terms, because larger individuals travel and cut faster, and load mass affects larger individuals less.) Individual workers could theoretically diminish the effect of queuing by cutting smaller pieces, effectively reducing their own delivery rate to reduce the waiting time of their nestmates and so boost their delivery potential. Figure 8.1.2 displays Burd et al.'s measurements of leaf fragment sizes in relation to these predictions. Workers cut smaller leaf fragments than predicted by individual rate maximization, and the observed fragment sizes more closely matched the predictions of colony rate maximization. Ydenberg and Schmid-Hempel (1994), Kacelnik (1993), and Roces and Nuñez (1993) provide more discussion of load size in leaf-cutter ants.

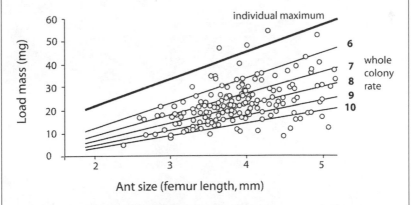

Figure 8.1.2. Load masses of leaf fragments cut by leaf-cutter ants (*Atta colombica*) from the tree *Tocoyena pittieri*. The line labeled "individual maximum" shows predictions based on maximization of individual delivery rates in the absence of queuing. The lines labeled "whole colony rate" show predicted load masses if ants maximize delivery to the colony taking queuing into account, the magnitude of which is given by the parameter λ. The "whole colony" lines lie below the "individual maximum" line and better match the data. (After Burd et al. 2002.)

(Box 8.1 continued)

In these two examples, interference and cooperation at resource collection sites both result in a tactical reduction of load size by provisioners. In other early studies, Martindale (1982) and Ydenberg and Krebs (1987) considered how territorial intruders affected provisioning tactics and found theoretical and empirical support for the idea that intruders cause a reduction in load size and patch residence time. Central place foraging models provide a simple framework for investigating the effects of social interactions on provisioning.

Bee-eaters feeding themselves or fledged young at these same sites (i.e., with no travel to the nest involved) ate many small prey, confirming that they must have been rejecting opportunities to deliver small prey in favor of waiting for larger items. Krebs and Avery used their field measurements to predict the critical travel times beyond which delivery of small items was no longer worthwhile and compared their predictions with their observational data (fig. 8.1).

"Multiple-prey loaders" face a different problem: they must decide how many prey items to collect before they return to the central place. Larger loads require increasingly long loading times, so multiple-prey models predict that foragers should collect large loads only when they must travel a long way from the central place. Kacelnik (1984) studied European starling (*Sturnus vulgaris*)

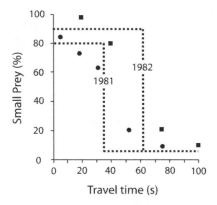

Figure 8.1. Measured and predicted composition (percentage of small prey) of prey collected for delivery by bee-eaters from capture sites distant from the nest in two different years. The dashed line shows the diet predicted by an energy gain–maximizing central place foraging model, which below the critical travel time should contain small and large prey in proportion to availability, and above it only large prey. The shaded bar shows the location of the best-fitting threshold, plus standard error, estimated from the data. (After Krebs and Avery 1985.)

parents collecting mealworms according to an experimentally controlled schedule and at manipulated travel times. Individual starlings clearly upheld the basic prediction that larger loads are a consequence of longer travel times (fig. 8.2).

Readers should understand that central place foraging is not synonymous with provisioning. The former uses the structure of repeated excursions from a central place to a site where some resource is collected. The essential feature of provisioning is the collection of a resource that does not fuel the provisioner's energy supply (e.g., nesting material or food for another). Many central place problems involve provisioning, but others, such as diving by air-breathing animals (Ydenberg 1988) or surface breathing by aquatic animals (Kramer 1988) clearly do not.

Currencies

What should central place foragers maximize? Kacelnik (1984) compared the load sizes that his European starlings collected with the predictions of four objective functions, or "currencies." The currency he called *delivery* is the total delivery of prey energy to the nest on each trip, divided by round-trip time. The currency called *yield* subtracts from the total delivery the amount of energy spent by the parent on each trip, all divided by round-trip time; while that called *family gain* further subtracts the energy spent by the young during each trip. These three closely related measures are all rates and are all expressed in units of watts (joules per second). The fourth currency is somewhat different: it takes the total delivery and divides by the energy expended by the parent. We call this currency *efficiency* (joules delivered to the nest per joule expended by the parent), and it has no units. Statistically, the family gain currency matched Kacelnik's observations best, but all four currencies made similar predictions, and he could not discriminate among them unambiguously.

Houston (1987) pointed out that all of these currencies combine the energy budgets of the parents and young in ways that do not accurately reflect who receives and who pays for the delivered energy. For example, yield subtracts the energy the parent expends from the energy delivered to the young, even though parents do not consume the prey they deliver to the nest. Field studies show that parents regularly consume prey items at the collection site, but always before beginning to collect a load for delivery (Brooke 1981; Kacelnik 1984; Krebs and Avery 1985). Central place foraging models simply ignore this self-feeding, and none of these studies accounted for it in making model predictions.

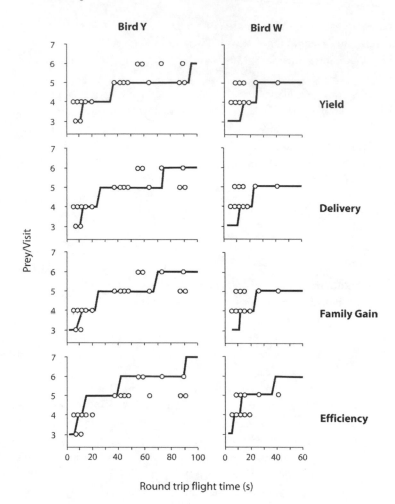

Figure 8.2. Number of prey (mealworms) collected for delivery to a nest by parent starlings from a feeding table at which the experimenter made prey available on a controlled schedule. The graphs show data (open circles) for two birds (Y and W) in relation to the predictions (solid line) of the four central place foraging currencies described in the text. Note that the data represented in the four panels for each bird are the same, but the prediction changes slightly. (After Kacelnik 1984.)

Provisioning Models

In a key step of the development of provisioning models from central place foraging models, modelers slowly recognized that they should account separately for the energy delivered to nestlings and the energy parents consume and expend (Ydenberg and Schmid-Hempel 1994). Only one central place foraging study published before Houston's (1987) paper recognized this key distinction. In a model of flight speed for parent birds delivering food to off-spring, Norberg (1981) separated parent and offspring accounts by requiring that provisioners spend some time acquiring the food needed to cover the costs of the trip. We can measure delivery as the amount of energy or material delivered (e.g., to offspring) over some period without confusing this with the provisioner's own energetics. So, a conceptually correct provisioning model must find the tactic that maximizes delivery, subject to the requirement that the provisioner (in this case, the parent bird) spend enough time to meet its own energy requirements. As Houston (1987, 255) says of the parent bird example, "the strategy that maximizes fitness is the strategy that maximizes the conversion of the parent's time and energy into energy for the young."

I call models with this explicit treatment of self-feeding "provisioning" models to distinguish them from central place foraging models. The differences are small but significant. Provisioning models keep the parent's energy budget separate from the energy delivered to the brood by measuring the parent's energy budget not in joules, but as the time the parent needs to find the food to balance its own books. This distinction means that we do not have to measure delivery in units of energy. We can consider the delivery of water to cool a wasp nest (e.g., Kasuya 1982), sticks to build a nest (e.g., McGinley 1984; Nores and Nores 1994), or any other material.

The Basic Provisioning Model

After delivering one prey item, a great tit must immediately turn around and fly back to find another. How fast should it fly to the foraging site? Faster flight, of course, reduces travel time, but it also increases the time that must be spent in collecting fuel for the trip. As Norberg (1981) noted in his original paper on the topic, the delivery-maximizing flight speed depends on the time that the provisioner must spend in feeding itself. The basic provisioning model analyzes this problem.

To find the solution, we assume that the provisioner can choose from a list of n behavioral tactics $i = 1, 2, 3, \ldots, n$. The tactics could be successively higher travel speeds, successively shorter patch residence times, successively smaller minimum prey sizes, or variations on any of the other tactics listed in table 8.1. When the provisioner uses delivery tactic i, it expends energy at rate c_i and delivers food at rate d_i. By convention, we arrange the provisioner's options in order of energy expenditure, so using option 1 costs the least per unit time, and using option n costs the most. The provisioning model finds the tactic (choice of i) that maximizes the total delivery over some time period (usually a day), called D_i. Typically the provisioner faces a trade-off because options that deliver food at higher rate also cost more to implement, and so require more self-feeding time.

Next, we divide a provisioner's time budget into time spent in delivery, self-feeding, and resting, so that total time $T = t_d + t_s + t_r$. The provisioner must allocate enough time to self-feeding, t_s, to maintain a positive energy balance. During self-feeding the provisioner obtains energy at rate b_s and expends energy at rate c_s (obtaining a net self-feeding rate of $b_s - c_s$). When at rest, the provisioner expends energy at rate r.

With estimates of the basic cost and delivery parameters, we can easily calculate how much delivery time each option allows, and so compute the total daily delivery. The provisioner must maintain a positive energy balance, and so the energetic gain while self-feeding must equal the energetic expenditure on all activities. To begin, we assume that nothing limits the provisioner's total energy expenditure, which means that the provisioner doesn't need to spend time resting. (We consider this assumption further below.) With this simplification, self-feeding at rate b_s for time t_s recovers the day's energy expenditure, so that $b_s \cdot t_s = t_d \cdot c_i + t_s \cdot c_s$. Solving for t_d yields the time available for delivery after accounting for the time that the provisioner must spend self-feeding:

$$t_d = \frac{t_s(b_s - c_s)}{c_i}. \tag{8.1}$$

To find the total daily delivery for option i, we multiply the time available for delivery [eq. (8.1)] by the rate of delivery (d_i):

$$D_i = \frac{d_i\, t_s(b_s - c_s)}{c_i}. \tag{8.2}$$

Equation (8.2) summarizes the relationships between the net self-feeding rate ($b_s - c_s$) and the provisioning tactics available. A heightened net self-feeding rate increases the time available for delivery. However, it may at the same time allow a higher-workload tactic (higher c_i) to increase the total delivery. Generally speaking, higher self-feeding rates permit the provisioner to sustain harder work, and the tactic that maximizes total daily delivery intensifies from lower-delivery to higher-delivery tactics (increasingly higher c_i) as the self-feeding rate rises. Figure 8.3 gives a worked example.

The role of the self-feeding rate in these predictions helps us resolve a puzzle in foraging theory. Central place foraging models generally use performance criteria such as "maximize the net rate of energy gain," but studies have sometimes found that efficiency maximization gives a better fit to the data

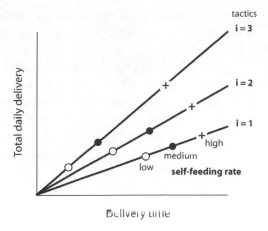

Figure 8.3. The dependence of total daily delivery on the self-feeding rate and the tactical options available, as described in equations (8.1) and (8.2). The three lines labeled $i = 1, 2, 3$ represent three successively higher-workload delivery tactics. For each tactic, open circles indicate the delivery time attainable if the provisioner adopts a low self-feeding rate; solid circles, an intermediate self-feeding rate; crosses, a high self-feeding rate. Along the line representing any tactical option, delivery time, and hence total delivery, increases with self-feeding rate, but at any self-feeding rate, working harder reduces the attainable delivery time. A shift to higher workloads with increasing self-feeding rates maximizes the total daily delivery.

(Ydenberg 1998). Provisioning models can explain this, because the predicted behavior depends on the self-feeding rate. The term d_i/c_i in equation (8.2) represents the efficiency of option i: at low self-feeding rates, the total delivery is determined largely by its value, and behavior (i.e., choice of i) should match that predicted by an efficiency (or efficiency-like) currency. As the self-feeding rate increases, it becomes possible to sustain a higher workload, and the measured behavior should approach the predictions of the three rate-maximizing currencies. McNamara and Houston (1997) give a general derivation and discussion of this important point. Thus, a provisioning model can accommodate rate-maximizing and efficiency-maximizing behavior within a single framework.

Few studies have tested this critical prediction. (Figures 8.1 and 8.2 show measured behavior as well as predictions about behavior based on central place foraging currencies, but provisioning predictions require an estimate of the self-feeding rate, which we do not yet have.) Waite and Ydenberg (1994a, 1994b) measured the deliveries of Canada gray jays (*Perisorus canadensis*) hoarding raisins. Birds came to a feeder where they could have one raisin immediately and obtain two more if they waited an experimentally controlled time. (Waiting at the feeder for the larger load is a lower-workload tactic because waiting is an inexpensive activity relative to flying and hoarding.) Obviously, jays can do better with three-raisin loads when the waiting time is short.

Waite and Ydenberg (1994b) showed that birds shifted abruptly from three-raisin to one-raisin loads as they increased the experimental waiting time. More importantly, each individual shifted to lower-workload tactics during the winter, when the self-feeding rate presumably falls (Waite and Ydenberg 1994a). Figure 8.4A summarizes these results.

A direct experimental test would manipulate the self-feeding rate and predict the effect on provisioning behavior. Palestinian sunbirds (*Nectarina osea*) feed insects to their nestlings (Markman et al. 1999), but feed themselves largely on nectar. Few bird species show such a marked difference in parental and nestling foods (but see Davoren and Burger 1999), so Palestinian sunbirds provide an opportunity to manipulate the provisioner's self-feeding rate. Markman et al. (2002) randomly assigned sunbird territories to low, medium, or high self-feeding rate groups, which they manipulated by varying the sugar concentration in feeders placed in the territory. Changes in sugar concentration caused a variety of behavioral changes. Parents worked harder when high sugar concentrations produced high self-feeding rates: they visited the nest more (fig. 8.4B) and reared larger nestlings. Although not designed to test a provisioning model (Markman placed his work in a life history framework), these results agree with the expectations of the provisioning framework.

Markman et al. controlled the self-feeding rate in their experiment, but in nature, provisioners can often make decisions about their self-feeding rate. For example, parent bee-eaters feed themselves on prey caught at the same sites where they capture prey for their nestlings. As each potential prey item flies by, they must decide whether to ignore it, catch and eat it, or deliver it to their nestlings. This decision process affects the self-feeding rate, and hence the achievable delivery rate. In general, a change in the self-feeding options alters provisioning behavior, even if the provisioning options do not change (Houston and McNamara 1999).

This central feature of provisioning models has wide-ranging implications. For example, students of avian breeding systems have assumed that the brood size of territorial birds increases with prey density because birds can find and deliver prey more easily. High prey densities could also mean that parents can achieve higher self-feeding rates, so that they can work harder at food delivery. We will need imaginative experimental work controlling both delivery and self-feeding opportunities to resolve this issue (e.g., Kay 2004). The possibility that different locations provide opportunities for self-feeding and food for delivery has interesting implications for provisioning; box 8.2 gives an example.

A.

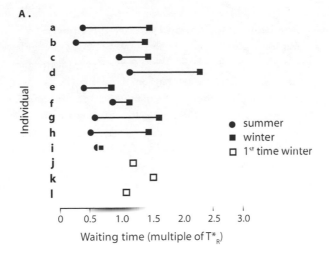

B.

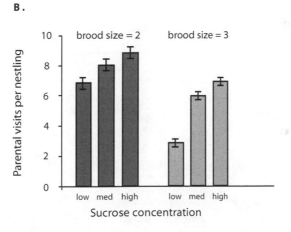

Figure 8.4. Harder work with higher self-feeding by (A) Canada gray jays hoarding raisins and (B) Palestinian sunbirds. (A) Measured threshold waiting times (plus 95% CI) relative to a standard for nine individual jays (a–i) measured in summer (solid circles) and again in winter (solid squares), when the self-feeding rate was lower. Individuals worked harder (waiting time was shorter) in summer when the self-feeding rate was higher. The waiting times for three jays measured for the first time in winter (j–l, open squares) indicate that an order effect cannot explain the observed difference. (After Waite and Ydenberg 1994a). (B) The number (with SE) of parental nest visits per nestling for Palestinian sunbirds with broods of two or three nestlings receiving low (0.25 M), medium (0.75 M), or high (1.25 M) sucrose concentrations in feeders on their territories.

8.4   Energy Metabolism and Provisioning Capacity

Davidson (1997; see also Kay 2004) found that tropical rainforest canopy ecosystems are dominated in both numbers and biomass by several hard-working ant species. These "high-tempo" species all feed on plant or homop-

BOX 8.2 **Provisioning and Spatial Patterns of Resource Exploitation**

Provisioning mason bees (*Osmia lignaria*) feed themselves nectar, but deliver both pollen and nectar to their nests. Williams and Tepedino (2003) placed nest boxes so that bees could fly in one direction to a patch of flowers with a high nectar content, but little pollen (*Heterophyllum capitatum*), or in the opposite direction to a patch with high-pollen but low-nectar flowers (*Salix* spp.). By examining pollen loads on returning bees, they concluded that the bees must have visited patches of both types of flowers on virtually every provisioning trip, in spite of the fact that this must have involved a lot of extra travel. How could this happen?

To model this situation, consider a provisioner with a central place located midway between two food patches. Patch 1 provides food for self-feeding and for delivery to the central place, while patch 2 provides food only for self-feeding. Are there any conditions under which a delivery-maximizing provisioner should visit both patches?

Each time the provisioner visits patch 1, it collects a load of size L for delivery, which requires time t_p. Round-trip travel to either patch requires time t_t. On each visit to patch 1, the provisioner must spend time t_{s1} self-feeding to cover the energetic costs of the trip. Alternatively, the provisioner could cover its costs by traveling occasionally to patch 2, where it can achieve a higher self-feeding rate, and spending time t_{s2} self-feeding. To balance its energy budget using this two-patch tactic, the provisioner must make x ($x < 1$) trips to patch 2 (on average) for each trip to patch 1.

When using only patch 1, the provisioner achieves a delivery rate of $d_1 = L/(t_t + t_{s1} + t_p)$. When it uses both patches, it can deliver at rate $d_2 = L/[(1 + x)t_t + x(t_{s1} + t_p)]$. The provisioner should use both patches if $d_2 > d_1$. Substituting our delivery rate expressions and simplifying gives

$$x(t_t + t_{s2}) < t_{s1}$$

In words, it makes sense to use both patches when the time that the provisioner must commit to self-feeding from patch 2 (which includes the travel time there and back) is less than the self-feeding time at patch 1. One can easily find combinations of parameter values for which this condition holds. Although they do not use this modeling format, Williams and Tepedino (2003) propose essentially this explanation for the puzzling provisioning behavior of mason bees.

teran exudates (honeydew)—substances that provide lots of energy but little protein. Workers consume these high-energy foods while they feed protein-rich prey (mostly arthropods) to the brood. The provisioning framework developed in section 8.3 emphasizes the relationship between a provisioner's energy metabolism (expressed as c_i, the energy expenditure rate of delivery tactic i), fueled by self-feeding at rate $b_s - c_s$, and the consequent delivery rate (called d_i). These parameters summarize many aspects of an organism's phys-iological ecology. This section explores some features of energy metabolism relevant to provisioning models and provides a framework for understanding how it is that some species can be so hard-working.

Consider again our stereotypical avian central place forager that works as hard as possible delivering prey to its offspring. Over the past three decades, many studies have sought to characterize energetic capabilities by measuring and comparing energetics in the field. Drent and Daan (1980) argued that free-living birds could not expend more than four times their basal metabolic rate (BMR). Kirkwood (1983) reached a similar value by comparing records of maximum food intake for various species. More recently, Hammond and Diamond (1997) compiled information on the maximum sustained metabolic scope (SMS), measured as the ratio of mass-adjusted sustainable metabolic rate (SusMR; kJ/gd) to resting metabolic rate (RMR; kJ/gd), and found that it ranges widely among species, but has a median value of about four. Similarly, there is great interspecific variation among social insect species in the speed and energetic cost of worker task performance (called "tempo" by Oster and Wilson 1978).

In the most comprehensive review of the avian literature, Williams and Vézina (2001) listed more than fifty field studies that measured the energetic expenditure of birds during reproduction, and concluded that our understand-ing of intraspecific variation remains rudimentary. The fact that two con-generic seabirds, the masked booby (*Sula sula*) and the gannet (*Sula bassanus*), show strikingly different work rates (1.6 BMR for boobies and 6.6 BMR for gannets) illustrates the depth of our ignorance. How does this variation arise? Why don't all provisioners work hard? Ecologists usually turn to life history theory to explain this variation, but this subsection considers purely energetic possibilities.

The development of equation (8.2) assumed no limit on the amount of energy that a provisioner can expend, but some sort of limit probably exists. Several things might limit daily energy expenditure, including access to food, the rate at which muscles can generate power to do work, or the rate at which fuel or oxygen can be assimilated or distributed to the musculature (Hammond and Diamond 1997). These limitations could influence provision-ers in several ways. If access to food limited the rate of energy expenditure, then a provisioner's time would all be occupied either by self-feeding or by delivering prey. When food is abundant, in contrast, self-feeding would allow a high rate of energy intake, but limitations on muscular activity or assimilation might limit the provisioner's energetic output. In the face of these processing limitations, a provisioner could either spend some of the day resting in order to avoid exceeding the maximum sustainable expenditure (as in some seabirds; Houston et al. 1996) or (if assimilation limits the conversion of input to energetic output) use reserves to increase the total daily energetic expenditure.

If the objective is to maximize total daily delivery, a time-limited provisioner should maximize delivery per unit time, and an observer would record performance matching the predictions of a rate-maximizing currency. However, when energy expenditure limits total daily delivery, a rate-maximizing provisioner would reach the expenditure ceiling before the end of the available time, and would have to spend the remainder of its time resting. In this scenario, it makes sense to maximize delivery per unit energy expenditure (i.e., maximize efficiency), and models show that less expensive options that deliver at a lower rate can achieve a higher overall rate of delivery in these situations. In this case, an observer would record behavior that matches the predictions of efficiency or modified efficiency currencies. These issues are treated more fully by Hedenström and Alerstam (1995), Ydenberg (1998), Houston and McNamara (1999), and Nolet and Klaassen (2005). Thus, provisioning behavior operates within an envelope bounded at one extreme by rate maximization and bounded at the other extreme by efficiency maximization. Self-feeding rates determine the predicted behavior: high self-feeding rates should shift provisioners toward higher workloads, while low self-feeding should have the opposite effect.

Why should some provisioners have high energy capacities while others have low energy capacities? Hammond and Diamond (1997) suggest that animals with high energy capacities need expensive metabolic machinery, including organs with high metabolic rates such as the liver, heart, and kidneys (Daan et al. 1990; see also section 5.3). Enhanced metabolic performance (e.g., Suarez 1996, 1998) can evolve under selection or develop on a physiological time scale within individuals, but it always comes at the cost of a metabolic machine that is more expensive to run. So why does it make sense for some animals to maintain this expensive machinery while others do not?

In the framework developed here, potential delivery depends not only on prey availability but also on the provisioner's capacity for hard work, which in turn requires fuel from self-feeding. If high delivery rates enhance fitness, then better self-feeding opportunities also allow increased metabolic capacity. For example, dominant individuals probably have better access to food (e.g., Hogstad 1988), and so can support higher metabolic rates. Bryant and Newton

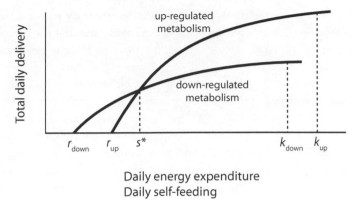

Figure 8.5. Hypothetical scheme of relationships between total daily delivery (on *y*-axis), resting metabolic rate (*r*), self-feeding rate, daily energy expenditure, and maximum daily energy expenditure (*k*) (all on *x*-axis). Total daily self-feeding and total daily energy expenditure must balance. As self-feeding increases, the provisioner can expend and deliver more. The lower curve shows the relationship for a "down-regulated" metabolism, and the upper curve for an "up-regulated" metabolism. Up-regulation gives higher deliver capacity, but also generates a higher resting metabolic rate, and so is more expensive to maintain. In the example shown, the provisioner benefits from up-regulated metabolism when the attainable self-feeding rate exceeds the rate labeled *s**.

(1994; see also Hogstad 1987) interpret the higher BMR of dippers (*Cinclus cinclus*) as a cost of dominance, but if high BMR translates into better provisioning, the chain of causation could be reversed: the higher BMR may be a benefit of dominance. Figure 8.5 shows these relationships diagrammatically.

Davidson (1997) suggests that some ant species dominate rainforest canopies because they can easily obtain high-energy exudates that favor the evolution of high tempos. These high-tempo ant species can deliver protein resources at a high rate and can vigorously defend territories, traits that in turn lead to the high reproductive rates that give these species their dominance. This example shows the broader ecological consequences of provisioning behavior.

Other social insect studies suggest similar relationships. Four honeybee species in the genus *Apis* show marked differences in worker mass-specific metabolic rate, colony metabolism, and the intensity of provisioning. The high-tempo species (*A. mellifera* and *A. cerana*) deliver more resources than the low-tempo species (*A. dorsata* and *A. florea*) and produce offspring at a greater rate, but have higher worker mortality (Dyer and Seeley 1991). We can examine the great interspecific variation in metabolic expenditure documented among birds in the same way. In fact, some investigators hypothesize that endothermy itself—a particularly expensive metabolic mechanism—has evolved to enhance aerobic capacity and support the vigorous exercise required in many forms of provisioning and parental care (Farmer 2000).

As these examples imply, close ecological relationships connect the energetically expensive activities of animals (such as provisioning), the foraging behavior that obtains the fuel, and the metabolism that powers the activity. Nonetheless, few studies have as yet considered these ideas and their implications.

8.5   Demand and Provisioning

An increase in brood size or an approaching period of shortage can increase the demand for delivered food. None of the provisioning models described so far consider variations in demand. Their assumptions and equations incorporate only "supply-side" parameters such as encounter rate (e.g., waiting time), travel distance, self-feeding rate, and the energetic cost of delivery. This section considers how demand may affect provisioning. What properties should we expect provisioning strategies to have for responding to demand?

A large number of excellent studies on birds and social insects demonstrate that provisioners respond to natural variation in demand, increasing the delivery of materials when demand rises and vice versa. We expect this, of course; demand varies in nature. Demand can vary predictably (e.g., nestlings grow), or unpredictably (e.g., bad weather). Experimentalists have used a variety of methods to manipulate demand. In studies of birds, the most common experimental technique manipulates clutch or brood size. Interest has been concentrated on the fitness consequences for parents and offspring, with little attention given to the tactics parents use to increase delivery. Social insect studies have usually focused on how colonies recover their populations and stored reserves after an experimentally imposed demand. A few studies have considered the behavior of individual workers (Cartar 1991; Schmid-Hempel et al. 1993).

Provisioners might use several basic tactics to meet increased demand. First, provisioners could simply spend more time provisioning. The longer hours of work will necessarily increase their daily energy expenditure. Second, a provisioner can work harder—for example, by flying faster. When experimenters remove pollen stores from a beehive, more workers collect pollen, and each individual worker works harder at the task (Eckert et al. 1994). Third, provisioners can use energy from body reserves to fuel extra delivery effort. And finally, provisioners can alter the selection of prey for delivery. I discuss fueling of delivery from reserves and prey selection in the next few paragraphs.

Powering Delivery from Stored Reserves

Provisioners can overcome the limits imposed by self-feeding on energetic expenditure by using reserves to fuel activity (Houston 1993). Many studies have reported reductions in the body mass of parent birds during periods of intense provisioning activity (e.g., Moreno 1989). Most researchers take this to indicate that energy expenditures exceed energy intake during provisioning. A parallel body of results describes the responses of social insect colonies, especially honeybees, to the challenges of imposed parasite loads (Janmaat et al. 2000) or removal of pollen (Fewell and Winston 1992). These studies commonly show that colonies reduce their reserves as they react to the manipulation and recover to their former state.

In analyzing whether provisioners should "dip into" their reserves to address an unexpected demand, most models assume that reduction of the provisioner's stores has life history costs, such as an increased risk of starvation. The provisioner must balance these costs against the advantages of increased delivery capacity. Nur's (1987) model provides a paradigmatic example. It seeks to explain patterns in the provisioning responses of songbird parents to (manipulations of) brood size. The model assumes that increased feeding frequency increases nestling weight and survival, but reduces parental weight and survival. Nur concludes that parents feed larger broods more frequently because the greater fitness value of the brood makes increased effort worthwhile (see also Beauchamp et al. 1991).

Swifts (*Apus apus*) delivered more food to experimentally enlarged broods, but each nestling received less, resulting in a lower mean chick mass (at age 12 days; Martins and Wright 1993b). In addition, Martins and Wright found that parents lost mass during the provisioning period, and they assumed that a reduction in self-feeding caused this weight loss. They argued that this weight loss imposes a risk on parental survival, but others have suggested that provisioning swifts may lower their body weights to reduce flight costs. Neither idea seems complete. Why is mass loss risky, given that parents can recover quickly? On the other hand, if lowering mass lowers flight costs, why don't all parents lose weight?

Thinking about the relationship between the use of a fat reserve and self-feeding suggests a slightly different explanation. A small reserve of fat could boost a parent's delivery rate, either by providing the power for a period of energetic expenditure above the sustainable limit or by reducing the need for self-feeding for a brief period. A provisioner could expend this fat reserve slowly and steadily on a programmed schedule to meet a predictable increase in demand (e.g., growing nestlings), or it could expend its reserve all at once to meet demand during unpredictably poor conditions (e.g., cold, rain). Under many circumstances, the provisioner would be able to "restock" its reserve when conditions improve. The key idea here is that expending the reserve need not endanger the provisioner's survival. The provisioner maintains the reserve (presumably at a small ongoing cost to provisioning capacity) in order to buffer provisioning capacity against fluctuations in prey availability.

Prey Selection

Provisioners can alter their selection of prey for delivery, and this gives them another way to address the problems of increased demand. Scores of papers, including many good experimental studies, reveal that provisioners change the size or type of prey delivered when the demand at the delivery point changes (e.g., Siikamäki et al. 1998; see Moore 2002 for a review). In some cases, parents deliver larger prey to larger offspring; this could occur for the simple reason that small offspring cannot swallow large prey. A more sophisticated hypothesis holds that parents change prey selection to boost the energy delivery rate. For example, when experimenters increase brood size, European starling parents deliver poorer but more easily obtained prey, increasing the delivery of energy at the expense of other nutrients (Wright et al. 1998).

A third possible hypothesis involves variance sensitivity (Stephens and Krebs 1986). (The ecological literature usually calls this concept "risk sensitivity," a term borrowed from economics, where it refers to variable returns on invested capital. Unfortunately, ecologists also use the term "risk" in other ways, as in "risk of predation." Substituting the terms "variance" or "danger," as appropriate, eliminates any potential confusion.) From their beginnings, central place foraging models focused on how changes in provisioning tactics (e.g., selectivity in a single-prey loader) affect the mean delivery rate. If there is any stochasticity in components of the provisioning process (e.g., capture time, handling time, prey size), there will also be variance about the mean delivery rate. The variance itself will also be affected by the tactic chosen. So, in the most general case, provisioning tactics affect both the mean and variance of delivery rates.

Variance becomes important when deviations above and below the mean delivery rate have different effects on fitness. The original development of variance sensitivity theory focused on starvation avoidance (Stephens and Krebs 1986). In these "shortfall-avoidance" models, falling below a requirement results in a different fitness outcome (starvation) than exceeding it (survival). However, the basic idea applies whenever the cost of falling below the mean differs from the benefit of exceeding the mean. Thus, we expect that provisioners will show variance-sensitive behavior when fitness increments resulting from delivery above and below the mean are unequal (Ydenberg 1994). This asymmetry could arise via falling short, as the early model imagined, but it could also arise if the growth (and hence fitness) of offspring shows diminishing returns with delivery.

Figure 8.6 gives a worked example to show how variance sensitivity can shape provisioning tactics. A parent common tern (*Sterna hirundo*) flies from its breeding colony to a lake, where it searches for a fish. Fish vary in size, and the tern encounters them sequentially as it flies over the lake. When highly selective, a tern spends more time searching for a suitable fish, but delivers larger fish than when it is less selective. As the tern becomes increasingly selective, the total daily delivery initially rises because it delivers larger fish, but the delivery rate falls if the tern becomes too selective, because it then spends too much time searching for a suitable fish. Selectivity also affects the variance in daily delivery: at low selectivities, the tern spends little time catching prey and most of its time ferrying small prey to the nest. This tactic leads to a total daily delivery rate with little variation. For a selective tern, on the other hand, the time to capture an acceptable item varies greatly, and the variation in total daily delivery rises.

Moore (2002) calculated the total daily delivery resulting from each possible prey selection tactic using a computer simulation. Moore's simulation created stochasticity by randomly drawing prey items from a given size distribution. Each time the simulated provisioner encountered a prey item, the computer applied a minimum acceptable prey size rule. When the item exceeded the minimum size, the provisioner delivered it to the nest; otherwise, the provisioner continued searching. The program computed total daily delivery for 1,000 simulated days, then calculated the mean and standard deviation of total daily delivery from this distribution. Moore (2002) used the "z-score" method of Stephens and Charnov (1982) to show that when the demand at the nest rises above the expected delivery, provisioners should adopt more variance-prone tactics, which in this case means becoming more selective (i.e., delivering larger prey). In field experiments, he manipulated the brood size of common terns and found that, as predicted, the mean size of prey delivered increased with brood size (see fig. 8.3.1). Moreover, natural variation in brood size and interannual variation in prey availability led to changes in prey selection that Moore could explain in the same way. Moore's study suggests that common tern parental provisioning repertoires regularly include variance-sensitive responses. Box 8.3 explores this idea further.

So far, I have illustrated how any one of several basic tactics can influence the delivery that a provisioner can attain, but this one-at-a-time analysis probably does not reflect reality. In nature, real provisioners must simultaneously decide what kind of prey to deliver, how fast to fly, and where to search. The empirical studies by Moore (2002) and Wright et al. (1998) on the common tern and starling provide the most complete pictures to date. Both studies found that in response to experimentally manipulated brood sizes, parents changed the amount of time they spent delivering and altered their

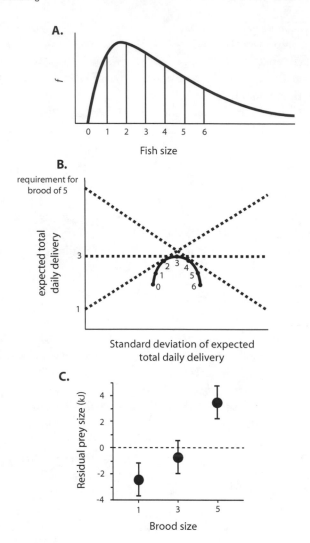

Figure 8.6. Variance-sensitive provisioning in the common tern. (A) Parents travel back and forth between the resource point and the breeding colony, capturing a single fish on each trip from the size distribution shown. Parents can capture any encountered prey (shown as tactic 0) or can be progressively selective for larger prey (tactics 1–6). (B) The expected means and standard deviations in total daily delivery resulting from each prey selection tactic are shown here, and the "z-score" method of Stephens and Charnov (1982) applied to find the tactic minimizing the probability of a shortfall for brood sizes of 1, 3, and 5, indicated by tangents. The model predicts that as demand (brood size) rises relative to the expected delivery, parents should become more selective. (C) Prey choice of parental common terns. The graph shows the residual (with annual differences taken into account) prey size (with standard error) delivered in relation to manipulated brood size. (Data from Moore 2002.)

selection of prey for delivery, as well as the amount of self-feeding, but did not change their flight speed (i.e., workload). Moore (2002) interpreted the observed changes in prey selection, delivery time, and self-feeding as tactics

BOX 8.3   Variance–Sensitive Provisioning

a

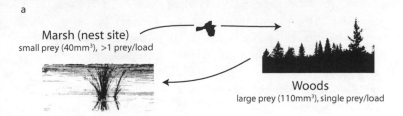

Marsh (nest site)
small prey (40mm³), >1 prey/load

Woods
large prey (110mm³), single prey/load

normal nestling diet 50/50 by volume

(% small prey delivered, by volume)

	Deprivation		Satiation	
	control	experiment	control	experiment
Before	46	51	41	56
After	44	79	46	28

c

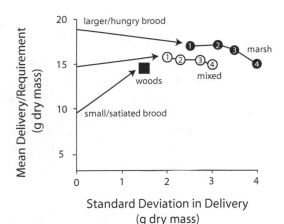

Figure 8.3.1. Variance-sensitive provisioning by red-winged blackbirds. (A) Parents choose loads of several small prey from the marsh surrounding the nest or single large prey from a distant woodland. (B) Whittingham and Robertson's (1993) experimental results show that parents normally deliver about 50% small and 50% large prey. When the investigators deprived broods, the blackbirds delivered more small prey, but they delivered more large prey when the investigators fed broods to satiation. (C) The shortfall-minimizing diet (small or large prey, calculated as in Figure 8.6B). Numbers inside the symbols represent the number of items loaded per trip. The model predicts that parents should deliver more *small* prey when demand increases, but should deliver large prey when demand decreases. (From Moore 2002.)

(Box 8.3 continued)

Whittingham and Robertson (1993) studied the provisioning of nestlings by red-winged blackbirds (*Ageliaus phoenicus*). Parents captured prey either in the marsh surrounding the nest, where they captured loads of small insects, or in distant woodlands, from which they delivered a single large caterpillar. Typically, parents delivered equal volumes of small prey and caterpillars. In a carefully controlled field experiment, Whittingham and Robertson deprived or satiated nestlings and measured the provisioning response of the parents. They found that after deprivation the proportion of small (marshland) prey increased, while after satiation it decreased. At first blush, this observation seems to disagree with the study of common tern provisioning described in the text. In that study, parents became more selective when feeding larger, more demanding broods, as our model of variance-sensitive provisioning predicted. Can variance sensitivity also explain a switch to smaller prey, as Whittingham and Robertson observed?

Variance sensitivity theory would predict that in this case, small prey represent the more variable option. When experimental deprivation of nestlings increases demand, parents become variance-prone (fig. 8.3.1) and so use the small-prey option more. Based on parameter estimates supplied by Linda Whittingham, Dave Moore (2002) used the simulation approach described for the common tern study, and found that marshland prey do indeed give more variable returns. The blackbirds spend most of the trip to the distant woodland in flight, and the large prey there vary little in size, so the delivery rate varies little per trip. However, when they forage close to the nest, small differences in prey size and in the rate of prey discovery greatly affect the rate of return. So Whittingham and Robertson's observations agree with the variance sensitivity hypothesis. In this case, however, small prey offer the more variable option, while larger prey vary more in the common tern case (see fig. 8.6).

that terns used when the brood requirement exceeded the expected delivery. He also suggested that terns would not benefit from increasing flight speed during provisioning, calculating that the extra self-feeding needed to fuel faster flight would have required more time than the increase in flight speed would save. These field studies confirm that real provisioners use an array of tactics to increase delivery during high-demand periods.

Social insects have additional tactics at their disposal because colonies can increase the number of foragers as well as the effort of each forager. For example, Janmaat et al. (2000) experimentally created different levels of demand in honeybee colonies by removing pollen reserves and creating infestations of parasites (*Varroa jacobsoni* mites). Both treatments had effects: the parasite treatment increased the pollen load collected by provisioners, while the pollen reserve treatment increased the number of provisioners that collected pollen. Both tactics boosted delivery rates, and colonies with high delivery rates converted pollen into new brood less efficiently than colonies with low delivery rates (for similar results, see Fewell and Winston 1992; Eckert et al. 1994).

No one has so far considered how provisioners should combine the various tactics at their disposal to meet heightened demand. For social insects with a large worker force to allocate to various tasks, Schmid-Hempel et al. (1993) argue that the increases in delivery resulting from more foragers as well as from greater effort by each forager both show diminishing returns. Therefore, moderate, linked increases in both tactics are better than a large response in only one. This simple approach could form the basis of an investigation of this problem in other systems.

8.6   Provisioning from a Life History Perspective

Provisioning is a dangerous business for many animals. Rhinoceros auklets (*Cerorhinca monocerata*), for example, are vulnerable to predators on land, and they are very cautious about coming ashore. Even in the forest at night on offshore islands, predators are a hazard for these birds. Harfenist and Ydenberg (1995) showed that parent auklets terminated provisioning sooner in areas of colonies in which eagles preyed on adults. They also showed that this early termination of parental care affected the mass at which offspring left the nest, which presumably had an effect on their survival prospects in turn. Thus, as with many other aspects of behavior, if we are to understand provisioning behavior, we must view it as part of the animal's life history, because decisions made at any one time (e.g., another delivery to the offspring) have consequences for both parent and offspring. Ideally, we would like a theory of provisioning behavior that integrates life history and provisioning perspectives.

Effort and Investment

Provisioning theory and life history theory share notions of "effort" and "investment" (Houston 1987). Much of the literature on parental care and provisioning, especially in altricial birds, treats effort and investment as synonyms, but this discussion will distinguish between them. Here the term "effort" will refer solely to energy expenditure. The term "investment" will refer to time, material, or effort devoted to the current delivery point that somehow reduces or compromises the provisioner's ability to give time, material, or effort to other delivery points (including future ones). By this definition, all investment has fitness costs as well as benefits, and hence selection will adjust investment ("allocation") patterns to maximize fitness.

Does the high energy expenditure of provisioning necessarily come at some life history cost? Specifically, does today's provisioning reduce a provisioner's survival or fecundity? Much of the literature on social insects makes little use of these concepts, but many avian studies have attempted to document future costs of reproduction, though with mixed results. Martins and Wright's (1993b) study of swifts provides an example. Like the parents in many other avian studies, swifts worked harder in response to increased demand, whether experimentally (brood size manipulation) or environmentally (bad weather) imposed. In some studies, avian parents sustained a higher workload for a considerable period. Social insects show similar responses when challenged in analogous experiments (Fewell and Winston 1992). Both groups of organisms often show short-term changes in body weight, condition, immune system traits, or the size of stored reserves in response to the challenge. But most studies have simply inferred the fitness consequences of provisioning effort, leaving us with little conclusive evidence. Even the study cited most often in support of the cost of reproduction (Gustaffson and Sutherland 1988) is open to an alternative interpretation. Gustaffson and Sutherland observed that female collared flycatchers (*Ficedula albicollis*) laid slightly smaller clutches in the season after their clutch sizes were experimentally increased. Lessells (1991) pointed out that this could occur if the females "recalibrated"—as if they had "overestimated" the clutch size they thought they could rear in the previous season because it proved to be too much work for them. Daan et al. (1996) offer the most compelling study. They manipulated the broods of European kestrels (*Falco tinnunculus*) and confirmed by recovering carcasses that adult kestrels with enlarged broods experienced a higher mortality rate.

This pattern of mixed results may be seen because we need large samples and prolonged field studies to measure the cost of reproduction accurately. In addition, if the cost of reproduction is small, it may be difficult to detect, especially over the range at which provisioners normally operate (Lindén and Møller 1989). However, the largest clutch manipulation study ever carried out (on the great tit, *Parus major*, reported in Pettifor et al. 1988) shows no trace of a cost of reproduction. Moreover, the failure of so many studies to detect an effect contrasts with the clear signal from similar studies that brood enlargement reduces the survival and recruitment of individual offspring (Lessells 1991).

One might also explain failures to document a life history cost of reproduction by arguing that evolutionarily recent environmental changes have reduced the costs of high effort. Such arguments have often been raised in the literature to explain, for example, why seabirds can rear enlarged broods. (Ironically enough, Wynne-Edwards's [1962] group selection "restraint" theory for seabird clutch size predicted that seabirds could raise extra young, exactly as many studies have found. Lack's [1968] clutch size theory based on individual selection predicted that seabird parents would be unable to do so, which made such apologia necessary. Ydenberg and Bertram [1988] provide a discussion of this interesting point.) For altricial birds, the absence of hawks and falcons since the late 1940s (see Ydenberg 1994 for a brief discussion) may have made provisioning of offspring much safer than it was when provisioning behavior evolved in this group. Studies have shown that the presence of raptors affects winter fat levels (Gosler et al. 1995) and fledging mass (Adriaensen et al. 1998) in breeding birds, so perhaps it is not too much to suppose that the absence of raptors over the past five decades has affected our ability to detect life history costs. Certainly no one can dispute the power of the idea of the cost of reproduction, or its importance in many groups of organisms, but three decades of study have failed to establish clearly just how life history costs play a role in provisioning behavior among birds.

The Costs of Provisioning Effort

Another way to explain why studies of avian provisioners seldom find evidence for the cost of reproduction lies in the distinction between effort and investment. The cost of reproduction represents only one reason why provisioners might make less effort than the maximum. Thus, experimentally induced higher workloads might not affect the provisioner's own survival or future fecundity. For example, a provisioner could maintain a small fat reserve, which reduces its delivery capacity. The provisioner would benefit because it could expend the reserve to increase its delivery capacity when demand unexpectedly escalates. An observer would measure this expenditure of the reserve as a loss of "condition," but this loss would not jeopardize the provisioner's survival: the provisioner maintains the reserve solely to buffer delivery capacity against shortages that occur from time to time.

Nest defense may also select for reduced provisioning effort. Dyer and Seeley (1991) explain how differences in nest architecture lead to differences in work tempo among various species of honeybees. High-tempo honeybee species nest in enclosed cavities, while the low-tempo species have open nests covered in a curtain of workers. Dyer and Seeley propose that open nests require more workers for defense and thermoregulation and, they argue, have therefore selected for reduced tempo because high-tempo workers experience high mortality rates, which makes it difficult to maintain a large worker population. Lack (1968) documented a similar correlation in altricial birds: ground-nesting species (presumably with the most vulnerable nests) have the smallest clutches and fastest growth rates. This observation suggests a relationship between nest vulnerability and provisioning tempo, as found among honeybee species, though the direction appears to be reversed. In both cases, we hypothesize, nest vulnerability has created selection on the metabolic capacity of the provisioner.

Some authors have argued that expending energy reduces longevity (e.g., Calder 1985). According to this view, a high workload reduces the life span, in the same way that overusing a flashlight more quickly leads to a dead battery. More recent authors offer more sophisticated interpretations. Some argue that energetic expenditure can reduce longevity by compromising an organism's immune response (Richner et al. 1995), while others argue that reproductive effort accelerates senescence (Gustaffson and Pärt 1990). If these arguments are correct, and if provisioning involves increased effort, then all provisioning entails a cost of reproduction.

Another possibility is that extra effort may expose provisioners to predators (Magnhagen 1991) or parasites (Richner et al. 1995). For example, Harfenist and Ydenberg (1995) found that parent rhinoceros auklets stopped provisioning sooner in areas where eagles posed a threat. Some authors argue that predation has played a role in the evolution of nocturnal provisioning and precocity in seabirds (Gaston 1992). Provisioning studies have barely scratched the surface of this problem; specifically, we still need experimental studies that document the responses of provisioners to the threat of predation. Studies of predation risk in other foraging situations illustrate the feasibility of these experiments (Lima and Dill 1990). Nonacs and Dill (1990) experimentally demonstrated that workers of the ant *Lasius pallitarsis* balanced the quality of the food available at a feeding site against the danger posed by a large predatory ant. Dukas (2001) discusses the potential effect of predation danger on insect pollinators (especially the many provisioning species such as bees) and notes a general lack of studies on the idea. Though it seems reasonable to suppose that each provisioning excursion carries some extra exposure to danger, we have, as yet, few good data with which we can evaluate how this exposure affects provisioning.

8.7   Summary

Models of provisioning have their origins in the central place foraging models of the late 1970s. In contrast to central place foraging models, provisioning models explicitly consider time for self-feeding. This small change provides a way to keep separate the energy budgets of the provisioner and those it provisions. It also reconciles some important and apparently conflicting experimental results with the theory, and it opens a window on the relationship between provisioning behavior and energy metabolism. In particular, it suggests ways to understand why some provisioners work hard while others work at a leisurely pace.

Many studies describe how conditions at the delivery point create demand that affects provisioning behavior, but central place foraging models do not take account of this. Integrating provisioning models with results from metabolic capacity studies and variance sensitivity models greatly expands the range of phenomena that our models can explain and suggests many interesting avenues for investigation. Nevertheless, important and long-standing questions about the maximum delivery capacity of provisioners, the relation of this capacity to clutch and brood size, and the possible significance of predation danger to provisioners remain unresolved.

8.8 Suggested Readings

The basic central place foraging models with which this account begins are best described and interpreted by Stephens and Krebs (1986). The ideas of my own that I present in this essay build on two previous accounts (Ydenberg 1994, 1998) that also describe provisioning models. Students interested in the subject should begin by mastering the material in these publications.

An overview of general ideas about parental care is given by Clutton-Brock (1991). Engrossing accounts of systems in which provisioning is important can be found in Gentry and Kooyman (1986) on fur seals and Winston (1987) on honeybees. Marvelous accounts of avian biology are too numerous to list; I suggest a book by Kemp (1995) on the hornbills, which have one of the more bizarre provisioning systems.

Even advanced students will find the books by Houston and MacNamara (1999) and Clark and Mangel (2000) challenging, but these works are indispensable for a modern and rigorous approach to models of provisioning.

Telander

Foraging in the Face of Danger

Peter A. Bednekoff

9.1 Prologue

A juvenile coho salmon holds its position in the flow of a brook. To conserve energy, it positions itself in the lee of a small rock. Distinctive blotches of color on its sides, called parr marks, provide effective camouflage. As long as it holds its position, it is virtually impossible to see. The simple strategy of keeping still hides it from the prying eyes of potential salmon-eaters. Kingfishers and herons threaten from above, and cutthroat trout, permanent residents of the stream, seldom reject a meal of young salmon. The threat posed by these and other predators is ever present.

The clear water flowing past the salmon presents a stream of food items: midges struggle on the surface; mayfly nymphs drift in the current. But, here's the rub: to capture a prey item, the salmon must dash out from its station, potentially telegraphing its position to unwelcome observers. When the salmon feels safe, it will travel quite a distance to intercept a food item, making a leisurely excursion to collect a drifting midge as far as a meter away from its location.

Detecting a predator changes the salmon's behavior. Depending on the level of the perceived threat, the salmon has several options. It may flush to deep water or another safe location. It may stop feeding altogether, but hold its position. It may continue feeding, but dramatically

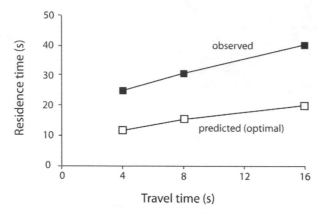

Figure 9.1. Patch residence time increases with travel time between patches (as predicted), but blue jays stay in patches much longer than the optimal residence time. Solid squares show observed residence times; open squares show the predicted optimal residence times. (After Kamil et al. 1993.)

reduce the distance it will travel to intercept food. This series of graded responses represents a sophisticated and often effective strategy to avoid predators. Sophisticated or not, all of these responses reduce the salmon's feeding efficiency. The salmon's problem is far from unique; virtually all animals face a trade-off between acquiring resources and becoming a resource for another.

9.2   Overview and Road Map

Resource acquisition is necessary for fitness, but it is not sufficient. Food is generally good for the forager, but not if the forager is dead. Danger affects animal decisions in many ways (see reviews in Lima and Dill 1990; Lima 1998). Animals often face a trade-off between food acquisition and danger: the alternative that yields the highest rates of food intake is also the most dangerous. A growing area of research focuses on this fundamental trade-off. This chapter examines how danger from predators affects foraging behavior.

Early theory assumed that fitness was highest when the net rate of foraging gain (i.e., net amount of energy acquired per unit time) was highest. Early empirical tests consistently showed that foragers are sensitive to foraging gain (see Stephens and Krebs 1986). As predicted, many animals spend more time feeding in each patch when patches are farther apart (Stephens and Krebs 1986; Nonacs 2001). Animals often stay in patches longer, however, than the time that would maximize their overall rate of energy gain (Kamil et al. 1993; Nonacs 2001; fig. 9.1). Tests suggested that early rate-maximizing models were partly right: foragers are sensitive to their rate of energy gain, but often do not fully maximize it (see also Nonacs 2001). This observation

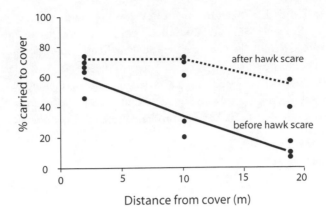

Figure 9.2. Black-capped chickadees are more likely to carry small food items to cover before eating them when cover is close and after a simulated hawk attack. (After Lima 1985a.)

suggested that some non-energetic costs must be important. By pointing out the importance of such costs, early tests of rate-maximizing models provided the springboard for the study of foraging and danger.

Black-capped chickadees often carry food items from a feeder to a bush before consuming them. They are more likely to carry larger items and are more likely to carry items if the feeder is closer to a bush (Lima 1985a; fig. 9.2). Carrying an item to a bush decreases a chickadee's rate of intake, but intake rate is decreased less with large items and close distances. Steve Lima hypothesized that chickadees carried food to cover in order to reduce their exposure to predators. He tested this hypothesis by flying a hawk model in the area during some trials. After having seen the hawk model, chickadees were more likely to carry food to safety (Lima 1985a; fig. 9.2). Thus, animals are willing to reduce their intake rate in order to reduce danger.

To begin this chapter, I examine why foraging gain and danger are generally linked, and I build a life history framework for modeling foraging and danger. I discuss how danger may change with the internal state of the animal, time, and group size. These topics lead to inquiries on how animals assess danger and whether they should overestimate danger. I close with my view of the prospects of the field. Within each section, I outline some principles, often with the help of simple models, and illustrate those principles with a sampling of examples.

9.3   Why Does Increased Foraging Lead to Greater Danger?

Animals often face alternatives that differ in both foraging gain and danger. Obviously, foragers should avoid options that combine poor feeding with great danger and choose options that offer good feeding with little danger. Most often, however, animals face difficult choices in which the options for better feeding also entail greater danger. Such difficult choices are ubiquitous for several reasons, and wherever one or more of these reasons applies, organisms face a trade-off between feeding and danger. After sketching out various routes to a trade-off, I return to a general conceptual approach because the many routes to a trade-off converge on the same basic consequences.

Time Spent Exposed

Guppies feed day and night when no predators are around, but only during the day if predators are around (Fraser et al. 2004). In response to indications of danger, many animals restrict their feeding time (Lima 1998, see especially table II). An animal that feeds part of the time can restrict its feeding to the safest period, but it must extend its feeding time into more dangerous periods in order to feed for longer. For example, small birds must extend their feeding time into the twilight periods around dusk and dawn, when they are less able to detect attacks in the low light and deep shadows (see Lima 1988a, 1988c; Krams 2000). For bats that feed on insects, darkness is safer, but emerging before nightfall may allow greater feeding (Jones and Rydell 1994). Feeding at night is also safer for minnows (Greenwood and Metcalfe 1998) and juvenile salmon (Metcalfe et al. 1999). In order to increase feeding, however, these fish have to feed during the more dangerous daylight period.

Habitat Choice

While actively foraging, animals often choose between habitats that differ in danger and productivity. For example, aquatic snails feeding on algae face a trade-off because more algae grows on the sunny side of rocks, but the tops of rocks are also more exposed to fish predators (Levri 1998). The basic ecology of exposure leads to the trade-off: exposure to sunlight allows more photosynthesis, but exposure often leaves foragers more vulnerable to predators. Similarly, sunfish can find more zooplankton to eat in the open-water portions of lakes because these areas produce more phytoplankton, which support the zooplankton. The open areas, however, provide no refuge from attack, whereas the weedy littoral zone provides refuge, but less food (Werner and Hall 1988). Animals switch between these two kinds of areas during growth because both foraging gain and danger change as they grow (Werner and Gilliam 1984).

In other cases, the attack strategy of the predator and the escape strategy of the prey combine to create the trade-off. In boreal forests, the swooping attacks of pygmy owls make the outer, lower branches of trees particularly dangerous (Kullberg 1995), and small birds avoid these branches unless competition or hunger forces them there (e.g., Krams 1996; Kullberg 1998b). Within a foraging group, individuals on the leading edge will first encounter new sources of both food and danger (Bumann et al. 1997). Animals often move to edge positions when hungry (Romey 1995) and to central positions when alarmed (Krause 1993). Habitat choice may involve another layer of compromise when foragers face conflicting pressures from different kinds of predators. For example, grasshoppers can reduce bird predation by staying low on a blade of grass, but they can minimize predation by lizards and small mammals by positioning themselves high on grass stems. When both kinds of predators are around, grasshoppers choose intermediate positions (Pitt 1999). As these examples emphasize, animals choose between habitats on small as well as large spatial scales, and both kinds of choices have ecological consequences.

Movement

Creatures great and small move less when predators are around (Lima 1998, table II). A forager actively searching for food can cover a greater area by moving faster. By covering a greater area, it is likely to encounter more feeding opportunities, and may also encounter more predators (Werner and Anholt 1993). Besides simply crossing paths with more predators, moving foragers increase the likelihood of an attack. Anaesthetized tadpoles are less likely to be killed by aquatic invertebrate predators (Skelly 1994), and tadpoles generally move less when danger is greater (Anholt et al. 2000). When movement is in short bursts, as in degus, greater movement may involve both faster speeds while moving and shorter pauses between bursts (Vasquez et al. 2002). I will consider movement in further detail below after developing a general model of foraging in the face of danger.

Detection Behavior

Most animals perform behaviors that increase their chances of detecting and escaping from predators. The best studied of these behaviors are pauses during foraging to scan the environment for potential danger (see Bednekoff and Lima 1998a; Treves 2000). Animals can raise their rate of food consumption by scanning less frequently, but at the cost of detecting attacks less effectively (e.g., Wahungu et al. 2001). Investigators have often operationally defined vigilance as raising the head above horizontal. While this operational definition works well for birds and mammals, animals with different body forms and lifestyles may require other operational definitions. For example, lizards basking with their eyes shut and one or more limbs raised off the substratum seem to be showing little antipredator behavior (Downes and Hoefer 2004). Overall, the varied postures and attention required for foraging probably affect predator detection in many organisms. For example, guppies react less quickly to predators when foraging than when not foraging, and even less quickly when foraging nose down (Krause and Godin 1996).

Depletion and Density Dependence

For a burrowing animal such as a marmot, safety comes from fleeing back to the burrow (Holmes 1984; Blumstein 1998). Marmots feed near their burrows, and so deplete food in the area (Del Moral 1984). Due to this depletion, a marmot can feed at a higher rate, but at greater danger, by venturing farther from the burrow. Thus, reactions to initial differences in danger produce differences in foraging. Many lizards also feed from a safe central place (Cooper 2000). Such lizards can find more prey farther out, but at a cost. The actions of a lizard also produce a gradient of food and danger for its potential prey. The grasshoppers the lizard preys on can find a richer, less depleted food supply near the lizard's perch, but obviously, feeding near the lizard increases the possibility of attack (Chase 1998). Thus, a spatial trade-off at one trophic level may have cascading effects on other trophic levels.

In a manner similar to food depletion, density dependence can produce a trade-off when potential prey congregate. By congregating, prey decrease one another's feeding rates through competition and also decrease one another's danger through safety-in-numbers advantages. When avoiding predatory perch, 92% of small crucian carp concentrate in the safer shallows, compounding the differences in food availability between shallows and open waters (Paszkowski et al. 1996). In theory, the outcome depends on the balance of competitive and safety-in-numbers effects and on how free predators are to choose habitats, but we may often expect habitats to be made either safe but poor or rich but dangerous by these mechanisms (Hugie and Dill 1994; Moody et al. 1996; Sih 1998).

9.4 Modeling Foraging under Danger of Predation

Foraging for a Fixed Time

Bluehead chubs alter their foraging in response to changes in energetic returns and danger from green sunfish. The best explanation for their behavior combines food and danger in a life history context (Skalski and Gilliam 2002). To build models of foraging under danger of predation, we start from the first principle of foraging theory—that food is good. We assume that higher foraging success leads to greater reproductive success in the future. To include danger, we need a second principle—that death is bad for fitness. Early research was uncertain on how to incorporate danger into foraging models (see box 1.1), perhaps because it is not obvious how to combine the benefits of foraging and the costs of predation. Because the costs and benefits are in different units, we need to translate both foraging gain and danger into some measure of fitness. A life history perspective is essential, and it leads to a simple solution that exists precisely because the costs and benefits of foraging under danger of predation are linked.

Decisions made under danger of predation are life history problems because, if predation occurs, the forager's life is history. In a life history, the basic currency to maximize is expected reproductive value, $b + SV$, where b is current reproduction, S is survival to the following breeding season, and V is the expected reproductive value for an animal that does survive to the next breeding season (see Stearns 1992). I concentrate here on foraging and fitness during a period without current reproduction ($b = 0$), so the measure of fitness is SV, the future benefits multiplied by the odds of surviving to realize them. I expect increased foraging to decrease survival to the time of reproduction, but to increase future reproduction if the animal does survive.

Death lowers expected future fitness to zero. Therefore, the cost of being killed is the reproductive success a forager could have had if it had survived. This linkage means that when we ask how much risk a forager should accept to produce one additional offspring, we need to know how many offspring it would produce otherwise. For example, a forager that would otherwise expect to produce one offspring might risk a lot to produce a second, while a forager that would otherwise expect to produce three offspring should risk less to produce a fourth, and a forager that would otherwise expect to produce a dozen should risk little to produce a thirteenth. This linkage of costs and benefits sets up an automatic state dependence: the potential losses from being killed increase with previous foraging success, so the relative value of further foraging is likely to be lower (see Clark 1994). Even if the fitness gains of foraging are constant, the costs should increase, since the expected reproductive value increases, and that entire value would be lost in death. In line with this logic, juvenile coho salmon are more cautious when they are larger, because larger individuals expect greater reproduction if they survive to breed (Reinhardt and Healey 1999).

Now I will repeat these arguments mathematically. For a nonreproducing animal, fitness equals the future value of foraging discounted by the probability of surviving from now until then, $W(u) = S(u)V(u)$, where u is a measure of foraging effort, W is fitness, S is survival, and V is future reproductive value.

Fitness, survival, and future reproductive value are all functions of foraging effort u. In general, we expect survival to decrease and future reproductive value to increase with foraging effort. More specifically, we expect survival to decrease exponentially with mortality, $S(u) = \exp[-M(u)]$, where $M(u)$ is mortality.

Mortality rate, $M(u)$, and future reproductive value, $V(u)$, could take various mathematical forms. For simplicity, I define foraging effort as a fraction of the maximum possible effort, so that u varies from zero to one and does not have units. This allows mortality, $M(u)$, and future reproduction, $V(u)$, to be given as simple functions of foraging effort.

Mortality is a function of the amount of time spent exposed to attack, the attack rate per unit time, and the probability of dying when attacked (see Lima and Dill 1990). Greater overall foraging effort could affect any of these components. For now I use a descriptive equation for mortality, $M(u) = ku^z$, where k is a constant and the exponent z gives the overall shape of the trade-off. Later we will examine two specific cases to see what k and z might mean biologically, but for now I will simply label k as the mortality constant and z as the mortality exponent. The general principle is that mortality should increase with foraging effort at a linear or accelerating rate; that is, $M(u) = ku^z$ with $z \geq 1$. If foragers exercise their safest options first, we expect an accelerating function because additional food comes from increasingly dangerous options. A mathematically convenient value for the exponent, $z = 2$, matches observed changes in behavior well enough (Werner and Anholt 1993), but other values are not ruled out, so I also examine a linear relationship ($z = 1$) as well as more sharply accelerating ones ($z = 3$ and $z = 4$). For all values, survival declines as foraging effort increases, but the contours of the decline depend on the exponent of the mortality function, z (fig. 9.3). As we

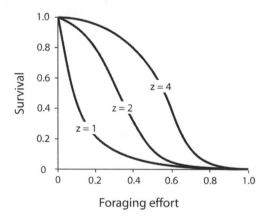

Figure 9.3. Survival declines with foraging effort. The swiftness of the decline varies with z, the exponent of the curve relating foraging effort to mortality.

shall see near the end of this chapter, the value of this exponent determines whether foragers should over- or underestimate danger.

For the relationship between foraging effort and future reproductive value, we will use $V(u) = \kappa u$. In this equation, the constant κ translates foraging effort into future offspring, and u is foraging effort. We expect future reproductive value to increase with total foraging effort. Studies have shown that greater foraging success leads to greater fitness in adult crab spiders (Morse and Stephens 1996), water striders (Blanckenhorn 1991), and water pipits (Frey-Roos et al. 1995). Particularly for any organisms that are able to grow, reduced foraging in the presence of predators can lead to considerable long-term losses of potential reproduction (Martin and Lopez 1999; see also Lima 1998, table III).

A linear relationship between foraging and future reproductive value is useful for its simplicity. Other relationships may occur in nature, and the relationship may differ between the sexes even within a species (Merilaita and Jormalainen 2000). I use a linear relationship here because it allows simple models with clear conclusions, even though these models may somewhat understate the effects of danger. The results of more complex models, in which future fitness is a decelerating function of foraging gain, strongly support the conclusions I reach using this simpler linear relationship.

To complete the modeling framework, assume that foraging effort must be greater than some required effort, R. This requirement, R, is the required rate of feeding divided by the maximum rate of feeding and so is a proportion without units. A forager starves if its foraging effort is less than the requirement, and avoids starvation as long as its foraging effort is greater than the requirement. A forager gains some amount of fitness, $V(R)$, by just meeting the requirement, but increases its future reproductive value by foraging at a rate higher than the requirement.

Assembling the pieces described above, we get the overall equation for fitness: $W(u) = S(u)V(u) = [\exp(-ku^z)][\kappa u]$. We can find the optimal foraging effort, u^*, if we differentiate $W(u)$, set the derivative to zero, and solve for u. We find that

$$u^* = \frac{1}{\sqrt[z]{kz}}, \tag{9.1}$$

so long as $u^* \geq R$.

The foraging effort that maximizes fitness (i.e., is optimal) decreases as the danger constant k increases. As the mortality exponent z increases, optimal foraging effort decreases less sharply with increases in k (fig. 9.4). Modelers sometimes assume that animals maximize survival during the nonbreeding season (see McNamara and Houston 1982, 1986; Houston and McNamara

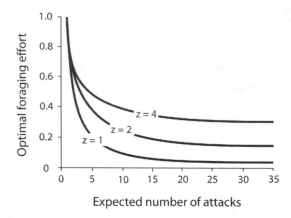

Figure 9.4. Optimal foraging effort declines with the expected number of attacks by predators. The swift-ness of the decline varies with z, the exponent of the curve relating foraging effort to mortality.

1999). This assumption is justified whenever the requirement is greater than the feeding rate that would otherwise be optimal, $R > 1/(\sqrt[z]{kz})$. Thus, a life history approach can converge on models that assume survival maximization even when future reproductive value increases linearly with foraging effort.

In order to examine our model further, we need to look more closely at the relationship between mortality and foraging effort, $m(u) = ku^z$. Mor-tality depends on the encounter rate with predators, time spent exposed to predators, and the probability of being killed per encounter. The relationship between mortality and foraging effort includes effects on any of these three components. I consider two situations here.

First, consider tadpoles encountering predatory dragonfly larvae. Tadpoles move while foraging, while dragonfly larvae sit and wait for prey. If a tadpole moves faster, it encounters both more food and more predators. In this case, the exponent, z, reflects changes in metabolic cost per distance moved and the constant, k, combines the relative encounter rate with predators and the probability of being killed in an encounter. This logic applies to any forager moving at different speeds with relatively immobile food and predators.

Second, consider birds hunted by *Accipiter* hawks, which move a great deal while hunting. Because the hawks seek them out, greater foraging effort will not cause foragers to encounter more predators, but it may make them more likely to be killed when they do encounter a predator. In this case, the constant k includes a constant attack rate, α, and the exponent, z, reflects how foraging effort increases the probability of being killed in an attack. This logic applies when predators move rapidly and foragers are relatively im-mobile.

The interpretation of the mortality function, $m(u) = ku^z$, depends on the biology of the predators and prey. I use the second scenario to gain some insight into models that maximize survival. Here an animal is exposed for a period T at an attack rate α. Thus, αT is the number of attacks the animal can expect while foraging. The optimal feeding rate will approach the requirement with only a modest number of attacks at a moderate level of requirement, particularly if the mortality exponent z is small (see fig. 9.4), and will approach any level of requirement if the expected number of attacks is large enough. Danger can cause animals to behave as if they are foraging to meet a set requirement.

Gathering Resources with No Time Limit

So far, we have considered foraging for a fixed time. We can modify our framework to address the classic problem of gaining a fixed amount of resources from foraging within a potentially unlimited amount of time (Gilliam 1982; see also Werner and Gilliam 1984; Stephens and Krebs 1986; Houston et al. 1993). Here a fixed amount of reproduction, V, occurs whenever sufficient resources, K, are accumulated. Thus, the reproductive value is fixed, but the time to reach it depends on gain, so $W(u) = \exp[-M(u)T(u)]V$, and $T(u) = K/u - R$. In this function, fitness will be maximized when $M(u)/(u - R)$ is minimized, which is another rendering of Gilliam's M/g rule (see box 1.1). Using our equation for mortality, $M(u) = ku^z$, we differentiate and solve for the optimal foraging effort,

$$u^* = \frac{zR}{z - 1} \tag{9.2}$$

The mortality exponent, z, is the key parameter; $u^* = 2R$ if $z = 2$, but $u^* = 4R/3$ if $z = 4$. In contrast to our previous results, here the optimal foraging effort decreases when mortality is a more sharply accelerating function of foraging effort (fig. 9.5). If the requirement, R, is large, foraging at the maximum rate ($u^* = 1$) may be the best option available to foragers. Notice also that the optimal effort does not depend on the constant, k, but only on the exponent z. This means that the shape of the trade-off is the key, while the exact level of danger is irrelevant. Animals in environments with different absolute levels of danger would have the same optimal behavior as long as their trade-offs between foraging and mortality followed the same basic function.

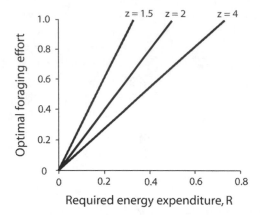

Figure 9.5. For growing animals, optimal foraging effort increases with the amount of energy required to stay alive, *R*, and decreases as the mortality function becomes more sharply accelerating.

9.5   Danger May Depend on State

Big fish may be better able to escape than small fish, and predators may attack large clams more often than small clams. Body size often influences danger, but most models of growth ignore this possibility. We would like to know if danger depends on attributes of the individual that we label as *state*. Many studies demonstrate that antipredator behavior depends on state (see Lima 1998; Clark and Mangel 2000), but that is not the same as demonstrating that danger depends directly on state because we expect antipredator behavior to depend on state whenever future reproductive value depends on state (Clark 1994; see section 9.4). Fatter voles might venture out less on moonlit nights for a variety of reasons. In order to say that they experience a higher risk, we must directly compare fat and skinny voles.

Examining the direct relationship between danger and state is difficult for a combination of theoretical and empirical reasons. In theory, the time course of behavior depends on whether the effects of behavior and state combine by addition or multiplication when we calculate mortality rate (Houston et al. 1993). Empirically, this suggests that we should compare several quantities that are difficult to measure. I will illustrate this problem with the example of fat reserves of small birds in winter. Theoreticians originally developed these ideas for birds weighing 20 g or less, but the same principles apply to any other animal for which starvation is a realistic threat.

In winter, small birds do not grow, but they do need large energetic reserves to survive the long, cold nights, plus any other periods of deprivation. Feeding more has value because it reduces the probability of starvation. Even in very harsh conditions, however, reserve levels are far lower than the reserves carried by long-distance migrants, suggesting that wintering birds could carry more reserves than they do. From this framework has grown the study of optimal energetic reserves for foragers that could die of either predation or starvation. This area has expanded rapidly (see chap. 7) and now possesses an impressive body of theory (Lima 1986; McNamara and Houston 1990; Bednekoff and Houston 1994a, 1994b; Brodin 2000; Pravosudov and Lucas 2000, 2001b) as well as a large collection of novel results that generally support the theory (e.g., Gosler et al. 1995; Bautista and Lane 2001; Thomas 2000; Olsson et al. 2000; Cuthill et al. 2000; see also Cuthill and Houston 1997).

Whether birds pay extra costs when carrying more fat reserves is an important, unsolved puzzle (see Witter and Cuthill 1993). Without such mass-dependent costs, the only cost of reserves is the foraging needed to acquire them (see box 7.3). If carrying reserves reduces the risk of starvation, then we would expect small birds to pay the acquisition costs for large reserves early in winter, unless carrying those reserves also imposes a cost (Houston et al. 1997). Some animals do fatten up for winter, but small birds seem to match their foraging to their daily demands and therefore end up foraging more intensely when days are short and cold and food is less abundant. In theory, this pattern makes sense with mass-dependent costs, but not without them (see Lima 1986; Houston et al. 1997). Mass-dependent costs in models make it uneconomical to forage in summer and fall and carry the reserves until needed in winter. These costs cause foragers to behave as if they are meeting a requirement over a fairly short time horizon (see Bednekoff and Houston 1994b).

Excess body mass might be costly in several ways. Extra mass might impair foraging performance, particularly while hovering or hanging from small twigs (Barbosa et al. 2000; Barluenga et al. 2003), or it might lead to increased energy expenditure (see Witter and Cuthill 1993; Cuthill and Houston 1997). These costs tax the value of reserves, but should not cause small birds to refuse "free" food when they encounter it. If possessing larger reserves leads to greater danger, however, this could make even free food too expensive to eat. Birds at feeders generally eat far less than they could, and willow tits may employ hypothermia at night even when ad libitum food is available to them during the day (Reinertsen and Haftorn 1983; see also Pravosudov and Lucas 2000). These observations strongly hint that mass-dependent predation may help explain the fat reserves of small birds.

Logic and hints are a great start, but in science, we require evidence to decide the issue. Unfortunately, we do not have the required evidence, and we are unlikely to get direct evidence from the field. It is difficult enough to observe any acts of predation; to also know the relative masses of the victims and control for differences in behavior may be too much to ask (see also Cuthill and Houston 1997).

Scientists have turned to indirect techniques. The first asks whether simulated exposure to predators causes birds to alter their reserve levels. Small birds have sometimes lost mass when chased occasionally (Carrascal and Polo 1999) or shown a model predator (Lilliendahl 1997, 2000; van der Veen 1999; Gentle and Gosler 2001), but in other tests they gained mass (Lilliendahl 1998; Pravosudov and Grubb 1998b). Warblers preparing for migration accumulated fat reserves faster, but attempted to leave at a lower mass, when in the presence of a simulated predator (Fransson and Weber 1997). This seems a sensible "exit strategy," but the increased reserves for residents contradict the predictions of current theory. In a more recent test, results seem to support theoretical predictions on long but not on short days (Rands and Cuthill 2001).

Another indirect way of looking for mass-dependent predation costs has been to measure flight performance. Aerodynamic theory suggests that mass must affect some aspects of escape flight (Hedenström 1992). Field observations suggest that performance on takeoff is likely to be critical (Cresswell 1994, 1996). Experimental results to date allow several interpretations. For zebra finches, small changes in mass have a large effect on flight speed when birds are taking off spontaneously (Metcalfe and Ure 1995), but little effect when birds are startled into flight (Veasey et al. 1998). Large fat reserves slow escape flights by startled birds (Kullberg et al. 1996, 2000; Lind et al. 1999) and also lower takeoff angle and maneuverability during flight (Witter et al. 1994). The diurnal changes in body mass between dawn and dusk, however, have little effect on escape flights for four species of small birds (Kullberg 1998a; Kullberg et al. 1998; van der Veen and Lindström 2000; see Lind et al., in press, for review). These results have been interpreted to mean that mass-dependent predation applies only to reserves above some threshold level (Kullberg 1998a; Kullberg et al. 1998; Brodin 2000). I suggest they are consistent with continuous but perhaps nonlinear functions. Either way, the effects of daily mass changes on escape performance seem small, though in theory small differences in flight speed might result in large differences in danger (Bednekoff 1996). While direct effects of mass have been hard to demonstrate, related studies show that reproductive states are likely to lead to greater danger: female starlings, zebra finches, blue tits, and pied flycatchers escape more slowly during reproduction, most probably due to a combination of carrying eggs and depleting their wing muscles to produce eggs (Lee et al. 1996; Veasey et al. 2001; Kullberg, Houston, and Metcalfe 2002; Kullberg, Metcalfe, and Houston 2002).

9.6   Danger May Change over Time

On moonlit nights, hunting owls can see gerbils easily, so gerbils may forgo foraging until a darker night. Besides the cycles of days, tides, and seasons, foraging conditions vary for a host of reasons. Many of these variations affect aspects of danger. If a forager faces a mix of situations that differ in danger, it should choose different levels of foraging effort to apply to each. Working through a set of assumptions (box 9.1), we find that the difference in danger has effects on foraging beyond the effect of the average amount of danger. Foragers are predicted to change their behavior more in response to variations in danger than to average rates of danger. Under some conditions, foragers might react only to variations in danger, not to average rates (see box 9.1). In other words, each individual may respond to the differences in danger that it experiences, while different individuals at different overall danger levels may behave similarly.

In Box 9.1, foragers using option i gather food at rate u_i and suffer mortality at rate $\alpha_i u_i^2$ under each option. The solutions for the two situations have equal M/g ratios, since $\alpha_i u_i^2/u_i$ reduces to $\alpha_i u_i$ and the solution stipulates that $\alpha_1 u_1 = \alpha_2 u_2$. Thus, Gilliam's M/g rule emerges from the problem, even though I did not formulate the problem like Gilliam's habitat choice scenario. We should not use the M/g rule as a starting point for models, because it ignores effects of time and state on fitness (Ludwig and Rowe 1990; Skalski and Gilliam 2002), but we should not be surprised when models produce solutions that relate to this surprisingly general rule (see also Houston et al. 1993; Werner and Anholt 1993; Lima 1998).

9.7   Danger Often Depends on Group Size

In one study, solitary male grey-cheeked mangabeys died at twelve times the rate of males in groups (Olupot and Waser 2001). Gathering into groups often yields benefits for both finding food and avoiding predation (see also chap. 10). For tropical birds, flocking correlates with predation pressure and survival, even when broad taxa and geographic areas are compared (Thiollay 1999; Jullien and Clobert 2000). How does grouping decrease danger? Individuals could benefit if being in a group decreases either the per capita attack rate or their individual danger per attack. Though larger groups should be easier for a predator to detect or encounter, on geometric principles, we expect this increase to be less than proportional to group size (Vine 1973; Treisman 1975). Investigators see this pattern for aggregations of aphids attacked by ladybugs (Turchin and Kareiva 1989) and for tadpoles attacked by fish (Watt et al. 1997).

BOX 9.1 **Allocation of Foraging Effort When Danger Varies
over Time**

Here I illustrate the effect of periods with differing danger on foraging
behavior. I assume that the two periods differ in their attack rate, α, but
have the same function for mortality per attack. Any function u^z with $z > 1$
will do, but I choose $z = 2$ to allow comparison with Lima and Bednekoff
1999b. Situation 1 occurs some portion of the time, p, and situation 2 the
rest of the time, $(1 - p)$. Foragers may have different foraging efforts (u_1,
u_2) in the two situations. Now our overall fitness is

$$W(u_1, u_2) = e^{-[a_1 p u_1^2 + a_2 (1-p) u_2^2] T} k_v \bar{u} T. \quad (9.1.1)$$

Here $\bar{u}$ denotes the average rate of feeding and equals $p u_1 + (1 - p) u_2$.
We can find the optimal feeding rates by taking the partial derivatives of
W for u_1 and u_2 and setting each to zero. After rearranging the results, we
find that

$$\alpha_1, \bar{u}_1 = \frac{1}{2\bar{u} T} \quad \text{and} \quad \alpha_2 u_2 = \frac{1}{2\bar{u} T}. \quad (9.1.2)$$

Therefore, $\alpha_1 u_1 = \alpha_2 u_2$, which means that the ratio of the feeding rates is
the inverse of the ratio of the attack rates. Substituting using this relationship yields the actual rates:

$$u_1 = \frac{1}{\sqrt{2\alpha_1 T \left(p + \frac{\alpha_1}{\alpha_2}(1 - p) \right)}} \quad \text{and}$$

$$u_2 = \frac{1}{\sqrt{2\alpha_2 T \left(\frac{\alpha_2}{\alpha_1} p + (1 - p) \right)}} \quad (9.1.3)$$

so long as $\bar{u} \geq R$.

The importance of this model is that allocation of antipredator behavior
to dangerous periods leads to greater changes in antipredator behavior than
we would expect in response to average levels of danger. For $V(u) = k_v u T$
and $M(u) = \alpha u^2 T$, without allocation, foraging rate changes with the square
root of attack frequency, whereas with allocation, they change proportionally. If $V(u)$ increases less than linearly, the relative effect of allocation is
greater. If we assume survival maximization, all changes would be the result of allocation of antipredator behavior (see Lima and Bednekoff 1999b).
As we saw earlier, we can assume survival maximization if the required
feeding rate exceeds the feeding rate that would be optimal otherwise.

Blue acara cichlids, however, attack larger schools of guppies at higher per capita rates (Krause and Godin 1995). Furthermore, predators differ. For example, sparrowhawks attack larger groups of redshanks at a roughly constant per capita rate, while peregrine falcons attack at a rapidly declining per capita rate (Cresswell 1994). On average, across predators and situations, attack rate probably increases less than proportionally with group size. Practically, however, attack rates are difficult to estimate. In many situations, we may have difficulty rejecting the possibility either that attack rate is independent of group size or that attack rate increases proportionally with group size. I consider each of these possibilities in developing the theory below.

In considering the danger per attack, we focus on one member of a group. If an attack occurs, what is the probability that our focal animal will die in the attack? If the predator can kill only one member of the group, we expect our focal animal to be the victim in $1/n$ of instances, where n is the size of the group. The odds improve if detection of an attack allows all group members to avoid it. Collective detection occurs if detection by one forager somehow increases the chances that others in the group will detect the attack. The classic models of antipredator vigilance assumed perfect and instantaneous transfer of information (Pulliam 1973; see also McNamara and Houston 1992). Under these assumptions, danger decreases very rapidly with group size (fig. 9.6). Collective detection, however, is far from perfect and instantaneous. For small birds that do not employ alarm calls, collective detection seems to occur only when two or more individual detectors leave the group in rapid succession (Lima 1994a, 1995a, 1995b; Lima and Zollner 1996). In other situations, one detector can put the flock to flight, but not as quickly as multiple detectors (Cresswell et al. 2000; Hilton et al. 1999). Individual detectors flee a considerable fraction of a second ahead of the rest (Lima 1994b; see also Elgar et al. 1986), and fractions of second could mean the difference between life and death (Bednekoff 1996).

Models of collective detection track the flow of information among group members through time. In box 9.2, I develop three of many possible special cases: one with perfect collective detection, one with no collective detection, and one in which collective detection requires that two or more group members detect the attack. Danger goes down with group size in all scenarios, though the exact shape of the decline depends on the effectiveness of collective detection (see fig. 9.6). The intermediate case (two-to-go rule) performs like no collective detection for small groups but is closer to perfect collective detection for large groups (see also Proctor et al. 2001).

The safety advantage of groups will depend on the shape of the decline in danger per attack with group size combined with the change in per capita

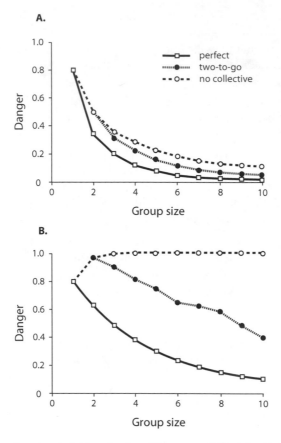

Figure 9.6. Danger depends on both information flow within groups and the relationship between attack rate and group size. (A) When attack rate is independent of group size, danger decreases sharply with group size and decreases somewhat with greater information flow. (B) When attack rate is proportional to group size, danger may increase with group size when information flow is faulty. ($f = 0.8$ throughout.)

attack rate with group size. We are still learning details about individual (Lima and Bednekoff 1999a) and collective detection, but these seem unlikely to negate any basic advantage in predator detection for larger groups. An enormous literature documents that vigilance rates decrease and feeding rates increase with group size (Elgar 1989; Roberts 1996; Beauchamp 1998; Blumstein et al. 1999).

Given any safety advantage for groups, if group size fluctuates, then danger fluctuates. By the logic of section 9.6, animals should forage intensely when they find themselves in the safety of a large group because soon enough they will be alone or in a small group (see Lima and Bednekoff 1999b). Many of the demonstrations of the group size effect on vigilance have relied on short-term

BOX 9.2   Three Models of Information Flow in Groups

In all three of the models presented here, I assume that group members independently scan for attacks and flee to safety as soon as they detect an attack. The predator chooses at random among those that do not flee. Each of these models assumes that each individual *fails* to detect an attack with probability f and feeds in a group of n individuals. (For continuity with earlier models in this chapter, I could present $f = u^z$, but this leads to a thicket of exponents that obscures the workings of the current models. Therefore, I stick with f for simplicity.)

Perfect Collective Detection

An attack succeeds only if no individual detects it, which happens with probability f^n, and all n group members share the risk equally:

$$\text{risk/attack} = \frac{f^n}{n}. \quad (9.2.1)$$

Two-to-Go Rule

Collective detection occurs if two or more individuals detect an attack. Therefore, an attack succeeds if no individual detects it or if only one individual detects it. The first possibility is the same as for perfect collective detection. Under the second possibility, the probability that any particular individual detects the attack is $(1 - f)$, and the probability that all other group members do not is f^{n-1}. The focal individual is in danger when it fails to detect the attack and one of the other $n - 1$ members of the group detects the attack. The detector flees the area, and the predator kills one of the $n - 1$ remaining individuals:

$$\text{risk/attack} = \frac{f^n}{n} + \frac{(n-1)(1-f)f^{n-1}}{n-1}$$
$$= \frac{f^n}{n} + (1-f)f^{n-1}. \quad (9.2.2)$$

No Collective Detection

The focal individual is in danger whenever it does not detect an attack. It could share the risk with zero to $n - 1$ other non-detectors. Summing across all possibilities looks daunting—

$$\text{risk/attack} = \sum_{i=0}^{n-1} \left(\frac{(n-1)!}{(n-1-i)!\,i!} \right) \frac{f^{n-1}(1-f)^i}{n-1}, \quad (9.2.3)$$

(Box 9.2 continued)

where i denotes the number of detectors—but eventually a simple solution emerges:

$$\text{risk/attack} = \frac{1 - (1 - f)^n}{n}. \quad (9.2.4)$$

The numerator of this solution is the probability that some individual will *fail* to detect an attack. The denominator is n, since each group member is equally likely to fail. Unless detection is very good (i.e., the failure rate, f, is small), this solution is nearly $1/n$.

---

changes in group size. Currently, we do not know whether the group size effect on vigilance diminishes when groups do not fluctuate in size.

So far, I have assumed that individuals within groups are identical, despite a rich literature on mixed-species groups (e.g., Bshary and Noe 1997; Dolby and Grubb 2000), in which advantages may come about because individuals differ in complementary ways. In addition, different positions with a group will generally bring different costs and benefits (Krause 1993; Romey 1995; Bumann et al. 1997). We expect different individuals to respond to their own costs and benefits, and therefore members of a group should adjust their vigilance to their local circumstances, which depend largely on the position and actions of neighbors, rather than group size per se. Though this logic is straightforward, it does not tell us how to measure local circumstances. In direct comparisons, vigilance is sometimes better predicted by a measure of local density (e.g., Blumstein 1996; Treves 1998) and sometimes by overall group size (Cresswell 1994; Blumstein et al. 1999). These mixed results may indicate that group size is sometimes a better correlate of animals' local circumstances than are our estimates of their local circumstances. Still, we would like to know what animals are assessing within groups. This brings us to our next topic.

9.8 How Do Foragers Assess Danger?

No living forager can really know its odds of being killed by a predator. Therefore, we should not expect foragers to make accurate estimates of danger, but the details of their estimates may have profound effects on their behavior and ecology (Sih 1992; Houtman and Dill 1998; Brown et al. 1999). What they might know and how they might know it are largely open questions. We can divide these questions into subquestions by remembering that predation involves encounter, attack, and capture stages. Since each stage requires the previous one, foragers generally experience many more encounters than attacks, and more attacks than captures. Here I present some examples of how different foragers may estimate the probabilities of encounter, attack, and capture.

Gerbils in the Negev Desert in Israel react to noises that indicate the presence of barn owls (Abramsky et al. 1996). Prey may know the general density of predators by detecting signs of predators. Part of this information comes directly from the inevitable sights, sounds, and smells that predators create. Furthermore, predators often betray their presence through their territorial behavior or other social interactions. Lions roaring and wolves howling are two examples that are obvious even to humans. Among seabirds, petrels limit their exposure after eavesdropping on the territorial calls of predatory skuas (Mougeot and Bretagnolle 2000).

When potential predators are nearby, foraging animals must assess the likelihood of attack. For fish and other aquatic organisms, chemical cues may provide detailed information on the capture of similar prey in the area (see Kats and Dill 1998; Wisenden 2000). Gerbils and other small rodents behave more cautiously on moonlit than on dark nights (Daly et al. 1992; Vasquez 1994). Although moonlight probably helps rodents detect attacks, it seems to help predators more in detecting prey. Moonlit nights are dangerous because of the increased probability of attack should predator and prey encounter each other.

Finally, how can a foraging animal assess its likelihood of escape if it were attacked? Likelihood of escape is intimately linked to detection behavior. We can even define detection to mean detection of an attack in time to avoid capture (see Bednekoff and Lima 1998b). Time needed to escape is a basic variable of escape that animals might assess. Time needed to escape depends on the distance and the structure of the habitat between the forager and safety (see Blumstein 1992). As mentioned earlier, small birds forage differently when farther from a refuge. The same applies to burrowing animals such as marmots, except that the refuge is a burrow rather than a bush or tree. Townsend's ground squirrels flee less quickly through shrub habitats than across open ground (Schooley et al. 1996), although shrub vegetation may also make predators harder to spot. Fox squirrels also react to escape substratum (Thorson et al. 1998).

Overall, foragers can respond to direct cues to attack rate or conditions that indirectly give cues to relative danger levels. Animals react to factors that are likely to affect their probability of escape. Differences in the probability of escape may drive many reactions to small differences in exposure. Because probability of escape after capture is difficult to estimate from experience, I

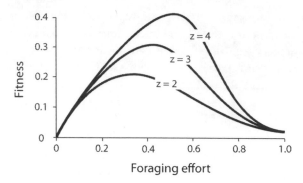

Figure 9.7. The symmetry of the overall fitness function with z, the exponent of the curve relating foraging effort to predator detection. This relationship determines whether prey should over- or underestimate predation risk. ($k = 4$, $\kappa = 1$.)

speculate that learning about escape probability is less important than learning about probabilities of encounter or attack.

9.9   Should Foragers Overestimate Danger?

Since a forager will never have full information on danger, is overestimating danger a more prudent course than underestimating it? Since a forager risks its entire reproductive value by foraging too much, but can increase it only by some fraction, it might seem that the costs of foraging too much are generally higher than the costs of foraging too little. Previous models have suggested that this intuition is sometimes correct, but not always (Bouskila and Blumstein 1992; Abrams 1994, 1995; Bouskila et al. 1995; Koops and Abrahams 1998). For the models used in this chapter, the fitness of foraging somewhat too much is often higher than the fitness of foraging slightly less than the optimum (fig. 9.7).

To examine this issue technically, we must check whether the third derivative of the fitness function is positive when evaluated at the optimum foraging rate; i.e., whether $W'''(u^*) > 0$ (see Abrams 1994). $W(u^*)$ is the peak of the function whenever we have an optimal foraging effort, so therefore the curve flattens out at the peak, $W'(u^*) = 0$, and curves down on either side, $W''(u^*) < 0$. The value of the third derivative tells us the shape of the downward curve on each side of the peak. For the model used in section 9.4—$W(u) = [\exp(-ku^z)]$ $[\kappa u]$—we find that $W'''(u^*) < 0$ when $z < 3$. The upshot is that prey should overestimate danger only if the costs of foraging are strongly accelerating.

Although the costs of foraging might accelerate rapidly, they are unlikely to do so in all situations. We can draw on observations that anuran larvae decrease foraging time and speed at approximately $1/\sqrt{}$ (change in food or predation) (Werner and Anholt 1993). If we back-calculate from equation (9.1), these results suggest that $z \approx 2$, which would suggest that anuran larvae might underestimate danger. Other empirical studies suggest that prey sometimes overestimate and sometimes underestimate danger. Moose without previous experience with bears or wolves are highly vulnerable when these predators recolonize an area, but rapidly adjust their behavior if they survive an initial encounter (Berger et al. 2001). On the other hand, New England cottontails seldom feed away from cover and are declining in the face of competition with bolder eastern cottontails now that predators are comparatively rare (Smith and Litvaitis 2000). In answer to my own question, it seems that foragers should not always overestimate danger, but might sometimes. For now, it may be enough to test how foragers estimate danger in the first place and leave questions of under- and overestimation for later consideration.

9.10   Prospects

I have three reasons for optimism about the future of this field. Two of the reasons are expansive, and one involves qualified optimism. The first reason for optimism is that the principles of foraging in the face of danger apply to an astonishing breadth of cases. Animals that do not face trade-offs between foraging and danger are probably exceptional, and may prove fascinating because they are exceptional. I believe that we will continue to apply these principles to important new cases. Novel opportunities and methods of study are likely to widen our understanding of the principles. For example, transgenic salmon are more willing to risk exposure to predators in order to obtain food (Abrahams and Sutterlin 1999), and both growth hormones and domestication affect antipredator behavior by brown trout (Johnsson et al. 1996). Opportunities to work on a large scale may be available due to population translocations and reintroductions (Blumstein et al. 1999), predator reduction campaigns (Banks 2001), and predator reintroduction and colonization (Berger et al. 2001). I am very optimistic that the field will continue to grow for the foreseeable future.

My second reason for optimism, albeit qualified, is that I believe we will make progress on tough quantitative questions using some highly tractable systems. The rapid expansion of the field has so far featured mainly qualitative studies (see Lima 1998). We have asked whether danger affects foraging, and these qualitative studies tell us that the answer is a clear-cut "Yes." We have identified several factors contributing to the observed effects. The next step is to compare these factors quantitatively. This task is far tougher. Models tell us that the exact shapes of the trade-offs between danger, foraging, and fitness could make a difference. Factors such as mass-dependent costs, allocation of antipredator behavior, and changes in attack rate with group size all influence the foraging-danger trade-off. Empirically, it is hard enough to distinguish values from zero and linear from nonlinear functions, much less to distinguish between types of nonlinear functions. Therefore, these issues should not be taken on lightly. I am confident, however, that we can gather data of exceptional quality and quantity using a variety of organisms, and even a handful of examples would go a long way toward establishing what general rules are likely.

Finally, I am excited about the continued application of and interaction with other fields. A large part of the ecological impact of predators may be behavioral (Schmitz, Beckerman, and O'Brien et al. 1997; Nakaoka 2000), and thus behavior can produce surprisingly powerful indirect effects in communities (e.g., Anholt et al. 2000; Peacor and Werner 2000). Adaptive behavior tends to destabilize population dynamics (Luttbeg and Schmitz 2000; McNamara 2001), but behavior based on imperfect information may stabilize predator-prey systems (Brown et al. 1999). The basic fact of antipredator behavior may have profound implications for population dynamics, species coexistence, and conservation (see chaps. 11–14). We can look forward to learning more in these areas.

9.11 Summary

The predictions of foraging theory depend on how feeding is related to fitness, and few things alter fitness as drastically as death. Foraging usually involves trade-offs between food acquisition and danger because options that increase the rate of foraging gain also increase the probability of predation. From a simple mathematical framework, conditions emerge that justify the assumptions of survival-maximizing models. Models often approximate Gilliam's rule of minimizing the danger of predation per net rate of foraging. State-dependent costs are likely, but measuring them has proved difficult. We have overwhelming evidence that animals change their foraging behavior in response to changes in predation risk, while our theories generally assume a constant risk of predation. In theory, larger foraging-predation trade-offs should occur when danger fluctuates. This suggests that much of our evidence for foraging-predation trade-offs depends on the fact that predation risk varies on scales to which foragers can respond. How animals sense changes in predation risk is largely an open question, and one with profound implications for ecology and conservation.

9.12   Suggested Readings

Current issues of journals in animal behavior and ecology provide many examples of the effects of danger on foraging. Two wide-ranging reviews summarize older examples (Lima and Dill 1990; Lima 1998). Two excellent books give details about the interplay between foraging and danger in groups (Giraldeau and Caraco 2000; Krause and Ruxton 2002). Cuthill and Houston (1997) investigate issues related to state dependence in greater detail. For anyone who wants to pursue the theory more deeply, I recommend one review (Houston et al. 1993) and two books (Clark and Mangel 2000; Houston and McNamara 1999).

Telander

Foraging with Others:
Games Social Foragers Play

Thomas A. Waite and Kristin L. Field

10.1 Prologue

On a bone-chilling winter night in the far north, a lone wolf travels through the boreal forest looking for his next meal. The half-dozen pack members in the adjacent home range howl periodically throughout the night. With each chorus, he resists the urge to howl in return. With each chorus, he feels the pull to cross over the ridge, descend into the cedar swamp below, and attempt to join the pack—to give up the solitary life. Suddenly, just before daybreak, he happens upon an ancient, arthritic moose. The chase begins. The moose flounders in the deep snow. Within minutes, the wolf subdues the moose, his tenth such success of the winter. He feeds beyond satiation and then rolls into a ball and sleeps. At first light, ravens arrive, gather around the carcass, and begin to feed. By midday, several dozen ravens are busily engaged in converting the carcass into hundreds of scattered hoards.

Later that winter, the same wolf travels through the adjacent home range, having recently become a member of the pack. Again, he happens upon a vulnerable moose. The chase begins. Within minutes, he and his new packmates manage to bring down the moose. As the newcomer in the pack, he must wait for his turn to feed. At first light, ravens begin to gather nearby and wait for their turn at the carcass. At midday, the ravens are still biding their time.

10.2   Embracing the Complexity of Social Foraging

The vast majority of carnivores live solitarily. Why, then, do wolves (*Canis lupus*) live in social groups? Surely, you might think, the advantages of social foraging must favor group living (sociality) in wolves. But the data suggest that wolves live in packs despite suffering reduced foraging payoffs (Vucetich et al. 2004). The data suggest that an individual wolf would often achieve a higher food intake rate if it foraged alone rather than as a member of a pack. So it appears that sociality persists despite negative foraging consequences. Why? Perhaps parents accept a reduction in their own intake rates if the beneficiaries are their own offspring (Ekman and Rosander 1992). But why would any individual stay in a pack if it could do better on its own?

In this chapter, we illustrate some theoretical approaches to analyzing such problems. We show that packs may form through retention of nutritionally dependent offspring, but we cannot readily explain why individuals with developed hunting skills belong to groups. This failure of nepotism as a general explanation prompts further analysis of the foraging payoffs. We incorporate a previously overlooked feature of wolf foraging ecology, the cost of scavenging by ravens. And voila! Predicted group size increases dramatically. Thus, it appears that benefits of social foraging favor sociality in wolves after all.

Throughout this chapter, we describe situations in which foraging payoffs depend not solely on an individual's own actions, but also on the actions of others. This economic interdependence means that the study of social foraging requires game theory (Giraldeau and Livoreil 1998). It also implies that animals may forage socially even if they never interact. Conventional foraging theory (Stephens and Krebs 1986) in effect assumes that foragers are economically independent entities. Until recently, the study of social foraging proceeded without a unified theoretical framework. Fortunately, Giraldeau and Caraco's (2000) recent book provides a synthesis of game theoretical models of social foraging that remedies this situation. The basic principle of such models is that the best tactic for a forager depends on the tactics used by others.

According to the classic patch model from conventional foraging theory (Charnov 1976b; see chap. 1 in this volume), a forager should depart for another patch when its instantaneous rate of gain drops to the habitat-at-large level. To illustrate the difference between conventional and social foraging, we examine how this patch departure threshold differs for solitary versus social foragers. Consider the following scenario (Beauchamp and Giraldeau 1997; Rita et al. 1997): An individual (producer) finds a patch, and forages alone initially, but then other individuals (scroungers) join the producer, arriving one at a time (cf. Livoreil and Giraldeau 1997; Sjerps and Haccou 1994). Each scrounger depresses the producer's intake rate by interfering with the producer's foraging. If interference is strong, the producer may leave immediately when the first scrounger arrives, even if it must spend a long time traveling to the next patch. Thus, a scrounger's arrival can lead a social forager to leave a patch much sooner than a solitary forager would. This scenario (see also box 10.1) emphasizes the basic theme that the economic interdependence of foraging payoffs shapes the decision making of social foragers.

BOX 10.1   The Ideal Free Distribution

Ian M. Hamilton

The Ideal Free Distribution (IFD; Fretwell and Lucas 1969) predicts the effects of competition for resources on the distribution of foragers between patches differing in quality, assuming that foragers are "ideal" (able to gauge perfectly the quality of all patches) and "free" (able to move among patches at no cost). The original IFD model assumed continuous input of prey and scramble competition. Under continuous input, resources continuously arrive and are instantly removed by foragers. Assuming equal competitive abilities and no foraging costs, the payoff of foraging in patch i is the rate of renewal of the resource, Q_i, shared among N_i foragers in the patch. At equilibrium, foragers will be distributed so that none can improve its payoff by unilaterally switching patches. In the original model, the ratio of forager densities between two patches at equilibrium matches that of the rates of resource input into the patches (i.e., $N_i / N_j = Q_i / Q_j$). This match in ratios is known as the *input matching* rule. At equilibrium, the fitness payoff to foragers is also equal in all patches. The input matching rule holds even for predators that do not immediately consume prey upon its arrival, so long as the only source of prey mortality is consumption by the predators (Lessells 1995).

There have been numerous modifications of the original model. Relaxing the ideal and free assumptions of the original model can result in *undermatching*, or lower use of high-quality patches than expected based on resource distribution (Fretwell 1972; Abrahams 1986). Undermatching is a common finding in tests of the IFD (Kennedy and Gray 1993; but see Earn and Johnstone 1997). Other modifications include changing the form of competition and the currency assumed in the model. In this box I briefly review these ideas. Extensive reviews of IFD models and empirical tests can be found in Parker and Sutherland (1986), Milinski and Parker (1991), Kennedy and Gray (1993), Tregenza (1995), Tregenza et al. (1996), van der Meer and Ens (1997), and Giraldeau and Caraco (2000).

(Box 10.1 continued)

Continuous Input, Unequal Competitors

If forager phenotypes differ in their abilities to compete for prey, and if their relative abilities remain the same in all patches, then there are an infinite number of stable distributions of phenotypes between patches (Sutherland and Parker 1985). However, all of these distributions are consistent with *competitive-weight matching*. If each individual is weighted by its competitive ability, the ratio of the summed competitive weights in each patch matches the ratio of resource input rates. At equilibrium, the mean intake rates are equal across patches.

If relative competitive abilities differ among patches, a *truncated phenotype* distribution is predicted (Sutherland and Parker 1985). Foragers with the highest competitive abilities aggregate in patches where competitive differences have the greatest effect on fitness payoffs, and those with the lowest competitive abilities are found where competitive differences have the smallest effect. Average intake rate is higher for better competitors.

Interference

Continuous input prey dynamics are rare in nature (Tregenza 1995). Interference models apply when prey densities are constant or gradually decrease over time and when the quality of patches to foragers reversibly decreases with increasing competitor density. There are several ways to model interference, which lead to different predicted distributions (reviewed in Tregenza 1995; van der Meer and Ens 1997). The simplest of these is the addition of an "interference constant," m (Hassell and Varley 1969), to the effects of forager density on patch quality, so that the payoff for choosing patch i is Q_i / N_i^m (Sutherland and Parker 1985). When $m < 1$, more competitors use the high-quality patch than expected based on the ratio of patch qualities. When $m > 1$, the opposite is predicted. When phenotypes differ in competitive ability, this model predicts a truncated phenotype distribution.

Kleptoparasitism

One form of interference that has been extensively investigated is kleptoparasitism, in which some individuals steal resources acquired by others. If kleptoparasitism does not change the average intake rate, but simply reallocates food from subordinates to dominants, no stable distribution is predicted (Parker and Sutherland 1986).

(Box 10.1 continued)

Models based on the transition of foragers among behavioral states, such as searching, handling, and fighting, have also been used to investigate the influence of kleptoparasitism on forager distributions (Holmgren 1995; Moody and Houston 1995; Ruxton and Moody 1997; Hamilton 2002). These models reach stable equilibria and predict greater than expected use of high-quality patches by all foragers when competitors are equal (Moody and Houston 1995; Ruxton and Moody 1997) and by dominant foragers (Holmgren 1995) or kleptoparasites (Hamilton 2002) when competitors are not equal.

Changing Currencies

The previous models all use net intake rate as the currency on which decisions are based. The IFD has also provided fertile ground for models exploring how animals balance energetic gain and safety (Moody et al. 1996; Grand and Dill 1999) and for empirical studies seeking to measure the energetic equivalence of predation risk (Abrahams and Dill 1989; Grand and Dill 1997; but see Moody et al. 1996). Hugie and Grand (2003) have shown how such "non-IFD" considerations as avoiding predators or searching for mates affect the distribution of unequal competitors (see above), resulting in a unique, stable equilibrium.

Some authors have also used IFD models to examine the interaction between predator distributions and those of their prey when both can move (Hugie and Dill 1994; Sih 1998; Heithaus 2001). These models predict that predators tend to aggregate in patches that are rich in resources used by their prey. If patches also differ in safety, prey tend to aggregate in safer patches, even when these patches are relatively poor in resources.

A recent model by Hughes and Grand (2000) used growth rate, rather than intake rate, as the fitness currency in an unequal-competitors, continuous-input model of the distribution of fish. In fish, like other ectotherms, growth rate is strongly influenced by temperature, and this model predicted temperature-based segregation of competitive types (body sizes) when patches differed in temperature.

This scenario also shows how social foraging can have both positive and negative consequences. Individuals may benefit from foraging socially because groups discover more food or experience less predation. In general, individuals may benefit by joining others who have already discovered a resource.

However, joining represents a general cost of social foraging. "Whenever some animals exploit the finds of others, all members of the group do worse than if no exploitation had occurred. The almost inevitable spread of scrounging behavior within groups and its necessary lowering of average foraging rate may be considered a cost of group foraging" (Vickery et al. 1991, 856). Recent work has revealed that foragers may sacrifice their intake rate to stay close to conspecifics (Delestrade 1999; Vasquez and Kacelnik 2000; see also Beauchamp et al. 1997). Other work has shown that social foragers may acquire poor information (i.e., about a circuitous, costly route to food) (Laland and Williams 1998). In the extreme, joining can lead to an individual's demise through tissue fusion (see section 10.5). These examples highlight the intrinsic complexity of social foraging.

This chapter reviews theoretical and empirical developments in the study of social foraging. Throughout, we explore joining decisions: When should a solitary individual join a foraging group? When should a group member join another member's food discovery? When should an individual join another through fusion of their peripheral blood vessels? We begin by exploring the economic logic of group membership. Next, we review producer-scrounger games, in which individuals must decide how to allocate their time between searching for food (producing) and joining other individuals' discoveries (scrounging). Finally, we review work on cooperative foraging.

10.3   Group Membership

Predicting Group Size

Stable Group Size Often Exceeds Rate-Maximizing Group Size

Many animals find themselves in a so-called aggregation economy, in which individuals in groups experience higher foraging payoffs than solitary individuals (e.g., Baird and Dill 1996; review by Beauchamp 1998). Peaked fitness functions are the hallmark of such economies (fig. 10.1; Clark and Mangel 1986; Giraldeau and Caraco 2000). By contrast, animals in a dispersion economy experience maximal foraging payoffs when solitary and strictly diminishing payoffs with increasing group size (e.g., Bélisle 1998). In an aggregation economy, the per capita rate of intake increases initially with increasing group size G. However, because competition also increases with group size G, intake rate peaks (at G^*) and then falls with further increases in group size. Clearly, this situation favors group foraging, but can we predict group size?

It might seem that the observed group size G should match the intake-maximizing group size G^*, at which each group member maximizes its fitness.

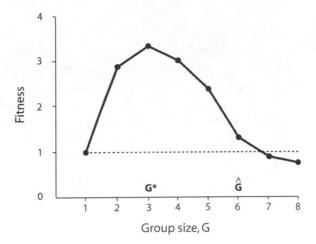

Figure 10.1. Hypothetical relationship between group size *G* and an unspecified surrogate for fitness (e.g., net rate of energy intake). This general peaked function is characteristic of an aggregation economy, in which individuals gain fitness with increasing *G*, at least initially. *G** (= 3) is the intake-maximizing group size. *G* may exceed *G** because a solitary individual would receive a fitness gain by joining the group. *G* may continue to grow until it reaches *Ĝ* (= 6), the largest size at which each individual would do better to be in the group than to be solitary. *G* is not expected to exceed *Ĝ* because a joiner that increases *G* to *Ĝ* + 1 would achieve greater fitness by remaining solitary.

Many studies, however, have found that G often exceeds G* (Giraldeau 1988). This mismatch is not unexpected. With a peak in the fitness function at G* (see fig. 10.1), the intake-maximizing group is unstable because a solitary forager can benefit from joining the group. A group of size G* will grow as long as foragers do better in that group than on their own, but it should not exceed the largest possible equilibrium group size Ĝ. At that point, solitary individuals do better to continue foraging alone than to join such a large group. Equilibrium group size may be as small as the intake-maximizing group size G* and as large as the largest possible equilibrium size Ĝ, depending on whether the individual or the group controls entry and on the degree of genetic relatedness between individuals (box 10.2).

Thanks to the development of this theory, it is no longer paradoxical to find animals in groups larger than the intake-maximizing group size G*. Yet the role of foraging payoffs in the maintenance of groups of large carnivores remains contentious (see Packer et al. 1990 for a fascinating case study). The wolves discussed in the prologue present a paradox, because pack size routinely exceeds the apparently largest possible equilibrium size Ĝ. Why would a wolf belong to a pack when it could forage more profitably on its own? Here we attempt to resolve this paradox while reviewing the theory on group membership.

BOX 10.2 Genetic Relatedness and Group Size

Giraldeau and Caraco (1993) analyzed the effects of genetic relatedness on group membership decisions. Consider a situation in which individuals benefit from increasing group size, and in which all individuals are related by a coefficient r. According to Hamilton's rule, kin selection favors an altruistic act (e.g., allowing an individual to join the group) when $rB - C > 0$, where B is the net benefit for all relatives at which the act is directed and C is the cost of the act to the performer. In the context of group membership decisions, both effects on others (E_R) and effects on self (E_S) can be either positive or negative, so we rewrite Hamilton's rule as

$$rE_R + E_S > 0. \quad (10.1.1)$$

Group-Controlled Entry

In some social foragers, group members decide whether to permit solitaries to join the group. Such groups should collectively repel a potential group member (i.e., keep the group at size G) when Hamilton's rule is satisfied. Here E_R is the effect of repelling the intruder on the intruder:

$$E_R = \Omega(1) - \Omega(G+1), \quad (10.1.2)$$

and E_S is the effect of repelling the intruder on the group:

$$E_s = G[\Omega(G) - \Omega(G+1)], \quad (10.1.3)$$

where $\Omega(1)$ is the direct fitness of the solitary intruder, $\Omega(G)$ is the direct fitness of each of G individuals in the current group, and $\Omega(G+1)$ is the direct fitness of each individual if the group decides not to repel the intruder. (As we highlight below, the group-level decision is based on the selfish interests of the individual group members.) Substituting these expressions for the effects of repelling the intruder on the intruder [E_R; eq. (10.1.2)] and on the group [E_S; eq. (10.1.3)] into equation (10.1.1) and dividing all terms by G, we see that selection favors repelling a prospective joiner when

$$\left(\frac{r}{G}\right)[\Omega(1) - \Omega(G+1)] + [\Omega(G) - \Omega(G+1)] > 0, \quad (10.1.4)$$

where we express both the indirect fitness (first term on the left-hand side) and the direct fitness (second term) of group members on a per capita basis.

By extension, group members should *evict* an individual from the group when $rE_R + E_S > 0$. Here the effect on the evicted individual E_R is

(Box 10.2 continued)

$\Omega(1) - \Omega(G)$, and the effect on the remaining group members E_S is $(G - 1)[\Omega(G - 1) - \Omega(G)]$.

Equation (10.1.4) indicates that repelling is never favored when $1 < G < G^*$, where G^* is the group size at which individual fitness is maximized, but repelling is always favored when $G > \hat{G}$, where $\hat{G}$ is the largest group size at which the individual fitness of group members exceeds that of a solitary. Thus, equilibrium (stable) group size must fall within the interval $G^* < G < \hat{G}$. Under group-controlled entry, the effect of increasing genetic relatedness is to increase the equilibrium group size. By contrast, if potential joiners can freely enter the group, genetic relatedness has the opposite effect.

Free Entry

Under free entry, group members do not repel potential joiners; thus, potential joiners make group membership decisions. Any such individual should join a group when Hamilton's rule is satisfied, where E_R is the combined effect of joining on all the joiner's relatives:

$$E_R = (G - 1)[\Omega(G) - \Omega(G - 1)], \quad (10.1.5)$$

and E_S is the effect of joining on the joiner:

$$E_S = \Omega(G) - \Omega(1). \quad (10.1.6)$$

Substituting, we see that joining a group of size $(G - 1)$ is favored when

$$r(G - 1)[\Omega(G) - \Omega(G - 1)] + [\Omega(G) - \Omega(1)] > 0. \quad (10.1.7)$$

An analysis of equation (10.1.7) reveals that, under free entry, the effect of increasing genetic relatedness is to *decrease* equilibrium group size. (For derivation of the expressions for equilibrium group size under both entry rules, see Giraldeau and Caraco 2000.)

Rate-Maximizing Foraging and Group Size

In wolf packs, group members control entry. Thus, pack size should fall somewhere between the intake-maximizing group size G^* and the largest possible equilibrium size $\hat{G}$ (see box 10.2). The data show that a group size of two maximizes net per capita intake rate and that individuals would do worse in a larger group than alone (i.e., $G^* = \hat{G} = 2$; see fig. 3 in Vucetich et al. 2004). Thus, this initial analysis cannot explain pack living.

Variance-Sensitive Foraging and Group Size

Our initial attempt might have failed for lack of biological realism. We assumed that each individual would obtain the *mean* payoff for its group size. However, in nature, the realized intake rate of an individual might deviate widely from the average rate. In principle, a reduction in intake rate variation with increasing group size could translate into a reduced risk of energetic shortfall. However, a variance-sensitive analysis indicates that an individual will have the best chance to meet its minimum requirement if it forages with just one other wolf (see fig. 4 in Vucetich et al. 2004). Its risk of shortfall will be higher in a group of three or more than alone. Thus, once again, foraging models fail to explain pack living.

Genetic Relatedness and Group Size

So far, foraging-based explanations seem unable to account for the mismatch between group size predictions and observations. Kin selection would seem to provide a satisfactory explanation (e.g., Schmidt and Mech 1997). After all, wolf packs form, in part, through the retention of offspring. However, kin-directed altruism (parental nepotism) does not account for the observation that pack size routinely exceeds the largest possible equilibrium group size $\hat{G}$. Although we expect group size to increase with genetic relatedness when groups control entry (see box 10.2), theory predicts that equilibrium group size cannot exceed $\hat{G}$, even in all-kin groups (Giraldeau and Caraco 1993).

Recalling that for wolves, the largest possible equilibrium group size $\hat{G} = 2$, kin selection cannot explain pack living. This does not mean, however, that group size should never exceed two. Consider immature wolves, which cannot forage independently. If evicted, they would presumably achieve an intake rate of virtually zero. Under this assumption, Hamilton's (1964) rule (see box 10.2) predicts group membership for nutritionally dependent first-order relatives (i.e., offspring or full siblings). However, individuals that can achieve the average intake rate of a solitary adult should not belong to groups, even all-kin groups (fig. 10.2). Thus, while kin selection offers an adequate explanation for packs comprising parents and their immature offspring, we still have not provided a general explanation for wolf sociality. How do we account for packs that include unrelated immigrants and mature individuals? Is there an alternative foraging-based explanation that has evaded us?

Kleptoparasitism and Group Size

Inclusion of a conspicuous feature of wolf foraging ecology, loss of food to ravens (*Corvus corax*), increases the predicted group size dramatically (fig. 10.3). Both rate-maximizing (fig. 10.3) and variance-sensitive currencies predict large pack sizes, even for small amounts of raven kleptoparasitism. Why does

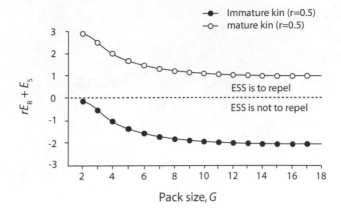

Figure 10.2. The application of Hamilton's rule to predict whether mature and immature solitary wolves should be allowed in packs of various sizes when the pack controls group entry (see also fig. 5 in Vucetich et al. 2004). The pack should repel any individual that attempts to increase the pack size from G to $G + 1$ when $rE_R + E_S > 0$ (i.e., above dotted line), where r is the coefficient of relatedness, E_R is the fitness effect on a repelled intruder, and E_S is the fitness effect of repelling the intruder on the current group members (see box 10.2). The points corresponding to $G > 2$ are based on the reciprocal exponential function for net rate of food intake (see fig. 10.1). Mature solitaries, assumed to have developed hunting skills, are assumed to achieve the average net intake rate of a solitary adult. Immature solitaries, with undeveloped hunting skills, are assumed to obtain no prey and to expend energy at $3 \times$ BMR ($=(3 \times 3,724$ kJ/d)/(6,800 kJ/kg) $= -1.6$ kg/d). A group comprising first-order relatives ($r = 0.5$) should accept an immature solitary with undeveloped hunting skills, but repel any mature solitary even if it is close kin.

including this cost shift the economic picture so dramatically? The key insight here is that individual wolves in larger packs must pay a greater cost in terms of food sharing with other wolves, but this cost is offset by the reduced loss of food to scavenging ravens. Such economic realities may commonly favor sociality in carnivores that hunt large prey and thus are vulnerable to kleptoparasitism (see Carbone et al. 1997; Gorman et al. 1998).

This case study highlights the value of applying formal theory. The failure of kin selection to explain wolf sociality prompted us to continue the search for a foraging-based explanation. Without modern theory on group membership decisions, we might have been satisfied to attribute large pack size in wolves to kin selection and unknown factors. Instead, our conclusions now lead us to ask why group members would prevent entry into the pack and why observed pack size is *smaller* than predicted (see fig. 10.3). The next subsection offers some perspective.

Recent Advances in the Theory of Group Membership

Recent theoretical studies have provided insights into the flexibility of group membership decisions. One such study used optimal skew theory to predict

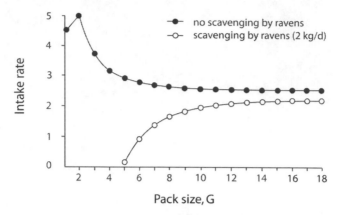

Figure 10.3. Relationship between pack size and average daily per capita net rate of intake assuming either negligible or minor scavenging pressure by ravens (see also fig. 6 in Vucetich et al. 2004). To assess how raven scavenging might affect the predicted relationship between pack size and intake rate, we first considered how pack size and rate of loss to scavengers (kg/d) affect the number of days required to consume the carcass of an adult moose (295 kg). For a given pack size and rate of loss, we calculated carcass longevity assuming a consumption rate of 9 kg/d/wolf. Then, to obtain kg/wolf/day as a function of pack size and number of ravens, we multiplied the kg/wolf/kill (a function of pack size and loss to scavengers) by the kills/day (a function of pack size).

group size (Hamilton 2000). This study modeled the division of resources as a game between an individual (recruiter) that controls access to resources and a potential recruit. If another individual's presence benefits the recruiter (fig. 10.4), the recruiter may provide an incentive to join or stay. The incentive may increase the recruit's foraging payoff, reduce its predation risk, or both. We restrict our attention to the simple case in which the incentive provides a foraging payoff. For joining to be profitable, this incentive must cause the recruit's payoff to equal or exceed the payoff it would obtain by remaining solitary.

This model predicts that the stable group size will fall between G^* (equal division of resources and group-controlled entry) and a maximum stable group size $\hat{G}$ (equal division of resources and free entry). Stable group size increases as the recruiter's control over resource division decreases (fig. 2 in Hamilton 2000). As this control decreases and the benefits of group membership increase, predicted group size G shifts from being transactional (i.e., where the recruiter provides an incentive) to nontransactional (i.e., where the joiner obtains a sufficient payoff without using any of the recruiter's resources) (see fig. 10.4). In transactional groups, the recruiter and joiners agree about group size because the stable size is the same for all parties. However, in nontransactional groups, there may be conflict over group size. Factors that reduce the recruiter's control (e.g., minimal dominance) or increase the benefits of group membership (e.g., large food rewards) will also increase the

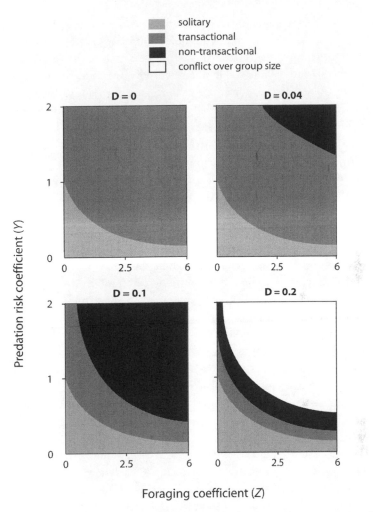

Figure 10.4. Numerical example of the joint effect of foraging (x-axis) and antipredation benefits (y-axis) favoring solitary versus social foraging. The panels represent situations in which the recruiter is assumed to have complete ($D = 0$, upper left panel) or varying degrees of incomplete ($D = 0.04$, 0.1, and 0.2) control over the division of resources. If the recruiter has complete control over the division of resources, all groups are transactional (i.e., the recruiter provides a joining incentive). Under incomplete control (e.g., $D = 0.2$, lower right panel), as the benefits of group foraging increase, groups switch from being transactional to nontransactional (i.e., the recruiter provides no joining incentive). If the benefits of group foraging are sufficiently high, the recruiter and joiners may be in conflict over group size (i.e., group size may exceed the optimum from the recruiter's perspective). (After Hamilton 2000.)

likelihood of conflict. In nontransactional groups, group size is likely to be stable only if joiners accrue no antipredation benefits. If joiners receive foraging benefits only, group size is likely to remain small (close to G^*) and under the control of the recruiter. However, if joiners accrue both antipredation and foraging advantages, group size is likely to be unstable. Predicted group size may increase to the maximum stable group size $\hat{G}$.

A compelling question remains: if models tell us that group size will equilibrate around some stable size, then why are observed group sizes so variable? A recent study used a dynamic model to address this question. Specifically, Martinez and Marschall (1999) asked why juvenile groups of the coral reef fish *Dascyllus albisella* vary in size (range: 1–15 individuals). They uncovered an explanation not only for why observed group size varies, but also for why it may often fall below the intake-maximizing group size G^*. Consider the natural history of *D. albisella*. Following a pelagic larval stage, these fish return to a reef, where they settle into juvenile groups. Martinez and Marschall modeled the joining decision as a trade-off between body growth (faster in smaller groups) and survival (better in larger groups), assuming that individuals reaching maturity by a specified date joined the adult population. When larvae encounter a group into which they may potentially settle, they must decide whether to join or to continue searching. By assumption, a larva settles only if the fitness value of doing so (i.e., the product of size-specific fecundity and probability of recruitment) exceeds the fitness value of further searching.

Rather than groups of a set size, Martinez and Marschall found that a range of acceptable group sizes arose from the fitness-maximizing choices of individuals. Their analysis suggests that, on any given day, fitness is maximized by settling in any encountered group that falls within the acceptable range. The policy for a larva settling early in the season is to settle in large groups ($G^* = 9$), which have high survival rates. By contrast, a small larva searching late in the season should settle as a solitary or join a very small group; otherwise, it will not grow fast enough to reach maturity. This dynamic joining policy creates persistent variation in group size, whereas conventional theory predicts that group size will equilibrate around a stable size.

The combination of this dynamic joining model with Ian Hamilton's recruiter-joiner model would allow new questions: Should current members provide a joining incentive to recruit new members? In the case of the coral reef fish *D. albisella*, would the size of this incentive depend on date, the recruit's body size, or current group size? Would increased foraging skew in large groups reduce the upper limit of acceptable group size earlier in the season? Would many more individuals choose to settle as singletons? Would the theory predict highly variable final group sizes? Under what conditions is group size stable? We expect Ian Hamilton's recruiter-joiner approach to play a key role in the development of group size theory, particularly in systems in which resource owners benefit from the presence of other individuals.

10.4   Producing, Scrounging, and Stable Policies

This section considers how animals should behave once they find themselves in a group in which some individuals parasitize the discoveries of others. This scrounging behavior is a pervasive feature of group foraging (Giraldeau and Beauchamp 1999). But should individuals always join others' discoveries? Doesn't scrounging become unprofitable if everyone does it? What is the optimal scrounging policy, and what factors affect the decision? Behavioral ecologists have analyzed these questions using two antagonistic approaches, information-sharing (IS) and producer-scrounger (PS) models. Here we briefly review these approaches and recent experiments that have tested them (see reviews by Giraldeau and Livoreil 1998; Giraldeau and Beauchamp 1999; Giraldeau and Caraco 2000).

Information-Sharing versus Producer-Scrounger Models

Information-sharing (IS) models assume that each group member concurrently searches for food and monitors opportunities to join the discoveries of others (Clark and Mangel 1984; Ranta et al. 1993). When a member discovers a food patch, information about the discovery spreads throughout the group, and by assumption, all members stop searching and converge on the patch to obtain a share. When individuals can search for food and for joining opportunities simultaneously, the only stable solution to the basic information-sharing model is to join every discovery (Beauchamp and Giraldeau 1996; but see extensions by Ruxton et al. 1995; Ranta et al. 1993, 1996; Rita and Ranta 1998; see also Ranta et al. 1998).

Producer-scrounger (PS) models, by contrast, assume that an individual cannot search simultaneously for food (the producer tactic) and for joining opportunities (the scrounger tactic) (Barnard and Sibly 1981). This incompatibility has important consequences for the optimal policy. Scroungers cannot contribute to the group discovery rate, so any increase in the frequency of scroungers reduces opportunities for scrounging. This relationship makes the payoff function for scrounging negatively frequency-dependent. When there are few scroungers, scrounging pays well. When everybody is a scrounger, there is nothing to scrounge, and producing pays well. The classic producer-scrounger game (box 10.3) predicts that foragers should adjust their scrounging frequency to a stable equilibrium (denoted by $\hat{q}$). At that equilibrium frequency, no one gains by switching from producer to scrounger or vice versa. In the terminology of game theory, this solution is a mixed evolutionarily stable strategy (ESS).

BOX 10.3 The Rate-Maximizing Producer-Scrounger Game

According to the classic producer-scrounger (PS) model (Vickery et al. 1991), each member of a social foraging group must decide how to allocate its time between two mutually incompatible tactics, producing (i.e., searching for food) and scrounging (i.e., searching for opportunities to exploit discoveries of others). The core assumption of the model is that individuals adjust their proportional use of the scrounger tactic to maximize their long-term rate of energy gain (but see Ranta et al. 1996). These adjustments lead to an equilibrium scrounger frequency at which producers and scroungers obtain the same payoffs and no individual can benefit from unilaterally altering its behavior.

At any moment, some proportion p of the G group members use the producer tactic, and the remaining $q = 1 - p$ individuals use the scrounger tactic. While using the producer tactic, an individual encounters food patches containing F items at rate λ. Upon each encounter, the producer obtains a items for its exclusive use before being joined by qG scroungers who "share" the remaining A food items ($F = a + A$) with the producer and one another. For an individual using the producer tactic, the expected cumulative intake I_p by time T is

$$I_p = \frac{\lambda T}{(a + A/n)}, \quad (10.2.1)$$

where $n (= qG + 1)$ is the number of scroungers joining the discovery plus the producer of the patch. For an individual using the scrounger tactic, the expected cumulative intake I_s by time T depends on the proportion $p (= 1 - q)$ of individuals using the producer tactic:

$$I_s = \frac{\lambda T}{[(1 - q)GA/n)]}. \quad (10.2.2)$$

Setting these two expressions equal to each other and rearranging yields an expression for the equilibrium frequency of the scrounger tactic:

$$\hat{q} = 1 - \left(\frac{a}{F} + \frac{1}{G} \right), \quad (10.2.3)$$

which implies that individuals should adjust their proportional use of foraging tactics in response to the finder's share (a/F) and the size of the group. This rate-maximizing PS model [eq. (10.2.3)] predicts that an

(Box 10.3 continued)

individual should reduce its proportional use of the scrounger tactic in response to an increase in the finder's share or a decrease in group size. However, neither the rate of encounter with patches (λ) nor the time horizon (T) influences the predicted producer-scrounger equilibrium (see "Testing the Variance-Sensitive Producer-Scrounger Game").

Theoreticians have modeled the producer-scrounger situation as both a rate-maximizing (Vickery et al. 1991) and a variance-sensitive game (Caraco and Giraldeau 1991; reviewed by Giraldeau and Livoreil 1998; Giraldeau and Caraco 2000). In the rate-maximizing game, the predicted equilibrium frequency of scrounging $\hat{q}$ decreases as a function of the finder's share of the food items (see box 10.3). In the variance-sensitive game, the scrounging frequency $\hat{q}$ depends on both the finder's share and the potential joiner's energetic requirement. The following discussion describes experimental tests of these two games.

Testing the Rate-Maximizing Producer-Scrounger Game

The rate-maximizing producer-scrounger game predicts that the proportional use of the scrounger tactic $\hat{q}$ increases with group size G and decreases as the finder's share increases (see box 10.3). Giraldeau and his colleagues tested the effect of the finder's share in a series of experiments using spice finches (*Lonchura punctulata*). These small seed-eating birds forage in flocks with nearly egalitarian social relationships. The spice finch's ground-feeding habit makes reasonable the assumption of incompatibility between searching for food and searching for joining opportunities. An early experiment revealed that the finder's share (a/F) was negatively related to the extent of food patchiness (Giraldeau et al. 1990). So in the experiments described below, Giraldeau and his colleagues manipulated food patchiness to test the predicted effect of finder's share on equilibrium scrounger frequency $\hat{q}$.

Giraldeau et al. tested spice finch flocks at three levels of food patchiness: very patchy, intermediate patchiness, and uniform. This procedure indirectly manipulated the average finder's share. As predicted, use of the scrounger tactic decreased as finder's share increased (fig. 10.5; see also Giraldeau et al. 1994). The observed use of the scrounger tactic matched the rate-maximizing scrounger frequency $\hat{q}$ reasonably well, but typically fell well below the basic information-sharing model's prediction. Thus, spice finches appear to balance

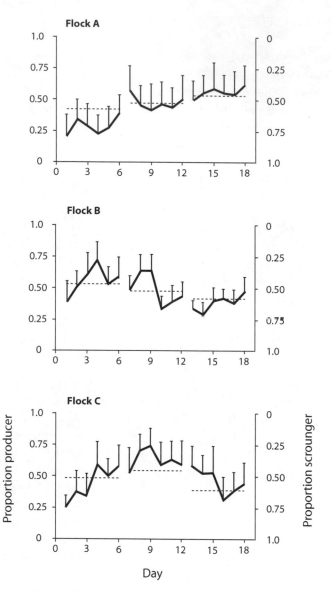

Figure 10.5. The observed (mean + 1 SE) proportional use of producer (left *y*-axis) and scrounger (right *y*-axis) tactics in three five-member groups of captive spice finches (*L. punctulata*). Each experimental group was tested using a unique series of three seed distributions (200 seeds distributed evenly among 10, 20, or 30 patches). By manipulating seed distribution, the experimenters indirectly manipulated the average realized finder's share (i.e., flock A: 0.20, 0.27, 0.33; flock B: 0.33, 0.27, 0.20; flock C: 0.27, 0.33, 0.20). As predicted, in all three flocks the proportional use of the scrounger tactic decreased as the average realized finder's share increased. The dashed horizontal lines indicate the predicted rate-maximizing behavior. (After Giraldeau and Beauchamp 1999; originally described in Giraldeau and Livoreil 1998; see also Giraldeau and Caraco 2000.)

producing and scrounging as the rate-maximizing producer-scrounger game predicts.

However, this study, like all previous studies, has several shortcomings. It failed to establish that producing and scrounging were truly incompatible or that the payoff for the scrounger tactic was negatively frequency-dependent. It also failed to establish whether the foragers converged on the equilibrium scrounging frequency $\hat{q}$, at which both tactics provide the same payoff. Fortunately, a recent study by Mottley and Giraldeau (2000) addresses each of these concerns.

Evidence for Negative Frequency Dependence

Mottley and Giraldeau designed an experimental apparatus that forced individuals to use either the producer or the scrounger tactic. To achieve this, they divided a cage into a producer and a scrounger compartment. An opaque partition prevented individuals from moving between the compartments. On the producer side, individuals could perch at any of twenty-two patches (half of which were empty) and pull a string to gain access to seeds. On the scrounger side, individuals could gain access to food only by sharing seeds produced by a bird on the other side.

Using this apparatus, Mottley and Giraldeau directly manipulated the frequency of tactics. For example, by placing all six subjects in the producer compartment, Mottley and Giraldeau could quantify the payoffs to producers in the absence of scroungers. By testing subjects in every permutation, they described the entire payoff curve for each tactic. Figure 10.6 shows the results. The payoff for the scrounger tactic decreased markedly as the frequency of scroungers increased, justifying the producer-scrounger game's assumption of negative frequency dependence.

The experiment used two patch conditions. In the uncovered condition, a producer's string-pulling action released seeds into an *uncovered* collecting dish that was easily accessible to the producer and any scrounger. In the covered condition, a partial cover limited scroungers' access to food. By varying the payoffs to producers and scroungers, this manipulation generated two predicted producer-scrounger equilibria that Mottley and Giraldeau could explore in a follow-up experiment.

Converging on Predicted Equilibria

To test whether group-foraging spice finches would converge on the predicted equilibria, Mottley and Giraldeau modified their apparatus to allow movement between the producer and scrounger compartments. Their results show that subjects converged first on the predicted scrounger frequency in

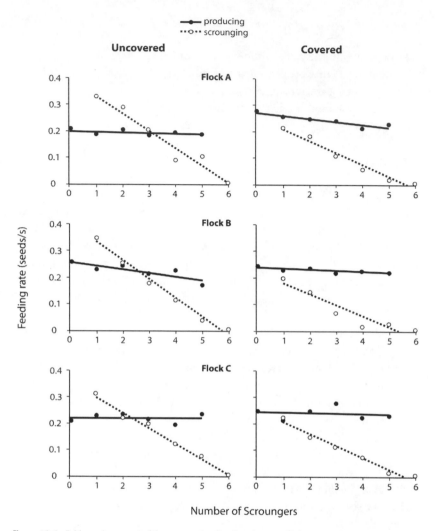

Figure 10.6. Evidence in support of the assumption that foraging payoffs for scrounging are negatively frequency-dependent (i.e., payoff for the scrounger tactic declines with increases in the proportion of individuals in the group using that tactic). Mean (+ 1 SE) observed food intake rates are shown for producing and for scrounging in three captive flocks (A, B, and C) of spice finches as a function of the number of individuals (out of six) scrounging. Subjects were tested under two patch conditions, covered and uncovered. In the covered patch condition, subjects experimentally constrained to use the scrounger tactic experienced reduced access to food. The purpose of these two patch conditions was to generate two distinct predictions for the equilibrium proportional use of the scrounger tactic. (After Mottley and Giraldeau 2000.)

the covered-patch condition and then on the higher predicted scrounger frequency in the uncovered-patch condition. These results constitute the best evidence to date that social foragers can adjust their scrounging frequency to predicted levels.

Testing the Variance-Sensitive Producer-Scrounger Game

Although the evidence just presented supports the rate-maximizing producer-scrounger game, it does not eliminate alternatives that minimize the risk of energetic shortfall (Caraco 1981, 1987; Caraco and Giraldeau 1991). To evaluate this possibility, Koops and Giraldeau (1996) exploited the fact that rate-maximizing and variance-sensitive producer-scrounger models make different predictions about the effect of patch encounter rate λ (and hence patch density) on equilibrium scrounger frequency $\hat{q}$. As box 10.3 shows, the rate-maximizing model predicts that patch encounter rate λ should not affect scrounger frequency, so a manipulation of patch density should not affect scrounger frequency. By contrast, variance-sensitive models predict that scrounger frequency should increase with patch density. Risk-sensitive foragers should adjust their scrounging in response to patch density for two reasons. First, the scrounger tactic yields a lower variance in expected payoff than the producer tactic. In addition, any increase in patch density increases mean intake rate. So, when patch density is high, variance-sensitive foragers should switch to the more conservative scrounging tactic. Koops and Giraldeau tested this prediction using captive European starlings (*Sturnus vulgaris*). As predicted, all eight subjects scrounged more when Koops and Giraldeau increased patch density. The rate-maximizing producer-scrounger model does not predict this flexibility. The results suggest that scrounger may be a variance-sensitive tactic, not strictly a rate-maximizing tactic.

Conclusions and Prospects

These results suggest that the producer-scrounger game provides useful insights into the dynamics of foraging groups. The best test to date (Mottley and Giraldeau 2000) forced individuals to play either producer or scrounger, so that the experimenters could unambiguously assign individuals to either producer or scrounger, and could be sure that the payoff to scroungers decreased with the frequency of scroungers, as the models require. With the assumptions of the producer-scrounger game satisfied, spice finches converged on the stable equilibrium frequency of scrounging. What remains unclear is whether natural foraging groups generally meet these assumptions. Future work should explore the incompatibility between producing and scrounging (Coolen et al. 2001) under natural conditions. Rather than viewing producer and scrounger as discrete alternatives, future theoretical work could consider the possibility that some individuals can search concurrently for food and scrounging opportunities, but that attentional constraints may limit performance (Dukas 1998b; Dukas and Kamil 2000). In systems in which individuals benefit from the presence of others (see fig. 10.3), the recruiter-joiner modeling approach may be appropriate (Hamilton 2000).

In addition to developing a general theory of the evolution of scrounging, future work should test elaborations of the producer-scrounger game (see Giraldeau and Beauchamp 1999). The study of joining policies appears to be in the beginning stages. Future workers should not restrict themselves to studying joining where fitness gains are straightforward, but should also pursue the more puzzling problem of cooperative joining, where individuals pay an apparent or real price in personal fitness. In the next section, we review some exciting developments in the study of the evolution of cooperative foraging.

10.5   Cooperative Games Unrelated Social Foragers Play

Up to now, we have focused on competition within foraging groups. The only exceptions have been group-membership games in which a cooperative individual has obvious selfish motives. First, through kin-directed restraint (see box 10.2), an individual that permits a genetic relative to join the group may gain through the indirect component of inclusive fitness. Obligately social animals such as ants and naked mole-rats provide extreme examples (Sherman et al. 1991, 1995). Second, Hamilton's (2000) recruiter-joiner model tells us that an individual may provide an incentive to a recruit, provided that recruit's presence increases the recruiter's fitness (for an example of shared paternity and egalitarian provisioning in the Galapagos hawk, see Faaborg et al. 1995). These routes to social foraging fit nicely within the "selfish gene" framework (Dawkins 1976). However, we find it more difficult to explain cooperative foraging between unrelated individuals, where the donor pays a cost. Some cooperative arrangements seem evolutionarily unstable because the donor could gain by "defecting." Here we offer a brief review of two evolutionary pathways—reciprocity and mutualism—to stable cooperation between unrelated individuals (see Reeve 1998 for a review of game theoretical models of cooperative kin groups). We then consider whether truly unselfish cooperative foraging can evolve through trait-group selection.

In this section, we adopt Dugatkin's (1997, 1998) definition of cooperation as "an outcome that—despite potential costs to individuals—is 'good' (measured by some appropriate fitness measure) for the members of a group of two or more individuals and whose achievement requires some sort of collective action." This definition implies that an individual can cooperate unilaterally. In other words, to cooperate, an individual need only perform an act that would achieve cooperation if other individuals also were to act appropriately. This definition of cooperation helps us quantify the payoffs in game theory matrices, because we can say that any given player "cooperated" if its opponent defected.

Reciprocity versus Mutualism

Here we describe the logic involved in using payoff matrices in a game theoretical framework. This approach entails specifying players, a set of behavioral options, and the consequences (payoffs) of these options, which depend on the actions of others. By making these assumptions explicit, one can predict when cooperative behavior should occur. To generate these predictions, one searches for the evolutionarily stable strategy (ESS).

Why Do Sentinels Cooperate?

We all know of situations in which humans take turns acting as sentries. Some group-living birds and mammals also "post" sentries (Bednekoff 1997; Clutton-Brock et al. 1999; Blumstein 1999; Wright et al. 2001). Sentinels position themselves in prominent positions where they can scan for approaching predators. When a sentinel detects a predator, it usually gives an alarm call. Group members often behave in a highly coordinated way, seemingly taking turns at sentinel duty. The protection provided by a single sentinel allows other group members to spend less time on vigilance and more time on foraging and other activities.

Why would any individual voluntarily engage in this seemingly dangerous, selfless behavior? The conventional answer has been that kin selection or reciprocity favors sentinel behavior. Before we outline reciprocity-based explanations of sentinels, however, we should acknowledge that sentinel behavior might not be dangerous after all. If sentinels are safe, then we can explain sentinels via mutualism, without the complex apparatus of kin selection and reciprocity (Bednekoff 1997).

Reciprocity and the Prisoner's Dilemma

Models of cooperation via reciprocity focus on the Prisoner's Dilemma game (Trivers 1971; Axelrod and Hamilton 1981). The ESS in the Prisoner's Dilemma is defection, even though mutual cooperation would yield a higher payoff. To see this, imagine two unrelated foragers faced with the prospect of cooperating (acting as sentinel) or defecting (refusing to act as sentinel). Under the payoffs in the matrix shown below, these players would be trapped in a Prisoner's Dilemma: The players face a dilemma because defection yields a higher payoff regardless of what the opponent chooses (i.e., $T > R$ and $P > S$), and yet two defecting players receive less than two cooperating players

		Player 2	
		Cooperate	Defect
Player 1	Cooperate	$R = 3$	$S = 0$
	Defect	$T = 5$	$P = 1$

$(P < R)$. In game theoretical terminology, the payoff matrix of the Prisoner's Dilemma game satisfies conditions $T > R > P > S$ and $R > (S+T)/2$, where T is temptation to defect, R is reward for mutual cooperation, P is punishment for mutual defection, and S is sucker's payoff.

For a single play of the game, we always predict defection, but repeated play of the game can make cooperation a rational choice. Axelrod and Hamilton (1981) confirmed this in a famous computer tournament in which they tested a range of strategies against each other. The winning strategy, tit for tat (TFT), cooperates on the first play and copies its opponent's move on each subsequent play. Tit for tat is evolutionarily stable if the probability of encountering the same player in the future is sufficiently high. Since animals with sentinel systems live in stable groups, tit-for-tat like reciprocity might explain sentinel behavior, but only if the payoff matrix really satisfies the conditions of the Prisoner's Dilemma game.

While it may be tempting to argue that reciprocity or kin selection explains sentinel behavior, any such argument would be speculative at best because no study has quantified the complete payoff matrix. Moreover, Bednekoff (1997) challenged this explanation, arguing that simple self-interest may explain sentinel behavior.

By-product Mutualism

Bednekoff argues that sentinel behavior may be a by-product mutualism. In a by-product mutualism, "each animal must perform a necessary minimum itself that may benefit another individual as a byproduct; these are typically behaviors that a solitary individual must do regardless of the presence of others, such as hunting for food" (Brown 1983). Thus, cooperative alliances may be favored simply because each individual benefits from other individuals' selfish actions.

The payoff matrix for a by-product mutualism will look something like this:

		Player 2	
		Cooperate	Defect
Player 1	Cooperate	$R = 5$	$S = 3$
	Defect	$T = 2$	$P = 0$

In this matrix, the players have no incentive to defect. By-product mutualism is the simplest, and perhaps the most common, pathway to cooperation. Unlike other pathways, it does not require relatedness (kin selection), cognitive abilities allowing scorekeeping (reciprocity), or population structure (Dugatkin 1997). Cooperators need not even be conspecifics.

By-product Mutualism among Safe, Selfish Sentinels?

What if sentinel behavior isn't dangerous after all? Bednekoff (1997) argued that sentinels might be safe and selfish rather than unsafe and selfless. He reasoned that even if they increase their risk of being the target of predators, an improved ability to detect and avoid predators might outweigh this risk. Even if sentinels expose themselves to minimal predation risk, they must pay an opportunity cost because they cannot forage and act as sentinel simultaneously. We might expect, therefore, that individuals will act as sentinels only when their energetic reserves are high. In addition, if a single sentinel provides an adequate early-warning defense, then even a well-fed individual will serve as a sentinel only when no one else is doing so. According to Bednekoff's model, sentinel behavior depends on both the prospective sentinel's nutritional state and the actions of others.

Combining these state-dependent and frequency-dependent aspects of sentinel behavior, Bednekoff developed a dynamic game to explore whether a coordinated sentinel system could emerge from the decisions of selfish individuals. In this game, each individual chooses forager or sentinel based on its energetic state and the actions of others. Provided group members share alarm information, a single sentinel greatly reduces everyone's predation risk, and additional sentinels add little protection from surprise attacks (fig. 10.7). Thus, an individual receives a large safety benefit if it acts as a sentinel when all other group members are foraging. However, an individual may not be able to forgo the foraging opportunity if its energetic state is too low. When no other individuals are acting as sentinels, Bednekoff's model predicts that a focal individual will serve as a sentinel even when its reserves are relatively low (fig. 10.8). However, when another individual is already acting as a sentinel, our focal individual is relatively safe, and so it should act as a sentinel only if its energetic state is near the maximum. The net effect is a sentinel system that appears highly coordinated even though simple selfishness guides the actions of each player. Thus, elaborate coordination and altruism emerge as a by-product of simple self-interested behavior.

Two recent studies support this model. First, meerkats (*Suricata suricatta*), a small social mongoose of South Africa, showed increases in various measures of sentinel activity in response to supplemental feeding (Clutton-Brock et al. 1999). Second, individual Arabian babblers (*Turdoides squamiceps*), a highly

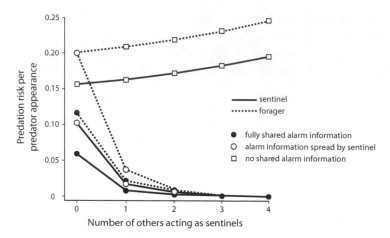

Figure 10.7. Per capita risk of predation for sentinels and foragers as influenced by the number of individuals acting as sentinels. Predicted predation risks are shown for three information-sharing scenarios: full alarm information sharing by any individual that detects a predator, alarm information sharing only by the sentinel, or no alarm information sharing. An individual may be killed if it fails to detect an approaching predator or if it receives no alarm from other group members. The predation risk faced by a particular individual depends on the actions of others (i e , whether they are foraging or acting as sentinel and whether they share alarm information). By assumption, each sentinel is three times more likely than a forager to be killed by an *undetected* predator, but is also four times more likely to detect an approaching predator. This combination makes a sentinel relatively safe, even if foragers fail to share alarm information. (After Bednekoff 1997.)

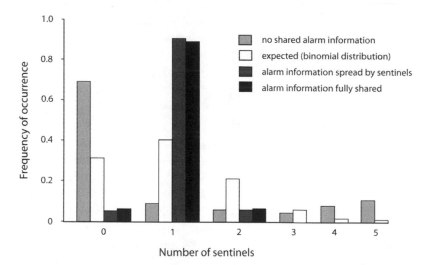

Figure 10.8. Predicted frequency distribution of numbers of sentinels under three information-sharing scenarios. White bars show expected numbers of sentinels (based on binomial distribution) assuming an individual's decision making is independent of the actions of other group members. (After Bednekoff 1997.)

social desert bird, devoted more time to sentinel behavior on days when experimenters provided supplementary food (Wright et al. 2001). Bednekoff's insight and the recent experimental work provide a strong challenge to kin selection and reciprocity as general explanations for sentinel behavior.

Testing the Prisoner's Dilemma versus By-product Mutualism

As our analysis of sentinels shows, the interpretation of cooperation depends on the payoff matrix. If the matrix shows a temptation to defect ($T > R$), then we need either reciprocity or kin selection to maintain cooperation; without a temptation to defect ($R > T$), simple selfishness can explain cooperative action. It may surprise the reader, therefore, to learn that numerous studies have "tested" reciprocity in Prisoner's Dilemmas without first establishing that a Prisoner's Dilemma exists. Tests generally assume that Prisoner's Dilemma conditions are met, but there is a weakness in this assumption in that the payoff matrix is rarely quantified. Doing so would require experimental control of the value for each cell in the matrix.

Clements and Stephens (1995) tested both the iterated Prisoner's Dilemma and by-product mutualism models of cooperation. The experiment allowed pairs of blue jays (*Cyanocitta cristata*) to peck on color-coded keys that represented cooperation or defection in precisely controlled two-player games. Food rewards for choosing each key corresponded either to Prisoner's Dilemma (PD) or by-product mutualism (M) conditions, specifically,

$$\mathbf{PD} = \begin{bmatrix} R = 3 & S = 0 \\ T = 5 & P = 1 \end{bmatrix} \text{versus } \mathbf{M} = \begin{bmatrix} R = 4 & S = 1 \\ T = 1 & P = 0 \end{bmatrix}.$$

The experiment's design exposed each pair of jays first to the Prisoner's Dilemma matrix (**PD**), then to the mutualism matrix (**M**), and then again to the **PD** matrix. On average, these tests took place over 60 days with about 200 plays per day. For each game, the two subjects reached a stable "solution" (i.e., in which at least 90% of trials involved mutual cooperation, mutual defection, or a combined strategy). All (three) pairs of subjects settled on mutual defection in the first Prisoner's Dilemma game, mutual cooperation in the by-product mutualism game, and reverted to mutual defection in the final Prisoner's Dilemma game. Thus, contradicting the dominant paradigm of nonkin cooperation, blue jays trapped in the Prisoner's Dilemma did not cooperate. Instead, Clements and Stephens saw cooperation only in by-product mutualism (i.e., when neither individual had any incentive to defect).

In follow-up work, Stephens et al. (2002) asked whether the jays' failure to cooperate in the Prisoner's Dilemma game could have been due to their strongly preferring immediate rewards rather than implementing an inappropriate strategy. Perhaps the jays failed to cooperate in the Prisoner's Dilemma game and thereby failed to maximize long-term gains because they strongly discount future rewards (see chaps. 1 and 2). Consistent with this idea, experimentally reducing discounting did lead to high levels of cooperation. However, it is hard to imagine natural conditions analogous to the highly contrived experimental conditions required to induce cooperation. Thus, both of these studies cast doubt on the widespread idea that animals faced with Prisoner's Dilemma conditions routinely engage in tit-for-tat-like cooperation.

Trait-Group Selection and Cooperative Foraging in the Extreme

The game theoretical models of cooperation considered above fit comfortably within the selfish gene framework. What appears at first glance to be altruistic turns out to be selfish. Here we invoke group selection in its nouveau form—trait-group selection—to explain how altruism could evolve in reproductively isolated populations, in which, in theory, the existence of even one selfish individual (S) would lead to rapid fixation of S and loss of the altruistic strategy (A).

In trait-group selection models (e.g., Wilson 1977; reviewed by Dugatkin 1997, 1998; Sober and Wilson 1998), cooperative traits differ in frequency across groups embedded within a larger interbreeding population. Reproduction takes place within these trait groups, but mixing allows groups to export surplus productivity resulting from cooperation. So, even though cooperative types tend to be selected against *within* groups, they may increase in frequency in the larger population. In general, trait-group selection favors cooperation when the within-group cost incurred by the altruist is outweighed by the between-group benefit of increased productivity.

Consider a foraging specialist that pays all of the costs of foraging while sharing the benefits with others. Inspired by a real-world example (described below), Wilson (1990) imagined a social insect in which multiple queens join together to found colonies. He assumed that in the primitive state all queens use the same selfish foraging strategy (S). He then asked, if a specialized (altruistic) foraging strategy (A) arises, can it increase in frequency in the metapopulation? The answer is yes. Trait-group selection favors A when $xy > 1$, where x (<1) represents the reduction in the altruist's probability of survival (scaled to S's probability of survival in all-S colonies) and y (>1) represents the positive effect of the altruist's behavior on her colony's probability of surviving at the expense of other colonies. In general terms, altruism can evolve even if the within-group cost is very large, provided that the between-group benefit is large enough to offset this cost. Next, we evaluate whether trait-group selection really favors this extreme form of cooperative foraging in nature.

Cooperative Colony Founding and Foraging Specialization in Ants

In some ant species, two or more queens may join forces to found a colony and rear their first workers, a phenomenon known as pleometrosis (reviewed by Choe and Perlman 1997). In nearly every such species studied to date, cofounding queens are not close genetic relatives (e.g., Rissing et al. 1989; Sasaki et al. 1996; but see Nonacs 1990), so we must invoke some form of selection other than kin selection to account for this extreme form of joining, in which one of the queens "loses." Proponents of trait-group selection contend that pleometrosis facilitates faster production of a larger worker force, which provides an advantage during intense intercolony aggression involving "brood raiding" (e.g., Mintzer 1987; Rissing and Pollack 1987, 1991; but see below and Nonacs 1993; Pfennig 1995).

Work on the desert leaf-cutter ant (*Acromyrmex versicolor*; Rissing et al. 1989; Cahan and Julian 1999) provides the most convincing evidence that trait-group selection favors cooperative colony founding. In many species with pleometrotic founding, only one queen survives after eclosion of the first workers (e.g., Balas and Adams 1996). But in *A. versicolor* and a few other species, the association develops into so-called primary polygyny, in which the queens coexist without antagonism (e.g., Mintzer 1987; Trunzer et al. 1998). Among such species, *A. versicolor* is exceptional in that foundresses forage outside the nest before the first brood of workers emerges, and they do so in a most unusual way. With no apparent coercion, a single queen takes on the role of foraging specialist. This individual provides all of the food for the growing larvae by collecting leaves for the young colony's fungus garden. This individual thus pays all of the foraging costs, while all queens produce workers. Thus, "personal" fitness gains cannot favor this cooperative system, unless the specialist produces a disproportionate number of workers.

For trait-group selection to favor the specialist forager, some between-group benefit must outweigh the within-group cost paid by the specialist. A recent experimental study found that the number of cofounding queens did not affect worker production in successful colonies (Cahan and Julian 1999). However, compared with three-queen colonies (5%), single-queen colonies (53%) often failed to establish a fungus garden. This failure led inevitably to starvation and death of the colony. This finding suggests that multiple-foundress colonies have a simple between-group advantage consistent with the claim that trait-group selection favors this extreme form of cooperative foraging.

Multipartner Urochordate Chimeras as Cooperative Foraging Units?

Even more dramatic than cooperative foraging in pleometrotic ants is the tissue fusion (chimerism) of genetically distinct conspecifics in various marine invertebrates. Recent studies of the cosmopolitan tunicate *Botryllus schlosseri* have explored some of the fitness consequences of this ultimate joining phenomenon (Rinkevich and Shapira 1999). In this species, two genotypically distinct colonies (partners) may fuse through their peripheral blood vessels, provided they share alleles at a single fusibility/histocompatibility locus (Weissman et al. 1990). Such bichimerical fusions yield no fitness gain for either partner (Rinkevich and Weissman 1992). In fact, a dramatic secondary allorecognition phenomenon usually follows fusion: the morphological elimination (resorption) of one of the partners (Rinkevich and Weissman 1987). This fact alone suggests a major fitness cost to the "loser" in this resorption. Group benefits may favor this astonishing form of self-sacrifice.

To examine this possibility, Rinkevich and Shapira (1999) measured fitness components as they experimentally manipulated partner numbers. Chimeras with more partners (colonies) had several advantages. Chimeras with four partners grew more than twice as fast as two-partner chimeras. These four-partner chimeras reached their peak size about a month sooner and contained over 50% more zooids (each producing an egg). In addition, whereas half of the two-partner chimeras fragmented soon after fusion, all of the experimental four-partner chimeras remained intact. Finally, one partner always resorbed the other in two-partner chimeras, but 69% of the subordinate partners in four-partner chimeras survived (although the group resorbed at least one partner in every case). Thus, the between-group benefit of larger alliances appears to be substantial. If trait-group selection favors large alliances, the between-group benefit must exceed the within-group cost. This seems likely, because the experiments revealed only one possible within-group cost: the number of zooids per partner decreased (nonsignificantly) with increasing numbers of partners. Overall, it seems that the between-group productivity advantage of larger chimeras outweighs any within-group cost.

These tantalizing results represent a good start toward understanding this extreme form of group foraging. One caveat, though, is that these results could be an artifact of the unnatural selective regime in the laboratory. Do fusions between unrelated colonies occur in nature? If so, what are the fitness consequences of chimera formation? Is fusion individually selfish after all? An exciting possibility is that the loser in the resorption process, while morphologically destroyed, may benefit by parasitizing the winner's germ or somatic cells. Beyond allowing us to scrutinize the possible role of trait-group selection in the maintenance of chimerism in this model system, chimerism presents many opportunities to study cooperative foraging.

An Epic Challenge for Social Foraging Theory

In their recent book on social foraging theory, Giraldeau and Caraco (2000) explored the implications of game theoretical models for population ecology (see also Sutherland 1996). We hope new workers will be inspired to scale up even further, toward a solution of the "tragedy of the commons" (Hardin 1968). This is no small challenge. Restraint that benefits others flies in the face of the selfish gene perspective. Recent advances in theories of cooperation offer hope in the form of new routes to cooperation and restraint. For example, recent theoretical (Nowak and Sigmund 1998) and experimental work (Wedekind and Milinski 2000; Milinski et al. 2002, 2006) suggest that *indirect* reciprocity, in which unrelated altruists and beneficiaries never even interact, may favor cooperation because altruists benefit from improved standing within the community. We remain optimistic that this and other advances (e.g., Nowak et al. 2000) will help us manage our own exploitive tendencies and help solve the biodiversity crisis (Wilson 2002).

10.6   Summary

This chapter introduces theoretical and empirical approaches to social foraging. We have emphasized the intrinsic complexity of decision making in a social context. Social foragers cannot act as if they were alone. Their decisions must reflect the actions of others. This economic interdependence often implies negative frequency dependence: the payoffs for a strategy decline when more individuals adopt it. Likewise, the payoffs for an individual foraging within a particular habitat patch decline when more individuals forage in that patch.

  We have reviewed both competitive and cooperative aspects of social foraging. In doing so, we have asked about a variety of joining decisions. When should a solitary individual join a foraging group? When should an individual in a group refrain from joining the food discoveries of others? When should unrelated foragers form cooperative alliances? Can joint foraging evolve despite fitness costs for the cooperator? These problems offer many exciting opportunities for additional work. Throughout, we have tried to emphasize specific challenges for future workers. To corrupt a metaphor (Giraldeau and Caraco 2000), we have just entered the social foraging patch and have not yet begun to experience diminishing returns.

10.7   Suggested Readings

Giraldeau and Caraco (2000) offer an excellent, timely, and comprehensive synthesis of social foraging theory. Galef and Giraldeau (2001) review how social environment influences foraging by biasing individual learning processes. Crespi (2001) considers the evolutionary ecology of social behavior, including cooperative foraging, in microorganisms. Sober and Wilson (1998) review evolutionary and psychological aspects of unselfish behavior in humans.

||||||||||||||||||||||||||||||||||||||||||||||||||||||||||||||||||||||||||||||||||||||||||||||||||||||||||||||||||||||||||||||||| **Part IV**

Foraging Ecology

Telander

Foraging and Population Dynamics

Robert D. Holt and Tristan Kimbrell

11.1 Prologue

Every ecology textbook tells the story of snowshoe hare cycles. The vaguely sinusoidal plots of hare densities wiggle across the bloodless page. The hare population traces out a complete cycle every 8 to 11 years; the difference between low- and high-population years can be as much as a hundredfold.

The on-the-ground reality of the hare cycle is anything but dry and academic. In peak years, the undergrowth of the boreal forest virtually quivers with hares. For the lynx, the cycle's peak means easy pickings and a distended stomach: a few short bounds, then a pounce, then a satiated lynx and blood on the snow. Many other predators also feed on hares in the peak years.

It is a different story in the low-population years. A world once spread with hare biomass is nearly empty. Without a ready supply of snowshoe hares, some lynx pursue red squirrels, which are smaller and harder to catch. Notwithstanding the difficulties, lynx adapt to squirrel hunting, and many continue to pursue squirrels even as the hare population recovers. The behavioral inertia of the lynx reduces the predation rate experienced by hares and helps the hare population return to its peak.

The connections between lynx foraging behavior and hare populations flow in both directions. The size of the hare population changes how the lynxhunt, but the hunting behavior of the lynx and other predators influences the growth of the hare population.

11.2 The Necessary Link

Fundamentally, ecologists want to understand patterns of variation in the distribution and abundance of species (Andrewartha and Birch 1954; Krebs 2001). The study of "population dynamics" represents an approach to questions of distribution and abundance that focuses on temporal and spatial variation in population size (box 11.1). Although the idiosyncrasies of natural history vary enormously among species, students of population dynamics find that most species follow one of a relatively small set of dynamic patterns (Lawton 1992). Some species have populations that persist indefinitely, whereas others regularly suffer local extinctions. For persistent populations, numbers may show small fluctuations around a well-defined equilibrium, or instead display large variations in abundance. These large variations may take the form of regular cycles or seemingly random variation. Species that regularly suffer local extinctions must have other populations nearby that can recolonize the empty patches, thereby forming a metapopulation that may persist indefinitely (Hanski 1999).

BOX 11.1   **Basic Concepts in Population Dynamics**

The essential data in population dynamics describe how numbers of individuals in a population vary through time and space. We focus here on a population at a single, spatially closed location. Population size is represented as $N(t)$, which could be either a function (continuous) or a variable with discrete values. Deterministic or probabilistic rules determine how $N(t)$ changes with time. For small populations, models must deal with the discreteness of individuals (for example, a population may contain 4 or 5 individuals, but not 4.78 individuals). This requires stochastic model formulations (which can be mathematically very difficult), because at the level of individual organisms, births and deaths are inherently probabilistic (Renshaw 1991). For large populations, we portray N as a defined function of time; this allows us to use simpler deterministic models. The "top-down" modeling approach described in the text usually involves deterministic models, whereas "bottom-up" individual-based models use stochastic rules. Foraging influences these rules because foraging decisions

(*Box 11.1 continued*)

influence birth rates (for consumers) or death rates (for both consumers and victims). Models may represent time, the independent variable, as a discrete or continuous variable. With discrete census intervals (e.g., an insect population in which one counts individuals in year 0, year 1, and so on), the difference equation $N(t + 1) = N(t)W(t)$ describes the population's dynamics, where $W(t)$ is the finite growth "rate" over one time unit. If, instead, we monitor the population continuously, a more natural formulation is $dN/dt = NF(t)$, where $F(t)$ is the instantaneous per capita growth rate. The quantities $W(t)$ and $F(t)$ describe the contribution an average individual makes to the next time step, and so are closely related to Darwinian fitness. Optimal foraging models all make assumptions about the relationship between foraging decisions and functions such as $W(t)$.

There are several basic questions that are perennial in the study of population dynamics. The most basic question one can ask about a population is, *does it persist, or go extinct* (i.e., does $N(t)$ go to 0 at large t)? One approach to this question relies on the most basic model of population growth, the exponential growth model $dN/dt = rN$, where r is the intrinsic growth rate (birth rate minus death rate). This model describes the dynamics of a continuously growing (or declining) population whose numbers are large enough that we can treat it deterministically. If $r < 0$, the population declines toward zero. Consequently, for a population to persist, it must have $r \geq 0$. In the text, we use this simple observation to conclude that optimal foraging may at times facilitate population persistence. Of course, if N is near zero, the assumption that abundance is a continuous variable breaks down. Stochastic birth-death models reveal that a population with expected births less than expected deaths at low N will surely go extinct, so the basic insight of the exponential model still holds; however, populations with positive r values may nonetheless go extinct at low N due to "demographic stochasticity," reflecting the inherent randomness of individual births and deaths (Renshaw 1991). These same models show that we cannot determine whether a population will become extinct using only the average values of births and deaths. For instance, if behavior A implies an average birth rate of 2 and an average death rate of 0, whereas behavior B leads to an average birth rate of 10 and an average death rate of 8, the two behaviors have the same expected fitnesses (as measured by expected growth rates), but they have different likelihoods of extinction (zero for A, because A strategists don't die, but greater than zero for B). This observation suggests that in very small populations we may need

(Box 11.1 continued)

a somewhat different measure of fitness to characterize evolutionarily persistence. At low densities, the effects of foraging decisions on deaths may be more important than numerically equivalent effects on births.

Given a closed, persistent population, one can ask, *why is it that the population does not increase in size indefinitely, but instead remains within some bound?* Broadly speaking, this requires that populations be "regulated," in the sense that their numbers decline when too high and increase when too low. In turn, such regulation must rest on density dependence experienced by individuals—average birth rates must decline, or average death rates must increase, when density increases. After many years of debate about whether we need population regulation and density dependence to explain population persistence, students of population dynamics now agree that density-dependent factors regulate populations (Royama 1992; Hanski 1999; Turchin 1999). Given that a population persists, it should have a frequency distribution of observed densities bounded away from zero (Turchin 1995).

Persistent populations may exhibit a wide range of dynamic behaviors, ranging from tight regulation near an equilibrium abundance to regular cycles to highly irregular fluctuations. *What causes unstable dynamics?* Ecologists continue to debate the relative importance of small-scale fluctuations versus the effects of climate versus direct within-species interactions versus interactions with other species in generating each of these dynamic outcomes (Bjornstad and Grenfell 2001). All of these processes have a role, but the relative contribution of each clearly varies among systems. A very active area of population ecology uses statistical time-series models to link observed population fluctuations to mechanistic population models (Bjornstad and Grenfell 2001; Kendall et al. 1999).

The three population questions we have just discussed roughly correspond to what Turchin (2001) suggests may be the basic laws of ecology, describing processes in all populations: (1) a propensity for populations without external constraints to grow exponentially, (2) the inevitability of density dependence and population regulation in a finite world, and (3) the likelihood of cycles arising when consumers exploit living resources. Royama (1992), Cappuccino and Price (1995), Hanski (1999) and Bjornstad and Grenfell (2001) provide useful reviews of population dynamics. Another layer of complexity in population dynamics that we have not discussed revolves around the importance of the internal "structure" of populations (for work on ages or stages, see Caswell 2001; for spatial structure, see Holt 1985; Hassell 2000; McPeek et al. 2001).

Foraging decisions are a fundamental driving force of population dynamics. The dynamics of a population arise entirely from four processes: births, deaths, immigration, and emigration (Williamson 1972). From a consumer's point of view, the resources it acquires while foraging govern its Darwinian fitness via effects on fecundity or survival, which translate into changes in population size. Because these effects may vary from one habitat to the next, decisions to disperse or change habitats also influence numerical dynamics via immigration or emigration rates. The relationships between foraging decisions and demographic rates thus link foraging theory and population dynamic theory. As the case of snowshoe hares indicates, the foraging decisions of one species may be tightly linked to the dynamics of many other species. For prey species, the foraging decisions made by predators can strongly influence mortality rates. For competing species that share a common food source, the foraging decisions of competitors can alter the environment. For all of these reasons, foraging decisions must profoundly drive or modulate population dynamics. Recent years have seen an upsurge of interest in this interface between behavior and population ecology, and a substantial literature on this topic now exists (e.g., Fryxell and Lundberg 1998; Abrams 1992, 1999; Abrams and Kawecki 1999; Krivan and Sikder 1999). The goal of this chapter is to provide an overview of the influences of foraging on population dynamics and the reciprocal influences of population dynamics on foraging.

11.3 "Top-down" versus "Bottom-up" Approaches Relating Individual Behavior to Population Dynamics

To understand phenomena such as the Arctic lynx-hare cycle discussed in the prologue, one needs population models. When abundances are great enough to be treated as continuous rather than discrete variables (see box 11.1), one uses differential equations (see also chap. 13), such as

$$\frac{dP}{dt} = P[bB(R) - m],$$ (11.1)

the predator half of a predator-prey model. The variables are P (predator density, say, of lynx [density is the number of individuals per unit area]), R (resource or prey density, say, of hares), and t (time). The expression dP/dt is the instantaneous rate of change in P; one can think of this as the difference in $P(dP)$ over a small time interval (dt). The biology enters into how one relates this change in density to foraging and other factors. The quantity $B(R)$ is a function describing the rate at which prey are captured and consumed (the predator's "functional response"; Holling 1959a) as a function of prey abundance. To relate foraging rates to predator population dynamics, one must determine how foraging affects predator birth and death rates. In this example, we assume that feeding influences births in a simple fashion, in that b is a conversion factor translating the rate of prey consumption by an individual predator into predator births. To finish this mathematical representation of predator demography, we also must account for deaths. Here we simply assume that predators die at a constant per capita mortality rate, m.

To complete the model, we need an expression for prey dynamics (e.g., hares):

$$\frac{dR}{dt} = G(R) - PB(R). \tag{11.2}$$

The quantity $G(R)$ describes how the prey population grows in the absence of predation. For instance, a hare population might grow according to the classic logistic expression $G(R) = rR(1 - R/K)$. The quantity r is the species' intrinsic growth rate (the rate at which it grows when rare enough to grow exponentially), and K is "carrying capacity," the prey abundance at equilibrium with births matching deaths. In describing the predator population, $B(R)$ expresses the rate at which an individual predator consumes prey as a function of prey abundance. Therefore, the total mortality imposed by predators on the prey population is $PB(R)$, which one must subtract from the prey's inherent growth to give the net growth shown in equation (11.2).

So far, we have said nothing specific about foraging. However, we can build assumptions about behavior into the detailed form of the functional response. Usually, $B(R)$ will increase with R, or at least not decrease; feeding rates typically rise with increasing prey numbers. (Sometimes this assumption does not hold, for example, if groups of prey defend themselves against predators, but we assume that this is not the case.) A classic predator-prey model arises if we make the following simplifying assumptions about foraging: that a predator searches at a constant rate a while foraging in a nondepleting patch, that each prey requires a fixed time h for the predator to handle it (during which other prey cannot be encountered), and that each consumed prey is worth a constant amount, b. Holling's "disc" equation describes the rate at which the predator consumes prey (Holling 1959a; Murdoch and Oaten 1975; Hassell 1978, 2000), which translates into a predator recruitment term of

$$bB(R) = \frac{baR}{1 + ahR}, \tag{11.3}$$

(the familiar saturating "type II" functional response). Figure 11.1 shows an example of this functional response in the context of the classic optimal diet model (see below and chap. 1). A crucial feature of this functional response is that predators become saturated with prey when prey numbers are large.

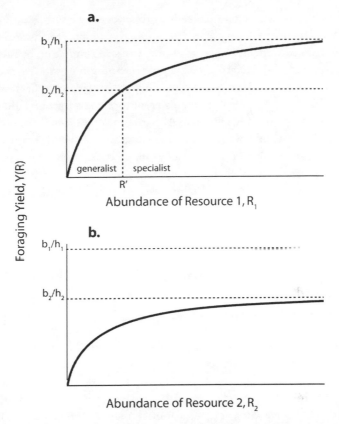

Figure 11.1. A graphical rendition of the classic optimal diet model, assuming sequential prey encounter and fixed handling times. The saturating curves represent the expected foraging yield of a consumer when it specializes on a particular resource (or prey type), of abundance R_i (resource 1 in A and resource 2 in B). The dashed lines represent the expected rate of yield resulting from having captured an item of type i (which equals the net benefit, b_i, divided by the handling time, h_i). The maximum gain rate from feeding exclusively on resource i (when it is very abundant) is b_i/h_i. Resource 1 is of higher quality than resource 2. (A) If resource 1 is sufficiently abundant, the expected yield from capturing and consuming a single item of type 2 is less than the consumer can achieve by ignoring that item and continuing to search for type 1; this implies that the consumer should specialize on resource 1 at abundances greater than the intersection shown and generalize at lower abundances of R_1. (B) Here, the consumer should always consume resource 1, because even the maximal foraging yield it can obtain from resource 2 is less than the yield it can obtain from a single encountered item of resource 1. As the graph shows, changing the abundance of resource 2 does not change this relationship.

With multiple prey types, all parameters are indexed by prey type $i = 1, \ldots,$ n. This n-prey-type extension of the disc equation produces the following harvest rate by a nonselective predator:

$$Y = \sum_{i=1}^{n} \frac{b_i a_i R_i}{1 + \sum_{i=1}^{n} a_i h_i R_i}. \tag{11.4}$$

A large theoretical literature takes this functional response as a given and uses it to analyze questions of predator-prey dynamics. For instance, saturating functional responses can permit prey to escape limitation by predators temporarily and can generate sustained predator-prey cycles such as the hare-lynx cycle. Model predators allowed to choose between prey types ("optimal foragers," for short) can exhibit very different functional forms relating feeding rates to prey density. For instance, Abrams (e.g., 1982, 1987) examined the functional responses of optimally foraging consumers for a wide range of ecological scenarios. Figure 11.2 shows an example of the nontraditional functional responses that can emerge when an optimal forager attacks two prey containing different ratios of two required nutrients (e.g., nitrogen and phosphorus). The rate of consumption of resource (prey type) 1 increases with the abundance of resource 1, but with abrupt thresholds between levels of feeding. Figure 11.2B shows how the rate of consumption of resource 1 varies with the abundance of the alternative resource. The functional response shows threshold responses, and despite an overall decline in attacks on resource 1 with increasing abundance of resource 2, some situations exist in which an increase in resource 2 leads to *increased* attacks on resource 1. These threshold responses, when integrated into a population model, would generate abrupt changes in population dynamics. Jeschke et al. (2002) provide a useful review of the wide range of functional response forms that ecologists have proposed and show how to incorporate digestive satiation as well as handling time constraints.

The above model of predator-prey dynamics illustrates a "top-down" approach to linking foraging and population dynamics (Schmitz 2001; Bolker et al. 2003). This approach takes an existing population model and refines one or more of its components in light of some idea about how the average consumer's foraging affects average birth or death rates. For instance, MacArthur and Pianka (1966) constructed a model of how predators select among prey when prey differ in caloric value and handling time. Several investigators interested in the effects of foraging on aspects of population dynamics have used the MacArthur and Pianka model to address issues such as indirect interactions between prey species (e.g., Holt 1977, 1983; Gleeson and Wilson 1986). Modelers call this approach "mean-field" modeling: the resulting equations describe how average (mean) predator foraging rates vary as a function of average (mean) prey densities, with a minimal specification of biological details. Mean-field models do not capture all of the complexity of real populations; because of their simplicity, however, they often generate very clear and testable predictions and clarify crucial conceptual issues.

Nonetheless, in many circumstances, considering average individuals ignores critical features of ecological systems, features that become apparent when one closely examines the behavior of individuals. Individual foragers

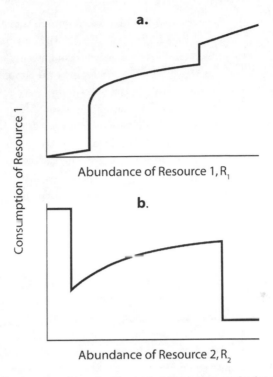

Figure 11.2. Functional response of an optimal forager exploiting two prey species that contain different mixes of two essential nutrients. (A) Consumption of resource (prey type) 1 with a fixed abundance of resource 2. (B) Consumption of resource 1 with a fixed abundance of resource 1. Abrams (1987) argues that consumer fitness should be an increasing function of the following quantity: minimum of $\{k_1 C_1 R_1 + k_2 C_2 R_2, \beta(1 - k_1)C_1 R_1 + (1 - k_2)C_2 R_2\}$, where R_i is population density of resource i, C_i is the attack rate on resource i, k_i is the proportion of nutrient a in resource i, $(1 - k_i)$ is the proportion of nutrient b in resource i, and β is the ratio of nutrients a and b required in the diet for the consumer to survive. Because the consumer needs both nutrients a and b, fitness can be assigned to the consumer only by determining which nutrient is limiting. To do this, the amount of each nutrient being consumed must first be compared, taking into account the ratio necessary for survival. The first term of the equation, $(k_1 C_1 R_1 + k_2 C_2 R_2)$, is the amount of nutrient a that the forager is consuming. The second term of the equation, $\beta(1 - k_1)C_1 R_1 + (1 - k_2)C_2 R_2$, is the amount of nutrient b the forager is consuming, but multiplied by β, which uses the ratio of nutrients necessary for survival to convert nutrient b into the equivalent units of nutrient a. Whichever amount is smaller is the nutrient limiting the consumer; therefore, the fitness of the consumer is the minimum of the first or second term in the equation. (After Abrams 1987.)

show considerable variation in encounter rates with prey, and this variation may influence overall population dynamics. Moreover, by focusing on individuals, one can explore the implications for population dynamics of features of foraging behavior such as sampling, learning, and state dependence (e.g., dependence of foraging decisions on hunger). Focusing on the behaviors of discrete individuals as a basis for developing population models is a "bottom-up" approach to modeling. The development of high-powered computers has allowed the ready exploration of models that incorporate the rich detail of individual foraging behaviors. Individual-based models have burgeoned in popularity (Grimm and Railsback 2005; Schmitz 2001). For example, Turner et al. (1994) modeled individual elk (*Cervus elaphus*) and bison (*Bison bison*) foraging in Yellowstone National Park. The landscape of the model was a grid with features matching spatially explicit data describing the Yellowstone landscape. The model tracked individual elk and bison as they foraged across the landscape under different winter conditions and fire patterns. These authors concluded that the proportion of elk and bison that could survive a severe winter depended on the spatial pattern of fire in the landscape, a conclusion that gives crucial guidance to park managers. Individual-based models made it easier to incorporate realistic spatial information about the landscape and details of individual foraging behavior. Individual-based models also permit investigators to explore the implications of alternative scenarios.

As with the bison and elk foraging model, most individual-based models begin with the investigator giving each individual a set of rules that define its behavior, position in space, and fate through time. These models typically represent space explicitly because each individual occupies a specific position. The computer takes these rules and applies them, individual by individual, to project the state of the system through time. Individual-based models commonly use probabilistic rules of individual behavior, which build stochasticity into the system automatically. We will describe an individual-based model for predator switching below.

Individual-based models do have disadvantages, however. To draw inferences from individual-based models, one must compute averages over large numbers of simulation runs; in complicated models, this makes it hard to survey the available parameter space thoroughly. In addition, the complexity of individual-based models makes it difficult to deduce which features of the system account for a particular observed outcome. Individual-based models can become so complex that they become a world unto themselves, requiring so much effort to understand that they distract from the model's original goals. It can be very useful to use a hybrid approach that combines "bottom-up" and "top-down" approaches. Several studies illustrate the benefits of such hybrid approaches (e.g., Keeling et al. 2000; Illius and Gordon 1997).

11.4   Implications of Population Dynamics for Foraging

Before discussing how foraging behavior can govern population dynamics, we will briefly consider how the dynamics of the resource base should influence how foraging behavior evolves.

Life in a Fluctuating World: Implications for Foraging Strategies

The magnitude and unpredictability of environmental variation strongly affects foraging strategies. The term "variance sensitivity" generally applies to decision making in the face of uncertainty (Stephens and Krebs 1986; Bateson and Kacelnik 1998). Unstable population dynamics in one species create a varying resource for any species that exploits it. In the next few paragraphs, we will discuss conceptual examples suggesting that temporal variability in the abundance of prey populations can change the relative fitnesses of alternative foraging strategies.

We will first consider the effect of temporal variation in the abundance of a preferred prey type on a forager's decision to be selective or opportunistic. Assume that a predator encounters two prey types sequentially. The predator feeds in accord with the classic diet model, so while it is handling an item of one prey type it cannot encounter any other prey. These assumptions lead to a prediction, described thoroughly in Stephens and Krebs (1986): the decision to be a generalist or specialist depends on the abundance of the higher-quality item (as measured by the b/h ratio). The model predicts an abrupt shift between specializing on the higher-quality item when it is abundant and eating both prey types at lower abundances of the preferred prey. Figure 11.1A shows this model graphically. A predator that consumes just the better prey, resource 1, has a type II functional response and a corresponding saturation curve (the solid line, $Y(R)$) describing the benefit it derives from foraging. The predator obtains a constant rate of return while consuming a single item of the less preferred species (the dashed line at b_2/h_2). If the benefit $Y(R)$ (the solid curve) exceeds b_2/h_2, the consumer will specialize on resource 1; if it falls below b_2/h_2, the consumer should also take resource 2 whenever it is available. The switch between the behaviors occurs at the resource level R', where the solid and dashed lines cross.

How does temporal variation in resource availability affect this switch point? Assume that the preferred prey has a constant abundance, but the less preferred prey varies greatly and unpredictably in abundance. Such variation does not matter to inclusion of the better prey in the diet because its inclusion does not depend on its abundance. By contrast, temporal variation in the abundance of the preferred prey can influence the predator's decision regarding the poorer prey, and in particular, makes indiscriminate consumption more likely. Let $R_1(t)$ be a function of time that describes the dynamics of the preferred prey about an arithmetic mean abundance of $\bar{R}$. Assume that predators can instantly and accurately assess resource abundance. Then, if $R_1(t) > R'$, the predator should specialize on resource 1; if $R_1(t) < R'$, it should generalize. If the predator can assess average foraging returns only over some

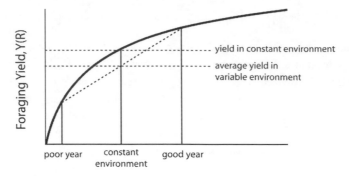

Figure 11.3. Unpredictable variation in resource abundance reduces expected foraging yield. The figure shows a typical type II functional response to a single resource. If we hold the abundance of the resource constant, the consumer achieves a higher foraging yield than it obtains from a variable mixture of good and poor years. If the environment varies between poor and good years, the average or expected yield is the midpoint of the straight line connecting the yields in each type of year. As shown, this average yield is lower than that in a constant environment with resource abundance equal to the average of the good and poor years. This is a case of Jensen's inequality. In the text, we suggest that this effect of nonlinearity may have implications both for diet choice and patch utilization.

long time period, then when faced with the decision to pursue encountered individuals of resource 2, the predator should compare b_2/h_2 to

$$\langle Y \rangle = \int_0^T \frac{Y(t)}{T} dt, \tag{11.5}$$

the long-term time-averaged foraging return. An algebraic rule called Jensen's inequality (Ruel and Ayres 1999) proves that $\langle Y \rangle$ is less than $\overline{Y}$, the foraging yield given a fixed $R = \overline{R}$ (see fig. 11.3 for a discussion of Jensen's inequality). We thus predict that an optimal forager is more likely to generalize in a fluctuating environment.

Instead of having the predator encounter both prey types in the same patch, we might imagine that the two prey types occupy different habitats. Temporal variation in the abundance of a prey species in a given patch (leaving the rest of the environment unchanged) should lower the expected foraging yield in that patch (given a saturating functional response) relative to other, unchanged patches. This should make it more likely that the predator will drop the patch from its foraging itinerary. In short, unpredictable variation in prey abundance tends to favor dietary generalization within habitats, but also may favor habitat specialization that leads to dietary specialization because predators may avoid habitats with variable prey abundance.

We have assumed that resources vary unpredictably in abundance. If instead resource abundance varies predictably (e.g., regular cycles), one expects learning mechanisms to evolve so that foragers can exploit such predictability. For instance, if prey abundance exhibits long-term cycles, and if predators make diet choice decisions by averaging over rather short time scales, predators may generalize when the population of the preferred prey reaches the low point of its population cycle and specialize when the preferred prey reaches a population peak. For instance, in the lynx-hare system, the lynx seems to specialize on hares when they are abundant but to attack a wider range of prey (red squirrels, etc.) when hares are scarce. This pattern may reflect the fact that hare cycles are quite predictable. If the predator must pay costs for such behavioral flexibility, then alternative foraging strategies may coexist in fluctuating environments. In section 11.6, we will discuss an example of coexisting foraging strategies involving predator switching when prey numbers vary through time.

Population Dynamic Constraints on Foraging Behavior

Changes in consumer abundance can alter the direction of selection on foraging. Guo et al. (1991) found that crowding in *Drosophila* cultures selected for increased feeding rates (more feeding "gulps" per minute). In the "scramble" competition experienced by larval fruit flies, larvae benefit from eating quickly, even if this means that they process food less efficiently. In another laboratory study with flies, Sokolowski et al. (1997) showed that high-density environments selected for a strategy that traveled farther ("rover"), whereas low-density situations selected for a strategy that traveled less ("sitter"). Rovers move more from patch to patch when feeding, whereas sitters concentrate their feeding in a patch. These examples show that changes in consumer abundance can alter how selection influences different facets of foraging (e.g., patch use strategies).

Consumer density may vary across a landscape because of chance or because consumers stay together in herds or other social groups (Giraldeau and Caraco 2000; see also chap. 10). If increased consumer density depletes patches more rapidly, why do some species aggregate? Obviously, the benefit of aggregating must outweigh the cost of increased competition. One hypothesis is that groups find new patches more quickly, and consequently, group members experience lower variance in consumption rates than individuals foraging alone. If variance in consumption reduces consumer fitness, then the decrease in variance due to grouping may overcome the cost of aggregation. Another hypothesis suggests that aggregation actually increases individual consumption rates. For instance, large herbivores often prefer to consume immature vegetation because it is more digestible and higher in protein. Aggregation may help to maintain vegetation at an immature stage; thus, an increase in forage quality due to higher consumption rates can overcome the cost of aggregation. Thus, the interaction of a group with resources can produce an "Allee effect"—so that over some range in density, individual consumer fitness increases with increasing density.

Investigators more commonly observe that fitness declines with increasing density (negative density dependence). Competition over shared resources, termed exploitative competition, can alter foraging tactics. Consider a predator that must select a diet from two prey types in depleting patches. Should it specialize on the higher-quality prey, eat prey indiscriminately, or modify its foraging rules as the patch is depleted? Game theoretical studies suggest that the answer depends on whether the predator has the patch all to itself or must share the patch with others (Holt and Kotler 1987; Mitchell 1990). If the predator is alone, it should generalize throughout each patch visit, as long as conditions at the time of patch departure favor generalizing. If, instead, a predator shares the patch with competitors, it should specialize early in patch visits and expand its diet as the patch is depleted. When many predators aggregate in a patch, the switch point occurs when consumption reduces the abundance of the preferred prey to the level predicted by the classic diet model (see fig. 11.1; Holt and Kotler 1987). The reason is that when generalists and specialists forage together, the generalists spend time handling low-quality items, while the specialists continue to deplete the high-quality items. Specialists thus achieve a higher overall foraging rate when they compete with generalists. However, if a predator has a patch to itself, it should maximize its rate of return over the entire time it occupies the patch. An isolated predator can reduce its cumulative search time by always attacking both prey types upon encounter. The likelihood of these two situations depends on predator abundance. When predators are rare and randomly distributed, we would expect to find them foraging alone, and so they should feed indiscriminately on both prey types; but when predators are abundant, it is likely that a predator will frequently share patches with other predators, and so initially in each patch, the predator should show diet selectivity.

Population Dynamics Can Indirectly Constrain Selection on Foraging Behavior

Because of human introductions, there is an increasing number of invasive species around the world. After their establishment, many non-native species remain rare and localized, but begin expanding their range after a long time lag. One hypothesis for this intriguing pattern is that initially these species are maladapted to the novel environment: the invasion lag may reflect a period of evolutionary adaptation (Holt et al. 2005). In some circumstances, evolutionary biologists believe that such maladaptation can persist. For instance, Ernst Mayr argued that recurrent gene flow from abundant populations at the center of a species' range into marginal, low-density populations can "swamp out" locally adapted genotypes and thus keep the marginal populations in a perpetual state of maladaptation—potentially including maladaptive foraging behavior.

The potential for such persistent maladaptation depends in part on population dynamics, which can influence the relative importance of selection versus nonselective evolutionary processes. In some populations, individuals may exhibit a mismatch with their environments because population dynamics magnify the importance of nonselective evolutionary processes such as genetic drift, hybridization, and gene flow (Holt 1987a, 1996a, 1996b; Crespi 2000; Lenormand 2002). Several population dynamic factors make nonselective effects more likely. Chronically small or highly variable populations may experience drift; populations low in abundance may also be particularly vulnerable to gene flow from other populations or hybridization with other species. Large environmental changes or long-distance dispersal may put species into circumstances they have not previously experienced, and if they persist, they may be initially maladapted. By contrast, one expects finely honed adaptations to local environments in large, stable populations that experience constant conditions for many generations.

If environmental conditions permit strong, persistent selection in abundant, stable populations, we expect realized foraging behaviors to be near predicted optima. Strong selection occurs when the adaptive peak relating fitness to phenotype curves steeply, so that a small deviation from the optimum has severe fitness costs. Weak selection occurs when the fitness peak is broad and flat, so that many phenotypes have fitnesses close to the optimum (fig. 11.4). The amount of deviation from local behavioral optima that one might find should depend both on the fitness costs of deviations from the optima and the demographic context of selection.

Spatial Variation

Consider a landscape where some habitats contain "source" populations producing an excess of births over deaths, with the excess forced into "sink" populations where deaths exceed births. We expect to observe a relatively good fit between phenotypes and environment in the source habitat, but a poorer fit between phenotypes and environment in the sink habitat. Models of adaptive evolution in sink habitats (e.g., Holt 1996a; Holt and Gomulkiewicz

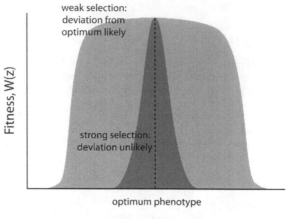

Figure 11.4. Fitnesses of phenotypes relative to an optimal phenotype. With a flat-topped fitness function, phenotypes far from the optimum may have almost the same fitness as the optimal phenotype; thus, selection acts only weakly against deviations from the optimum. If the fitness function is sharply peaked, phenotypes close to the optimum have fitnesses much lower than the optimal phenotype. The text argues that the degree of maladaptation one might observe depends both on the shape of the fitness function and on the demographic context of selection.

1997a, 1997b; Kawecki 1995) suggest that it may be difficult for selection to sculpt adaptations to sink environments, particularly harsh ones, at least when most mutations have small fitness effects. Recent environmental change (e.g., human-caused habitat degradation) makes this sort of maladaptation likely. When a trade-off exists between performance in the sink and source habitats, foragers will remain adapted to the source habitat, and local maladaptation in the sink can persist indefinitely. Moreover, a behavior may be "maladapted" when examined in one local environment, but "well-adapted" when evaluated over the entire range of environments a population experiences (Brown and Pavlovic 1992).

A similar phenomenon emerges even with ideal free habitat selection, which implies that in a stable environment, no individuals will occupy sink habitats. Assume that in each generation, individuals settle in one of two habitats that differ in their carrying capacity. Assume also that they use rules of thumb that create an ideal free distribution. When the population reaches demographic equilibrium, each habitat will be at its respective carrying capacity, so the fraction of individuals in habitat 1 will be $p = K_1/(K_1 + K_2)$. Now imagine that a mutant arises, and that this mutant increases fitness in habitat 1 by a small amount, at a life stage after habitat selection occurs. The probability that an individual bearing this mutant is in habitat 1 is p. If $K_1 \ll K_2$, then selection will be very weak, simply because the probability is low that the mutant will reside in the habitat where it can express its fitness advantage. This implies that drift will often eliminate weakly favored mutants that improve fitness in the rare habitat. Similarly, selection acts only weakly against deleterious mutants that reduce fitness in the rare habitat. In effect, selection will emphasize adaptation to the higher-K habitat at the expense of adaptation to the lower-K habitat.

Temporal Variation

In the lynx-hare system, populations fluctuate in abundance by factors of 100 or more. One could, in principle, compare the precision of foraging adaptations by lynx in years when lynx are common (which often are years following peaks in hare abundance) with that in years when lynx are rare (e.g., after a hare crash). A simple model suggests that when numbers are fluctuating, selection may act more effectively in years with large populations. Consider a species with discrete generations for which the equation $N(t + 1) = N(t)W(t)$ describes the population's dynamics. $W(t)$ represents the finite growth "rate" in generation t (technically, $W(t)$ is not a "rate," because the units are dimensionless, but this usage is customary in ecology). When populations vary in growth over time, we find long-term fitness by calculating the geometric mean of fitness (denoted by W_g) in consecutive generations. (We calculate the geometric mean of a string of n numbers by multiplying those numbers and taking the nth root. This method is relevant to population growth, which involves a multiplication of successive growth rates over time; see box 7.3.) The finite growth rate (W_i in year i) can vary as a function of population size and environmental factors. These can all vary with i, as well as with the value of phenotypic parameters such as body size or movement rates. Let us assume that the growth rate in each year depends on a single parameter, α (e.g., body size or speed of movement during foraging). We assume that α varies genetically within the population and that reproduction is clonal. If an increase in α increases geometric mean fitness, then selection should favor a clone with a higher value of α. The sensitivity of W_g to α (the derivative of W_g with respect to α) governs the direction and rate of evolution. One can show that this sensitivity is

$$\frac{\partial W_g}{\partial \alpha} = \left\langle \frac{1}{W_i} \frac{\partial W_i}{\partial \alpha} \right\rangle, \tag{11.6}$$

where the brackets denote an arithmetic average, in this case taken through time. Notice the term $1/W_i$ that appears on the right-hand side of equation (11.6). This term will be large in generations when the realized growth rate is small. This means that years of low fitness disproportionately influence the relationship between long-term fitness and our α parameter. If a trait has different effects in good and poor years, selection will emphasize the poor-year effects.

Typically, a population fluctuating between upper and lower (nonzero) bounds generates a negative statistical correlation between reproductive rate and population size (Royama 1992, 31), regardless of the exact mechanism regulating the population. In other words, on average, over time, animals reproduce or survive less well in years when they are most abundant! Thus, selection tends to emphasize years of high abundance because populations then experience lower fitness. This may seem counterintuitive, because one often assumes that population size reflects environmental plenitude and that low growth rates mean small populations. This view often holds true when we consider variation in space; we do observe larger populations in regions of plenty, but this intuition leads us into error when we consider variation in population size over time.

11.5 Implications of Foraging for Population Dynamics

Optimal Foraging and Population Persistence

As noted in box 11.1, the most basic question one can ask about a population's dynamics is, will the population go extinct or persist? A simple argument suggests that optimal foraging behavior, all else being equal, promotes long-term persistence in changed environments (Holt 1987a). Imagine that bad weather or bad luck has reduced the populations of two species to the point at which direct density dependence and frequency dependence have little effect. Each species exists in heterogeneous environments and must make foraging choices (e.g., should a forager be selective or opportunistic when faced with several prey types?) In species 1, all individuals make optimal decisions, whereas in species 2, many individuals make suboptimal decisions. By definition, the reproductive fitness associated with an optimal behavioral choice exceeds that associated with a suboptimal choice. Hence, when rare, species 1 has a larger average rate of population growth than species 2. Optimal foraging decisions should enhance a population's ability to rebound from dangerously low densities.

Optimal Foraging and Population Size

At high densities, however, we cannot make such a simple claim about how optimal foraging will affect population dynamics. Optimal foraging can indirectly depress population size in several situations. For instance, optimal foraging may permit predators to overexploit their prey. In almost any predator-prey model, overexploitation can arise; an increase in per predator capture rates may reduce the number of predators a prey population can sustain. In the basic predator-prey model presented in section 11.3, the prey population equilibrates when the rate of prey recruitment matches the rate at which the predator consumes prey (in symbols, $G(R) = PB(R)$). The predator population equilibrates when the predator birth rate equals the predator mortality (in symbols, when $cB(R) = m$, or when $B(R) = m/c$). We combine these two expressions to find that that the predicted abundance for the predator at equilibrium (the asterisk indicates equilibrium) is $P^* = G(R^*)/B(R^*) = G(R^*)c/m$. In words, the equilibrium density of the predator equals prey recruitment, times a conversion factor translating consumed prey into predator births, divided by predator mortality. If a predator exploits prey more efficiently (higher c), it needs to consume fewer prey to match its losses to mortality. For simplicity, assume that prey consumption is $B(R) = aR$, where a is the "attack rate"; this says that consumption increases proportionately with prey numbers. In this case, the predator equation leads to $R^* = c/am$. By working with these expressions, one can show that if predators interact via exploitative competition for prey, then selection favors the foraging behaviors that permit persistence at the lowest possible abundance for the prey. Biologically reasonable models of self-reproducing prey always lead to a hump-shaped relationship between prey abundance and prey recruitment ($G(R)$), with maximal total growth at a prey density ($\tilde{R}$) well above zero. If predators forage efficiently enough to push the prey density to below this critical value ($R < \tilde{R}$), then optimal foraging will reduce the number of predators sustained, because a high rate of foraging reduces the number of prey and so depresses total prey recruitment (and thus indirectly suppresses predator numbers).

Ideal Free Distribution

Another interesting way in which optimal foraging can reduce consumer population size is via patch use strategies. One goal of population dynamics is to understand patterns of abundance of species across space. A fundamental prediction about spatial patterns in abundance arises from foraging. Consider the following scenario: in one habitat patch, food items arrive at a rate of ten items per hour; in another, poorer habitat patch, five items per hour arrive. How should a group of foragers allocate themselves to these two patches? The ten items per hour patch is definitely better, but if everyone forages there, competition will reduce its value to the point at which some individuals can do better by foraging in the poorer five items per hour patch. The classic ideal free distribution addresses this problem (Fretwell and Lucas 1969; Fretwell

1972; see box 10.1). The ideal free distribution predicts that the distribution of foragers (the number of foragers in patch A divided by the number of foragers in patch B) should match the ratio of input rates. So in this example, we expect twice as many foragers in the ten items per hour patch as in the five items per hour patch. Another way of putting this is that, given an ideal free distribution, one expects an equalization of fitness across space (so that no individual benefits from moving).

As a population grows, it experiences a lower birth rate or a higher death rate because of density dependence (e.g., due to exploitative competition). In a constant environment, the population increases until it reaches a level at which births just match deaths (so that average fitness is unity). Modelers call the abundance at which this demographic balance occurs the "carrying capacity." Now assume that our population exploits several distinct patches, each with its own resource renewal rate, leading to density dependence in each patch. If the population can achieve an ideal free distribution, and is in demographic balance, then we know two things: overall fitness is unity, and all patches have the same fitness. For this to be true, we can infer that local birth rates match local death rates in each patch. Thus, each patch equilibrates at its own carrying capacity. The overall carrying capacity of a landscape for an ideal free forager is just the summed carrying capacities over the separate habitat patches.

Before pointing out how these dynamics may lead to a lower population size (compared with a population of nonideal foragers), some other population dynamic consequences of ideal free behaviors are worth noting. To understand population dynamics, one needs to characterize how populations behave when perturbed from equilibrium, which often requires unraveling the mechanisms that generate density dependence in local fitness. Ideal free behaviors can lead to transient patterns of habitat use or equilibrium distributions quite different from those expected with random habitat choice. For instance, if population numbers are low, all individuals should occupy the habitat providing the greatest individual fitness, but as numbers rise, there comes a point (because of density-dependent reductions in fitness) at which individuals may start exploiting alternative habitats. Thus, we expect the degree of habitat specialization to vary as population size fluctuates.

Density dependence in consumer fitness can arise for many reasons (see box 10.1). The most general may be that with exploitative competition, resource supply declines with increasing consumer numbers. The quantitative expression of this density-dependent decline requires one to understand how exploitation drives the dynamics of resource populations. Holt (1984) shows how one can use models of predator-prey population interactions within patches (similar to the model in section 11.3) to generate expressions relating predator fitness to local predator numbers. The bottom line of this mechanistic approach is that high carrying capacity for a consumer in a patch can be associated with either strong or weak density dependence in the patch. This is what allows ideal free behavior to generate reductions in total population size.

There are many reasons why one might observe deviations from an ideal free distribution leading, for instance, to undermatching of consumer numbers to resource inputs (see box 10.1). In general, these deviations imply that the carrying capacity of the population will not match the summed carrying capacities across the separate habitat patches. Holt (1985) gives examples in which an ideal free distribution greatly depresses the size of a population below the level expected if foragers were utilizing habitats at random. This is particularly likely if density dependence is strong in patches with a high carrying capacity. The degree to which a forager fits the ideal free assumptions thus matters greatly in determining its carrying capacity and in determining how its habitat breadth should vary with changes in its numbers.

11.6   The Interplay of Population Structure, Foraging Behavior, and Population Dynamics

Population dynamics and foraging can intersect in other ways. For instance, the prey "types" that standard foraging theory considers may not correspond to taxonomic species. Osenberg and Mittelbach (1989) studied pumpkinseed sunfish feeding on snails and found that they had to incorporate selectivity for size classes of snails to understand pumpkinseed diets. Additionally, transitions between life history stages (e.g., tadpole to frog) can create time lags in population growth. If a predator selectively attacks a particular life history stage, we may not observe the dynamic consequences of this predation until later. These time lags can lead to oscillations in abundance.

When individuals cannibalize their own species, foraging decisions may have particularly dramatic dynamic effects. Many invertebrate predators practice cannibalism (Polis 1981; Elgar and Crespi 1992). Cannibalism should reduce population growth—the principles of thermodynamics tell us that a cannibal cannot convert each conspecific meal into an equivalent newborn! Cannibalism often increases with density because increased density means there is more for a cannibal to eat. Detailed models of cannibalism typically include age or size structure. For instance, Claessen et al. (2002) developed a dynamic model of cannibalism in Eurasian perch, in which juveniles outcompete adults (because they survive at lower resource levels), but adults attack juveniles in a certain size range. The size range of juvenile vulnerability determines whether the population is stable or cycles. Adult foraging decisions implicitly determine the threshold size of juveniles included in the diet, and this, in turn, determines the dynamics of the population. Claessen et al. tailored their model to several well-studied fish systems and showed that it helps to explain dramatic observed patterns in population dynamics and size structure. For instance, their model predicts that cannibalistic "giants" will emerge in populations with unstable dynamics, while such classes of foraging specialists are less likely in stable populations. This prediction matches observed patterns of cannibalism in perch and other fish species.

Stability

As noted above, optimal foraging can increase the growth rate of populations at low numbers. If optimal foraging creates time lags in the effects of density dependence, then high growth rates can generate unstable dynamics. This instability arises because a moderately sized population can generate so many offspring that they exceed the environmental carrying capacity, so that the population crashes in the next generation. Depending on the details, this instability can generate either sustained population cycles or chaotic dynamics. Alternatively, as noted above, optimal foraging strategies may "buffer" populations from fluctuations. If individuals leave local habitats when foraging opportunities suddenly disappear, their behavior may moderate fluctuations in overall population size. So it seems reasonable that foraging decisions could stabilize populations in fluctuating environments.

The effects of foraging decisions on population dynamics make a compelling connection with community ecology (i.e., interactions between species; see chap. 12). To illustrate this connection, we will consider the "switching" of generalist predators among alternative prey species. Before discussing switching per se, however, we will briefly revisit the influence of functional responses on population dynamics. Consider a predator population at a fixed density. The functional response describes the relationship between predator attacks and prey abundance. The simplest, type II functional response predicts that the prey per capita mortality rate will decline as prey abundance increases. Since predation becomes less effective as prey numbers increase, predation alone cannot regulate the prey. A similar result holds if a predator feeds on two prey species in accord with the optimal diet model (Fryxell and Lundberg 1994). Fryxell and Lundberg (1994, 1998) suggest that constraints on optimal diet choice (leading to partial preferences) may facilitate population stability. As noted in section 11.3, real predators can also have more complex functional responses. For instance, if optimal foragers engage in nonforaging activities when prey are scarce, the attack rate per prey may increase with increasing prey density (at low prey densities). This leads to a sigmoid (or S-shaped) functional response (usually called a "type III") in which predation responds strongly to increases in prey density; this in turn may keep prey numbers in check (Murdoch and Oaten 1975).

For many years, the conventional wisdom was that type III responses were stabilizing. We now know that this is not always true. For instance, in the temperate zone, insect hosts and parasitoids (e.g., a moth and a parasitic wasp that attacks its caterpillar) can be locked in strong interactions that keep host densities well below the level set by available food. When one incorporates discrete, synchronized generations for such species into coupled population models and assumes constant attack rates, the typical outcome is large-scale oscillations in abundance that in reality would surely imply extinction for both species. Yet such strongly interacting host-parasitoid systems do persist. Much creative work has gone into trying to understand the factors that permit their persistence. One avenue that seemed reasonable was to focus on the detailed nature of the parasitoid's functional response. However, a type II functional response for the parasitoid turns out to enhance the inherent instability of the system (Hassell and May 1973). Holling (1959b) surmised that a type III response would stabilize the system, but Hassell and Comins (1978) later showed that this functional response could never (by itself) stabilize the interaction because the one-generation time lag between changes in parasitoid abundance and the resulting effects on host mortality simply overwhelm the potential stabilizing influence of the within-generation behavioral response (Bernstein 2000).

More recently, ecologists have argued that the persistence of intrinsically unstable predator-prey systems may depend on the interplay of population dynamics, foraging, and spatial structure. This idea has implications for how one views the evolutionary dynamics of foraging behaviors. As section 11.5 explained, the evolution of optimal predator foraging can lead to overexploitation in a closed predator population. Several authors (Gilpin 1975; Pels et al. 2002) have argued that dividing a population into partially isolated subpopulations (as often happens in nature) may reduce the evolutionarily stable level of predation and prevent the overexploitation predicted by local optimization.

Another potential influence on population stability comes from considering in more detail the community context of pairwise interactions. Up to now, we have largely looked at the interplay of foraging and population dynamics for single populations and for coupled predator-prey dynamics. We will now turn to considering the community context of optimal foraging (see also chap. 12) as we examine switching (Murdoch 1977) by generalist predators among multiple prey species.

Predator Switching

In the North Pacific Ocean, populations of seals, sea lions, and sea otters have sequentially collapsed over the last several decades (Springer et al. 2003). Scientists initially thought that physical changes in the ocean or competition with fisheries were to blame, but it now appears that killer whales are responsible. Killer whales usually consume great whales (such as sperm whales and bowhead whales), but great whale numbers were significantly decreased after World War II by human whaling. With their primary food source reduced, killer whales "switched" to consuming seals. When their consumption reduced the availability of these prey, killer whales then switched to sea lions and finally to sea otters. The result is the sequential collapse in pinniped and other populations observed by scientists (Springer et al. 2003).

Formally, we say that a predator "switches" if its relative attack rate on one of two prey species increases faster than the relative abundance of that species (Murdoch and Oaten 1975). The multi-prey functional response defined in section 11.3 provides a null model against which switching responses can be measured; if N_i' is the number consumed of prey type i, then this model implies that $N_1'/N_2' = (a_1/a_2)\,N_1/N_2$, so that the relative frequency of prey in the diet faithfully reflects (up to a constant) relative prey abundances.

Ecologists have traditionally believed that predator switching stabilizes both predator and prey populations and permits the coexistence of species that otherwise might exhibit competitive exclusion (Roughgarden and Feldman 1975). In the prologue, we discuss how lynx switch to consuming red squirrels when snowshoe hare numbers are low, a behavior that may allow snowshoe hare populations to recover more easily (at which point lynx switch back to consuming hares) and may also sustain the lynx during troughs of low hare numbers. However, as the collapsing pinniped populations in the killer whale example show, predator switching may not always lead to system stability.

Van Baalen et al. (2001) created a top-down model that illustrates how predator switching can have both stabilizing and destabilizing effects. They modeled a special case of predator switching in which the preferred prey type occurred in one patch and an alternative prey type occurred in a second patch. They further assumed that the alternative prey type existed at a fixed density, and that predators knew the optimal preferred prey density at which to switch between patches and could do so instantaneously and without cost. They found that the predator and preferred prey did not equilibrate to a fixed point, but instead formed a limit cycle. Predator switching did not produce a stable equilibrium because switching predators alternated between the preferred and alternative prey patches as the density of the preferred prey changed. This behavior destabilized population dynamics because switching released the preferred prey when their numbers were low, and this reproductive response permitted the prey to increase. The predator then switched back to the preferred prey, whose population was still growing, and continued to grow because there was a lag in the growth response of the predator population as a whole. The instability emerges from the interplay of individual behavioral responses and the time-lagged responses that are always inherent in the responses of populations to changes in the environment.

Van Baalen et al. (2001) also examined other cases in which predators were "nonoptimal" foragers. In this context, "nonoptimal" means only that the foraging behavior deviated from the expectations of the idealized model. For instance, predators may switch between patches gradually. Van Baalen and colleagues modeled this behavior by altering the proportion of time the predators spent foraging in the alternative prey patch. This gave the predators a sigmoid functional response, and a stable population equilibrium arose. When the switching threshold was replaced with a sigmoidal curve, predator switching acted to stabilize the populations. Thus, time dependence in prey switching made the system more stable. One possible reason for such time dependence is that the predator does not adjust to changes in prey numbers instantaneously, but instead has a learning curve that takes time. If gradual switching reflects the need for learning, these results suggest that slow learning may lead to more stable dynamics. However, if learning is so slow that predators in effect move randomly between patches, the system again becomes unstable, and predators and prey go extinct. The reason is that predators that switch between patches randomly do not necessarily leave the preferred prey patch when prey numbers are low, and they can drive the preferred prey to extinction.

This result illustrates the importance of considering complementary models of a given problem. Kimbrell and Holt (2004, 2005) have explored an individual-based computer model based on the mathematical model of van Baalen et al. (2001). Kimbrell and Holt's model took a "bottom-up" approach. Predators moved on a grid consisting of two patches and chose which of the two patches to exploit. When all predators foraged optimally, or when predators switched gradually between patches, Kimbrell and Holt's model agreed with van Baalen et al.'s results. However, when predators randomly switched between patches, Kimbrell and Holt found that populations of predator and prey could sometimes persist. This difference in results arises from the spatially explicit and stochastic nature of the bottom-up approach. The van Baalen et al. model assumed mean-field dynamics and thus assumed that all predators in the preferred prey patch consumed prey with a fixed probability. The Kimbrell and Holt model, however, explicitly modeled space in each patch and assumed that predators moved stochastically within each patch. Thus, at low predator numbers, the stochastic movement of predators left small

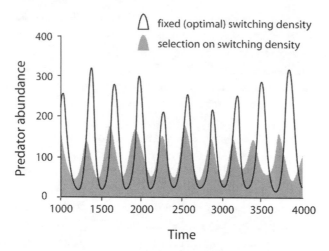

Figure 11.5. Time series of predator abundance from an individual-based model comparing a predator population that faced selection on prey switching density with a predator population that had a fixed "optimal" switching density. A homogeneous population of predators that switch between patches optimally (according to the criterion derived by van Baalen et al. 2001) produces large cycles in both predator and prey numbers. Permitting mutation in the predator's switch point leads to genetically heterogeneous populations, with a mixture of predators switching at different prey densities. The overall effect of evolution in foraging behavior in this example is to reduce the amplitude of predator-prey cycles without making them disappear. (After Kimbrell and Holt 2005.)

portions of the preferred prey patch without predators. These predator-free islands provided temporary refugia where the preferred prey could increase when rare. Thus, a bottom-up approach that explicitly included space and stochasticity yielded a different result than a top-down approach to the same problem with mean-field dynamics.

The bottom-up approach of Kimbrell and Holt (2005) also permitted them to explore evolution's influence on individual foraging behavior. They allowed the prey density at which predators switched to evolve by natural selection in the predator population. They then compared the resulting population dynamics with those of a predator population that switched between patches optimally (as given by the top-down prediction of the van Baalen et al. model). They found that incorporating evolution and individual differences in foraging behavior had large effects on the magnitude of population oscillations (fig. 11.5).

A mean-field model by Richards and de Roos (2001) found that random foragers moving over several discrete patches persisted with optimal foragers who paid a slight cost of habitat assessment. Kimbrell and Holt (2005) likewise found that random foragers could persist with optimal foragers that had slightly higher metabolic costs. When predators with different foraging strategies coexist, prey populations often experience smaller fluctuations than when a predator with a single switch point attacks them. Thus, as biologists begin to examine the evolution of foraging strategies, the ability of alternative foraging strategies to coexist in a patchy environment may have interesting implications for population dynamics.

This example of predator switching illustrates several themes of this chapter. First, as the van Baalen et al. (2001) model showed, optimal foraging can destabilize population dynamics. Yet persistence may nonetheless be ensured if populations are buffered from excursions to low density; thus, optimal foraging may enhance persistence, even if it is not stabilizing in the classic sense. Second, population stability may emerge as a by-product of constraints within which optimal foraging occurs. For instance, Kimbrell and Holt's (2004, 2005) simulations of switching constrained the movement of predators within patches, thus permitting some prey to escape in transient refuges. Third, the combination of strategic top-down models and more detailed individual-based simulations provides a useful way to analyze the relationship between foraging and population dynamics. Agreement between multiple modeling approaches can confirm the generality of results, while disagreements highlight the need for further investigations.

11.7   Optimal Foraging and Species Coexistence

In the previous section, we discussed how predator switching behaviors could regulate population dynamics. One reason this may be important is that predator switching can potentially promote the coexistence of prey species (Murdoch and Oaten 1975; Roughgarden and Feldman 1975). If so, then foraging ecology may have profound consequences for maintaining species diversity. In this section, we will explore some potential implications of predator foraging for the structure of prey communities, including the nature of indirect interactions among prey species.

A useful schema in community ecology is to characterize interspecific interactions by how the population size (or fitness, or other measures) of species i would be expected to change, qualitatively, if one were to manipulate the abundance of species j. So if we have two species, the pairwise interaction between them might be (0,0) (no effect), or $(-,-)$ (e.g., each species competes with the other), or $(+,-)$ (e.g., one species benefits, whereas the other suffers, from the interaction), or $(+,+)$ (e.g., each species benefits the other). If we consider two prey species, we can then ask how foraging decisions by a predator might influence the qualitative pattern of interactions between those prey.

To address species coexistence and persistence, we must ask whether each species in a community can increase when rare. Generalist predators can affect the coexistence of prey species in several ways, depending on the predator's foraging behavior and other features of the system (Holt and Lawton 1994). Consider first a spatially closed system in which factors other than prey availability limit predator numbers in a single homogeneous habitat. Alternative prey species may still interact via the predator's functional response. If the generalized disk equation (11.4) describes predator feeding, then as one prey species increases in abundance, predation on the alternative prey is relaxed. If a predator follows the classic diet model, the lower-quality prey species enjoys an additional benefit because the predator should ignore it when the preferred prey are abundant. When this happens, variation in the abundance of the less preferred prey has no effect on the preferred prey. In other words, one expects to observe $(+,+)$ or $(+,0)$ interactions between the prey species. Given a generalist predator with fixed numbers, an increase in the abundance of either prey relaxes predation on the other prey. But if the preferred prey species becomes sufficiently numerous, the alternative prey may enjoy a respite from predation because the predator becomes selective on just the preferred prey species.

Holt and Kotler (1987, see also Holt 1987b) argued that in spatial open systems, in which predators can move freely between patches, a richer array of possibilities may arise. To summarize their results, when predators choose prey optimally (in accord with the classic diet model) within patches, and also use a marginal value theorem criterion (Charnov 1976b) for entering and leaving patches, the two prey species will affect each other negatively (a $(-,-)$ interaction). Having two prey types together in a patch enriches that patch and makes it more valuable for predators to aggregate there. Thus, one prey species can indirectly enhance predation pressure on an alternative prey species. If predators show a strong reproductive response to their prey, so that predator numbers in a patch increase greatly with increasing prey density in that patch, this time-lagged demographic response strengthens the negative effects of one prey type on the other. Ecologists call this indirect interaction "apparent competition" (Holt 1977, 1997; Holt and Lawton 1994) because the prey species have negative effects on each other even though they do not compete directly.

If predators have suboptimal diets, but use a marginal value criterion to leave patches, $(+,-)$ interactions between alternative prey can occur (Holt and Kotler 1987). These effects arise when predators encounter both prey types within the same patch and when predators are constrained to be opportunistic when the diet model predicts they should be selective (e.g., when predators can assess prey quality only during consumption). At the other extreme, if predators find different prey types in different patches, Holt (1984) argued that optimal foraging by the predator should lead to an ideal free distribution for the predator population across patches, which in turn implies that the predation pressure on each prey species should be independent of the alternative prey.

The theoretical expectations discussed here apply most forcefully when we hold other aspects of population dynamics constant. If we embed foraging behavior into full-blown population dynamic models, then an even richer set of outcomes can occur. Hamback (1998), for instance, in a theoretical study of herbivore effects on plant species coexistence (including direct competition between the plants), compared the effect of optimal foraging on prey coexistence in seasonal versus nonseasonal environments. Food-limited consumers with fixed diets led to apparent competition and species exclusion in both kinds of environments. Comparable effects arose for consumers with optimal diets in the nonseasonal environment, but in the seasonal environment, the large variations in abundance led to annual diet shifts that promoted coexistence.

In the system explored by Hamback, optimal foraging facilitated stability and species coexistence. By contrast, Schmitz et al. (1997a) developed a model for an optimally foraging consumer, a herbivore, feeding on two plant species but constrained by time and digestive capacity. They parameterized their model using published foraging data for several taxa (an insect, a small mammal, and a large mammal), and in all cases found that optimal foraging led to unstable dynamics. If these populations experienced large oscillations, they could become extinct. Abrams (1999; see also Abrams and Kawecki 1999) likewise has argued that adaptive adjustment of foraging rates (e.g., with predators moving between patches) can destabilize prey populations, and that given such instability, switching can reduce the minimum density of one or both prey species, and so make coexistence less likely. Krivan (1997) examined a Lotka-Volterra predator-prey model with an optimally foraging, switching predator and, by contrast, concluded that the optimal foraging behavior promoted persistence. The difference between these results reflects differences in the researchers' assumptions about the temporal dynamics of the behavior itself; Krivan (1997) assumed that predators optimized their behavior at each instant, whereas Abrams assumed that because of factors such as constraints on learning, predators took time to shift their preferences. The contrast between these theoretical results highlights the potential importance of subtle differences in foraging behavior for population dynamics and species coexistence. An open challenge is relating these theoretical possibilities to real systems.

11.8 Prospectus

We could explore many other aspects of the interface between foraging and population dynamics. For instance, we have largely focused on how predators' foraging decisions translate into dynamic consequences either for themselves or for their prey. We have not considered how the risk of predation can change the foraging decisions of prey, or how a predator's predators might change the predator's foraging decisions (these topics are taken up in chap. 13). Interlocking behavioral rules can lead to long chains of rapidly transmitted indirect interactions between species, and such interactions can have strong and counterintuitive effects on population dynamics (Abrams 1993; Bolker et al. 2003). For instance, the interaction of adaptive behaviors can lead to unexpected effects, such as situations in which an increase in the predator density leads to an increase in the population growth rate of its prey! This can occur if the prey feeds on a biotic resource and tends to overexploit that resource if left to its own devices. If a predator reduces the prey's effectiveness at capturing its own resource, then the presence of predators can reduce the prey's tendency to overexploit that resource, so an increase in predator abundance can enhance prey growth rates. We expect that a major focus of future work on foraging and population dynamics will revolve around "foraging games" between species and how populations play these games against the background of environmental variability, as well as webs of shifting direct and indirect behavioral interactions among species in complex food webs. A general insight that emerges from this chapter's material is that a given foraging strategy can affect population dynamics in different ways in different situations; for instance, predator switching can stabilize population dynamics in some circumstances, but be destabilizing in others. We expect that an important direction of growth in population ecology will be the explicit incorporation of insights from foraging theory.

11.9   Summary

Foraging strategies have profound consequences for population dynamics because they can determine population growth rates in both predators and prey and ultimately influence average abundances as well as the pattern of fluctuations in abundance. The Arctic lynx-hare system shows how foraging behaviors link the population dynamics of multiple species together. Approaches to integrating foraging with population dynamics range from refinements of traditional population models to individual-based simulations incorporating detail about individual behavior. Population dynamics can influence the evolution of foraging behaviors by shifting the relative fitnesses of alternative strategies, which can lead to situations in which maladaptation is expected (e.g., sink environments). In turn, foraging behaviors can determine population persistence and dynamic patterns such as population cycles. Predator switching provides an example of a foraging behavior that can either stabilize or destabilize population dynamics, depending on the circumstances, and can also influence species coexistence. The interplay of foraging and population dynamics is likely to continue to be an important intellectual theme as ecology continues to mature.

11.10   Suggested Readings

Abrams and Kawecki (1999) provide a nice theoretical example of how adaptive foraging behavior can be destabilizing. They examine a parasitic wasp attacking two prey species. Holt (1983) and Holt and Kotler (1987) provide a detailed discussion of how different assumptions about predator foraging imply qualitatively different consequences for prey interactions. Fryxell and Lundberg (1998) provide a monographic overview of the integration of behavioral and community ecology, touching on many topics we have discussed in this chapter. Ovadia and Schmitz (2002) provide an excellent empirical study of how trade-offs between foraging and predator avoidance can have major consequences for species' abundances in a food chain interaction involving plants, grasshoppers, and spiders. Fryxell et al. (2005) use models and data on foraging, grass growth, and movement to suggest that adaptive foraging is required for Thomson's gazelles to persist in the temporally and spatially variable landscape of the Serengeti Plains in East Africa.

Telander

Community Ecology

Burt P. Kotler and Joel S. Brown

12.1 Prologue

Two species of gerbils, the 24 g Allenby's gerbil and the 40 g greater sand gerbil, live together on sand dunes in the Negev Desert. These species are very much alike. They eat mostly seeds (Bar et al. 1984), they are nocturnal, they live in burrows, they are caught by the same predators, and they compete intensively with each other (e.g., Mitchell et al. 1990). They invite a central question of community ecology: What promotes the coexistence of close competitors? How do these two species escape competitive exclusion?

Perhaps the answer has to do with their use of habitats. The two species use the varied substrata of the sand dunes differently. Allenby's gerbil predominates on sand dunes stabilized by vegetation, while the greater sand gerbil predominates on less stable sand dunes (Rosenzweig and Abramsky 1986). Habitat segregation intensifies at higher population densities (Abramsky and Pinshow 1989; Abramsky et al. 1990, 1991). Foraging theory suggests that habitat selection is based on the costs and benefits of habitat use (Fretwell 1972; Rosenzweig 1981). For this to explain species coexistence, each species must have a habitat that it uses and exploits better than its competitor (Brown 1989b). That is, Allenby's gerbils should use the stabilized sand more because they forage more efficiently there, and greater sand gerbils should forage more efficiently on the looser substratum. Experiments show, however, that Allenby's gerbils forage more efficiently in both habitats (Brown, Kotler, and Mitchell 1994). Habitat selection resulting from the costs and benefits of foraging evidently does not provide the necessary conditions for the gerbils' coexistence.

So, did foraging theory fail? We think not, and in this chapter, we hope to show how the use of foraging theory helped us discover and test for the mechanisms underlying the gerbils' coexistence, and to understand the emergent pattern of habitat selection.

12.2   Introduction

Community ecologists want to understand the mechanisms that determine the abundances, numbers, types, and characteristics of species found living in the same place. They study niches and how organisms that differ from one another partition those niches. Foraging theory helps us understand how the abilities and liabilities of animals determine where and when they can forage profitably and how much they profit under different circumstances. Understanding how each species' fitness changes with the density and frequency of other species will illuminate community ecology.

In previous chapters, we have seen that foragers at high densities select prey opportunistically, and that competition can restrict the numbers of habitats used by individuals of interacting species (often called the compression hypothesis; Schoener 1969). Even in these simple cases, foraging matters. Foraging both responds to and reveals aspects of intra- and interspecific interactions.

In this chapter, we will examine the community consequences of foraging from the perspective of niches and niche partitioning. Much of an animal's niche involves where it lives and how it feeds. Foraging theory connects the characteristics and behavior of organisms with population dynamics, species coexistence, and community dynamics. It provides the tools for revealing the mechanisms by which species coexist and by which communities are structured through the behaviors of the individuals. Foraging theory provides a window into the evolutionary ecology of communities, from the coadaptation of morphologies and behaviors to coevolution and speciation.

12.3   Species Coexistence

Two species that occur at the same time in the same place coexist when their population densities are dynamically stable, or at least bounded away from zero. Dynamic stability occurs when a system at equilibrium returns to its equilibrium point following small perturbations (i.e., has a stable equilibrium point). For pairs of interacting species, dynamic stability arises when intraspecific interactions are stronger than interspecific ones (e.g., May 1973). Mutual invasibility can also be a condition for species coexistence. Two species can coexist if each can increase when rare within a stable or persistent population of the other. Chesson (2000) provides an outstanding review of these mechanisms.

Species coexistence can be promoted by resource partitioning (when species utilize different food types), frequency-dependent predation (when the rate at which predators kill individuals of different species depends on their relative abundances; Holt 1977; Holt et al. 1994), nonlinear competition combined with resource variability (when the per capita growth rate of a competitor species increases nonlinearly with resource availability; Armstrong and McCohu 1980) and storage effects (when temporal variation leads species to be more successful in some seasons or years than in others; Chesson 1990, 2000). These mechanisms can stabilize communities whenever intraspecific interactions are stronger than interspecific ones. They are typically modeled as mass action models in which individuals come together and interact almost like molecules in an ideal gas. Mass action models do not explicitly consider behavior, or if they do, they do not allow behaviors to vary. However, foragers do behave, and their behavior often varies with population density, resource availability, and environmental conditions. Thus, behavior, especially foraging, can create and shape the stabilizing effects that promote species coexistence.

We can introduce foraging behavior into mass action models via functional responses (see chap. 5). Adding foraging decisions to these models generally affects community stability. Functional responses sometimes destabilize communities (e.g., Gleeson and Wilson 1986; Fryxell and Lundberg 1994; Krivan 1996), but they can stabilize communities when predators avoid or ignore prey species that are at low population densities. Patch use decisions and constraints on digestion or handling time can both stabilize communities in this way (Holt 1983; Schmitz, Beckerman, and O'Brien 1997). Here we examine how feeding behaviors shape species interactions and coexistence from the ground up and in greater depth by applying foraging theory.

12.4   Behavioral Indicators and Behavioral Titrations

Building community models in which species interactions emerge from the foraging decisions of individuals requires an understanding of how behavior influences fitness. Testing such models requires methods that lead animals to reveal aspects of their fitness through their behavior. Such methods are based on the costs and benefits of foraging when the forager experiences diminishing returns.

For example, Kotler and Blaustein (1995) examined microhabitat selection and patch use in the gerbils of the prologue, Allenby's gerbil and the greater sand gerbil (*Gerbillus andersoni allenbyi* and *G. pyramidum*, respectively). They asked how much richer open and dangerous microhabitats had to be for gerbils to value them equally with safer microhabitats under bushes. Kotler and Blaustein conducted their experiment in a large aviary where gerbils could forage on artificial patches (trays filled with seeds mixed into sand) placed in bush and open microhabitats. The gerbils experience diminishing returns while foraging in these trays, so the density of seeds left in a tray after a night of foraging, the giving-up density (GUD; Brown 1988; see box 13.2), reflects the forager's harvest rate when it leaves the patch. A forager exploits the patch until the harvest rate falls to a value equal to the cost of foraging (see chap. 13). A higher giving-up density signifies higher costs.

The experiment used barn owls (*Tyto alba*) to manipulate the danger level. In response to the owls' presence, the gerbils showed higher giving-up densities in the open than under bushes, revealing that owls pose a greater threat in the open. Then Kotler and Blaustein added seeds to the open trays until the gerbils were harvesting the same amount of seed from open and bush trays. *G. pyramidum* needed 4 times and *G. a. allenbyi* needed 8 times as much initial seed in the open trays to make the open microhabitats of equal value to the bush microhabitats (fig. 12.1).

A similar experiment studied guppies (*Poecilia reticulata*) foraging in the presence of predaceous cichlids (*Cichlasoma* sp.) and gouramids (*Trichogaster leeri*) (Abrahams and Dill 1989). The study was based on the idea that foragers should distribute themselves according to an ideal free distribution (see box 10.1). The experiment offered guppies a choice between two patches differing in danger (one side of the aquarium contained a predator). Most guppies avoided the dangerous side in favor of the safe side, leaving those fish willing to take the risk with higher feeding rates. The resource supply rate in the dangerous habitat was then increased to the level required to equalize the number of guppies on each side.

We call studies like these "behavioral titrations" (Kotler and Blaustein 1995). Foraging theory tells us that a forager should perform an activity (feeding, hiding) so long as the marginal benefit it derives from this activity exceeds its marginal cost. A forager should continue with the activity until the marginal benefit falls to equal the marginal cost. When choosing which activities to perform, a forager should allocate more time to activities with

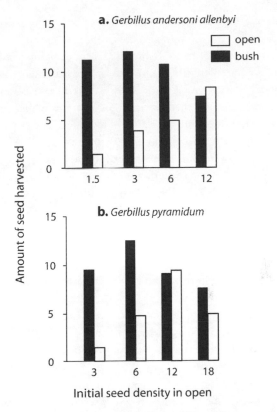

Figure 12.1. Behavioral titration. Total amounts of seed harvested from bush versus open microhabitats for (A) *Gerbillus andersoni allenbyi* and (B) *G. pyramidum*. Resource trays in the bush microhabitat contained a constant amount of seed from night to night, but trays in the open microhabitat varied. Bars of equal height for bush and open habitats indicate that gerbils place the same value on the two microhabitats. (After Kotler and Blaustein 1995.)

higher net marginal values and reduce time spent on activities with lower net marginal values. Hence, a forager's optimal allocation of time among activities should equilibrate the marginal values of the activities. Behavioral titration experiments provide a window into this equilibration. Researchers can take advantage of the animal's natural tendency to perform fitness titrations by conducting titrations of their own involving total value, total effort, and so on. In titration experiments, we use a quantifiable dimension of quality, such as food abundance, to measure the fitness value of another, more difficult to quantify dimension, such as predation risk. Titrations carried out in this manner form the basis for behavioral indicators that reveal a forager's perception of costs and benefits. Titrations can be used to test models of species interactions that involve foraging behaviors.

12.5   Behaviorally Mediated Indirect Effects

Tadpoles of two species of frogs, bullfrogs (*Rana catesbeiana*) and green frogs
(*R. clamitans*), live together with the predatory dragonfly larva *Anax junius* in
Michigan ponds. Werner and Anholt (1996) studied this system experimentally, manipulating the presence of caged *Anax* larvae while simultaneously
manipulating the densities and size classes of tadpoles. The caged *Anax* larvae
could not, of course, eat the tadpoles, but their presence did change the tadpoles' behavior: in general, the tadpoles moved more slowly, which affected
their feeding, mortality, and growth rates. Some of the effects were surprising. The growth rates of green frog and small bullfrog tadpoles were reduced,
but those of large bullfrog tadpoles were enhanced, and more large bullfrogs
completed metamorphosis in the presence of *Anax*! This happened because
while large and small bullfrogs compete strongly, *Anax* has a greater effect on
small bullfrogs. So, from the large bullfrogs' point of view, the presence of
Anax reduced competition from small tadpoles, allowing the large bullfrog
tadpoles to feed and grow faster.

In the terminology of community ecology, *Anax* had a behavioral indirect
effect on large bullfrog tadpoles via their interaction with small bullfrog tadpoles. In our example, the effect of *Anax* on the behavior of small bullfrogs
shaped the way in which small bullfrogs competed with large bullfrogs. Students of indirect effects typically focus on effects mediated through changes in
population densities and population growth rates, but one can consider other
traits, including activity times, foraging speeds, and individual growth rates.
When changes in behavior cause an indirect effect (e.g., as in our example
with *Anax* and *Rana*), we call it a behaviorally mediated indirect effect (Miller
and Kerfoot 1987; Werner 1992).

Indirect effects can cause what community ecologists call trophic cascades,
in which a predator reduces the density or foraging activity of its herbivore
prey, which in turn allows greater numbers of plants to grow (see chap. 13).
Indirect effects can result in higher-order interactions wherein the intensity
of the per capita effects of one species on another is altered by the presence
of a third (Kotler and Holt 1989). In our example, the *Anax* scare the small
bullfrog tadpoles, which move less, eat less, and grow more slowly. Because
the small tadpoles eat less, each one has less of a negative effect on both
its competitors and its periphyton food. Reduced feeding by small tadpoles
allows for greater periphyton density. The effect of predators on the tadpoles
thus "cascades" down to lower trophic levels.

To see how behaviorally mediated indirect effects can affect community
structure and coexistence, consider an environment with two equally productive

habitat types. One habitat provides more protection from potential predators. Two species that share a common predator and a common resource live in this hypothetical environment. The two species compete for the limited resource, but one is more vulnerable to predation than the other. In the absence of the predator, we expect the two species to compete intensely in both habitats, depleting all the available resources. We expect coexistence only if the two species differ in their resource-harvesting abilities in the two different habitats or in their relative energetic costs of foraging in the two habitats. Otherwise, the most efficient forager will win.

With the predator present, things change. Now, one habitat offers safety but little food, and the other offers more food that comes at a cost (recall our discussion of behavioral titrations in section 12.4). As the foragers balance the costs and benefits of each habitat and adjust their activities and habitat use accordingly, competition intensifies in the safe habitat, but weakens in the dangerous habitat. The predator indirectly affects the competitive interaction between the two prey species by influencing their behavior, so we have a behaviorally mediated indirect effect. In addition, we have a higher-order interaction because the predator's presence reduces the per capita effect of one competitor on the growth rate of the other. The presence of the predator and its effects on the habitat choices of the prey promote species coexistence, provided that the better competitor is more affected by the predator.

Werner and Anholt (1993) modeled key aspects of the tadpole-*Anax* system. They sought to understand how the individual decisions of foragers combined to create the observed behaviors that led to the indirect effects. They had their model foragers select swimming speed and proportion of time spent active so as to minimize the ratio of mortality risk to harvest rate. Increasing these parameters increased risk of predation and rates of resource depletion. Hence, these decisions permit the forager to determine its mortality risk, harvest rate, and individual growth rate. In general, both competition for resources and predation risk lead to slower optimal foraging speeds, lower activity levels, and slower growth rates. These effects in the context of interacting competitors yield indirect effects like those observed with the tadpoles.

Experiments by Peacor and Werner (1997) showed that the behaviorally mediated indirect effect predicted by theory and observed in the simple tadpole-*Anax* food web applies to more complex food webs, too. Peacor and Werner placed the same numbers of green frog and small bullfrog tadpoles in each of several experimental ponds. They then varied the densities of large bullfrog tadpoles and two classes of odonate predators (free-ranging *Tramea lacerata*; caged *Anax junius* and *Anax longipens*). Caged *Anax* led green frogs and large bullfrogs to reduce their activities. This treatment gave rise to three behaviorally mediated indirect effects, due mostly to the nonlethal effects of the *Anax*:

1. Large bullfrogs increased the movement of the smaller tadpoles (via interference and reduced resource levels), increasing *Tramea* predation on green frogs and small bullfrogs (an indirect effect spanning three trophic links).
2. Caged *Anax* reduced green frog activity, decreasing *Tramea* predation on green frogs.
3. Caged *Anax* increased the competitive advantage of small bullfrogs over green frogs, because green frogs responded more strongly to predation risk and thus spent less time active and grew more slowly (another indirect effect spanning three trophic links).

This example demonstrates how behavioral responses to predators can alter competitive interactions and even interactions among predators (see Schmitz 1998 and Wootton 1992 for similar studies with different taxa).

12.6   Habitat Selection

The world is heterogeneous. Resource density, cover from predators, foraging substratum, and types and numbers of competitors and predators are just some of the things that can vary in space or time. Specializations that increase a forager's ability to exploit particular conditions often come at the expense of decreasing its ability to exploit others. Consequently, selection can favor the ability of a forager to direct its activity to situations where it profits most. This coadaptation of ability and behavior can affect species interactions and community structure. For example, habitat selection can reduce competition if two species select different habitats. In fact, the strengths of species interactions emerge from the optimal behaviors of the interacting individuals. Box 12.1 explains two graphical tools (isodars and isolegs) that reveal properties of habitat selection as well as community organization based on habitat selection. The following examples apply these tools.

In the Rocky Mountains of southern Alberta, pine chipmunks (*Tamias amoenus*) coexist with deer mice (*Peromyscus maniculatus*) and red-backed voles (*Clethrionomys gapperi*) across a range of conditions differing in aspect and plant community, from xeric open meadow to mesic fir forest. Chipmunks are diurnal, forest-dwelling ground squirrels that larder-hoard seeds and nuts. Deer mice are nocturnal caching omnivores that climb well, while red-backed voles are terrestrial herbivores that are active day and night and eat seeds and

BOX 12.1   Isolegs and Isodars

The ideal free distribution (IFD) of Fretwell and Lucas (1969) provides the basis for understanding how individuals should distribute themselves among habitats in response to habitat quality and population density. The IFD is described in box 10.1. Isodars (Morris 1988) and isolegs (Rosenzweig 1981) link the habitat choices of individuals with the dynamics of populations and communities.

Isodars

The ideal free distribution assumes that foragers can change habitats without cost. Individuals choose the habitat that offers the highest fitness, and individuals can enter a habitat on an equal basis with those already there. Furthermore, the ideal free distribution assumes that fitness (per capita population growth rate) in a habitat declines with the habitat's population density (fig. 12.1.1). For example, the relationship between density and fitness may be linear:

$$\left(\frac{1}{N_A}\right)\left(\frac{dN_A}{dt}\right) = r_A - b_A N_A, \quad (12.1.1)$$

where N_A equals population density in habitat A, r_A equals maximum per capita population growth rate in habitat A, and b_A is the strength of density dependence in habitat A.

Consider two habitats, A and B. If habitat A offers higher fitness at low population density, then all individuals should choose habitat A at low density. As density in A increases, fitness decreases for each individual there. Eventually, fitness in habitat A drops to the point at which fitness in a crowded habitat A equals fitness in an unoccupied habitat B. At that point, individuals should be indifferent to habitat choice because both habitats offer equal returns. As population density grows further, individuals should distribute themselves such that fitnesses across the two habitats are equal:

$$\left(\frac{1}{N_A}\right)\left(\frac{dN_A}{dt}\right) = \left(\frac{1}{N_B}\right)\left(\frac{dN_B}{dt}\right), \quad (12.1.2)$$

which is equivalent to

$$A_A - b_A N_A = A_B - b_B N_B.$$

(Box 12.1 continued)

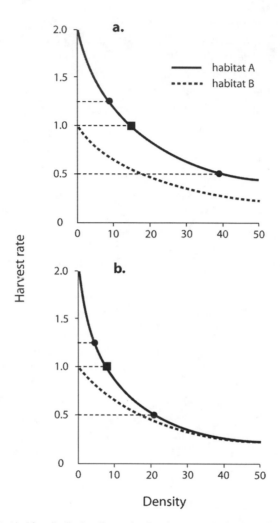

Figure 12.1.1. Ideal free distribution. The graphs show how per capita fitness declines in each of two habitats with each habitat's population density. At low population sizes, all individuals crowd into the preferred habitat A, as it provides a higher fitness reward than habitat B (shown by the upper solid circle emanating from the highest horizontal lines). At a critical population size in habitat A (shown by the solid squares), unoccupied habitat B offers the same reward as habitat A. At this critical density, individuals should be indifferent to the choice between habitat A and habitat B. At total population sizes above this critical density, individuals should spread themselves between habitats A and B such that expected fitnesses are the same for A and B, as shown by the solid circles emanating from the lowest horizontal equal fitness lines. (A) Habitat A has twice the productivity of habitat B. (B) Habitat B offers resources that are twice as easy to encounter as those in habitat A. (After Brown 1998b.)

(Box 12.1 continued)

Morris (1988) noted that this equation can be rewritten as

$$N_A = (A_A - A_B)/b_A + (b_B/b_A)N_B. \quad (12.1.3)$$

This equation specifies an *isodar*: the relationship between population densities (N_A and N_B) in two habitats for animals following the ideal free distribution (Morris 1988; fig. 12.1.2). We define an isodar as *all combinations of population densities in habitats A and B such that both habitats offer the same fitness reward.*

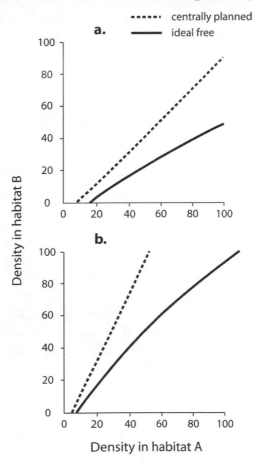

Figure 12.1.2. Isodars. The solid lines show the relationship between the numbers of individuals in habitat A and in habitat B such that individuals experience the same fitness in each habitat. (A) Habitat A offers twice the productivity of habitat B (same parameters as in fig. 12.1.1A). (B) Habitat B offers twice the ease of encountering prey as habitat A (same parameters as in fig. 12.1.1B). The dashed line ("centrally planned") represents the distribution that maximizes total productivity, rather than fitness. (After Brown 1998b.)

(Box 12.1 continued)

We can construct isodars from census data (e.g., Morris et al. 2000) by plotting estimated density in habitat A against estimated density in habitat B. By convention, we plot the density of the habitat with the higher productivity on the y-axis. The isodar's intercept $[(A_A - A_B)/b_A]$ gives the difference between the habitats in per capita growth rate at low population densities (i.e., in the productivities of the habitats). Morris refers to differences in habitats revealed by nonzero y-intercepts of the isodar as quantitative differences. The isodar's slope is the ratio of the terms that describe the intensity of density-dependent effects in habitats A and B (often due to differences in risk of predation). Morris refers to differences in habitats revealed by slopes different from 1 as qualitative differences.

We can extend isodars to examine species interactions. If two species, 1 and 2, share habitats A and B, then we can rewrite equation (12.1.3) as follows:

$$N_{1A} + \alpha N_{2A} = [C + \beta(N_{1B} + \beta N_{2B})],$$

where $\alpha = b_{11A}/b_{12A}$ and gives the average competitive effect of one individual of species 2 on species 1 in habitat A; $C = (A_{1A} - A_{1B})b_{1A}$ and gives the quantitative differences between the two habitats; and $\beta = (b_{12B}/b_{12A})$ and gives the average competitive effect of one individual of species 2 on species 1 in habitat B. Or, more conveniently, we can rewrite equation (12.1.3) as

$$N_{1A} = C - \alpha N_{2A} + \beta(N_{1B} + \beta N_{2B}). \quad (12.1.1)$$

We can use multiple regression to estimate the parameters in this relationship [eq. (12.1.4)]. Isodar analysis accurately detects exploitative competition (Morris 1988), but may fail to detect interference competition (Ovadia and Abramsky 1995).

Isolegs

Isolegs provide a different perspective on habitat selection (Rosenzweig 1981) (fig. 12.1.3). Again, the ideal free distribution provides the conceptual foundation. Isolegs give combinations of population densities at which two habitats provide equal fitness. Again, consider two species, 1 and 2, that share habitats A and B. The two species can either show a shared preference for the same, best habitat (say, A), or they can do better in different habitats (say, species 1 does best in A and species 2 does best in B) and show distinct

(Box 12.1 continued)

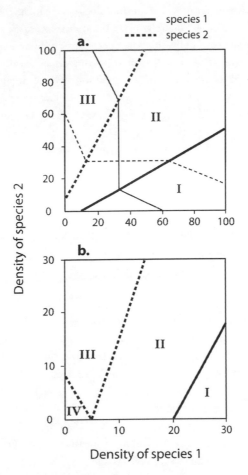

Figure 12.1.3. Isolegs and isoclines (A) The isolegs and isoclines for distinct-preference, two-species, density-dependent habitat selection. Below species 1's isoleg (solid, positively sloped line), species 1 resides in both habitats, while above its isoleg it occupies habitat A only. Below species 2's isoleg (dashed, positively sloped line), species 2 resides in habitat B only, while above its isoleg it occupies both habitats. Each species' isocline (thinner lines) has a negative slope in region I (species 1 is opportunistic and species 2 is selective), a vertical (species 2) or zero (species 1) slope in region II (both species are selective on their preferred habitat type), and a negative slope in region III (species 1 is selective and species 2 opportunistic). The point where the two isoclines cross in region II indicates the ghost of competition past—neither species appears to have a negative effect on the other at the equilibrium point. (B) Isolegs for shared-preference habitat selection where species 1 is the superior competitor in the preferred habitat. Species 1 and 2's isolegs have the same interpretation as in part A, with the addition of a second isoleg for species 2 (the short negative line). Inside this second isoleg, species 2 is selective on habitat 1. This creates a fourth region in the state space, IV, where both species are selective on habitat A and absent from habitat B. (After Brown 1998b.)

(Box 12.1 continued)

habitat preferences. Assume that species 1 does best in habitat A, and species 2 does best in habitat B. There are two important isolegs, one for species 1 and one for species 2. The isoleg for species 1 maps where species 1 goes from being selective on its best habitat (to the left of the isoleg) to being opportunistic in its use of both habitats (to the right of the isoleg). This is simply the effect of density dependence that we have seen previously in chapter 10, in the ideal free distribution (see box 10.1), and above. The other isoleg maps the same for species 2.

Consider the problems of species 1 without species 2. At low density, all members of species 1 select their preferred habitat A. As population density increases, fitness in A drops to the same level as fitness in B. This gives the x-intercept of species 1's isoleg. At this point, individuals can choose either habitat with the same consequences, and they should be indifferent. If density increases beyond this point, foragers should choose habitats opportunistically. Thus, the isoleg separates a region of selectivity (species 1 resides only in habitat A) from a region of opportunism (species 1 occupies both habitats A and B).

But what if species 2 is also present? At low density, individuals of species 2 will select their preferred habitat B. With some individuals of species 2 in habitat B, it now takes more individuals of species 1 in habitat A to reduce the value of habitat A to equal that of habitat B. The point at which fitnesses equilibrate now occurs at a higher density of species 1 (in A), and the isoleg moves up and to the right: as species 2 increases in B, the point where species 1 switches from being selective on A to being Isolegs and isoclines. opportunistic occurs at ever higher densities of species 1. This results in an isoleg that intercepts the x-axis and has a positive slope. We use a similar argument to find the species 2 isoleg, which also has a positive slope, but intercepts the y-axis. The result is a system of two isolegs, both with a positive slope, that separate the state space of N_1 and N_2 into three regions (fig. 12.1.3A). Above species 2's isoleg (region III in fig. 12.1.3A), species 1 selects habitat A and species 2 is opportunistic; between the isolegs (region II), both species select their own best habitat; to the right of species 1's isoleg (region I), species 2 selects habitat B and species 1 is opportunistic.

As optimal habitat selection behavior changes across these three regions, the intensity of competition also changes. The two species compete most intensely in the upper and lower regions (I and III), where one species selects its preferred habitat and the other occupies both habitats opportunistically. In the central region (II), however, the two species do not compete, because

(Box 12.1 continued)

the two species avoid each other by selecting their own preferred, best habitats. If population densities typically fall in this "no competition" region, the two species may evolve fixed habitat selection behavior that no longer responds to density. When this occurs, not even removal experiments can detect the interspecific competition that produced each species' habitat specialization. Rosenzweig (1991) calls this phenomenon "the ghost of competition past."

Zero population growth rate isoclines give the combinations of densities of each species at which the population growth rate for a species is zero. These isoclines reveal the dynamic stability properties of the ecological system of two interacting species and can show the ghost of competition past (see fig. 12.1.3A). The resulting changes in optimal habitat selection behavior in the different regions also change the intensity of competition between the species there. The isoclines change slope as they pass from one region to the next. This results in isoclines that kink as they cross the behavioral isolegs. The isoclines are vertical or horizontal between the isolegs and have negative slopes elsewhere (see fig. 12.1.3A). The kinking of the isoclines can produce a stable equilibrium point where one otherwise would not exist. Thus, the magnitudes of the competition coefficients emerge from behavior, and in fact, change as behavior changes (compare this with the models of mass action in which competition coefficients are givens).

In other cases, two species may prefer the same habitat (fig. 12.1.3B). Assume that both species prefer habitat A, but that species 1 is more despotic and specialized while species 2 is more tolerant across habitats. There can be three isolegs in this system. The dominant species has a single isoleg that, as in shared preference habitat selection, has a positive slope, and for the same reason. At low density, species 1 will inhabit habitat A exclusively, but increasing population density will eventually reduce fitness in habitat A to the level of habitat B, so species 1 will become opportunistic and begin to use habitat B. The presence of species 2 decreases the quality of alternative habitat B and leads to a positively sloped isoleg. The subordinate species has up to two isolegs. One separates the lower densities at which the subordinate species selects the preferred habitat from the higher densities at which it becomes opportunistic. For species 2 by itself, individuals will select habitat 1, and as its density rises, there will come a point where fitness in habitats A and B are equal. This point forms the isoleg's y-intercept. Below this point, species 2 selects habitat A; above this point, it chooses opportunistically. However, species 2's isoleg has a negative slope: increases in the density of

(Box 12.1 continued)

species 1 (also inhabiting habitat A at low density) will decrease the quality of habitat A and lower the point where habitats A and B are of equal quality. Species 2's isoleg will intercept the *x*-axis at or below species 1's isoleg, and that is why we can assume that there will be at least some members of species 2 in habitat B when we calculate the species 1 isoleg.

Finally, another isoleg for the subordinate species may exist above the first. At sufficiently high densities of species 1 (which mostly uses habitat A and may interfere with species B there), species 2 may choose to avoid the best habitat altogether due to intolerable costs of interference from the dominant species and instead select habitat B. This creates the new species 2 isoleg (to the right of the original) that separates opportunistic choice of the two habitats from a region of high species 1 density where species 2 should select the poorer habitat. This isoleg has a positive slope because adding more species 2 individuals to habitat B reduces its quality and makes habitat A more attractive.

The three isolegs create four regions with different combinations of optimal habitat selection behaviors (see fig. 12.1.3B). In region IV at the bottom left, both species select the best habitat, A. In region III, species 1 selects habitat A, but species 2 is opportunistic. In region I, species 1 chooses opportunistically, while species 2 shows an apparent preference for habitat B. The species compete most intensely in this region because both species occupy both habitats and population densities are high. And finally, in region II, species 1 chooses habitat A, but species 2 selects the poorer habitat. As in the case of distinct preference, shared preference habitat selection causes the zero population growth rate isoclines to kink as they pass from one region to the next.

We derive isodars and isolegs from the ideal free distribution, and we use them to reveal aspects of population growth, population regulation, species interactions, and community organization. Although they both explore habitat selection, notice that they consider different quantities. When we plot isodars, we plot density in habitat A versus density in habitat B; when we plot isolegs, we plot density of species 1 versus density of species 2. We can find both isodars and isolegs from simple census data. Additionally, experiments that give foragers a choice between habitats at different competitor densities can reveal the zero population growth rate isoclines of the system (e.g., Rosenzweig and Abramsky 1997). Thus, the ideal free distribution forms the basis for a comprehensive analysis of populations and ecological communities.

vegetation. Figure 12.2 shows the isodars for each species (Morris 1996). The isodars reveal a habitat generalist (chipmunk) and two habitat specialists (xeric habitat: deer mouse; mesic habitat: vole). Habitat selection responds to intraspecific density only, though the opportunism of the chipmunk occurs at a fine scale, and the habitat selection of the deer mouse and red-backed vole occur at a coarse scale. Theory suggests that a generalist and two specialists can coexist if the generalist experiences the environment as relatively fine-grained (Brown 1996), as do these rodents.

Morris et al. (2000) calculated isodars for two competing herbivorous rodents from the wet heathlands of eastern Australia. The heathlands are seasonally dry and burn frequently. There, the swamp rat (*Rattus lutreolus*) co-occurs with the eastern chestnut mouse (*Pseudomys gracilicaudatus*) in habitats that vary in age and edaphic conditions. The eastern chestnut mouse is especially common in recently burned sites, but is gradually replaced by the swamp rat as the effects of fire recede. In intermediate-aged stands, the two species co-occur across the range of edaphic conditions. The isodar analysis confirmed the asymmetric competitive dominance of the swamp rat over the eastern chestnut mouse in both wet and dry heath habitat, with stronger effects in drier sites. Isodars also revealed the superiority of *P. gracilicaudatus* in recently burned areas. Applying principles of density-dependent habitat selection corroborated the results of previous removal experiments that revealed much the same information, at much greater cost and effort (Higgs and Fox 1993).

Although isodars can reveal aspects of community organization, they are better suited for studying intraspecific behavior. In contrast, isolegs are defined only for two or more interacting species in heterogeneous environments. We can use experimental manipulations of population densities to find isolegs. The isoleg for a species gives all combinations of the densities of two (or more) species such that the species is indifferent in its use of the two habitats. Usually this isoleg considers the point at which a species goes from being selective on one habitat to being opportunistic on two habitats. There is a separate isoleg for each species. The isolegs exist in the same state space of species densities as the population growth rate isoclines from ecology (see box 12.1).

Abramsky, Rosenzweig, and colleagues manipulated the densities of gerbils in 1 ha field enclosures where two gerbil species, Allenby's gerbil and the greater sand gerbil, could choose between stabilized and semi-stabilized sand dunes within a mosaic of habitats (Abramsky et al. 1990, 1991; Rosenzweig and Abramsky 1997). The results supported the shared preference model (see fig. 12.1.3B) and the existence of a single isoleg for the dominant species, but two isolegs for the subordinate species. More importantly, the investigators deduced the general shapes of the zero population growth rate isoclines (indicators of the dynamic stability of the system, i.e., whether the two species

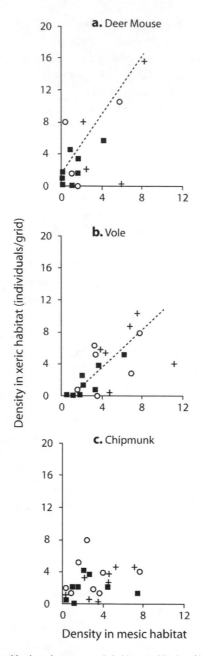

Figure 12.2. Population densities in xeric versus mesic habitats and isodars (dashed lines) for three species of montane rodents in the Rocky Mountains of southern Alberta, Canada: (A) deer mouse, (B) red-backed vole, and (C) pine chipmunk. Isodars are based on the ideal free distribution and are obtained by regressing population densities in one habitat versus the other. Isodar intercepts that differ from 0 reveal quantitative differences between habitats, and slopes that differ from 1 reveal qualitative differences (see box 12.1). For the deer mouse, the xeric habitat is both quantitatively and qualitatively superior; for the vole, the mesic habitat is quantitatively superior; for the chipmunk, the habitats are equally valuable. Symbols refer to different trapping sessions: first ■, second o, third +. (After Morris 1996.)

coexist) through the application of the ideal free distribution. They did so by connecting pairs of enclosures with gates. By allowing only one species to pass through the gates, Abramsky and Rosenzweig could fix competitor densities in the two connected enclosures while allowing the target species to adjust its distribution and activity. Using this technique, Abramsky and Rosenzweig measured the effect of the species with a fixed density on the level and distribution of foraging activity of the species that could move freely between enclosure halves. In this way, the species that is free to move reveals the effect of competition with the other species on it (the competition coefficient) through its habitat selection behavior. By repeating this treatment over a range and combination of competitor densities, Abramsky and Rosenzweig could render the shape of the isoclines. Remarkably, their data support the nonlinear isoclines that foraging theory predicts (Abramsky et al. 1991, 1994; Abramsky, Rosenzweig, and Subach 1992; fig. 12.3).

The data from these experiments can also be examined with isodar analysis (Ovadia and Abramsky 1995). The isodars confirm shared preference habitat selection for the semi-stabilized habitat, with G. pyramidum experiencing the stabilized and the semi-stabilized sand as qualitatively similar, but G. a. allenbyi experiencing the stabilized sand as qualitatively superior. The isodars reveal a flip-flop in the habitat preferences of G. a. allenbyi. At low population densities it prefers the semi-stabilized sand habitat, but at high densities it prefers the stabilized habitat. The isodars also revealed resource competition between the two species, but failed to detect interference. Abramsky and Rosenzweig's ability to set conditions in different enclosures and then allow the animals to perform their own titrations made this a successful experiment.

We can use this approach to address other questions in community ecology. Abramsky et al. (2000), for example, used it to measure the energetic cost of interspecific competition. They established four G. pyramidum individuals in one of two connected enclosures, along with 40 or 50 G. a. allenbyi individuals. The G. a. allenbyi could move freely between the two enclosures (through species-specific gates); the G. pyramidum could not (as in the above experiments). G. a. allenbyi individuals adjusted their enclosure-specific activities in response to the differing competitive regimes in the two enclosures. More G. a. allenbyi activity occurred in the enclosure without the competitor. Next, Abramsky et al. carried out an experimental titration, adding seeds to the enclosure with G. pyramidum until G. a allenbyi was equally active in both enclosures. Adding 4.5 g of seeds to each of 24 trays balanced the effect of four competitors. To date, similar titrations have measured the benefits of habitat selection, the cost of temporally partitioning the night, and the cost of apprehensive foraging under predation risk (Abramsky et al. 2001, 2002a, 2002b).

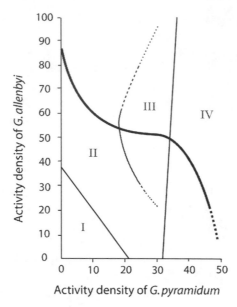

Activity density of *G. pyramidum*

Figure 12.3. The density-dependent habitat selection isolegs for *Gerbillus andersoni allenbyi* (lines) and *G. pyramidum* (light curve) and the isocline of *G. a. allenbyi* (heavy curve) drawn in a state space of activity densities (i.e., activity as measured by tracking plots) of the two species. The isolegs separate regions of optimal behavior. In region I, both species prefer semi-stabilized sand dunes; in region II, *G. pyramidum* still prefers the semi-stabilized habitats, but *G. a. allenbyi* opportunistically exploits both the semi-stabilized and stabilized habitats; in region III, *G. a. allenbyi* exhibits apparent preference for the stabilized habitats, and *G. pyramidum* continues to prefer the semi-stabilized habitats; in region IV, *G. a. allenbyi* continues to exhibit apparent preference for the stabilized habitats, and *G. pyramidum* uses both habitats. The zero population growth rate isocline changes slope in the different regions as habitat selection behavior changes, and with it, the intensity of competition. Note (1) the strong interactions between the gerbils when the subordinate *G. a. allenbyi* and the dominant *G. pyramidum* occur at low densities and both species forage mostly in the preferred semi-stabilized habitat; (2) the strong interactions when *G. pyramidum* is at high densities and using habitats more opportunistically; and (3) the less intense interactions at intermediate densities when the dominant *G. pyramidum* is still selective on the preferred semi-stabilized habitat, but the subordinate *G. a. allenbyi* already favors the stabilized habitat. (After Abramsky et al. 1991.)

12.7   Optimal Behavior and Consumer-Resource Models

Coexisting species often differ in body size, but such differences do not always lead to coexistence based on food size selection. Coexisting species of granivorous desert rodents often differ in body size (e.g., Brown 1975), yet may overlap almost completely in the sizes of the seeds that they consume (e.g., Lemen 1978). In contrast, coexisting species of Darwin's finches may show distinct differences in both their beak sizes and the seed sizes in their diets (Grant 1986; see section 12.8). Can foraging theory illuminate the causes for such different outcomes?

Imagine two consumer species that compete for two types of food resources. The two species may differ in many respects, including the rates at which they encounter the resource types, the values of the different resource types, and the handling time needed to consume a resource item. Encounter rates, values, and handling times all affect rates of energy gain and determine diet and patch use decisions (see chap. 1). The species compete through their effects on resource density. The foraging aptitudes of individuals and the foraging choices that they make determine their effects on the resources and their energy gains. What are the conditions for species coexistence that emerge from the species' optimal behaviors? The answer depends on the distribution of the resource types, the nutritional relationship between them, the rates at which the resources are renewed, and the rates at which consumers harvest them. Thus, coexistence in these circumstances is at heart a foraging problem (MacArthur 1972; Tilman 1982).

In such consumer-resource systems, coexistence depends on the resources. If two competitors exploit a single resource, the species whose individuals can subsist on the lowest density of that resource typically outcompetes the other. The threshold density of a resource at which a consumer species can just survive is referred to as R^*. Above R^*, the consumer species harvests enough of the resource to have a positive population growth rate, and vice versa when the resource is below R^*. The expectation is that a population of consumers will grow or decline until its density promotes a resource abundance of R^*. The consumer species with the lowest R^* outcompetes the other under equilibrial consumer-resource dynamics.

For two consumer species to coexist in these models, there must be more than one resource. Vincent et al. (1996) examined coexistence on two resource types for optimal foragers that can choose both their diet and their habitat use. Their models varied two things: two resources occurred together in the same habitat or in separate habitats, and the resources could be either essential (the resource in shortest supply determines fitness) or perfectly substitutable (both resources contribute additively to fitness). Vincent et al. factorially combined these properties of the environment and resources to create four cases. To find conditions for coexistence, they examined the zero net growth isoclines and the resource depletion vectors.

Zero net growth isoclines are plotted in a state space of resource densities. They represent all combinations of the densities of two resources such that a forager has a zero population growth rate (Tilman 1982). The zero net growth isocline represents the two-resource equivalent of R^*. When resource abundances lie above the isocline, the consumer species has a positive growth rate, and vice versa when resource abundances lie below the isocline. At equilibrium, the consumer species should deplete resources to some point along the zero net growth isocline.

The shapes of zero net growth isoclines depend on the resources' nutritional quality and spatial distribution and on the optimal foraging behavior of the consumers. For instance, consider a consumer species opportunistically harvesting two perfectly substitutable resources that occur together. In this case, the consumer species' zero net growth isocline is linear with a negative slope. The R_1 and R_2 intercepts (densities of resources 1 and 2, respectively) represent the original R^*s for the situation in which there is only one resource.

Two perfectly substitutable resources can instead occur apart in different habitats. In this situation, the consumers can seek only one or the other resource at a time. In a situation analogous to the ideal free distribution, the consumers now seek the habitat that offers the highest harvest rate of resources. At equilibrium, the abundance of each resource will be driven down to its R^*. Hence the zero net growth isoclines are the horizontal and vertical lines emanating from the R_1 and R_2 intercepts of the preceding example with two resources occurring together.

For essential resources, growth is limited by the resource in shortest supply. Regardless of whether resources occur together or apart, the isocline resembles an L. The level of each leg of the isocline is set by the density of that resource that yields a harvest rate equal to the foraging costs. Each of the isoclines emerges from the properties of the environment (foods together versus foods apart), the nutritional properties of the foods (substitutable versus essential), and the foraging behavior of the consumers. The consumers in these systems adopt a feeding strategy of opportunism, partial selectivity, or complete selectivity so as to maximize their fitness.

The population sizes of consumers and their foraging strategies result in resource depletion. The harvesting of resources and the renewal of resources result in a new equilibrium abundance of resources that is lower than it would be in the absence of harvesting. The depletion vector of a consumer species gives all combinations of equilibrium abundances of two resources that will occur as the population size of that consumer increases. The depletion vector starts at high values for R_1 and R_2 when the population of consumers is zero, then declines as the number of consumers increases. It is positively sloped if the consumers harvest some of both resources. It is horizontal if the consumers harvest only R_1, and it is vertical if the consumers harvest only R_2. Depletion vectors can be plotted in the same state space as zero net growth isoclines (Tilman 1982).

As with zero net growth isoclines, the distribution and characteristics of the resources and the foraging behavior of the consumers determine the position and shape of the depletion vectors. For substitutable resources occurring in the same habitat, the slope of the depletion vector is determined by the consumer's rates of encountering the two resources, a_1 and a_2, and the abundances of the two resources, R_1 and R_2. The slope at any point in the state space of R_1 and R_2 is given by $a_2 R_2 / a_1 R_1$. For substitutable resources that occur in separate habitat patches, the optimal habitat selection behavior of the consumers becomes paramount in determining the shape of the depletion vector. In this case, the consumers should balance their activity between habitats so that the fitness values of their harvest rates are equal (assuming equal costs of foraging between habitats): $e_1 a_1 R_1 / (1 + a_1 h_1 R_1) = e_2 a_2 R_2 / (1 + a_2 h_2 R_2)$. This behavior produces a linear depletion vector whose slope is influenced by the consumer's energetic gain from the resource, e, rate of encountering the resource, a, and handling time for the resource, h. For essential resources, the ratio of the contribution of each resource to the consumer's fitness determines the slope of the depletion vector (Vincent et al. 1996).

The intersection of a consumer species' zero net growth isocline with its depletion vector determines the equilibrium abundance of resources at the equilibrium population size of that consumer species. We can find the conditions for coexistence by combining the zero net growth isoclines and depletion vectors of two different species. As a first condition, coexistence requires that the zero net growth isoclines cross. If not, one species (the one with the zero net growth isocline closest to the origin) can always outcompete the other by depleting resources to a point where the second species can no longer exploit them profitably. When zero net growth isoclines cross, different resources limit each species. Coexistence also requires that each species consume more of the resource that most limits its own growth; that is, the species with the shallower zero net growth isocline must have a depletion vector that increases less steeply (fig. 12.4).

Traits that affect foraging aptitudes also help determine the zero net growth isocline and the depletion vectors, and hence conditions for coexistence. Tra-offs in those traits among the consumer species cause zero net growth isoclines to cross. Zero net growth isoclines typically include the coefficients for encounter rate (a), handling time (h), and conversion efficiency (e) of resources, but depletion vectors often have only one of these coefficients. Hence, it is often the trade-offs among the coefficients in the depletion vectors that determine when two consumer species can and cannot coexist. When substitutable resources co-occur in the same patch, the relevant trade-off for coexistence requires differences between the two consumer species in their rate of encountering each resource. One consumer must have a higher rate of encountering resource 1, while the other consumer species must have a higher rate of encountering resource 2. For coexistence on essential resources, the two consumer species must have a trade-off in their conversion efficiencies (es). In this case, the

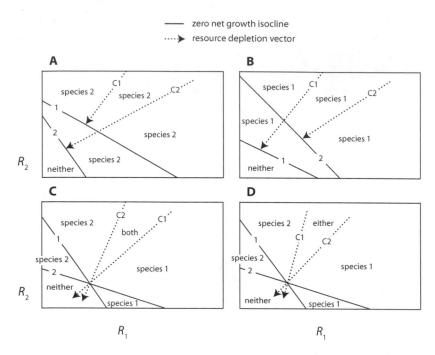

Figure 12.4. Zero net growth isoclines and resource depletion vectors for two species, 1 and 2, plotted in a state space of resource density (R_1, R_2). Consumers are limited by the resource in shortest supply. Typically, for coexistence, zero net growth isoclines (labeled 1 and 2 for species 1 and species 2, respectively) must cross, and the resource supply point (the maximum amount of the two resources in the absence of consumption) must lie in a region bounded by the two resource depletion vectors (labeled C1 and C2 for consumption by species 1 and species 2, respectively). The regions of each panel where both species coexist or where one species or the other wins out in competition when the resource supply point lies within that region are labeled. (After Vincent et al. 1996.)

species with the lower ratio of conversion efficiency of resource 1 relative to resource 2 should also leave the higher amount of resource 1 when it stops foraging (highest R_1^*). In contrast, when resources occur in separate habitats, coexistence can result from trade-offs between the two consumer species in encounter rate (a), handling time (h), or conversion efficiency (contributing to e). Thus, for organisms following the rules of optimal diet and habitat selection models, the distribution and nutritional quality of resources limits the kinds of trade-offs and mechanisms that promote species coexistence. In general, habitat selection offers more opportunities for coexistence than opportunistic feeding on co-occurring foods because a larger suite of trade-offs satisfy the conditions for coexistence under habitat selection than under overlapping diet choice.

How, then, does this apply to the desert rodents and the finches? In both cases, the coexisting species consume seeds of various sizes that co-occur in patches. They are most likely exploiting substitutable resources that co-occur.

In this situation, coexistence by diet choice requires a trade-off in encounter rates with the different food types. The ability to encounter large seeds must come at the expense of the ability to encounter small ones. The desert rodents often forage on buried seeds that are encountered by olfaction. Any characteristic that improves their ability to smell large seeds also probably improves their ability to smell small seeds, so the required trade-off does not exist. We must look elsewhere for a mechanism of coexistence (see section 12.8). For the finches, encounter rates with small versus large seeds do not appear to vary between small- versus large-beaked birds. But small-beaked foragers cannot generate enough force to crack open large seeds with thick coverings. Effectively, it is as if they do not encounter such seeds, resulting in a trade-off of encounter rates according to beak size and seed size and providing the necessary conditions for coexistence.

12.8   Mechanisms of Species Coexistence of Optimal Foragers

Consider the two gerbils, G. pyramidum and G. a. allenbyi, discussed previously. Recall that these species show distinct patterns of habitat selection, but they do not coexist due to habitat selection. Might the foraging abilities of the two gerbil species and salient features of their environment reveal the mechanism by which they coexist? In regard to the foraging abilities of the gerbils, the same field experiments that showed the smaller G. a. allenbyi always to be a more efficient forager than the larger G. pyramidum also suggested that G. pyramidum often arrives at resource patches first. In addition, G. pyramidum can handle food items more quickly and feeds faster at high seed densities (Kotler and Brown 1990). Isoleg analysis suggests that G. pyramidum dominates G. a. allenbyi via interference competition (Abramsky et al. 1990). On the other hand, G. a. allenbyi has evolved an especially low metabolic rate (Linder 1987) that should reduce its energetic costs of foraging. In regard to the gerbils' environment, predictable afternoon winds redistribute seeds and renew seed patches daily (Ben-Natan et al. 2004). The aptitudes of the gerbils and the daily renewal of seeds open the possibility that these gerbils partition resource variability, with G. pyramidum using its ability to interfere and harvest seeds quickly to monopolize and deplete rich resource patches early in the night. Later, G. a. allenbyi, by virtue of its especially low energetic cost of foraging, can forage profitably on what remains (Brown, Kotler, and Mitchell 1994). The result is temporal partitioning.

To test this mechanism, Kotler, Brown, and colleagues conducted two experiments. In the first, Kotler, Brown, and Subach (1993) looked for temporal partitioning. They set out groups of six seed trays at the beginning of

Table 12.1   Temporal partitioning in two gerbil species, *Gerbillus andersoni allenbyi* and *Gerbillus pyramidum*

Species	First visit	Last visit	Average visit	Average patch	Giving-up density
G. a. allenbyi	2.61 h	8.95 h	5.82 h	0.75 g	0.506 g
G. pyramidum	1.63 h	6.54 h	4.02 h	1.42 g	0.734 g

Source: After Kotler, Brown, and Subach 1993.
Note: The table lists the average times of the first visit by each species in resource patches, the last visit of the night, and average time of a visit (all in units of hours after sunset). The table also lists the average value of a patch (in units of grams of millet seeds in the patch) and the giving-up density (in units of grams of millet seeds left in the patch).

the night and collected one tray for analysis every 90 minutes. This technique provided snapshots of gerbil activity and patch depletion during the night. The faster forager, G. *pyramidum*, started foraging earlier in the night than G. *a. allenbyi*, but G. *pyramidum* also stopped foraging earlier (table 12.1). More importantly, G. *pyramidum* encountered richer resource patches, on average twice as rich as those G. *a. allenbyi* encountered. In contrast, the more efficient G. *a. allenbyi* extracted 0.25 g more seeds from each patch (table 12.1; Kotler, Brown, and Subach 1993). Thus, this pair of gerbils partitions nightly seed resources temporally. Each species biases its activity toward times of the night and resource densities in which it is the superior competitor.

In the second experiment, the researchers used fenced enclosures to test for interference (Ziv et al. 1993). They created experimental communities that differed in the presence or absence of G. *pyramidum*, and then recorded the intensity and timing of gerbil activity using sand tracking. When both species were present, the gerbil species showed dramatic temporal partitioning (fig. 12.5), with G. *pyramidum* dominating the early hours of the night and G. *a. allenbyi* the hours toward dawn. In the absence of the larger G. *pyramidum*, however, G. *a. allenbyi* expanded its activity to include all hours of the night. Indeed, the level of activity it achieved early in the night without a competitor was higher than previously observed in the presence of G. *pyramidum*. Thus, the small gerbil species compensated for all the "missing" G. *pyramidum* activity caused by the larger species' absence.

The following picture emerges. When the gerbils emerge from their burrows in the evening, they find rich patches of seeds created by afternoon winds, but as the night wears on, their foraging necessarily reduces the quality of these patches. G. *pyramidum* feeds quickly and aggressively outcompetes G. *a. allenbyi* in the rich patches of the early evening. The more energetically efficient G. *a. allenbyi* can extract more from each patch and outcompetes G. *pyramidum* in the depleted patches that occur late at night.

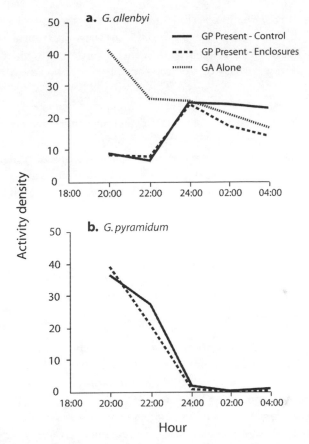

Figure 12.5. Timing of nightly activity for two species of gerbils, (A) *Gerbillus andersoni allenbyi* and (B) *G. pyramidum*, in experimental communities. *G. pyramidum* always forages early in the night, when seed resource patches are rich. When these species co-occur, they display temporal partitioning, with *G. a. allenbyi* foraging later in the night than *G. pyramidum* and thereby experiencing poorer resource patches ("control" communities and "GP present—enclosures" communities). When *G. a. allenbyi* lives without *G. pyramidum* ("GA alone" communities), it expands its time of activity to include the earlier hours of the night. (After Ziv et al. 1993.)

The study described above represents a mechanistic approach to communities, in which mechanisms for the coexistence of competitors are sought in the costs and benefits of adaptive behaviors. Foraging theory can be applied to reveal the mechanisms by which the species of a community coexist. Optimal foragers reveal their preferences, aptitudes, and handicaps through their foraging decisions. When faced with several options, an optimal forager should choose the one that yields the highest marginal value in terms of fitness (i.e., makes the largest contribution to per capita population growth rate). Thus, the abilities and the liabilities of the individuals determine the foods individuals exploit, the times, habitats, and microhabitats they utilize, their vigilance, and so on. The foragers' abilities, in the context of the environment, determine where and when they can forage profitably. Foraging theory allows us to quantify their behaviors, measure their costs and benefits of foraging, and test possible mechanisms of coexistence. This approach makes it possible to identify salient features of the environment and relevant characteristics of the organisms that allow species coexistence just by asking the animals.

Hence, a mechanism of species coexistence has two necessary ingredients: an axis of environmental heterogeneity or niche axis along which the species can segregate, and an evolutionary trade-off such that each species has a part of the axis at which it profits more than any of its competitors (Brown 1989a, 1989b). *G. pyramidum* and *G. a. allenbyi* can coexist because patch quality in their environment varies during the night. This nightly variation forms the necessary "axis of environmental heterogeneity" for our gerbils, but axes of environmental heterogeneity come in many forms: differences in food size, resource density, temperature, cover, or even predator type or density. In addition, our gerbils specialize on different parts of the patch quality axis because they have responded differently to the evolutionary trade-off between foraging speed and efficiency, primarily through body size. More generally, coexistence requires that each organism profit more than its competitors along some part of the niche axis. This usually happens via specialization: a jack-of-all-trades is a master of none (MacArthur 1972).

When both conditions hold, even when one competitor is at its carrying capacity and exploiting all of its profitable opportunities, the other species can still find profitable opportunities. The heterogeneity must be great enough, and the trade-off severe enough, that a competitor at low density can obtain more than the resources it needs for its maintenance and replacement. If so, then the species will coexist because each can increase when rare and invade a community that its competitors dominate. This mutual invasibility criterion provides a sufficient condition for coexistence (Chesson 2000). We can find more precise conditions for coexistence using game theory (e.g., Chase et al. 2001), but simpler mutual invasibility criteria provide valuable assays for empirical testing.

Researchers have identified several potential mechanisms of species coexistence, and future work will probably uncover many more. Table 12.2 lists six important coexistence mechanisms, along with the axis of environmental heterogeneity and the corresponding trade-off that promotes coexistence in each case. The mechanisms range from resource partitioning to habitat selection. Habitat selection can include partitioning of time or space, or partitioning of spatial or temporal variation in resources or hazards.

Table 12.2   Axes of environmental heterogeneity and evolutionary trade-offs that permit niche partitioning for six major mechanisms of species coexistence

Axis	Trade-off
1. Food resource partitioning	Foraging efficiencies on different food types that may vary in encounter rates, handing times, energetic content, nutrients, toxins, gut passage rates, etc. Each species must have a food type on which it profits more than its competitor.
2. Bush/open microhabitat selection	Foraging efficiencies in bush vs. open microhabitats based on differences in energetic cost of foraging, harvest rates of resources, or risk of predation. Each species must have a microhabitat in which it has the lowest giving-up density.
3. Habitat selection in a mosaic	Foraging efficiencies in different habitats based on differences in energetic cost of foraging, harvest rates of resources, or risk of predation. Each species must have a habitat in which it has the lowest giving-up density.
4. Spatial variation in resource abundance	Foraging versus traveling efficiencies.
5. Temporal variation in resource abundance (daily or annual)	Foraging versus maintenance efficiency or foraging efficiency at high versus low resource abundance.
6. Temporal variation in foraging costs (daily or annual)	Foraging costs and efficiencies during different time periods. Each species must have a time period in which it has the highest foraging efficiency.

Source: After Brown, Kotler, and Mitchell 1994.

This mechanistic approach has been applied to other communities. One example involves seed-eating rodents of the Sonoran Desert (Brown 1989b). Here, a kangaroo rat (*Dipodomys merriami*), a pocket mouse (*Perognathus amplus*), an antelope squirrel (*Ammospermophilus harrisii*), and a ground squirrel (*Spermophilus teretecaudus*) all coexist. The kangaroo rat, the pocket mouse, and the ground squirrel coexist via a seasonal rotation of foraging efficiency wherein each species has a time of year during which it is superior to its competitors. Differential susceptibilities of the three species to a seasonally changing array of predators drive the rotation. At the same time, the kangaroo rat coexists with the antelope squirrel via a mechanism involving spatial variation in resource abundance and a trade-off between foraging costs within patches versus the costs of traveling among patches. Effectively, the larger antelope squirrel uses its superior speed to move among rich patches and skim off the "cream," while the smaller kangaroo rat uses its relatively low metabolic costs to forage more efficiently within patches on the remaining "crumbs."

More than one mechanism can operate in a community, and a single species can be involved in more than one mechanism within and across communities. *D. merriami* provides a good example. In the example above, its predator avoidance abilities help it coexist with pocket mice and ground squirrels through a mechanism involving seasonal rotation of foraging efficiencies, but its small body size and low metabolic costs help it coexist with antelope squirrels through a mechanism involving spatial variation in seed densities. In different communities, its predator avoidance abilities again come into play, but this time in promoting bush versus open microhabitat partitioning with still more energetically efficient pocket mice (Kotler 1984). *D. merriami* occurs in communities containing at least 88 different combinations of coexisting species (Brown and Kurzius 1987) that vary in their numbers of species and their characteristics. The mechanisms by which *D. merriami* coexists in all of these situations must vary. As the environmental conditions change from location to location, so too will the axes of heterogeneity and the relevant trade-offs among the species that allow for their coexistence.

We can combine the heterogeneities outlined here to generate still further, unique mechanisms. One such example involves larger, more arboreal red squirrels (*Tamiasciurus hudsonicus*) and smaller, more efficient eastern chipmunks (*Tamias striatus*) in Quebec, Canada. Quebec experiences strong seasonality and offers a range of forest types ranging from coniferous to mixed deciduous forests (Guerra and Vickery 1998). Red squirrels have exclusive access to resources during the winter, when chipmunks hibernate. Measurements of giving-up densities reveal that red squirrels forage more efficiently in spring in coniferous forest, while chipmunks forage more efficiently in all other forest types. Squirrels and chipmunks coexist via a combination of habitat selection in time and in space. Different forest types and different seasons provide the necessary environmental heterogeneity, and differences in body size, torpor strategies, and arboreal abilities provide the necessary trade-offs.

Studying the foraging behaviors of two or more coexisting species often suggests the mechanisms of coexistence underlying the community's biodiversity. Two examples include the interactions of tropical nectar-feeding hummingbirds (Feinsinger and Colwell 1978) and of Darwin's finches (Grant 1986). Hummingbirds may partition flower species according to dispersion and nectar reward. Tropical hummingbird species can be categorized by the length of their bills. Among short-billed hummingbirds, some species have higher wing disc loading than others (wing disc loading is the ratio of body mass to the area swept out by a wing beat and indicates the power needed for hovering). Species with high wing disc loading have short, broad wings that provide greater maneuverability and good interference ability, but high wing disc loading also makes flight more expensive (Feinsinger 1976). Hummingbirds with high wing disc loading use their fighting ability to defend territories with large clumps of moderately rewarding to rich flowers. Hummingbirds with low wing disc loading have longer wings, lower flight costs, and reduced interference ability. They cannot defend territories, but can forage profitably on dispersed or poor flowers. Finally, long-billed hummingbirds with low wing disc loading and large body sizes need very rewarding flowers to forage profitably. These hummingbirds are particularly apt at harvesting nectar from flowers with long corolla tubes (which exclude the short-billed hummingbird species) and with wide dispersions (precluding territorial hummingbird species). Hummingbird species arrange themselves across communities along axes of flower density and corolla length, based on trade-offs of body size and wing size that influence flight costs, flight speed, maneuverability, and interference ability.

Darwin's finches (*Geospiza*) partition seeds according to seed size based on their beak depth (Grant 1986). Birds with larger beaks can open larger and harder seeds than those with smaller beaks. Larger beaks also permit faster handling of larger seeds. Birds with smaller beaks can handle smaller seeds more quickly, but cannot open many large seed species. So, large-beaked finches profit most from the largest seeds, while small-beaked finches can exploit smaller seeds most efficiently. In the field, finches specialize on the seeds they can harvest most efficiently and coexist by resource partitioning according to seed size.

Mechanistic approaches to the study of ecological communities based on foraging theory hold much promise. So far, advocates of this approach have examined only a handful of communities, identifying only a tiny subset of co-existence mechanisms. We look forward to a much larger sample before we can answer even simple questions such as "Do coexistence mechanisms vary more within or between continents?" or "How do mechanisms of species coexistence change along clines of species diversity?" Our ability to answer such questions may help us conserve biodiversity, manage natural and artificial ecosystems, and meet the challenge of global climate change.

12.9   The Evolutionary Ecology of Communities

Two types of three-spined sticklebacks live in Paxton Lake, British Columbia. One form feeds on the lake bottom near the shore on an array of large aquatic invertebrates. The other feeds on small zooplankton in open water, although it feeds near the shore during the nesting season. The two types vary morphologically as well. The bottom feeder has a deep body, a big mouth (for its larger prey), and a small number of gill rakers. The open-water feeder has a slender body, a narrow upturned mouth, and many long gill rakers. Schluter and McPhail (1992) recognize these forms as separate species recently descended from a common ancestor via sympatric speciation. Schluter and McPhail have not formally described the two species, so following their practice, we call the bottom feeder the benthic species and the open-water feeder the limnetic species.

The two species choose habitats adaptively. The limnetic species captures more food per strike and has a higher energetic intake rate than the benthic species in open water. The benthic species captures more food per strike and has a higher energetic intake rate than the limnetic species in the benthic habitat (Schluter 1993). Within species, the benthic species has a higher food capture rate in the benthic habitat than in open water; the limnetic species has approximately equal feeding rates in both habitats. Coadaptations between morphology and behavior contribute to the species-specific performances in the two habitats. In open water, the fish lunge at prey using characteristic "S-start" strikes; in the benthic habitat, they take mouthfuls of sediment. The limnetic species' slender body makes it much better at "S-start" strikes, while the wide mouth of the benthic species allows it to take bigger mouthfuls of sediment. These differences in foraging performance translate into differences in individual growth rates. The benthic species grows about twice as fast as the limnetic species when both species feed in the benthic habitat; the limnetic species grows twice as fast as the benthic species when both feed in open water (Schluter 1995). Interestingly, hybrids have intermediate characteristics and thrive in the laboratory, yet they grow only 73% as fast as either parent species in the wild (Hatfield and Schluter 1999).

In addition to their behaviorally flexible food capture strategies for each habitat, the sticklebacks' morphology exhibits adaptive phenotypic plasticity (Day et al. 1994). When Day et al. fed each species its competitor's diet, it developed morphological features that more closely resembled those of the competitor, especially in the length of the gill rakers and in head depth. The limnetic species showed greater plasticity, consistent with its more opportunistic habitat use, and less skewed habitat-specific feeding rates. Yet, even when fed the competitor's diet, the two species remain morphologically distinct, suggesting that many of the differences between them are fixed and heritable. *Reaction norm* is a term often used to describe the interaction between genes and environment in determining an organism's phenotype. It formalizes the idea of phenotypic plasticity. While both sticklebacks exhibit appropriate and similarly directed phenotypic plasticity, they each exhibit this plasticity according to a distinct and species-specific reaction norm.

Overall, each species' heritable reaction norm, foraging ability, foraging tactics, and diet choice represent a coadapted syndrome in response to habitat variation (benthic versus limnetic), food type, and food availability. The recent divergence of these species, together with the fact that they have always lived together in Paxton Lake, suggests sympatric speciation driven by differences in optimal diet and optimal habitat use. The reaction norm of an animal feeding benthically produces a morphology that enhances aptitude within that habitat at the expense of aptitude in the limnetic habitat. Once a fish possesses this morphology, it is more likely to direct its foraging behavior toward the benthic habitat than a fish of the same species that has moved along the reaction norm toward a more limnetic morphology. Once fish exhibit directed foraging behavior based on their phenotypes, the possibility exists for natural selection to favor an exaggeration of these morphological differences by selecting for a divergence of reaction norms. Eventually, the coadaptation of morphology and feeding strategies produces a community with two species.

The resulting two species become defined by their foraging behaviors and the form of their reaction norms. Empirically, the superior performance of each species on its characteristic diet and in its characteristic habitat, along with the inferior fitness of intermediate types, attests to strong disruptive selection. It appears, then, that the distinct ecological opportunities offered by the two habitats to a phenotypically plastic species led to the speciation of an intermediate species into two daughter species with more extreme reaction norms. Foraging behavior is key to this process. Varied feeding strategies select for phenotypic plasticity, the coadaptation of behavior and morphology selects for divergent reaction norms, and divergent reaction norms define the new species.

An intriguing question in community ecology concerns whether communities evolutionarily take species or make species (Kotler and Brown 1988; Wilson and Richards 2000; Chase et al. 2001). To what extent does a given community represent that selection regime that shaped the characteristics of the species coexisting within it, or to what extent did the species currently coexisting within a community evolve characteristics in response to other circumstances that exist elsewhere in the species' ranges? If a community is primarily a species taker, then invasions from a regional species pool filter through mechanisms of coexistence to assemble communities. If a community is a species maker, then interactions within the community act through natural selection to shape the diversity and characteristics of the species within the community. The Paxton Lake sticklebacks appear to conform to a "species-maker" scenario. The current bird community of the Hawaiian Islands, with its preponderance of introduced exotic species, clearly conforms to a "species-taker" scenario. The formation and composition of species in most communities probably result from the joint action of species-taking processes (in which selective forces occur elsewhere) and species-making processes (in which selection acts in the community on the community). With time, both processes work together to shape ecological communities. Mechanisms of coexistence provide the species, and evolution helps shape morphologies and behaviors of individuals to fit the community context and may further promote speciation. As a result, coexistence often represents an ESS (evolutionarily stable strategy) (Wilson and Richards 2000).

At an ESS, a species' heritable phenotype maximizes fitness given the circumstances. In this case, it is the fitness of the evolutionary strategy that is optimized, and it is optimized over all of the circumstances in which individuals possessing the strategy find themselves. Generally, the scale at which the phenotype of a species represents an ESS is probably larger than the scale over which mechanisms of coexistence operate. It is this disjunction of the evolutionary scale of optimal phenotypes and the scale of ecological contingencies relating to coexistence that gives a central role to flexible feeding behaviors (and reaction norms) in revealing and influencing community organization (whether the community represents a coevolved ESS or not). The next subsection examines the tools used for modeling the coevolution of communities within the context of evolutionarily stable strategies.

Models of Evolution in Communities

Models of evolution in communities show that the coevolution of interacting populations can lead to speciation and place limits on species diversity. Mitchell (2000) modeled a community in which individuals move around a landscape, randomly encountering habitat patches. These patches represent a continuum of habitat properties in which habitat type varies continuously from stressful to benign. Based on the habitat properties of a patch, a forager can choose to exploit the habitat or move on to another patch (this model can be modified into a diet model in which patches are food items instead of habitat patches). The foragers possess an evolutionary strategy (heritable phenotype) that determines their ability to exploit habitats according to stress. Often, ecological models with a habitat continuum permit the coexistence of an unlimited number of species (Abrams 1988; Tilman and Pacala 1994), each one specialized at a point along the continuum. Mitchell's model, in which the species can evolve, results in a discrete set of species at the ESS despite the habitat continuum.

In Mitchell's model, foragers pay travel costs when moving between habitat patches and foraging costs when exploiting a habitat patch. Regardless of evolutionary strategy, all forager species have their lowest costs in the least stressful habitat type. Foraging costs increase with habitat stress. Foragers with a stress-intolerant strategy have very low foraging costs in benign habitats, but their foraging costs increase rapidly with habitat stress. Foragers with a stress-tolerant strategy have relatively high foraging costs in benign habitats (relative to the stress-intolerant strategy), but their foraging costs increase much more slowly with habitat stress. Depending on their evolutionary strategies, foragers will have ranges of habitat stresses at which they are relatively superior to their competitors and make larger foraging profits. For simplicity, Mitchell defined a species as the population of individuals possessing the same value for the evolutionary strategy. Given that a species enjoys an absolute advantage when it is in its best habitat type, should it always select that habitat type?

Travel costs affect the cost of habitat selection. The more selective a forager is, the farther it must travel to reach the next suitable patch, and this increases the cost of habitat selection. So, optimal habitat selection often predicts something less than strict selectivity. The species' stress tolerance strategy will affect its optimal habitat selection behavior. At the same time, an animal's habitat selection behavior and that of others will affect the optimal value of its stress tolerance strategy. As with the hummingbirds and the sticklebacks, we see a coadaptation between optimal feeding behaviors and heritable phenotypes. And the strategies and behaviors of others influence the optimal combinations of behaviors and morphology. We need game theory to analyze the evolution of stress tolerance because the profitability of a patch depends on the condition in which other foragers have left it. Mitchell's model combines this logic by finding the behavioral ESS for habitat selection given the interacting individuals' stress tolerance, while finding ESS values for stress tolerance for individuals that choose habitats optimally.

When foragers experience high travel costs, the ESS results in a single species with a stress tolerance strategy that utilizes a wide range of habitats (fig. 12.6A). The species cannot afford to restrict itself to its best habitat because travel costs greatly reduce the value of being picky. Furthermore, the ESS population of foragers reduces the profitability of the "preferred" habitat. When the species has a non-ESS value for its stress tolerance strategy, two processes produce Darwinian evolution toward the ESS value: natural selection can favor variants that more closely resemble the ESS, and immigrants with strategies closer to the ESS can invade and displace resident values farther from the ESS. As the community slowly approaches evolutionary equilibrium, it can support two species at ecological equilibrium, one on either side of the ESS. Eventually, the community will contain a single strategy, the ESS. If we equate strategies with species, then travel costs set limits on the numbers and characteristics of the species that the community can contain, even before reaching the ESS. In Mitchell's model, the community away from the ESS

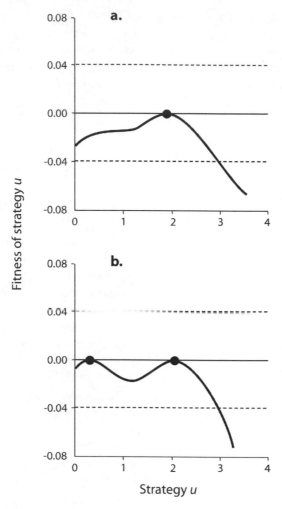

Figure 12.6. The frequency-dependent adaptive landscape plotting fitness for strategy u for a range of values for u in an environment containing a continuum of habitat types that vary in stress. (A) Foragers experience high travel costs. The result is a single ESS. The high travel costs make specialization for the best habitats too costly. (B) Foragers experience lower travel costs, and greater habitat selectivity is now possible. The ESS community now contains a greater number of species. (After Mitchell 2000.)

can produce up to twice as many species as the community at the ESS (see Cohen et al. 1999; Vincent and Brown 2004).

At lower travel costs, the number of species at the ESS grows from 1 to 2, and then from 2 to 3 (fig. 12.6B). So long as there is a finite travel cost, the number of forager species at the ESS will always be finite. All of these model communities represent the interplay between foraging behavior, phenotypic evolution, and community structure.

This model considers only a single mechanism of coexistence: habitat selection. Many more types of heterogeneity influence real-world communities, so we can expect actual ESS limits to be greater. Nonetheless, this model demonstrates that competition among locally adapted organisms can promote a fixed and finite number of species within a community. While regional processes and the size of the regional species pool set the rate at which new species arrive in a community, local ecological and evolutionary processes determine the characteristics and numbers of the coexisting species.

12.10 Summary

Foraging theory gives us unique insights into the coexistence mechanisms and the forces that structure and shape assemblages of species. For a species to coexist with its competitors, members of the species must experience positive fitness at some point; that is, the strategy of an individual in the population must lead to a positive per capita growth rate. The fitness of an individual depends on its foraging profit. Thus, the characteristics that really matter for the community are the characteristics that really matter for foragers. There are many such characteristics, including properties of the forager such as encounter rates and energetic costs; properties of prey such as handling times, energetic value, nutrient content, the bulk of various food items (digestion time), and search time; and properties of predators such as mortality risk. This chapter shows how parameters like these determine the intensity of species interactions, conditions for and mechanisms of species coexistence, and even the characteristics of coevolved species in an ESS community.

12.11 Suggested Readings

Chesson (2000) provides an excellent review and synthesis of the theory of species coexistence. Other important theoretical treatments of species coexistence include the consumer-resource models of Holt et al. (1994) for a pair of competitors that share a common predator and of Vincent et al. (1996) for optimally foraging competitors exploiting resources that may differ in quality or in spatial distribution. Mitchell (2000) shows how coevolution among optimal foragers can lead to communities whose species are shaped and determined by the ESS conditions and in which species interactions set local limits on species diversity. Morris (1988, 1996) provides theoretical explanations and empirical examples of the application of isodars, and Rosenzweig and Abramsky (1997) do likewise for isolegs.

Foraging theory provides the tools for understanding a community in depth, as demonstrated in gerbils. The behavior of individuals of the constituent species (Kotler et al. 1991), the salient features of the environment and the species that promote coexistence (Kotler, Brown, and Subach 1993; Ziv et al. 1993; Brown et al. 1994), and even the resolution of the foraging game played among competitors and their predators (Kotler et al. 2002; Kotler, Brown, and Bouskila 2004; Kotler, Brown et al. 2004) can be understood by applying foraging theory. The article by Rosenzweig and Abramsky noted above, which concerns the gerbils, provides an excellent summary of the application of isolegs, isoclines, and isodars to better understand this community.

Telander

Foraging and the Ecology of Fear

Joel S. Brown and Burt P. Kotler

13.1 Prologue

The reintroduction of wolves in 1995 changed Yellowstone National Park. Riparian habitats have seen a marked increase in willows and aspen. The streams running through these willow thickets meander more. Wetlands have reappeared. Birds and butterflies have increased in the taller and more complex galleries along the riparian stretches, and they breed more successfully than before. Can wolves really have such restorative power?

Wolves reshaped the Yellowstone ecosystem through their effects on elk. Without wolves, elk could forage anywhere with impunity. They browsed their way through every aspen and willow grove and prevented regeneration. The riparian galleries gradually disappeared, which in turn led to the near-extinction of beavers. Without beavers, streams ran faster and eroded more, and the marshy wetlands impounded behind beaver dams and diggings were lost.

Things changed when the wolves came back. Of course, wolves devour elk, but much more importantly, they scare them. Frightened elk spend more time vigilant and less time feeding. They bunch up more, which lowers their feeding efficiency. Most of all, fearful elk avoid dangerous habitats such as thickets. Frightened elk released the willows and the aspen, which formed thickets with tall canopies that created new habitat for birds and brought about a recovery of beavers and their activities. Streams slowed down and returned to their earlier meandering form. Fear can be a powerful ecological force.

13.2   Introduction

Predators kill prey. With this in mind, Schaller (1975), in his classic book *The Serengeti Lion*, documented just how many prey lions kill. Although lions kill large numbers of wildebeests and zebras, the number killed represents only a small fraction of the prey population. Schaller reasonably concluded that lions contribute little to the regulation of their prey's population sizes. Lions kill too few individuals to regulate prey populations.

Another feature of Serengeti grazers is their apparent restraint in grazing their pastures. Compared with domestic grazers such as goats, sheep, and cattle, the Serengeti's natural grazers seem to leave a lot of food uneaten. Perhaps wild grazers are more sophisticated, prudently leaving some vegetation uneaten to generate new fodder for tomorrow. In domestic grazers, centuries of artificial selection for productivity have reduced vigilance and increased consumption (see chap. 6), a luxury that wild grazers cannot afford. However, fear, rather than prudence, probably drives the Serengeti grazers' restraint. Gustafsson et al. (1999) ran domestic and wild-type pigs (*Sus scrofa*) through an identical foraging challenge. The domestic pigs won. The researchers noted that the wild pigs seemed distracted and not fully attentive to their foraging tasks.

Lions and other predators are important to their prey's ecology more for the fear they instill than the mortality they cause directly (Sinclair and Arcese 1995). Death by a predator makes the threat credible, but the threat itself is enough to leave an indelible mark on the ecology of prey and predators.

Fear induces prey to forage more tentatively, in fewer places, in larger groups, or at restricted times. Fear by prey induces behavioral countermeasures on the part of their predators—predators use stealth, boldness, and habitat selection to manage fear in their prey. The prey species' altered feeding patterns cascade down the food chain to affect the prey's resources—the vegetation of the Serengeti would be radically different in the face of fearless grazers. Fear not only strongly affects the foraging behavior of prey (see chap. 9), but also affects the foraging behavior of predators (predator-prey foraging games), the population dynamics of predator and prey (see chap. 11), the food of the prey (via trophic cascades), community interactions among prey and predator species (mechanisms of coexistence; see chap. 12), coadaptations between behaviors and morphologies (coevolution), and the conservation and management of natural areas (see chap. 14). All these topics fall under the ecology of fear. Box 13.1 considers a mechanistic approach to fear, outlining the endocrine correlates of stress and the interplay between stress and starvation avoidance.

BOX 13.1   Stress Hormones and the Predation-Starvation Trade-off

Vladimir V. Pravosudov

Animals usually elevate their levels of glucocorticoid hormones in response to stress. This response, which is considered a homeostatic mechanism (Wingfield et al. 1997; Silverin 1998), is an important adaptation to short-term changes in the social and physical environment that directs behavior toward immediate survival. Long-lasting stress, however, can cause chronically elevated levels of glucocorticoid hormones that produce many deleterious side effects, such as wasting of muscle tissue, suppressed memory and immune function, neuronal death, and reduced neurogenesis in the hippocampus (Sapolsky 1992; Wingfield et al. 1998; McEwen 2000; Gould et al. 2000).

Stress and stress responses are relevant to the study of predation-starvation trade-offs. Experiments increasing predation risk, for example, have recorded effects on energy management (e.g., Witter and Cuthill 1993; Pravosudov and Grubb 1997), but in some cases individual birds reduced their body mass, while in others birds actually increased their mass after exposure to a model predator (e.g., Pravosudov and Grubb 1997, 1998; van der Veen and Sivars 2000). To interpret these results properly, it is important to understand the hormonal mechanisms underlying mass change.

Cockrem and Silverin (2002) recently demonstrated that captive great tits (*Parus major*) responded to the presentation of a stuffed owl with increasing corticosterone levels, whereas free-ranging tits exposed to a stuffed owl did not. These results suggest that studies of captive animals may not accurately reflect the response of free-ranging birds to heightened risk of predation. Animals confined to small laboratory spaces may show longer or stronger stress responses in response to a predator stimulus than the same stimulus would produce in the wild. For example, small rodents exposed to an owl call in a restricted laboratory space immediately showed elevated levels of glucocorticoid hormones (Eilam et al. 1999), but that does not mean that these animals would do so in natural conditions, or that the elevated levels would persist as long.

(Box 13.1 continued)

In fact, much of the research on energy regulation in birds has been carried out in captivity (e.g., Witter and Cuthill 1993; Pravosudov and Grubb 1997). The concentration of plasma corticosterone may increase not only as a result of experimental treatment, but also as a result of stressful conditions in captivity. For example, Swaddle and Biewener (2000) reported that additional exercise in captive starlings (*Sturnus vulgaris*) resulted in reduced flight muscle mass. They concluded that birds strategically reduce muscle mass to reduce flight costs. However, it seems possible that the experimental birds could have perceived the experimentally induced exercise as stressful and responded with elevated corticosterone levels, which are known to result in loss of protein from flight muscles (Wingfield et al. 1998). Sadly, the birds' corticosterone levels were not measured in this study, and the question becomes whether natural increases in flight activity would also result in corticosterone elevation. The possibility that the experimental birds were stressed because of the treatment in captivity means that we must be careful in interpreting the results of such an experiment.

With this caveat in mind, we should nevertheless recognize that short-term responses to predator exposure that increase an individual's chances of escape—for example, by helping to mobilize energy reserves (Wingfield et al. 1998; Silverin 1998)—could be adaptive. Glucocorticoid hormones may also mediate other important antipredator behaviors, such as alarm calls and vigilance (Berkovitch et al. 1995). To understand how stress hormones can mediate antipredator tactics, we need to study the entire chain of events (stimulus → hormones → behavior), and it is especially important to establish experimentally the link between perception of predation risk and glucocorticoid hormones.

The risk of starvation may serve as a stressor, either through hunger effects or through the perception of food unpredictability. Avian energy management tactics such as fat accumulation and food-caching behavior have been studied intensively (e.g., Witter and Cuthill 1993; Pravosudov and Grubb 1997). This work shows that birds accumulate more fat and cache more food when environmental conditions become unpredictable. For birds, higher fat loads increase flight costs and, importantly, reduce maneuverability, thus increasing an individual's vulnerability to predation. Much theoretical and empirical research has studied this trade-off between the risks of starvation and predation (e.g., Lima 1986; McNamara and Houston 1990; Macleod et al. 2005). Unfortunately, the literature on fat

(Box 13.1 continued)

regulation in birds has paid little attention to the mechanisms regulating fattening processes. This is unfortunate, because many factors known to affect birds' fattening decisions also affect birds' physiology. Unpredictable weather and limited food supplies are well known to affect levels of glucocorticoid hormones, which appear to strongly influence birds' behavior (e.g., Wingfield et al. 1998). Furthermore, several studies have demonstrated that elevated corticosterone levels result in increased fat deposits and loss of protein from flight muscles (Wingfield and Silverin 1986; Silverin 1986; Gray et al. 1990). It seems likely that stress responses form a central part of this mechanism, and measures of corticosterone levels will undoubtedly add an important dimension to our understanding of how animals manage their energy reserves.

Studies have documented a variety of effects. Limited and unpredictable food supplies affect levels of glucocorticoid hormones (Marra and Holberton 1998; Kitaysky et al. 1999; Pravosudov et al. 2001; Reneerkens et al. 2002). Reneerkens et al. (2002) suggested that elevated corticosterone levels may induce more exploratory behavior. Moderately elevated levels of glucocorticoids caused by limited and unpredictable food supplies could result in improved spatial memory and cognitive abilities (e.g., Pravosudov et al. 2001; Pfeffer et al. 2002). For example, data presented by Pravosudov and Clayton (2001) and Pravosudov et al. (2001) suggest that corticosterone may be mediating seasonal changes in spatial memory performance in food-caching birds. It has often been suggested that high levels of stress and high levels of stress hormones have a negative effect on memory performance and the hippocampus (McEwen and Sapolsky 1995; McEwen 2000), but in fact not much is known about the effect of moderately elevated levels of glucocorticoid hormones. Diamond et al. (1992) showed that, below a certain threshold, there is a positive correlation between hippocampal neuron firing rate and corticosterone concentration, and a negative correlation above that threshold. These results suggest that moderate elevation of baseline corticosterone may result in improved spatial memory performance. Similarly, Breuner and Wingfield (2000) showed that Gambel's white-crowned sparrows (*Zonotrichia leucophrys gambeli*) increase their activity with moderately increased corticosterone levels, but after the concentration of corticosterone exceeds a certain threshold, activity strongly decreases. In food-caching mountain chickadees (*Poecile gambeli*), individuals with corticosterone implants designed to maintain moderately elevated corticosterone levels over more than a month demonstrated

(Box 13.1 continued)

enhanced spatial memory in addition to caching and consuming more food than placebo-implanted birds (Pravosudov 2003). Thus, it appears that chronic but moderate elevations in baseline levels of glucocorticoid hormones might effect several important changes, such as improved cognitive abilities, increased exploratory, feeding, and food-caching behavior, and maintenance of optimal fat reserves, which all could be important adaptive responses to prevailing foraging conditions rather than "stress."

It also seems that corticosterone may be mediating cognitive tasks beyond spatial memory. For example, greylag goslings (*Anser anser*) that successfully solved a novel foraging task had higher levels of fecal corticosterone than unsuccessful goslings (Pfeffer et al. 2002). The meaning of this intriguing finding can at the moment only be speculated upon, and much work is needed to establish the role of glucocorticoid hormones in memory and cognition in particular, and the mediating role of hormones as a mechanism within the general framework of starvation-predation trade-offs in general.

This chapter considers fear as a cost of foraging, the ecological consequences of animals using time allocation to ameliorate predation risk, the ecological consequences of vigilance behaviors, fear responses and population dynamics, and foraging games between clever predators and fearful prey. Throughout, the chapter combines concepts from foraging theory with concepts from population and community ecology. Its goal is to show how ideas from the study of foraging under predation risk can help us understand predator-prey interactions and the role of predators in ecological communities.

13.3 Fear and the Predation Cost of Foraging

Fear as a noun describes "an unpleasant emotional state characterized by anticipation of pain or great distress and accompanied by heightened autonomic activity; agitated foreboding . . . of some real or specific peril." The definition goes on to describe fear as "reasoned caution" (*Webster's Unabridged Dictionary*, 3rd edition, G. & C. Merriam, 1981). Is fear merely an organism's assessment of risk, or does it involve more? We will argue that fear combines an organism's assessment of (1) danger, (2) other benefits and costs associated with the dangerous activity or situation, and (3) the fitness loss to the organism in the case of injury or death. We define fear as an organism's perceived cost of injury or mortality.

When foraging under predation risk, the organism can and should treat predation risk as a cost of foraging (Brown and Kotler 2004). Combat pay or hazardous duty pay in human occupations reflects an attempt to place a monetary value on risk. Similarly, animals place an energy value on predation risk. Titration experiments with ants (Nonacs and Dill 1990), tits (Todd and Cowie 1990), desert rodents (Kotler and Blaustein 1995), and fish (Abrahams and Dill 1989) all show that a higher harvest rate or food reward can coax an animal into accepting a riskier feeding situation. Several foraging models (Brown 1988, 1992; Houston et al. 1993) have triangulated on the form of this predation cost of foraging. If we define fitness as the product of a survivor's reproductive success, F, and the probability of surviving to enjoy that success, p, then we can write the following equation for fear as a foraging cost:

$$P = \frac{\mu F}{(\partial F/\partial e)},$$
(13.1)

where P is the predation cost of foraging (units of joules per unit time), μ is the forager's estimate of predation risk (units of per unit time), F is survivor's fitness (unitless as a finite growth rate), and $\partial F/\partial e$ (units of per joule) is the marginal fitness value of energy, e. Note that p does not appear in the expression, as it cancels out (see Brown 1988, 1992).

According to equation (13.1), an animal's sense of fear can rise in three ways. First, an animal should be more fearful in a risky (high μ) than a safe situation (low μ), all else being equal. Second, an animal with a lot to lose (high survivor's fitness, F) should be more fearful than one with less to lose (Clark [1994] has referred to this phenomenon as the asset protection principle). Third, an animal that gains less from an additional unit of energy (lower marginal value of energy, $\partial F/\partial e$) should be more fearful than one that has much to gain. In human experience, when a well-intentioned friend warns you against an activity because "it's dangerous," this often reveals the worrier's judgment that the activity offers a "pointless risk": some danger with little benefit.

For the ecology of fear, the predation cost of foraging has two useful properties. First, it reveals more than just predation risk. It integrates other aspects of the forager's condition; namely, its current state or prospects (F) and the contribution of additional energy to those prospects $(\partial F/\partial e)$. Second, it shows how food and safety behave as complementary resources in the sense that safety is valuable only if the organism has something to live for, and having excellent prospects is valuable only if the organism survives to realize this potential. Formally, food and safety are complementary because increasing the energy state of an organism (giving it more food and increasing *e*) increases the marginal rate of substitution of energy for safety (increasing *e* probably increases F and decreases $\partial F/\partial e$).

A species of sparrow, the dark-eyed junco, reveals these aspects of the cost of predation in its foraging behavior. Lima (1988a) fed one group of juncos and withheld food from another before releasing them to feed on a complex of artificial habitats. Consistent with the idea of predation risk as a cost of foraging, the juncos biased their feeding effort toward the safer habitat, which for these small birds lies closer to cover into which they can escape. Consistent with the complementarity of food and safety, the hungry juncos spent more time feeding in dangerous habitats away from cover.

We (Brown, Kotler, and Valone 1994) estimated the size of the predation cost of foraging to desert rodents foraging for seeds. We did this by measuring the giving-up density of free-living rodents in standardized experimental food patches. Using laboratory measurements of the rodents' gain curves, we could convert giving-up densities into quitting harvest rates (joules per minute). Subtracting estimates of the metabolic cost of foraging (adjusted for ambient temperature and activity intensity) from the quitting harvest rate leaves an estimate of the predation cost of foraging. For a kangaroo rat (*Dipodomys merriami*) and ground squirrel (*Spermophilus tereticaudus*) inhabiting a creosote-bush habitat in Arizona's Sonoran Desert, we estimated that predation costs were roughly three times higher than the metabolic costs of foraging. For two gerbil species (*Gerbillus pyramidum* and *G. andersoni*) inhabiting sand dunes in the Negev Desert, similar studies found that the costs of predation were four to five times higher than metabolic costs. While one would like to have many more studies for many more species, these studies support the idea that predation risk represents a considerable cost of foraging.

The cost of predation does not necessarily have to correlate with actual mortality caused by predators (Lank and Ydenberg 2003). The predation a species experiences has already been filtered through the lens of antipredator behaviors. If cautious behavior pays big dividends in safety, then cautious animals may pay a relatively high cost of predation in lost food gains even while experiencing little actual mortality. Brown and Alkon (1990) saw this with the Indian crested porcupine (*Hystrix indica*). Its spines bespeak antipredator morphology, and indeed, the porcupine is virtually impervious to predation by the leopards, wolves, hyenas, and jackals inhabiting its environment in the Negev Desert. However, measures of its foraging behavior showed that the porcupine paid a high predation cost of foraging when active on moonlit nights or in habitats free from perennial shrub cover. How can we reconcile the observation of little mortality due to predators with the observation of

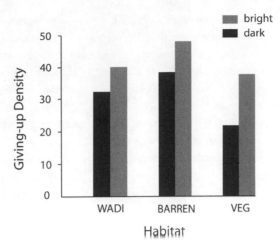

Figure 13.1. The giving-up densities of porcupines (*Hystrix indica*) in experimental food patches set in the Negev Desert, Israel. A high giving-up density suggests a high perceived cost of predation. Food patches began with 50 chickpeas mixed into 8 liters of sifted sand. The porcupine's perceived cost of predation increases with moonlight, and decreases with the amount of perennial shrub cover. The authors observed higher giving-up densities (shown as the mean number of chickpeas left behind in a food patch) on moonlit nights (bright) than on nights with less than a quarter moon (dark). Giving-up densities were highest in a habitat without any perennial shrub cover (BARREN), lowest in a habitat with ca. 12% shrub cover (VEG), and intermediate in the habitat immediately adjacent to the porcupine's burrow (< 5% shrub cover, WADI). (After Brown and Alkon 1990.)

a very high predation cost of foraging? Two factors probably contribute to this pattern: harassment from predators and the need for the porcupine to respond to this harassment. On moonlit nights or in open habitats, predators may easily spot porcupines. Furthermore, it may pay predators to deviate from their path and challenge encountered porcupines—an ill or otherwise incapacitated porcupine may be vulnerable. To deter the unwanted attentions of a predator, a healthy porcupine may be obliged to raise it quills and take up a defensive posture. In this way, predators represent more of a harassment cost than a mortality cost to the porcupines (fig. 13.1).

In Aberderes National Park, Kenya, the black rhinoceroses suffer harassment from spotted hyenas, and many exhibit missing tails from such encounters. However, we know of only one instance in which hyenas killed a black rhinoceros. In this case (reported by a ranger in 1998), a pack of hyenas set upon the rhino when it became mired in wet clay. Before killing the rhino, the hyenas dehorned it. These hyenas had probably never killed a rhino before. However, their experience harassing rhinos, and the rhinos' responses to this harassment, suggest that the hyenas had ample experience with rhinos and their defensive tactics. In response to hyena harassment, rhinos perceive a lower foraging cost of predation in the more open habitats of the forests and glades of Aberderes. In these habitats, they have more room to maneuver. Berger and Cunningham (1994) reported that dehorning of black rhinoceroses in Namibia to discourage poaching led to attacks by hyenas on mothers and their young. The speed of the hyenas' response suggests that the hyenas and rhinos had considerable behavioral experience with each other's tactics. A tension exists between rhinos and large carnivores even though the carnivores almost never kill rhinos. It is unlikely that any organism, regardless of taxon, is free from a foraging cost of predation.

Even top predators experience a foraging cost of predation. They probably have two sources of predation-like costs. First, top carnivores often inflict injury or death on one another in the form of direct interference. The claws and teeth that make predators dangerous to prey also make them dangerous to one another. Examples include dragonfly larvae attacking each other, the susceptibility of venomous snakes to conspecifics' venom, and the posturing and fighting within groups of mammalian carnivores. Great-horned owls may raid the nests of red-tailed hawks, and vice versa. Lions steal the captures of spotted hyenas, and spotted hyenas reciprocate by harassing or killing lone lionesses or their young. The presence of conspecifics or other predator taxa can increase the foraging costs of an individual predator.

Second, prey can injure carnivores. If oblivious to injury or pain, a mountain lion can probably kill a North American porcupine easily. However, a muzzle or paw full of quills may incapacitate and starve a lion. Sweitzer and Berger (1992) found that mountain lions increased their consumption of porcupines during an extreme winter with deep snow. J. Laundré (personal communication) found porcupine quills embedded in several dead mountain lions retrieved during a period of low mule deer abundance. A predator faced with the risk of injury while capturing prey should add a cost of "predation" to its other hunting costs. A predator down on its luck (in a low energy state or with a high marginal value of energy) should be willing to broaden its diet to include higher-risk prey or to take on bolder hunting tactics that simultaneously increase the probabilities of success and injury.

More generally, one can think of the predation costs of foraging as the opportunity costs a forager pays while trying to avoid a catastrophic loss. This catastrophic loss can emerge from the risk of mortality or injury from predators, amensals, prey, competitors, combatants, and even accidents. The giving-up density of raccoons increases with height in a tree (Lic 2001), presumably as a consequence of the greater risk of falling from increasing heights.

The examples developed here show the importance and pervasiveness of the predation costs of foraging. The next step in our analysis considers how animals respond to these costs. Three classes of responses can affect the organism's ecology, the ecology of its predators, and the ecology of its own resources: time allocation, vigilance, and social behaviors. The next two sections explore some of the ecological consequences of time allocation and vigilance (chap. 10 deals with social foraging).

13.4 Ecological Consequences of Time Allocation

Animals should balance the conflicting demands of food and safety (see chap. 9). In terms of time allocation, this balancing can occur in the context of patch use (small-scale habitat heterogeneity in food availability and risk) or habitat selection (large-scale heterogeneity). Within a depletable food patch, a forager should stop foraging when

$$H = C + P + O, \tag{13.2}$$

where H is the quitting harvest rate, C is the metabolic cost of foraging, P is the predation cost of foraging [as given in eq. (13.1)], and O is the missed opportunity of not spending the time at other fitness-enhancing activities (Brown 1988). Each of these terms can have units of energy per unit time, nutrients per unit time, or resource items per unit time, although for any given application of equation (13.2) we must express all four elements of the equation in the same units. Box 13.2 explains how giving-up densities can be used to estimate the costs of predation.

BOX 13.2   Giving-up Densities

Joel S. Brown

When a goose is grazing, it does not eat entire grass plants. A part of each leaf is torn away, and a part is left behind. Nor does a browsing moose eat all the twigs and leaves from each bush. Foragers at depletable patches do not consume all of the contents. We call the amount of food that a forager leaves behind the "giving-up density," or GUD.

    Even humans exhibit GUDs. An "empty" drink can or bottle is not actually empty—there are dregs left that could be had with enough dexterity, patience, and perseverance. The same goes for eating pieces of chicken. Some do indeed eat all—meat, cartilage, marrow, and bone. But generally,

(Box 13.2 continued)

most humans leave some of the chicken uneaten at the end of a meal. This remainder is also a GUD.

What do GUDs tell us about the forager, its environment, and its opportunities and hazards? The marginal value theorem conceptually anticipates GUDs. In most food patches, the forager's harvest rate declines as the food is depleted, and there is a positive relationship between the patch's current prey density and the forager's harvest rate. Since the GUD is simply the current prey density when a forager quits the patch, the GUD provides a surrogate for the forager's quitting harvest rate. The predictions of the marginal value theorem can be recast in terms of GUDs. A forager should have a higher quitting harvest rate (higher GUD) in a rich than in a poor environment; and a forager should have a higher quitting harvest rate (higher GUD) as travel time among patches declines.

Two studies, one with bees (Whitham 1977) and one with tiger beetles (Wilson 1976), empirically anticipated GUDs. Whitham asked why honeybees left dregs of nectar behind in flowers. He suggested that bees may be unable to access all of the flower's nectar, or that it might not be worth the effort. This latter interpretation sees the flower as a depletable food patch, and sees the dregs as a GUD reflecting the costs and benefits of harvesting the flower. Wilson examined the consumption of insect prey by tiger beetles as influenced by the tiger beetles' habitat of origin. Tiger beetles from habitats rich in prey consumed a much smaller proportion of the offered prey than tiger beetles from habitats poor in prey. He suggested that partial prey consumption may be analogous to the use of patches where the tiger beetles' harvest rate declines as the prey is consumed. The GUD of the tiger beetles corresponded to the beetle's habitat quality as predicted.

How thoroughly should a forager use a food patch when there may be predation risk, activity-specific metabolic costs, and numerous alternative activities to consider, or when the patch itself may become depleted as a consequence of the forager's activities? We will start by defining some terms. Let predation risk, Φ (units of per time), be the forager's instantaneous rate of being preyed upon while engaged in some risky activity. Let the reward from foraging, f (items or joules per unit time), be the instantaneous or expected harvest rate of resources while foraging under predation risk. Let a forager have a number of alternative foraging choices that vary in risk, Φ, and reward, f. With depletable food patches, we assume that patch harvest rate, f, declines as resources are harvested. The effect of predation risk on the cost of foraging depends on how risk and resources

(*Box 13.2 continued*)

combine to determine fitness. Let $F(e)$ be survivor's fitness. It gives fitness in the absence of predation (expressed as a finite growth rate). Assume that F increases with net energy gain, e. Let p be the probability of surviving predation over a finite time interval. This probability is influenced by the cumulative exposure of the individual to risky situations. As more time is allocated to risky situations, p declines; as more time is allocated to safer situations, p increases.

Consider four fitness formulations. Each of these formulations shares a time constraint such that the time devoted to all activities must sum to the total time available:

1. Max p subject to $F > k$
2. Max F subject to $p > k$
3. Max $(F - 1) + p$
4. Max pF

The first model considers an organism attempting to maximize the probability of surviving over some time interval with the requirement of maintaining a certain energy state, k. This model can be appropriate for animals surviving through a juvenile or larval stage to adulthood, or for animals that must survive through a nonbreeding season. The second model considers an organism that attempts to maximize its state while maintaining a threshold level of survivorship, k. Given that survivorship is really a component of fitness, rather than a constraint, this model seems less applicable. This safety constraint can provide an approximation for fitness maximization when the modeler wants the objective function to merely be net energy gain. The third model closely fits classic predator-prey models in which fitness is the difference between population growth in the absence of predation and the predation rate. This model applies where there is either a rapid conversion of energy gain into offspring or where there is communal raising of young or full compensation by the surviving partner so that the death of a parent or helper does not jeopardize the current state and investment in offspring. The fourth model, in which an organism's fitness is its survivor's fitness (or net reproductive value in dynamic programming models; see Houston et al. 1993) multiplied by the probability of achieving that fitness, is probably most applicable to food-safety trade-offs. In this case, a forager must survive over some finite time period before realizing its fitness potential.

The optimal patch use strategy (Brown 1992) shows that in all cases, a food patch should be left when the benefits of the reward rate, H, no longer

(Box 13.2 continued)

exceed the sum of metabolic, C, predation, P, and missed opportunity, O, costs of foraging: $H = C + P + O$. In the following equations (one for each fitness formulation), the term on the left-hand side is H, and the terms on the right-hand side are C, P, and O, respectively:

$$\text{Model 1}: f = c + \frac{\boldsymbol{\mu p}}{\boldsymbol{\Phi_F(\partial F/\partial e)}} + \frac{\Phi_t}{\Phi_F(\partial F/\partial e)}$$

$$\text{Model 2}: f = c + \frac{\boldsymbol{\mu p \Phi_P}}{\boldsymbol{\partial F/\partial e}} + \frac{\Phi_t}{\partial F/\partial e}$$

$$\text{Model 3}: f = c + \frac{\boldsymbol{\mu p}}{\boldsymbol{\partial F/\partial e}} + \frac{\Phi_t}{\partial F/\partial e}$$

$$\text{Model 3}: f = c + \frac{\boldsymbol{\mu F}}{\boldsymbol{\partial F/\partial e}} + \frac{\Phi_t}{P(\partial F/\partial e)}$$

In these models, $\partial F/\partial e$ is the marginal value of energy, and Φ_t is the marginal fitness value of time from relaxing the time constraint. In model 1, Φ_F is the marginal survivorship value of relaxing the energetic state constraint. In model 2, Φ_P is the marginal value of relaxing the survivorship constraint.

In all of these models, the cost of predation (shown in boldface in each of the above equations) has units of energy per unit time or resources per unit time. The currency of risk, μ, is converted into the currency of reward, f, by multiplying the predation risk by the marginal rate of substitution, MRS, of energy for safety. The MRS depends on the fitness formulation. For instance, in model 4, the MRS is the ratio of survivor's fitness to the marginal value of energy. Hence, in model 4, the energetic cost of predation is the predation risk multiplied by survivor's fitness divided by the marginal fitness value of energy: $\mu F/(\partial F/\partial e)$ (Houston et al. 1993 derive this cost of predation for dynamic programming models).

The forager's quitting harvest rate upon leaving a patch should be influenced by all of the parameters associated with the costs and benefits of foraging. From the perspective of the predation cost of foraging,

1. GUDs should be higher in a risky than in a safe habitat or scenario.
2. GUDs should be higher for a forager with a higher energy state or survivor's fitness, F.
3. GUDs should be lower for a forager with a higher marginal value of energy, $\partial F/\partial e$.

Measuring natural GUDs poses challenges in terms of accurately quantifying initial and ending resource abundances and identifying the quality

(Box 13.2 continued)

and quantity of the resource as perceived by the foraging animal. Olsson et al. (1999) measured the natural GUDs of lesser spotted woodpeckers in Sweden. Upon a woodpecker leaving a branch, the branch was collected and X-rayed to determine the number of food items removed (empty cavities) and the number of food items remaining (cavities containing a beetle larva). GUDs have generally been measured by making an experimental food patch that includes a container, a substratum (this increases search time and encourages diminishing returns), and food. For seed-eating rodents and birds, a common practice has been to mix 1–5 g of millet seeds into 1–5 liters of sifted sand or dirt. This mix is then poured into a shallow plastic or metal tray. The GUD is measured by sieving the remaining seeds from the sand following foraging and weighing or counting them.

GUDs have been measured for ungulates such as ibex (Kotler et al. 1994) and mule deer (Altendorf et al. 2001) by using wooden boxes filled with plastic chips as a substratum. For such animals, the food can be alfalfa pellets or other animal chow. GUDs have been measured for the Indian crested porcupine (*Hystrix indica*) by burying 20-liter metal cans in the ground and filling them with sand and chickpeas (Brown and Alkon 1990). Mealworms pressed into moist or dry sand have provided useful food patches for measuring the GUDs of European starlings and North American robins (Olsson et al. 2002; Oyugi and Brown 2003). Korb and Linsenmair (2002) developed a food patch for measuring the GUDs of termites (*Macrotermes bellicosus*) in a savanna and forest habitat of Ivory Coast. Morgan (1994), who measured the GUDs of woodpeckers and nuthatches, used PVC pipes drilled with holes as the receptacle, wood chips as the substratum, and mealworms or sunflower seeds as the food.

The next issue in measuring natural GUDs concerns the identity of the forager. In many cases, just a single species will forage from the experimental patches, as in the case of the crested lark (*Galeria cristata*) at a Negev Desert site (Brown et al. 1997). In other cases, several species may use the patches, as in the case of two nocturnal and two diurnal rodent species at a Sonoran site (Brown 1989b). The identity of the species can sometimes be determined by footprints in the substratum or other telltale sign, direct observations, camera traps, or more recently, PIT tags. Sometimes individuals from more than one species may use the same food patch during the course of the day or night. In this case, it may be of interest to know the sequence of visits (Ovadia and Dohna 1998), the last species in the patch

(Box 13.2 continued)

as a measure of foraging efficiency, and the first species in the patch as a measure of priority or interference competition (Ziv et al. 1993).

Work with GUDs has verified most of the relationships described here. For endotherms, colder temperatures can increase thermoregulatory costs. In accord with this expectation, gerbils in the Negev Desert exhibit an inverse relationship between temperature and GUDs (Kotler, Brown, and Mitchell 1993). Experimental manipulations of temperature (exposure of trays to solar radiation versus shade for gray squirrels and American crows in winter; Kilpatrick 2003) resulted in the expected change in GUDs.

The foraging substratum strongly influences the ease of finding food. Gerbils have higher harvest rates on millet harvested from sand than from loess. As expected, gerbils have a lower GUD in trays with sand than trays with loess (Kotler et al. 1999). Food quality should also influence GUDs. Schmidt et al. (1998) soaked sunflower seeds in distilled water, tannic acid, or oxalic acid. Relative to the control food, the GUDs of fox squirrels were a tiny bit higher on tannic acid and substantially higher on oxalic acid.

Olsson et al. (2002) compared, in aviaries, the GUDs of starlings from a good environment and from a poor environment. The starlings from the poor environment had lower GUDs than those from the good environment. Under natural conditions, white-footed mice from higher-quality environments had higher GUDs than those from lower-quality environments (Morris and Davidson 2000). Similarly, lesser spotted woodpeckers with higher-quality territories exhibited higher GUDs (Olsson et al. 1999, 2001). More studies have used GUD titrations to show how decreasing the marginal value of energy increases GUDs. In these experiments, animals in aviaries (e.g., gerbils, Kotler 1997; starlings, Olsson et al. 2002) or free-living animals (fox squirrels, Brown et al. 1992) are given a food augmentation that is assumed to reduce their marginal value of energy. Food augmentation increases GUDs, and this increase is often more pronounced in risky than in safe microhabitats (Brown et al. 1992; Kotler 1997) and in the presence of predators (Kotler 1997).

The information state of a forager may leave a diagnostic "fingerprint" on the forager's GUDs across a variety of food patches that vary only in their initial prey density (Olsson and Holmgren 1998). To diagnose the forager's information state and patch use strategy, the researcher needs to know the distribution of patch qualities and needs to measure GUDs as influenced by the patch's initial prey density. Valone and Brown used the simple notion of over- versus underutilization of rich and poor patches

(Box 13.2 continued)

to determine when desert granivorous birds and rodents conformed to a prescient information state (exact knowledge of the current patch's initial and current prey density), fixed time state (no information on the current patch's actual value), and Bayesian assessment. This initial application of GUDs to information state has been expanded and refined by modeling and empirical work on woodpeckers in Sweden (Olsson and Holmgren 1998; Olsson et al. 1999). With an application to a shorebird in the Netherlands (the red knot), Van Gils et al. (2003) provide a guide to using GUDs, initial prey densities, and giving-up times to determine the patch use strategy and information state of foragers facing uncertainty about the initial prey density of patches.

The largest application of GUDs has been to investigate habitat variation in predation risk. When the same forager has access to similar food patches across the habitats of its home range, those food patches should offer the same metabolic and opportunity costs of foraging. Differences in GUDs will then reflect differences in perceived predation risk. In aviary experiments with direct (owls) and indirect (lights) cues of predation risk, GUDs for desert rodents were consistently higher on nights with owls or lights than on nights without (Brown et al. 1988; Kotler et al. 1991). By far the most frequent result is for microhabitats near cover ("bush") to have lower GUDs and to be perceived as safer than microhabitats away from cover ("open"). Besides the examples discussed above, examples with rodents include Namib desert gerbils (Hughes and Ward 1993), the pygmy rock mouse (Brown et al. 1998), multimammate mouse (Mohr et al. 2003), degu (Yunger et al. 2002), white-footed mouse (Morris and Davidson 2000), common spiny mouse (Mandelik et al. 2003), laboratory rat (Arcis and Desor 2003), deer mouse (Morris 1997), chipmunk (Bowers et al. 1993), fox squirrel (Brown and Morgan 1995), and gray squirrel (Bowers et al. 1993). In birds, bobwhites had lower GUDs in bush than in open microhabitats (Kohlmann and Risenhoover 1996).

Safety in cover is not a rule, however. Refreshing counterexamples in which GUDs are lower in the open than in covered habitats include kangaroo rats faced with predation risk from rattlesnakes (Bouskila 1995), crested larks on sand dunes in the Negev Desert (Brown et al. 1997), and mule deer in southern Idaho (Altendorf et al. 2001). Rattlesnakes lie in ambush under shrubs, presumably making the bush microhabitat more dangerous than the open. Foraging under and near shrubs may handicap the crested lark, whose escape tactics include jumping into the air and

(Box 13.2 continued)

taking flight. Mule deer experience predation risk from mountain lions that ambush them either in forest patches (Douglas fir) or along forest-open (sagebrush) habitat boundaries.

GUDs can reveal both within- and between-habitat heterogeneities in predation risk. In general, animals in higher-risk habitats should show even sharper responses to microhabitat or temporal variation in risk (Lima and Bednekoff 1999b; Brown 2000). For fungus-rearing termites, higher GUDs in the gallery forest suggested that it is the higher-reward, higher-risk habitat relative to the savanna habitat. When the researchers simulated predation events near food patches, GUDs increased in the savanna habitat while, as an extreme response, foraging ceased in the forest (Korb and Linsenmair 2002).

Illumination makes owls more lethal predators on rodents (Kotler et al. 1988; Longland and Price 1991), and many nocturnal rodents use illumination as an indirect cue of increased predation risk (Brown et al. 1988; Kotler et al. 1991; Vasquez 1994). GUDs increased with moonlight in the Indian crested porcupine (Brown and Alkon 1990) and the Namib desert gerbil, *Gerbillurus tytonis* (Hughes et al. 1995). Other studies have found very small (South American desert rodents; Yunger et al. 2002) or more complex relationships between patch use and moonlight (Bouskila 1995; Mandelik et al. 2003). GUDs have also been used to examine the effects of predator odors on small mammal foraging behavior (Pusenius and Ostfeld 2002; Thorson et al. 1998; Herman and Valone 2000).

In conjunction with patch use theory, GUDs become a concept that can be used to estimate foraging costs, measure predation risk, and link individual behaviors with population- and community-level consequences (see chap. 12). In behavioral studies, GUDs complement other measures of feeding behaviors such as patch residence times, giving-up times, and measures of vigilance behavior. In population and community studies, GUDs complement measures of population sizes and habitat distributions. In conservation biology, GUDs can provide a behavioral indicator of habitat suitability and population status. The opportunity and challenge of using GUDs is to make appropriate measurements and appropriate interpretations.

Under many circumstances, we can rearrange equation (13.2) to generate the μ/f rule of Gilliam and Fraser (1987), where μ is the instantaneous risk of predation and $f = H - C$ is the net feeding rate (Brown 1992). According to the μ/f rule, a forager should direct its foraging to the patch or habitat with the lowest ratio of risk to feeding rate, or in a depletable environment, the forager should leave each patch when this ratio rises to a threshold level. Regardless of whether one uses giving-up densities or a μ/f rule to express patch departure rules, the threshold level of patch acceptability should rise with predation risk and with the state of the forager, but decline as the marginal value of food to the forager increases.

The Landscape of Fear

If predation risk varies in space and among food patches, then the forager should adjust its giving-up density to the variation and particulars of the food patches. Foragers should extract more food from patches (have a lower giving-up density) in safe areas and should extract less (have a higher giving-up density) in risky areas. Spatial variation in predation risk produces a landscape of fear (*sensu* Laundré et al. 2001). The landscape of fear describes how the animal's foraging cost of predation varies in space. This term refers to a spatially explicit landscape in which position with respect to refuges and ambush sites, escape substrata, sight lines, and possibly other landscape properties influences the foraging cost of predation.

Van de Merwe (2004) used giving-up densities in experimental food patches to measure the landscape of fear in the Cape ground squirrel (*Xerus inauris*). Van de Merwe measured landscapes using an 8×8 grid of food patches spread throughout an area of 1 ha. By converting the observed giving-up densities into quitting harvest rates on grain, Van de Merwe specified lines of equal foraging costs (in units of J/min) on a map of the landscape. Although the real landscape seemed flat and homogeneous, Van de Merwe's calculated landscape of fear showed striking peaks and valleys, with some areas well below 500 J/min, but other areas with fear-induced foraging costs above 6,000 J/min. Obstructions created by shrubs raised foraging costs, while proximity to burrows lowered costs.

Over time, variation in the availability and composition of resources should come to reflect the landscape of fear. As a general rule, we expect a positive relationship between foraging opportunities and a forager's predation risk. Consequently, foragers should find higher standing crops of resources in riskier places. A forager's response to its landscape of fear may alter the species composition of its prey. For instance, the plant species in a community may experience a trade-off between competitive ability and resistance to herbivory. Thus, we would expect to find strongly competitive plant species dominating areas of high predation risk and herbivore-resistant species dominating areas of low predation risk to the herbivore. The herbivore's landscape of fear should influence the herbivore's use of space, spatial heterogeneity in the plant community, and the predator's likelihood of capturing the herbivore.

For the prey, a positive relationship between areas of high food supply and predation risk influences both its energy state and its sources of mortality. A forager in a lower energy state has a lower predation cost of foraging than one in a higher energy state. According to the asset protection principle, a forager that is down and out should forage in riskier food patches and reduce each patch to a lower giving-up density. Even in the same environment, an individual in a lower energy state perceives a flatter landscape of fear than one in a higher energy state. Like Lima's (1988a) juncos, a forager in a lower than average energy state should adopt a riskier and more profitable patch use strategy than one in a higher than average energy state. Consequently, individuals in a lower energy state can and should accrue resources more rapidly than individuals in a higher energy state. By the end of the day, all of Lima's juncos may have converged on the same energy state. In a highly varied landscape of fear, individuals in poor body condition can feed in riskier but more rewarding locations—effectively converting safety into body condition—while those in good body condition can feed in safer, less rewarding locations—effectively converting body condition into safety.

The landscape of fear should also influence patterns of mortality. Desert granivorous rodents rarely appear to be in poor body condition, and they do not seem to die from starvation. Yet food addition experiments verify that food limits their population sizes (see Brown and Ernest 2002). If the predation cost of foraging exceeds the metabolic costs of foraging, that means that these rodents usually extract much less from food patches than they could. If starvation threatens, a desert rodent can obtain food quickly by exploiting riskier patches. As an animal's energy reserves decline, starvation becomes certain, whereas predation risk always has a probabilistic element. Better to play Russian roulette with the predators than to starve. As the energy state of the animal declines toward zero, the cost of predation also declines toward zero (as $F \to 0$, $P \to 0$). Hence, a starving animal should always be willing to forage in a food patch that covers its metabolic costs of foraging. If the landscape of fear varies dramatically from one location to the next, most foragers should succumb to predation rather than to starvation.

We can use foraging theory to predict the likelihood of mortality sources to the forager when food and safety vary temporally. Increasing predation risk or increasing food availability can actually cause a shift in mortality away from predation and toward starvation (McNamara and Houston 1987b, 1990). This can happen because a forager with higher expectations of food may be willing to take more chances with its energy state, and an animal that takes such a gamble experiences a greater risk of starvation. Similarly, under higher predation risk, a forager may take more chances with its prospects for food in exchange for greatly reducing its exposure to predation risk. Because both food and safety act as partially substitutable resources, actual causes of death may reveal little about the magnitudes of predation and food supply as limiting factors.

The landscape of fear can also provide mechanisms of coexistence for both the prey and the predators. As discussed in chapter 12, trade-offs among forager species in energetic foraging efficiency versus susceptibility to predation can provide a mechanism of coexistence. The foraging specialist may have the lower giving-up density in safe food patches, whereas the antipredator specialist may have the lower giving-up density in risky food patches.

Predator Facilitation and the Landscape of Fear

Charnov et al. (1976) recognized that two predators, seemingly competing for the same prey, can actually help each other by promoting fear in their shared prey. Consider the case of desert rodents responding to owl and snake predation. In deserts, islands of shrubs sit in seas of open space (Brown 1989b; Bouskila 1995; Kotler et al. 1992). Owls capture rodents more effectively in the open (Kotler et al. 1988; Longland and Price 1991). In response to owls, rodents bias their foraging toward the shrub microhabitat. Snakes can exploit this fear response by ambushing rodents under shrubs. In response to snakes, rodents bias their foraging toward the open (Kotler, Brown, Slotow et al. 1993). As an indirect effect, owls kill gerbils that would otherwise go to feed snakes, and vice versa. As a behavioral indirect effect, owls make it easier for snakes to kill gerbils, and vice versa. Throughout deserts, the differing fear responses of desert rodents to owls and snakes may promote these two predators' coexistence.

Time Allocation and Trophic Cascades

The predation cost of foraging produces a behavioral analogue to trophic cascades within exploitation ecosystems. In a standard trophic cascade, predators kill prey that consume resources (Hairston et al. 1960; Oksanen et al. 1981; Oksanen 1990). The presence of the predator depresses the abundance of the prey, and hence increases the abundance of the resources. This represents a positive indirect effect of predators on a prey's resource. Because of its influences on patch use and the predation cost of foraging, the predator does not even need to kill the prey to benefit the prey's resources. The mere threat of predation causes the prey to harvest fewer resources from patches. The prey will leave risky food patches at higher giving-up densities than safe patches. In this way, African lions may, by the mere threat of predation, discourage zebras and wildebeests from overgrazing.

The effect of predators on a resource's population growth rate or population size via the fear responses of the prey has been given several names: a higher-order interaction ("lions discourage zebras from hurting grasses"), behavioral indirect effect, or trait-mediated effect (Werner 1992; Wootton 1993). These terms emphasize different aspects of the problem. The phrase "higher-order interaction" recognizes that increasing the abundance of lions reduces the magnitude of the interaction coefficient between zebras and grass. (The interaction coefficient gives the effect of changing the population size of zebras on the population growth rate of grass.) The phrase "behavioral indirect effect" recognizes that changes in behavior can affect populations in the same way that products of interaction coefficients produce indirect effects. Lions discouraging zebras from hurting grasses and lions killing zebras that hurt grasses can have similar consequences for the resource's population size and population dynamics.

In fact, the fear responses that predators cause may affect prey populations more than the direct mortality effect of predators on their prey. Schmitz, Beckerman, and O'Brien (1997) created enclosures in which spiders threatened grasshoppers that fed on vegetation. In one treatment, the spiders could frighten and kill grasshoppers. In the other, glued mouthparts allowed the spiders to frighten, but not actually harm, the grasshoppers. The experiments measured the dynamic changes in the grasshopper populations. As predicted based on fear and direct mortality, the population sizes of grasshoppers were lower in both treatments than in treatment with predators absent. Interestingly, grasshopper population sizes fell about the same amount in the fear only and the fear + predation treatments. In both treatments, the grasshoppers greatly restricted their use of space within the enclosures, thus reducing the resource base from which they fed (see Beckerman et al. 1997; Schmitz and Suttle 2001).

Many wonderful examples show how changes in fear can change the abundance of the prey's food, even if the prey's density stays the same. We will consider just two of them. In montane meadows, Huntly (1987) observed the effects of herbivorous pika (*Ochotona princeps*) on surrounding vegetation. By harvesting food near their rocky refuges, pika created a gradient in food quality and abundance: low near refuges, high far from refuges. Huntly created rock piles for pika in the meadows away from existing refuges. The pika immediately began using these temporary refuges. Soon, the vegetation near the new refuges began to resemble the vegetation near the original refuges.

These results suggest that the pika's cost of predation and giving-up densities increased with distance from a refuge.

Abramsky, Shachak, et al. (1992) also created rock pile refuges, but in a different context. In the rocky habitats of the Negev Desert, Israel, large numbers of snails subsist on the soil alga that grows from frequent dewfall. Most dawns provide a dewy period in which algae and snails have a brief window of activity. During the remainder of each day and night, the snails roost conspicuously on rocks or shrubs. Mightn't they provide easy pickings? Indeed, spiny mice (*Acomys cahirinus*) do consume snails in these habitats, but seem to have little effect on the snail population. Abramsky et al. created grids of rock piles that greatly reduced the mean distance to refuge for a foraging spiny mouse. Within days, the snails on these grids disappeared as their chewed and broken shells appeared at the edges of the mice's new refuges. As marginally nutritious foods, snails reside safely below the mice's threshold of acceptability because a snail isn't usually worth the predation risk.

13.5   Ecological Consequences of Vigilance

Foragers can use vigilance to reduce predation risk while they continue to feed. When a forager shifts its attention away from foraging to detect predators, its feeding rate within a food patch will decline. We will use the variable u to represent vigilance and consider how the trade-off between feeding rate and safety shapes the optimal vigilance strategy.

Many models consider vigilance (e.g., McNamara and Houston 1992; Lima 1988b, 1995a), but most of them focus on the relationship between vigilance and group size. We will develop a simple model that shows how ecological variables can influence the optimal level of vigilance. Using this model, we can explore the ecological consequences of vigilance.

We will let the instantaneous predation rate, μ, be given by the following expression:

$$\mu = \frac{m}{(k + bu)}, \tag{13.3}$$

where m is the encounter rate of prey with predators, k is the inverse of predator lethality in the absence of vigilance, and b is the benefit of vigilance. We will assume that vigilance can vary from $u = 0$ to $u = 1$, and hence predation risk can vary from m/k to $m/(k + b)$. Equation (13.3) expresses the same risk of predation as equation (13.1), except that we have now broken predation risk into four components, one of which is vigilance. It can be instructive to substitute equation (13.3) into equation (13.1) for μ and see how predator density, lethality, and vigilance influence the cost of predation, P:

$$P = \frac{mF}{[(\partial F/\partial e)(k + bu)]}.$$

Under the assumption that equation (13.3) describes the relationship between vigilance and predation risk, a forager maximizes fitness, pF, by adopting a vigilance level, u^*, of (Brown 1999)

$$u^* = \left\{ \frac{mF}{[bf_{\max}(\partial F/\partial e)]} \right\}^{1/2} - \frac{k}{b}. \tag{13.4}$$

The optimal level of vigilance behaves (mostly) as one would expect. Vigilance increases with the prey's encounter rate with predators, its survivor's fitness (yet another form of the asset protection principle; Clark 1994), and predator lethality. Vigilance declines with net feeding rate and with the marginal value of energy. Optimal vigilance exhibits a hump-shaped pattern when plotted against the value of vigilance. If vigilance reduces predation risk effectively, the forager needs very little vigilance. If vigilance has little effect on predation risk, then there is no point in being vigilant.

In this formulation, vigilance reduces feeding rate according to $f(u) = (1 - u)f_{\max}$, specifying the cost of vigilance in units of reduced feeding rate. Here $f_{\max}$ gives the forager's feeding rate in the absence of vigilance, so that $f_{\max} = f(0)$. The rate at which feeding rates decline with vigilance sets the exchange rate between food and vigilance (cf. Gilliam and Fraser's [1987] tenacity index).

Vigilance and Trophic Cascades

Vigilance, like the cost of predation, sets off a behavioral cascade that influences both the forager's prey and its predator (Kotler and Holt 1989). In a typical trophic cascade, a predator inflicts mortality on a forager species, and the reduced forager population inflicts less mortality on its prey. Hence, the presence of a third trophic level (the predator) raises the standing crop of the first trophic level (the forager's prey; Hairston et al. 1960). At the extreme, trophic cascades can lead to the paradox of enrichment (Rosenzweig 1971). Imagine a system with three trophic levels characterized by exploitative competition only. Increasing the productivity of the plants (via precipitation, nitrogen, temperature, etc.) will paradoxically cause no increase in the number of herbivores, because the predators increase in numbers so as to just

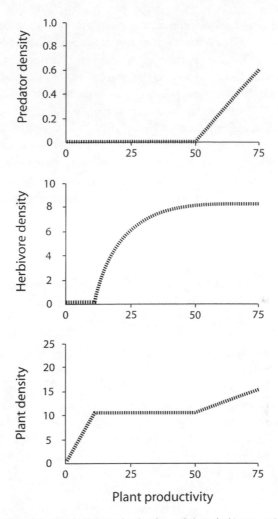

Figure 13.2. The effect of plant productivity on the abundance of plants, herbivores, and predators. The model assumes that herbivores and predators compete only through exploitative competition. Below a threshold productivity level, the abundance of plants cannot support any herbivores. As productivity increases, so does the plant abundance. Above a threshold level of productivity, the system can support herbivores. In this region of productivity, the abundance of plants remains constant with productivity, while the abundance of herbivores increases with productivity. Above a second threshold in productivity, enlarged herbivore populations can now support a predator population. Above this productivity threshold, the abundances of both plants and predators increase with productivity, while the abundance of herbivores remains constant with productivity (see Oksanen and Oksanen 1999).

match the increased productivity of the herbivores. The extra productivity goes straight up the food chain, through the prey and to the predators. The increased productivity of the plants causes an increase in plant biomass, no change in herbivore abundance, and an increase in the number of predators (fig. 13.2; see Oksanen and Oksanen 1999).

A similar phenomenon happens when predators frighten prey into increased vigilance. The predator's presence reduces the herbivore's feeding rate on plants. With fear in the system, we need to distinguish between mortality-driven (N-driven) and fear-driven (μ-driven) population interactions. Obviously these are endpoints of a continuum, but the distinction is useful. Holt (1977) anticipates this distinction in his "r/a" measure of apparent competition ability. The term r is the prey's intrinsic growth rate; it measures how fast the prey population can grow in the face of predation (N-driven component). The term a is the predator's rate of encountering the prey; it measures the predator's ability to catch the prey (μ-driven component). A prey species with a higher r/a has an advantage over other prey species in terms of persisting in the face of a common predator.

In an N-driven system, predators have little effect on the behavior of prey. The predator influences the dynamics and abundance of its prey through direct mortality. Classic predator-prey models (Rosenzweig and MacArthur 1963) fall into this category. Current interpretations of the lynx-hare cycle and weasel-vole cycle fall into the category of N-driven predator-prey systems. Despite some behavioral responses, the populations of hares (Krebs et al. 1995) and voles (Hanski et al. 2001) rise and fall with the tide of lynx and weasels, respectively, as these highly effective predators inflict high and often unsustainable mortality on their prey.

In a μ-driven system, the predators do not appear to control their prey's population through mortality. A casual examination may even suggest that the predators have little or no effect on prey mortality. However, the predators may strongly influence prey behavior, population dynamics, and population sizes by inducing increased vigilance in the prey. In both N- and μ-driven predator-prey interactions, the predator reduces the prey's per capita population growth. In N-driven systems, this occurs via direct mortality (the prey feed the predator). In μ-driven systems, this occurs more subtly via reduced prey fecundity (forgone opportunities to convert resources into offspring) or indirect increases in prey mortality from other sources (increased likelihood of starvation, exposure, or death by other predators and pathogens).

Zebras and lions on the Serengeti and the Indian crested porcupine and its suite of Negev "predators" probably represent μ-driven predator-prey interactions. Fox squirrels (*Sciurus niger*) and gray squirrels (*S. carolinensis*) may represent a μ-driven predator-prey interaction in which fear underlies the mechanism of coexistence. In the midwestern United States, fox squirrels occupy wood margins and gray squirrels occupy the interiors of many of the same woodlots and forests (Brown and Yeager 1945; Nixon et al. 1968; Brown and Batzli 1984). It seems that gray squirrels are better at interference and exploitative competition, but are more sensitive to predators, than fox squirrels (Stapanian and Smith 1984; Lanham 1999; Steele and Wiegl 1992). This trade-off forms a mechanism of coexistence in which the fox squirrels occupy the riskier habitats and gray squirrels the safer habitats.

Two observations suggest that the coexistence of these squirrels is a μ-rather than an N-driven process. First, it is unlikely that the squirrels actually provide the prey needed to support the predators that promote the squirrel species' coexistence. For instance, in Illinois, abundant populations of voles, white-footed mice, chipmunks, and cottontail rabbits support the hawks, owls, foxes, and coyotes that generate the riskier and safer habitats. Hence, the level of predation risk results from apparent competition (Holt 1977; Holt and Lawton 1994) in which voles and rabbits indirectly interact with the squirrels via their common predator. Second, this diverse and sometimes abundant population of predators kills very few squirrels. Red-tailed hawks, for example, have difficulty catching fox squirrels and gray squirrels (Temple 1987), and the predators' diets contain relatively few squirrels. In a bizarre twist of community ecology, predators that the squirrels do not support and predators that mostly frighten rather than kill the squirrels dictate many aspects of the squirrels' foraging behavior (Brown et al. 1992; Bowers et al. 1993; Brown and Morgan 1995) and promote their coexistence.

Vigilance and Predator Facilitation

Vigilance can produce predator facilitation and promote the coexistence of predators (Sih et al. 1998). Imagine a forager facing two predators in which u_1 provides effective vigilance against predator 1 and u_2 provides effective vigilance against predator 2. We can extend the vigilance model to recognize two sources of predation risk:

$$\mu_i = \frac{m_i}{(k_i + b_i + b_i u_i)}, \qquad (13.5)$$

where $i = 1, 2$.

The forager must now choose its optimal level of vigilance in response to its encounter rates, m_i, with each of the predators. The optimal level of vigilance for each predator still satisfies equation (13.4) for u^*. But now the presence of the first predator not only influences u_1^*, it also influences u_2^*. Increasing the abundance of predator 1 causes an increase in m_1 that increases u_1. This increase in u_1 lowers the average feeding rate of the forager, which decreases its state, F, and increases its marginal value of energy, $\partial F/\partial e$ (if there are diminishing returns to survivor's fitness from net energy gain). Either fitness effect will reduce u_2 and make the forager more vulnerable to predator 2. As in the case of owls, snakes, and gerbils, the presence of predator 1 can make it easier for predator 2 to catch the prey. Predator-specific vigilance strategies promote coexistence among predator species if paying attention to one sort of predator causes the forager to be less attentive to another predator species (Sih 1998; Sih et al. 1998).

13.6   Fear and Population Dynamics

Foragers can pay the cost of predation either by directly feeding their predators (in N-driven systems) or by changing their behavior and thereby reducing their fecundity (in μ-driven systems). Here we examine how one can incorporate fear into predator-prey dynamics and how μ-driven systems determine the shape of predator and prey isoclines (Holt 1983; Abrams 1982, 1984; Holt et al. 1994; Holt and Lawton 1994).

We start with a typical model with three trophic levels:

$$\frac{dR}{dt} = rR(K - R)/K - f_N(R)N,$$

$$\frac{dN}{dt} = N(\beta f(R) - d_N) - f_P(N)P,$$

$$\frac{dP}{dt} = P(\gamma f_P(N) - d_P),$$

where r is the plant's intrinsic growth rate, K is the plant's carrying capacity, $f_N(R)$ is the functional response of herbivores on plants, β is the conversion rate of plants consumed by herbivores into new herbivores, d_N is the herbivore's density-independent mortality rate, $f_P(N)$ is the predator's functional response on herbivores, γ is the conversion rate of herbivores consumed by predators into new predators, and d_P is the predator's density-independent mortality rate. The variables R, N, and P give the population sizes of plants, herbivores, and predators, respectively.

Type I and Type II Functional Responses with No Fear

The simplest isoclines emerge if both herbivore and predator have type I (linear) functional responses ($f_N(R) = a_N R$ and $f_P(N) = a_P N$). In this case, the isoclines are all linear planes (fig. 13.3). Doubling the number of plants doubles the herbivore's harvest rate, and doubling the number of herbivores doubles the predator's harvest rate. This model produces the trophic cascades seen in models of exploitation ecosystems exhibiting the paradox of enrichment (Oksanen and Oksanen 1999).

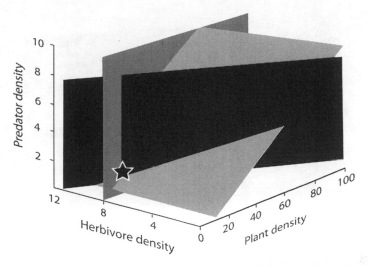

Figure 13.3. The isoclines of plants, herbivores, and predators in a system where herbivores and predators have type I functional responses. All isoclines are planes. The herbivore's isocline (in the foreground) rises along the plant and predator axes—increasing the standing crop of plants increases the number of predators that the herbivores can support and still maintain zero population growth. The plant's isocline declines linearly along the herbivore and plant axes. As the number of herbivores increases, the system can sustain a smaller standing crop of plants while maintaining zero population growth. The predator's isocline is a plane that is independent of predator and plant abundances. It emerges from a subsistence abundance of herbivores. Above a threshold abundance of herbivores, the predators have a positive population growth rate, while below this threshold they experience a negative population growth rate. The star indicates the equilibrium point ($R^* = 20$, $N^* = 8$, $P^* = 1$). The following parameter values were used: $r = 1$, $K = 100$, $a_N = a_P = d_N = \beta = \gamma = 0.1$, $d_P = 0.08$.

Nonlinear isoclines introduce considerable complexity. Isoclines become nonlinear when herbivores must spend time handling the plants they consume (h_N) and predators must spend time handling the herbivores they consume (h_P). The handling time that predators must devote to consuming a captured prey creates a type II (decelerating) functional response. A type II functional response offers the prey safety in numbers and can produce a hump-shaped prey isocline. Figure 13.4 shows such isoclines (surfaces) when both herbivore and predator have type II functional responses to their respective foods. The prey achieve safety in numbers because predators cannot attack a new victim while they are handling a current victim. This safety in numbers weakens the stability of equilibrium points, tends to make the predator-prey dynamics less stable, and increases the likelihood of an unstable equilibrium with nonequilibrial dynamics (as in lynx-hare cycles and weasel-vole cycles). But, even with its nonlinear isoclines and its more complex suite of dynamic outcomes, the predator-prey interaction is still completely N-driven. Furthermore, the system will continue to exhibit classic trophic cascades (assuming equilibrial population dynamics) and the paradox of enrichment (Rosenzweig 1971).

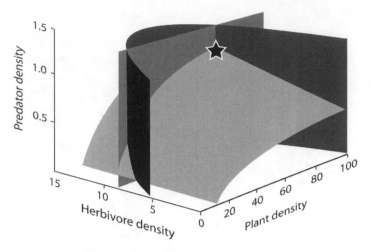

Figure 13.4. The isoclines of plants, herbivores, and predators in a system where the herbivores and predators have a type II functional response. The predator isocline remains the same as in figure 13.3 (a vertical plane that is independent of plant abundance or predator abundance). The herbivore isocline rises at a decelerating rate along the plant, herbivore, and predator axes. With either more plants or more herbivores (safety in numbers), the herbivores can support more predators. The plant isocline takes on a hump shape in the plane defined by the herbivore and plant axes. At low plant abundances, safety in numbers from herbivores dominates the slope (rising portion), whereas at high plant abundances, competition among plants dominates (negatively sloped portion). The star indicates the equilibrium point: $R^* = 83.2$, $N^* = 8$, $P^* = 0.748$. The following parameter values were used: $r = 1$, $K = 100$, $a_N = a_P = 0.166$, $h_N = h_P = 0.5$, $d_N = \beta = \gamma = 0.1$, $d_P = 0.08$.

Increasing plant density will simply result in more predators with no change in the equilibrium abundance of herbivores.

Introducing Prey Fear Responses

What happens to these systems if the herbivores respond to predators with vigilance? We replace $f_N(R)$ with $(1 - u^*)\, f_N(R)$, where u^* is the herbivore's optimal level of vigilance; this incorporates the idea that vigilance reduces prey capture. We replace $f_P(N)$ with $m'N/(k + bu^*)$, where m' gives an individual predator's encounter rate with its prey to incorporate the idea that the prey's vigilance reduces the predator's effectiveness. From the prey's perspective, $m = m'P$. These substitutions give the model a μ-driven component. The herbivores' vigilance reduces their feeding rate, their fecundity, and their mortality due to predators. The more effectively vigilance reduces predation risk, the more the system will behave as μ- rather than N-driven. Figure 13.5 shows the isoclines for such a system.

The herbivore's vigilance strikingly changes the plant and predator isoclines. In the N-driven model, plant isoclines were independent of predator density. Now, an increase in predators increases the number of herbivores that a given plant population can tolerate without declining. This pattern indicates a behavioral trophic cascade in which predators make the herbivores less effective at killing plants. The herbivores strongly influence the plant isocline. Because of vigilance, the herbivores, at low density, have a decreasing effect on plant fitness as predators increase. Plants can tolerate extremely high herbivore abundance so long as there are sufficient predators. The hump of the plant isocline shifts to smaller numbers of herbivores and disappears completely at high predator densities. Herbivores can magnify the negative effect of plants on other plants. Increasing the number of plants increases inter-plant competition and reduces herbivore vigilance, and this reduction in vigilance makes herbivores more effective harvesters. These combined effects warp the contours of the plant isocline into convex folds along the predator-plant axis and along the predator-herbivore axis (fig. 13.5A).

Vigilance affects the herbivore isocline little, except that it increases more steeply along the predator and plant axes. This steepness occurs because vigilant herbivores can manage higher predator numbers as the herbivores exploit increases in plant abundance (fig. 13.5B).

With vigilance, the predator isocline increases with predator density and comes to depend on plant density—in contrast to the N-driven model, in which plant density did not affect the predator isocline. Herbivore vigilance means that predators adversely affect one another because the additional predators make the herbivores more vigilant and less catchable. Therefore, as the number of predators increases, they require a higher standing crop of herbivores to sustain themselves. In addition, the predator isocline increases with plant density. Increasing the abundance of plants makes the herbivores less vigilant and easier to catch, and hence the predators require a lower subsistence number of herbivores. Because of these behavioral effects of vigilance, the predators have a negative direct effect on themselves (which tends to be stabilizing), and the plants have a positive direct effect on the predators, just as the predators have a positive direct effect on the plants (fig. 13.5C).

To an ecologist, a three-trophic-level system with herbivore vigilance would appear to exhibit both "top-down" and "bottom-up" regulation. Increasing plant productivity causes a large increase in plants, a large increase in the numbers of herbivores, and proportionately smaller increases in the number of predators. The predators face a larger herbivore population composed of less catchable individuals. On the other hand, the seemingly less important predator population (recall the lions and zebras) exerts considerable top-down control via the herbivore's feeding rate and vigilance. Remove the predators and the herbivores will overgraze, not because their population size increases (N-driven), but because each less vigilant herbivore now feeds more (μ-driven).

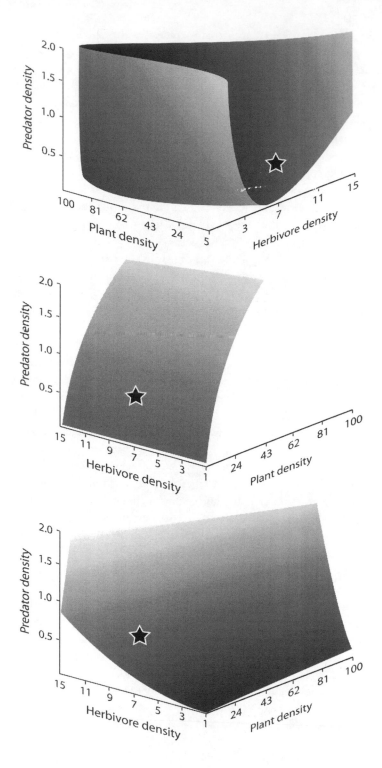

To an ecologist, a fear-driven system also looks like "ratio-dependent predation"—the idea that predators experience a zero growth rate at some fixed ratio of prey to predator abundance (see Abrams 1997; Akcakaya et al. 1995). In the classic predator-prey model, the predators merely require a fixed standing crop of prey, independent of the numbers of predators. But when vigilance decreases catchability in response to predation, each predator may require a fixed number of prey (at least as a first-order approximation) to maintain a stable predator population.

Prey fear responses become crucial to debates over Oksanen's exploitation ecosystems (the three-trophic-level model without vigilance and with exploitative competition), top-down versus bottom-up regulation of ecosystems, and ratio-dependent models of predator-prey interactions. When herbivores have effective fear responses toward their predators, exploitation ecosystems become intricate. Top-down and bottom-up effects become flip sides of the same vigilance-induced *direct* effects of predators on plants and vice versa, and approximate ratio-dependent predation becomes the expected outcome of the positively sloped predator isocline (with respect to the predator axis).

Prey vigilance and the arrangement of isoclines in figure 13.5 may resolve a paradox of the classic Rosenzweig and MacArthur (1963) predator-prey model (these authors in their paper fully anticipate this resolution!). Here's the paradox: When predators capture prey very efficiently, they require a low prey density for subsistence. The predator's vertical isocline intersects the prey's isocline in a region where the prey's isocline increases with prey density (positive density dependence of the prey on themselves). The intersection of these isoclines creates an unstable equilibrium that can lead to limit cycles, prey extinction followed by predator extinction, or the extinction of the predator. When predators capture prey very inefficiently, they require a high density of prey for subsistence, which creates a stable equilibrium point because the predator's isocline intersects the prey's in a region of negative slope (negative density dependence of prey on themselves). But the predator is now highly susceptible to environmental fluctuations that take the prey population below the subsistence level. Paradoxically, efficient predators produce intrinsic instability in predator-prey dynamics, while inefficient

Figure 13.5. The isoclines of (A) plants, (B) herbivores, and (C) predators for a model in which the herbivores can use vigilance to manage predation risk. This figure shows the isoclines in separate graphs to prevent confusion and to illustrate how strongly herbivore fear responses change the plant and predator isoclines. The model used here directly extends the model used in figure 13.3. This model assumes that herbivores and predators have zero handling times on their respective prey. The star in each graph shows the equilibrium point: $R^* = 15$; $N^* = 7.847$, $P^* = 0.722$. The following parameter values were used: $r = 1$, $K = 100$, $a_N = 0.15$, $d_N = \beta = \gamma = 0.1$, $d_P = 0.08$, $m' = 1$, $k = 0$, $b = 30$. Hence, $\mu = m'/bu$ and $F(\partial F/\partial e) = 1/\beta = 10$.

predators produce vulnerability to extrinsic variability in prey population numbers.

Fear responses by prey can break this paradox. At low predator abundances, the predator can efficiently catch unwary prey. At higher predator numbers, the predator becomes less efficient as the prey become increasingly wary and uncatchable. The high efficiency of predators at low predator numbers buffers the predator from environmental stochasticity, while the inefficiency of predators at higher predator numbers promotes a stable equilibrium and reduces intrinsically unstable dynamics. Rosenzweig and MacArthur (1963), in addition to their "classic" predator-prey model, anticipate these stabilizing effects of prey that respond behaviorally to their predators.

Self-regulation also provides a hypothesis to explain "why big fierce animals are rare" (Colinvaux 1978). Carnivores may represent a sufficient threat to one another to keep their densities low. It seems to take a large prey base to support carnivores (10,000 kg of prey to support 90 kg of carnivore, for instance; Carbone and Gittleman 2002). The word "fierce" suggests a role for the prey and their fear responses. Fierceness is a property of the prey rather than the predator: a predator is fierce because it induces fear in its prey. And, in a highly μ-driven system, the prey's fear responses produce a system in which their fierce predators can and must be rare. From the perspective of ecological energetics, N-driven predator-prey systems support higher densities of predators. The prey compensate for predation risk by higher fecundity that sends energy up the food chain as they feed the predators. In μ-driven systems, the prey respond to the presence of predation risk by forgoing fecundity, reducing mortality, and thus sending less energy up the food chain. Fear contributes to the length and to the transfer efficiency of food chains.

13.7   Foraging Games between Prey and Predator

"The deer flees, the wolf pursues" (Bakker 1983). We have considered how prey react to predators with fear responses, but predators need not be passive partners in this interaction. Predators can anticipate and respond to the prey's fear responses. Clever prey and clever predators produce a foraging game of fear and stealth. The abilities of prey and predator to respond to each other contribute to the character and stability of the predator-prey interaction (Abrams 2000). Although relatively few studies have addressed this problem, some recent work has done so (Lima 2002).

Habitat Selection Games

In the conventional ideal free distribution (IFD; see box 10.1), which measures a forager's fitness in units of energy gain, animals should distribute themselves among habitats in a way that equalizes per capita net energy gain (Fretwell and Lucas 1969; Fretwell 1972; Rosenzweig 1981). If habitats vary in food and safety, animals should still equalize the fitness value of net energy gain, but they should discount net energy gain by the cost of predation as given in equation (13.1). In other words, a forager occupying a risky habitat must have a higher harvest rate than one occupying a safer habitat (Moody et al. 1996; Brown 1998; Hugie and Grand 1998; Grand 2002).

Hugie and Dill (1994) expand on the ideal free distribution under predation risk by allowing both prey and predators to select among habitats. The ESS (evolutionarily stable strategy) distribution of prey and predators creates a spatial paradox of enrichment. In the absence of predators, more productive habitats will harbor more prey as the prey equalize their feeding opportunities. In the presence of predators, however, the prey must balance both food and safety, and the predators must equalize their feeding opportunities among habitats. If predators catch prey with equal ease in all habitats, then equal opportunities for the predators require that each habitat possess the same density of prey. This, in turn, means that the prey in more productive habitats harvest more resources. So, for the prey to have equal opportunity among habitats, there must be more predators in the more productive than in the less productive habitats. The ESS condition on the predators tends to equalize prey abundances independently of habitat productivity for the prey. And the ESS condition on the prey means that predators must bias their activity or distribution toward more productive habitats (fig. 13.6).

We can extend this model by permitting vigilant prey. With vigilant prey, we find a new ESS that requires a higher abundance of more vigilant prey in productive habitats, and a slight shift of predators from the more productive to the less productive habitats. At the new ESS, the more productive habitats have both more prey and more predators than the less productive habitats. In a more productive habitat, the prey are more vigilant, less efficient foragers, and harder to catch. In terms of ecological energetics, the less productive habitats actually have higher ecological efficiencies of transferring energy from one trophic level to the next.

Hugie and Dill's foraging game (1994) can also occur in time (Brown et al. 2001). Instead of productivity or resource availability varying among habitats, we imagine prey resources varying in time. The owls preying on gerbils in the Negev Desert represent such a system (see chap. 12). Afternoon winds redistribute sand and seeds every night, creating a resource that is renewed

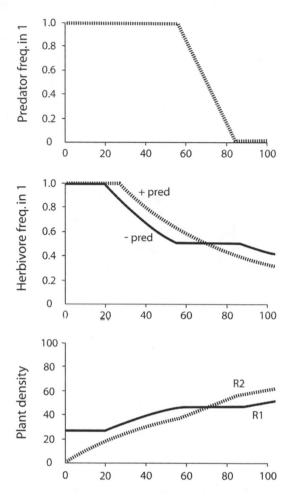

Figure 13.6. Density-dependent habitat selection when both predator and herbivore distribute themselves according to an ideal free distribution. In both habitats, plant, herbivore, and predator population dynamics follow the three-trophic-level model with type I functional responses for herbivores and predators to their respective prey. In these graphs, we hold the quality of habitat 1 fixed. The plants in habitat 1 have a carrying capacity of 70. We allow plant dynamics to go to equilibrium, but hold the total numbers of predators ($P = 1$) and herbivores ($N = 6$) fixed. The graphs show the distributions of predators (frequency in habitat 1), herbivores (frequency in habitat 1), and plants (density of plants in habitat 1 and habitat 2: R_1 and R_2, respectively) as the carrying capacity of habitat 2 is increased from 0 to 100. For the herbivore, one line shows its distribution when there are no predators (− pred.) and the other shows its distribution when predators are present in the system (+ pred.). Except for the habitat-specific carrying capacities and the fixed values for N and P, the parameters in both habitats are set to the same values as those in figure 13.4.

nightly (Ben-Natan et al. 2004). The gerbils begin the night with the greatest abundance of seeds, and then they deplete these seeds throughout the night (Kotler et al. 2002). Without predators, clever gerbils would become active after dusk, deplete the seeds to the point at which they could no longer profit energetically, and then return to their burrows and remain dormant. If the owls "knew" this, their activity would track the gerbil activity pattern. But, if owls were most active at dusk, then clever gerbils might want to find a different time for their peak activity. In fact, the ESS distribution of activity by gerbils and owls follows an ideal free distribution in time.

The owl ESS requires a constant level of gerbil activity throughout the night (assuming the gerbils do not vary their vigilance during the night). The gerbil ESS requires an equalization of opportunity throughout the night. In accord with Gilliam and Fraser (1987), the ratio of risk to net feeding rate (μ/f rule) must remain constant throughout the night. This requires higher owl activity early in the night and less as the night progresses. Over the course of the night, owl activity should track the level of seed resources for the gerbils, and gerbil activity should remain relatively constant. If the gerbils also vary their vigilance during the night, then early gerbils will be more numerous and more vigilant. To the owls, this part of the night will offer more, but less catchable, gerbils. In accord with these expectations, gerbil activity does decline as the night progresses, and gerbils behave more apprehensively early in the night (Kotler et al. 2002).

The gerbil-owl foraging game produces distinctive predator and prey isoclines (fig. 13.7). The prey isocline descends steeply from an asymptote along the predator axis. This typically happens in predator-prey models in which the prey have refuges. In this model, the prey have a behavioral refuge. Below a certain level of gerbil activity, it is not profitable for the owls to be active at all. Hence, the gerbils can sustain any number of inactive owls. At gerbil densities beyond this threshold, the gerbil isocline declines almost linearly. Because additional gerbils attract greater owl activity, the gerbils experience greater danger as their numbers increase. The owl isocline rises vertically from the subsistence abundance of gerbils that an owl requires. When there are no owls, the gerbils are most active and most catchable. As the number of owls increases, the owl isocline takes on a positive slope as the gerbils respond by becoming less active and less catchable. The intersection of the gerbil and owl isoclines produces a stable equilibrium, as both gerbils and owls have negative direct effects on themselves.

Patch Use Games

In traditional patch use models, the forager seeks patchily distributed and unresponsive prey. Patches are depleted because foragers harvest prey. Charnov et al. (1976) recognized that a predator could also depress patch quality if prey become harder to catch or if prey escape the area. This behavioral resource "depletion" can occur in unusual ways. A Neotropical tree frog lays its eggs out of harm's way in foliage above a pond, but a vine snake can still consume

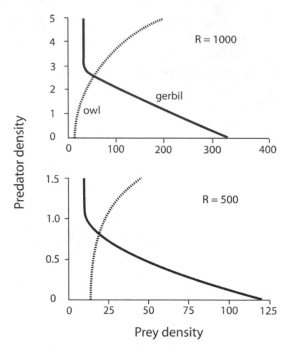

Figure 13.7. The predator and prey isoclines for the gerbil and owl foraging game. The size of the initial pulse of resources (*R*) in the bottom panel is twice that of the top panel. In both systems, the gerbil isocline declines from an asymptote and then become almost linear as it approaches the prey axis. The owl isocline rises vertically from the prey axis (at this point the prey have no fear) and then slopes sharply to the right as the presence of additional predators makes each gerbil less active and catchable. Increasing the size of the resource pulse (bottom panel) increases the equilibrium number of both gerbils and owls.

the eggs. Warkentin (1995) showed that the tadpoles, after a certain stage of development, perceive the vibrations caused by an approaching and feeding snake. The tadpoles respond by hatching prematurely. Even as the snake depletes the patch of eggs through consumption, the patch also depletes itself as the tadpoles hatch and escape. When pea aphids (*Acyrthosiphon pisum*) attack a broad bean (*Vicia fava*) (Guerrieri et al. 1999), the plant secretes pheromones that attract parasitoid wasps (*Aphidius ervi*). The pea aphids' patch quality declines both through their own herbivory and through increased risk of parasitism.

This kind of predator-prey patch use game occurs most commonly when a mobile predator has a larger home range than its prey. Mule deer populations exist in fragmented forest patches in the western montane habitats of North America. Their predator, the mountain lion, requires a home range that encompasses several forest patches. In this game, the mountain lion must decide how long to remain in any given forest patch, and the mule deer must choose vigilance levels in response to their perception of the mountain lion's whereabouts. The prey's information state becomes a critical feature of a game of fear and stealth. The deer would like complete information on the current whereabouts of the mountain lion (prescience), while the lion would prefer to keep the deer ignorant of its whereabouts. The deer's information strongly influences the behavioral ESS and the subsequent stability of the predator-prey dynamics in this system (Brown et al. 1999).

In the mountain lion's absence, perfectly informed deer can be totally at ease (vigilance, $u = 0$). When a lion arrives, they can immediately adjust their vigilance to the u^* appropriate for the lion's encounter rate and lethality [eq. (13.4) for the optimal level of vigilance] (fig. 13.8). At the other extreme, ignorant deer must adopt a fixed level of vigilance appropriate for the average level of lion proximity. More realistically, the deer's information state lies between these extremes of prescience and total ignorance. Imperfectly informed deer can never be sure when a lion is nearby, but the longer a lion is present (or absent) in their area, the more aware of its presence (or absence) they become. While never completely sure, the deer continuously update their expectation of encountering a lion based on time and direct or indirect cues given by the lion.

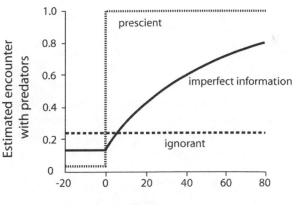

Time before and after predator

Figure 13.8. Three types of learning curves for prey becoming aware of a predator's presence. The x-axis shows the time since the predator's arrival (i.e., the predator arrives at time zero, and negative time refers to time before the predator's arrival). A prescient prey animal knows the predator's whereabouts. Hence, its expectation of encountering the predator is 0 prior to the predator's arrival and then jumps immediately to 1 (standardized to represent one predator within the prey's area) upon the predator's arrival. An ignorant prey animal that never actually knows the predator's whereabouts has a constant expectation of encountering a predator. An imperfectly informed prey animal never actually knows where the predator is, but it can use cues to modify its expectation of a predator encounter. Its baseline expectation of predator encounter is higher than the prescient prey's and lower than the ignorant prey's. (After Brown et al. 1999.)

The prey's information state influences the predator's patch use behavior (how long it stays in a given area) and the prey's vigilance tactics. We can describe the value of a patch to the predator by multiplying the number of prey by their catchability. With prescient or ignorant prey, prey catchability stays the same while the predator occupies a patch; it's just that the predator catches ignorant prey more easily than prescient prey at their ESS levels of vigilance.

When the deer have imperfect information, they must select their baseline level of apprehension (Brown et al. 1999)—the optimal level of vigilance when available information suggests that there is no lion around. This baseline vigilance level acts as the set point for raising vigilance when the deer detect a lion's arrival. If the deer set an excessively high baseline, they waste foraging opportunities when lions are actually absent. If they set too low a baseline, they risk death by reacting too slowly to the arrival and presence of a lion. The deer's ESS baseline level of vigilance will fall somewhere between 0 (the optimal baseline for the prescient prey) and the fixed u^* of the ignorant deer. When faced with deer with imperfect information, the lion reduces its chances of capturing a deer simply by spending time in the patch. When the lion first arrives, the deer are using their baseline vigilance level and are hence at their most catchable. As the deer begin to notice the lion, they increase their vigilance and become less catchable (fig. 13.9). As soon as the lion arrives in the patch, patch quality begins to decline for the lion. If patch quality declines to some threshold (set by the overall quality of the environment), then the lion should leave the patch and seek another. A lion should spend less time per patch (higher giving-up threshold) in a rich than in a poor environment. A lion should spend more time per patch when the deer have a low baseline level of apprehension.

The deer's information state and the resulting ESS behaviors for deer and lions influence the predator-prey dynamics and behavioral indirect effects among deer and lions. Prescient prey can result in unstable predator-prey population dynamics (van Balaan and Sabelis 1993, 1999; Brown et al. 1999; Luttbeg and Schmitz 2000). Adding more deer reduces their feeding rate and encourages them to be even more vigilant when a lion is present. This contributes to safety in numbers and to the destabilizing of behavioral feedbacks. Increasing the number of lions produces few behavioral indirect effects because deer do not become more vigilant; instead, they endure more frequent periods of lion presence. The predator isocline remains roughly vertical, but actually has a slight negative slope. Perfectly informed prey also create a repelling isocline that occurs at higher prey densities than the regular predator isocline. Consequently, the deer-lion model has two unstable equilibria (both interior solutions) and one stable equilibrium point (a corner solution). The corner solution has no lions and the deer at their carrying capacity. Another solution

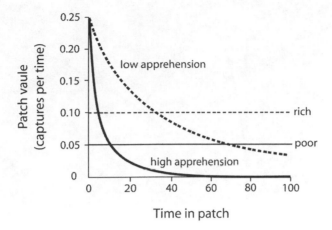

Figure 13.9. Patch use strategy of a predator that depresses patch quality by frightening its prey. Prey catchability is highest upon the entry of the predator into the area. The longer the predator remains in the area, the less catchable the prey become as they become aware of the predator's presence. Once prey catchability declines to a threshold, the predator should abandon the area and seek another. The predator's threshold of acceptability should be higher in an environment rich in prey than in one poor in prey. Furthermore, prey catchability declines faster with time when the prey begin with a high baseline level of apprehension. A predator will spend less time in an area where prey have a high baseline level of vigilance, and will spend more time in patches where prey have a low baseline level of apprehension. (After Brown et al. 1999.)

involves oscillatory dynamics around the lower, unstable interior equilibrium point.

When the deer have imperfect information, the deer-lion foraging game produces very different isoclines. The cost of fear responses can offset the safety in numbers that deer gain from the dilution effect. Increasing the number of deer makes them more catchable as they lower their baseline vigilance level. The prey isocline still follows a humped shape, but deer fear responses shift the peak to the left. Increasing the number of lions makes the deer less catchable because the deer increase their baseline vigilance level. This produces a predator isocline with a positive slope. The resulting stable equilibrium point occurs in a region where both deer and lions have negative intraspecific direct effects (fig. 13.10). When the deer have imperfect information, the deer-lion model (Brown et al. 1999) produces isoclines and population dynamics similar to the gerbil-owl foraging game model (Brown et al. 2001). In all cases to date, predator-prey foraging games produce isoclines and dynamics that would be hard or impossible to predict without some understanding of fear-stealth games.

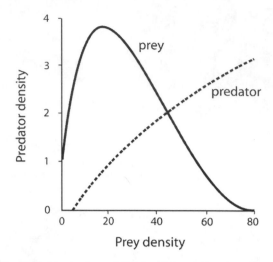

Figure 13.10. The isoclines of the deer-mountain lion foraging game when the deer have imperfect information on the lions' whereabouts. The hump in the prey isocline occurs at low prey densities. The predator isocline has a positive slope. The positive slope occurs because the presence of more predators renders the prey less catchable. Hence, with more predators, the predators require a higher prey density to subsist. The resulting equilibrium point is stable. (After Brown et al. 1999.)

13.8   Summary

Predation creates a special foraging cost because it poses a catastrophic risk: the forager risks losing everything through injury or death. Foraging animals should and do respond strongly to this risk via vigilance behaviors, habitat selection, and group size. These nonlethal effects of predators have implications for the ecology of prey species, of their predators, and of their resources. The ecology of fear examines the ecological and evolutionary implications of foragers responding behaviorally to their predators and vice versa.

In most situations, the richest foraging opportunities also carry the greatest predation risk. Because predation risk represents a cost of foraging, foraging animals should titrate food and safety as they seek and deplete foraging opportunities. Foragers should deplete safe patches more thoroughly than risky patches, thus creating a correlation between predation risk and foraging opportunities. Ecologically, this relationship between predation risk and foraging opportunities means that a forager can improve its energy state if it is willing to risk the possibility of predation, or it can purchase some safety by forgoing the best foraging opportunities.

Feeding animals should distribute their foraging efforts in response to both resource productivity and the landscape of fear. When predation risk varies in space (the landscape of fear), a population's distribution will not match resource productivities, as simple resource-matching models predict. In general, foragers will underutilize risky areas and overutilize safe areas.

From the perspective of population dynamics, prey can cope with predators in two ways. They can increase fecundity to offset losses to predators (N-driven systems), or they can forgo fecundity and increase their survivorship (μ-driven systems). In N-driven systems, the foragers harvest more food, turn over faster, and feed the predators. The foragers pay the predation cost of foraging directly to the predators, and forager mortality feeds and supports the predators. In μ-driven systems, the prey forgo foraging opportunities, forgo fecundity, and deny food to the predators. N-driven systems enhance the transfer of energy up the food chain, while μ-driven ones stifle this transfer of energy.

Trade-offs in fear responses that affect habitat selection or vulnerability to predators create nonlethal effects that increase opportunities for species coexistence. In conjunction with habitat selection, each prey species will thrive in the habitat within which it feels safest. Alternatively, one prey species may be able to exploit resources more competitively under safe conditions while the other does so more competitively in the face of predation risk. Predator-induced shifts in foraging behavior and predator-induced mortality can influence community dynamics in similar ways.

When both prey and predator alter their behavior in response to each other, the predator-prey interaction becomes a foraging game. In a foraging game, the prey can alter their vigilance, habitat selection, activity patterns, activity times, and group size in response to predator numbers and behavior. The predators can adjust their habitat selection, stealthiness, and boldness in response to prey abundance and behavior. These responses can stabilize or destabilize predator-prey dynamics. Regardless of its effects on the dynamics, this game increases the number and intensity of direct and indirect effects relating predators to their prey. For instance, if predators frighten their prey, then the mere presence of an individual predator can reduce the success of other predators. When the prey have excellent information about predator numbers and whereabouts, the predator-prey foraging game tends to destabilize predator-prey dynamics. When the prey have imperfect or poor information, it tends to stabilize population dynamics.

The ways in which predators frighten their prey may be more important than the number of prey they kill (Kotler and Holt 1989; Schmitz, Beckerman, and O'Brien 1997). Foraging theory in the guise of the ecology of fear permits an integration and appreciation of the full impact of predators on ecological systems. In fact, in the absence of foraging theory, it may be impossible to predict and understand most predator-prey systems.

13.9   Suggested Readings

Rosenzweig and MacArthur (1963) underpin current textbook models of predator-prey dynamics, and they anticipate in advance the significance and consequences of predator-prey foraging games. Lima (2002), using part review and part synthesis, makes the case for incorporating fear responses into classic approaches to predator-prey interactions. Holt (1983) and Abrams (2000) provide inspiration and some clear paths to incorporating optimal foraging behaviors into models of population interactions, as well as discussing the consequences of adaptive behaviors for population dynamics and stability. Sih (1998) reviews and contributes novel ideas for placing predation in the context of game theory. Grand (2002) provides an introduction to and formalisms for how predation risk contributes to the coexistence of habitat selectors. Hugie and Dill's model (1994) deserves careful study as the first explicitly game theoretical model of a predator-prey foraging game. Laundré et al. (2001) for elk and wolves, Schmitz, Beckerman, and O'Brien (1997) for grasshoppers and spiders, and Abramsky et al. (1998, 2002a) for gerbils and owls provide experimental and empirical demonstrations of the crucial roles that fear plays in the ecologies of predator and prey.

Telander

On Foraging Theory, Humans, and the Conservation of Diversity: A Prospectus

Michael L. Rosenzweig

14.1 Prologue

The Tertiary is over. The world of our remote ancestors has nearly vanished. No nostalgia can save it; no yearning can restore it. We have entered the geological era of *Homo sapiens*. Like it or not, we are the boss.

We take what we want where we want it. We take land and sea, water and air. We corral a stupendous fraction of the earth's productivity and mineral resources (Vitousek et al. 1997). With clever apparatuses, we adapt to an unprecedented variety of environmental conditions, turning them all into a semblance of the semiarid tropical climate in which our physiologies evolved. Where we have not yet learned to live, we dream of living. No previous era in the history of life has seen our ilk.

We have not eradicated in ourselves the basic, acquisitive nature that natural selection insists upon in all successful life forms. That was the real flaw of Marxist thought: it dreamed that Man without unfulfilled needs would become generous. But, while a competitive and exploitative Mankind may confound socialist economics and disappoint theologians and moralists, it looms as a death warrant for every ecosystem whose resources we expropriate. The rest of life can do little to thwart us.

But *we* can do something. We can abstract. We can contemplate what we are doing. We can even predict the consequences. And we can find alternatives. Our plans have already restructured the world of life un-intentionally. Why should they not do so on purpose? And who is to say whether that purpose need be malevolent or malicious?

Fortunately, evidence indicates that we would rather share our world with other species, conserving at least patches of it as relics of our environmental heritage (Kellert and Wilson 1993; Wilson 1984). We have developed a world-wide network of set-asides—national parks, wildlife refuges, nature reserves, and the like. We restore ourselves in them, spending prodigious quantities of money and time. We join and support organizations devoted to them and to the preservation of specific species in them. As much as we can afford to, we surround ourselves with nature (Orians 1998). We install parks in our cities and towns. We tend our lawns; plant herbs, trees, and shrubs; and pay extra for property that allows us to do so.

14.2   Introduction

This chapter assumes that we humans do care about preserving natural di-versity. It will explore the ways in which foraging theory and studies of for-aging may improve our ability to make a difference. Much of it will be a call for focused research, rather than a synthesis or a review of what has already happened.

The chapter has several themes. It views human beings as sophisticated products of natural selection. We ourselves are optimal foragers. In that con-text, it asks how we should go about setting the rules for set-asides. It also wonders about what people really want from nature. It notes the promise of studies of foraging and habitat selection. These studies can reveal the under-lying relationships among species, and they can also provide environmental indicators and tools for further study. And the chapter calls attention to a relatively new strategy for conservation, reconciliation ecology. Reconcilia-tion ecology makes use of sophisticated methods for natural history research in order to develop new habitats in which humans and the natural world can coexist (Rosenzweig 2005).

14.3   Human Beings as Optimal Foragers

Can anyone imagine that selection has refined the foraging abilities of insects and fish, spiders and reptiles, birds and mollusks—not to mention mammals—but not of *Homo sapiens*? Yet I have sat on committees with first-rate minds in various human-oriented sciences, and I have heard their well-meaning lips deny that human behavior has any genetic roots. Certainly, their opinions stem from goodwill, from a determination to see that genetics is never again used to oppress people. However, their reticence to view people as products of natural selection can actually hurt people by negating the good that our institutions and understanding can do. On the other hand, if we admit that people do have innate tendencies toward certain behaviors, then we and our world stand to gain.

Recent evidence presented by Morris and Kingston (2002) strongly reinforces the notion that people exhibit behaviors consistent with a long history of selection to improve foraging abilities. Morris's work depends on Fretwell and Lucas (1969), who pointed out that individuals, when faced with choices of habitats (see box 10.1), will distribute themselves and their activities so that no individual can gain an advantage by unilaterally changing its habitat choice. Their work established the connection between population size and habitat selection because as population grows in a habitat, the advantage gained from foraging there declines. Sometimes optimal habitat choices result in what Fretwell and Lucas termed "ideal free distribution." Conforming to the ideal free distribution often means that more individuals use the richer habitat.

Human Isodars

Isodar plots, invented by Morris (1987), help us to compare the properties of different habitats (see box 12.1). In an isodar plot, each axis is the population size of a species in a specific habitat. Each point is the set of a species' habitat-specific populations at a single time. The line fitting those points is the isodar.

Human population distributions conform to an isodar (Morris and Kingston 2002). Urban and rural populations form its axes. In 1995, in 154 nations, large and small, rich and poor, authoritarian and free, people lived in urban and rural habitats in proportions that follow it. Of course, there is statistical noise in the relationship, much of which can be accounted for by subdividing nations into high and low per capita CO_2 emissions. In the 76 nations with emissions below the median, more people lived in rural habitats than in urban ones. In the 73 nations above the median, about half the people lived in rural habitats.

The point is that the human isodar exists. People follow innate rules of density-dependent habitat selection that manifest themselves in all societies. No one claims that the isodar proves people achieved an optimal habitat distribution in 1995. The isodar of 1995 may reflect conditions of a past era and be quite inappropriate for 1995, but it exists.

Adjusting Costs and Benefits of Nature Reserve Exploitation

In yesterday's world, people made their living by harvesting resources from the bounty of environments resembling today's set-asides. Thus, today's nature reserves seem, to the very core of the human psyche, to be patches of beckoning abundance in a sterile world. Morris's isodar comes to remind us that our evolved psyches urge us to not let them lie unexploited!

Sometimes such urges afflict very rich individuals. The very rich may visit a set-aside and find it releasing passions in them that perhaps they never knew they had. Beyond better education and strict law enforcement, there's not much we can do to tame their atavistic selfishness.

Sometimes the urges are collective, infecting rich organizations of people hell-bent on taking the last 1% of something. Although they are already making lots of profit, simple institutional greed moves them—probably reinforced by groupthink (Janis 1972). And what they do is rarely illegal; they buy legality with their profits. Harnessing the power of foraging theory cannot stop them directly, although it may create a world in which their behavior loses its profitability by virtue of an excessive cost in the courts of public opinion.

But sometimes very poor people, who happen to live nearby, threaten set-asides. This scenario applies to many of the world's richest set-asides. Exploiting such set-asides could make a great deal of difference in the lives of their poor neighbors, at least for a time. In these cases, we must understand people as foragers, which is to say, as rational beings behaving intelligently to improve their lot. The set-aside is a resource-rich patch next to an impoverished one. It will attract foragers in substantial numbers.

Policymakers and conservationists know full well what they must do to protect their country's set-asides. They need to develop incentive-compatible systems for reconciling human behaviors with conservation efforts (Gadgil and Seshagiri Rao 1995). That is the strategy. Its tactics involve adjusting the cost and benefit parameters of those behaviors. But many policymakers have shown little imagination. Ignoring benefits, they act only to increase the costs. Fines and prison terms for poaching go up. More wardens enforce the restrictions, with increased powers to injure, and even to kill, suspected violators.

Sadly, the proponents of such policies have greatly underestimated the value of the contraband to poachers. So such policies generally fail, except in rich countries where poachers gain comparatively little by their activities. Escalating the cost of poaching usually leads poachers to increase their prices—an enhanced reward to compensate them for the greater risks and higher costs. Most perversely, such increases could even increase poaching, because people, acting like perfectly sane foragers, ought to shift their activities to a resource that has become more lucrative. (Ask yourself, how many narcotics dealers would there be if greengrocers sold hemp and coca leaves at the price of cabbage?) Hence, increasing the cost of poaching may also increase its benefits and nullify some, all, or even more than the increase in costs.

Consider the following case from Zimbabwe (Muchapondwa 2002b):

> Rhino . . . cause minimal damage to agriculture. . . . The virtual elimination of the black rhino [in Zimbabwe] is due to the high value of the horn. . . . [Despite] the imposition of a complete embargo on trading in rhino parts and derivatives . . . the illegal trade has flourished. The [government] had increased its surveillance.

Anti-poaching operations assumed the proportions of moderately intensive anti-insurgency warfare, employing the same tactics and equipment, including automatic weapons, sophisticated radio and intelligence networks, vehicles, boats, helicopters, and fixed-wing aircraft. Law enforcement was, however, tackling the effect rather than the cause of the problem. Poaching was motivated by the high price of rhinoceros horn on the illegal market, which had been handed a monopoly by the prohibition on legal trade. Protecting wildlife by giving it a value benefits landholders, often the rural poor, whereas trade bans, if they are effective, destroy this benefit.

In a few cases, cost-increasing tactics have eliminated the benefit of wildlife almost entirely. This negative tactic can work if it prevents the sale of the resources. When no one, not even a Russian nobleman or a Park Avenue matron, may own a sea otter coat, sea otter populations rebound. When international traffic in ivory becomes illegal, as does the sale of ivory artifacts, elephants stand a chance.

Nevertheless, in the past 10 or 20 years, a fundamentally new kind of policy has surfaced. Instead of increasing the costs or reducing the benefits of poaching, this policy seeks to increase the benefits of alternative non-poaching activities to people who live close to set-asides. It replaces bureaucratic regulations with rewards (Gadgil and Seshagiri Rao 1995).

Residents may train to become wardens themselves, or they may learn how to participate in managing the set-aside. Often hunting or ecotourism provides the rewards. Residents become guides or involve themselves in the supporting industries, such as food and lodging. Conservation can be profitable (Daily and Ellison 2002).

Yet, for all their benefits, ecotourism and trophy hunting are limited industries. To make reserves successful in the long term, we must reject the idea that we can manage reserves as hermetically sealed ecosystems. Instead, we must learn how to integrate set-asides with other means for humans to earn their livelihoods.

In that regard, David Western's approach in Kenya has been particularly fruitful (Western 2001). It uses the set-asides to enrich economic opportunities in surrounding areas. Outside the reserves, people engage in a wider variety of legal and profitable activities than inside. Yet residents understand that profits outside the reserve depend on the creatures within it. The result: areas around reserves receive overflows of wildlife from the reserves themselves, actually extending the ranges of the species in the reserves.

Policymakers can succeed if they take into account the intimate connections that nearby residents have to set-asides and to the conservation of wild species. Again, consider the lessons learned in Zimbabwe (Muchapondwa 2002b): Because they received money from wildlife exploitation, the Mahenye community agreed to move some of its villages away from a portion of its land, a small, fertile patch of excellent wildlife habitat. Most of the wildlife were elephants (*Loxodonta africana*). The Mahenye got more from selling the right to hunt an elephant than they lost through the crop losses they incurred by the move. The community used the money for local infrastructure: a school, a road, a borehole, and a grinding mill. As the community's earnings grew because of elephant conservation, it allocated more land to wildlife. Then the community itself started to control poaching. People were reluctant to kill wildlife even to protect their crops. Finally, the community began to use some of the wildlife profits to compensate its members for crops losses. It had decided to use the wildlife to increase its income.

Now, the people of Zimbabwe are not crass materialists. Indeed, they are poor, but they mix their respect for the profitability of elephants with a love for them. Elephants are a destructive nuisance to them, yet they are actually willing to pay something to preserve the elephants near them (Muchapondwa 2002a). Indeed, we must never expect people to be cold-hearted optimal foragers. They will always combine their implicit foraging calculations with a little bit of inexplicable mystery and aesthetics.

14.4   People as Bayesian Foragers: Shifting Baselines

As nature retreats, people rapidly accustom themselves to whatever nature remains. They cannot imagine what they are missing and rarely even try.

The depauperate environments with which we have surrounded ourselves during the past few centuries have deeply eroded our horizons. We expect to see nothing more than house sparrows and a few house plants. When, at last, we do take a trip to a national park or reserve, most of us depend on its wild things being in predictable places at predictable times. We have disconnected ourselves from the world of nature and have learned to prefer it that way. Nature makes us uneasy, even fearful.

But primarily, nature no longer holds promise for us. No promise of abundance. None of sustenance. Having conquered nature, we have lost both our esteem for her and our faith in her wealth. We have stopped believing in her robustness because we can no longer remember it.

Daniel Pauly (1995) calls this failure of intergenerational memory "the shifting baseline" syndrome. He illustrates it with a story about the grandfather of one of his colleagues. In the 1920s, the grandfather was a fisherman, drawing up his catch of mackerel from the waters of the Kattegat, an arm of the sea between Denmark and Sweden. Poor grandfather, it seems, was plagued by numerous, economically useless bluefin tuna that entangled themselves in his nets. Today, of course, bluefin tuna are rarely, if ever, seen in the North Sea. In those few places in the world's oceans where their dwindled schools do remain, experts meticulously monitor their population biology, and nations carefully apportion the right to fish them. Their flesh sells for a fortune.

Jeremy Jackson (2001) found evidence of our shifted baseline in the Caribbean. Once, great maritime powers added remote island systems to their far-flung empires because the abundance of turtles supported by those islands helped to provender their sailing ships. In the Caribbean, a few hundred years ago, green turtles were so abundant that ships struck vast shoals of them and sank! Green turtles, manatees aplenty, and teeming multitudes of man-sized herbivorous fishes kept Caribbean sea grasses closely cropped. Today, sea turtles of all species are rare or threatened.

Our baseline expectations have shifted in fresh waters, too. Consider an edible mussel, the giant floater, *Pyganodon grandis*. Living on the bottoms of some North American freshwater streams and lakes, it can quickly grow to be about 25 cm long. Ten generations ago, it was so abundant that in many places in the middle of the continent, it was a staple food. Brandauer and Wu (1978) estimate its population densities to have been six to twelve per square foot. In contrast, today's populations of giant floaters, like the majority of North American freshwater animals, have nearly vanished. In the waters of Colorado, they exist at population densities of less than one per hundred square feet in the few sites where they still survive at all (Liu et al. 1996). Thus, they are more than a thousand times scarcer than they were a century or two ago.

And then there is the principal pigeon of North America, the passenger pigeon. The last one died in the Cincinnati Zoo in 1914. Her ancestors had numbered in the billions just a century before (Schorger 1955). Professional pigeoners shot them in hordes and supplied the cities of the eastern seaboard with fresh pigeon meat for a century. But now they are gone, and their world is gone, and we can never imagine what it was like. That is the point. Our ancestors lived on an earth where they took nature's abundance and diversity for granted. We live on one where we take her fragility and poverty for granted. We simply have not experienced enough to know how different she could be.

And, indeed, we cannot bequeath our memories to our children. If they could but see what we saw when we were younger, they might be outraged at what they have lost. Thus, the human species as a whole is like a Bayesian forager, updating its expectations and estimates, generation after generation, of the probabilities of coming across habitats of each type and quality.

From the perspective of natural selection, it makes as little sense to defend a habitat that has ceased to exist as it does to search for one that contains an abundance of a perfect, but imaginary, resource. No conservation strategy can have long-term success if it merely tries to restore what a few doddering older members of our species recall with fondness. A truly victorious conservation plan will find a way to up the ante, to shift the baseline in the positive direction.

14.5   Reconciliation Ecology

Gordon Orians has dubbed the world we are creating "The Homogocene." Orians chose this word to reflect the breakdown in barriers between biogeographic provinces (Mooney and Cleland 2001). Nevertheless, it is an apt designation for our new world. The Homogocene threatens to be a time of mass, persistent loss of diversity, and not just because the world is losing its biogeographic boundaries—a minor threat in my view (Rosenzweig 2001a). To maintain diversity, we shall need to promote a sea change in our strategy of conservation.

Our current strategy is "reservation with restoration." We set aside what we can as reserves, and we attempt to repair degraded environments until they support some semblance of the natural flora and fauna (Rosenzweig 2003a). The most sophisticated of these efforts—the hotspot tactic—recognizes that not all areas of the world are equally valuable as set-asides. Some contain many more species than others; some contain species found nowhere else. The world's 25 silver-bullet hotspots constitute only 1.4% of its land area, but 44% of its vascular plant species and 35% of its vertebrate species are contained entirely within that 1.4% (Myers et al. 2000). (One guesses that they also contain a large proportion of its invertebrate species, but that proportion is unknown.)

Reservation with restoration has slowed the bleeding. But it relies on a static view of habitats and their distributions. Global warming may vitiate all current reserves. And even if we do somehow manage to get that problem under control, both current biogeography and paleobiogeography leave no doubt that area is as fundamental a property of an ecoregion as its precipitation and temperature. Shrunken ecoregions can preserve species diversity only in direct linear proportion to their size. In other words, lose about 95% of an ecoregion and expect to lose about 95% of its species diversity (Rosenzweig 2001b).

However, not all species require set-asides. The German language calls those that do, *kulturmeider* (culture avoiders), and those that do not, *kulturfolger* (culture followers). A new strategy of conservation biology, reconciliation ecology, seeks to convert *kulturmeider* into *kulturfolger*. As the first step in this process, reconciliation ecologists study the habitat and resource requirements of species. Next, they design new human-occupied habitats that offer the requirements for these species to thrive (Rosenzweig 2003b). Reconciliation ecology bulges with opportunities for the behavioral ecologist.

Not Your Grandfather's Natural History

Research for reconciliation ecology often begins with natural history. Not old-fashioned natural history, but natural history informed by modern techniques, modern theory, and conservation priorities. When reconciliation ecologists target a species for preservation, they study it carefully with a view toward determining what it needs to succeed in the natural world. Finally, they alter a human habitat in accordance with those needs. Notice: reconciliation ecology alters the habitat, rather than setting it aside in a reserve. Two examples should serve to illustrate how easy this can be in some cases, and how difficult in others.

The easy case is a bird, the loggerhead shrike (*Lanius ludovicianus*). Populations of this species and many others of its family are declining and disappearing over much of their range (Yosef and Lohrer 1995). Yet one imaginative person quickly discovered a way to reverse the trend.

Ruven Yosef began by observing the natural feeding behavior of the loggerhead shrike in southern Florida. From its perch on a fence or in a cabbage palm, a foraging shrike would scan its immediate surroundings for the large invertebrate prey that form the bulk of its diet. It would pounce only if that prey lay within a certain restricted distance (6.5 m from a fence; 9.3 m from a palm) (Yosef and Grubb 1992). One may speculate that this attack distance reflects a well-adapted forager. Foraging from farther away might give the targets so much time to react that they would too often escape. Marginal benefit would fall beneath a critical threshold, and natural selection would force the shrikes to ignore prey beyond the critical distance. Whether or not this speculation proves true, the foraging behavior is real, and the biologist can work with it (see also Cresswell and Quinn 2004 for the hunting tactics of sparrowhawks on redshanks).

Yosef mapped shrike perches on a working cattle ranch and discovered that, despite abundant prey, much of it was unavailable to shrikes because it was beyond their attack distance (Yosef and Grubb 1994). Yosef installed simple wooden posts in the patches of pasture that lacked perches. The shrikes responded immediately. Within the first spring, territories with the extra posts shrank an average of 77%, and the loggerhead shrike population increased 60%. The smaller territories also helped nestlings survive. Parent birds in smaller territories had 33% more successful clutches than controls, and raised 29% more chicks per successful clutch (Yosef and Grubb 1994).

South African ecologists have successfully applied Yosef's method to fiscal shrikes (Devereux 1998). German biologists used a similar method to restore a population of great grey shrikes (*Lanius excubitor*), in which small piles of rock took the place of the posts (Schön 1998). Van Nieuwenhuyse (1998) is applying a similar approach to populations of red-backed shrikes (*Lanius collurio*). Adding hunting perches to land already used for agriculture tweaks the habitat only a bit and does nothing to reduce its use by humans. It is also cheap.

The natterjack toad (*Bufo calamita*) in England proved a more difficult case. To learn how to rescue this threatened species, a veritable company of some 50 researchers and their assistants spent 25 years refining their understanding of natterjack toad natural history (Denton et al. 1997).

This team first focused on characterizing the natterjack's niche. The natterjack is a pioneer amphibian. It lives in open vegetation surrounding eutrophic pools of coastal dunes or oligotrophic pools of inland heaths. Unlike its chief competitor, *Bufo bufo*, it burrows in sand. When foraging at night, it operates at a body temperature 1.4°C higher than *B. bufo*, and it loses weight if forced to forage in dense, cooler vegetation. This helps to explain why its population declines when tall vegetation—such as birch, gorse, and bracken—begins to invade and shade its habitat. The increased shade also lowers the water temperature of the pools, slowing the development of natterjack tadpoles and subjecting them to damaging competition from *B. bufo*.

The company studied many other aspects of natterjack ecology. They looked at a unicellular gut parasite, *Prototheca richardsi*, and at predation by salamanders, Odonata, water beetles, water bugs, and Notonecta larvae. They studied pond chemistry and water quality (chlorides, sulfates, orthophosphates, ammonia, iron, sodium, potassium, calcium, magnesium, alkalinity, conductivity, color, and turbidity). They even studied pond depth and the contour of pond slopes.

Simply knowing the detailed natural history of *B. calamita* has supported the reestablishment of many healthy natterjack toad populations (Denton et al.

1997). The toad biologists increased grazing to maintain the early stages of succession. For the same reason, they cleared dense vegetation. They fought acidification by adding $Ca(OH)_2$ to natterjack ponds every year or two, or by scraping the sulfate-rich silt from their bottoms. They removed some *B. bufo* to give the natterjacks a fair start. And they built some 200 new ponds, not too deep—for that would have encouraged invertebrate predation—and not too steep, and sometimes lined with concrete to fight acidification. They used old bomb craters and active golf courses. At all sites with new ponds, *B. calamita* used at least one and usually most within a year or two. The new ponds reestablished, rescued, or increased natterjack populations at two-thirds of the sites.

Sophisticated Preference Studies

We can do better than to discover what a species needs. Using sophisticated foraging studies, we can actually discover what a species wants. Many ecologists continue to believe that a proper way to do this is to develop a utilization distribution for a species; that is, to accumulate information about the resources and habitats used by individuals and then rank these according to the intensity of their use. A slightly more sophisticated version of this method involves comparing these intensities with the proportions available in the natural environment. For example, habitat used 10% of the time by a species and extending over only 5% of its range would be viewed as twice as beneficial as a habitat used 50% of the time that covers 50% of the range.

But foraging ecology teaches us that we can rely on none of the above methods (Rosenzweig 1981). Observing where a species lives and what it does depicts merely its realized niche. Its fundamental niche may be quite a bit larger. Moreover, competitors and predators can profoundly affect the proportional use of habitats within the realized niche, so proportional use data may be an unreliable guide to the relative value of habitats. In the worst-case scenario, a habitat that is heavily used by a species may not even be part of its fundamental niche. It may instead harbor only sink populations of that species. The largest populations of the annual *Cakile edentula* grow in such sink habitats (Keddy 1981).

Foraging ecology allows us to compare and rank a species' habitats and resources more reliably. If one assumes that natural selection produces competent individuals, then one can study their choices to discover what benefits them most. Microeconomists often use such an approach to human behavior, calling it the study of "revealed preference" (see chap. 6). Students of animal behavior may estimate revealed preferences using the worldview of the ideal free distribution (Fretwell and Lucas 1969). They may equally well use patch use theory, which Charnov (1976b) introduced to ecology. Charnov asked how long a forager should remain in a patch—its giving-up time. Brown (1988) extended patch use theory by asking what foraging conditions in a patch should prompt a well-adapted forager to leave it—the giving-up density (GUD) of its resources (see box 13.2).

My colleagues and I use these ideas to structure our research programs, and our results have often surprised us. If a species experiences strong interspecific competition, then members of the species may use secondary habitats almost to the exclusion of primary habitats (Abramsky et al. 1990; Reynoldson 1983). Using extensions to ideal free distribution theory (see chap. 12), we can compare disparate rewards and threats, determining, for example, that the advantage of foraging without the threat of predation from barn owls has about ten times the effect on a foraging gerbil's behavior as does the very important advantage of foraging in semi-stabilized sand (Abramsky et al. 2002a, 2002b). Previously, with similar field experiments, colleagues had shown that the advantage of semi-stabilized sand was very subtly linked to time (i.e., sunset to midnight) as well as to space (Kotler, Brown, and Subach 1993; Ziv and Smallwood 2000). Reconciliation ecologists will need just this sort of knowledge to do their jobs.

We need to find out about our own (that is to say, human) habitat preferences, too (Orians 1998). Layer upon layer of civilization may obscure human preferences, but they are nonetheless real; we must take them into consideration when tinkering with the habitats in which we live. Human behavioral ecology remains an inchoate science, and I cannot foretell the extent to which optimal foraging principles will be useful to it. Perhaps the answers are already available, but lie embedded in the literatures of marketing, landscape architecture, and interior decoration.

In any case, reconciliation ecology does not seek to impose habitats on people. For its designs to be successful, people will have to appreciate them. It is one thing to establish a forest of shrike perches in a cow pasture, but quite another to do the same thing in someone's front lawn.

Studies of Guild Organization

Most of the reasons why conservation biologists need to know about community organization are perfectly clear. One cannot be a good steward without being aware of the potential hazards to one's charges. I believe that the great difficulty with the hotspot strategy is that it pays too little attention to this principle of stewardship. It tacitly assumes that things will always be the way they are. Yes, we must save hotspots. But we cannot rely on them staying hot without understanding what makes them that way. As I have already pointed out, a species may be present in an area—even abundant there—despite having no source populations in it. In addition, a species may be absent because it interacts with another species in some negative way. Thus, conservation of diversity calls for intensive study of how population dynamics and species interactions determine the geographic ranges of species.

Methods relying on foraging theory have done better than any others in elucidating interspecies relationships (see chap. 12). First, they have enabled us to predict the behavioral and dynamic consequences of several forms of guild organization. At least one form of these predicted dynamics is baroque and unique (the so-called "ghost of competition past" model). Yet the behavior of gerbils in field experiments supports it (Rosenzweig and Abramsky 1997). This suggests that studying foraging behavior in the field may actually help us diagnose population interactions and may help to reveal how guilds of similar species are organized.

The tactic of measuring giving-up densities (see chap. 13) has also proved invaluable in dissecting guild organization (Brown 1989b). It helps us to compare habitat qualities. It determines the relative tolerances and efficiencies of species. It provides an alternative way to look at the influence of predation threats. Finally, GUDs have even reached across taxonomic boundaries, revealing the intimate interaction between gerbils and a common species of lark (Brown et al. 1994).

Using Guild Organization

In addition to helping us save species in reserves, understanding guild organization may tell us which species need reserves in the first place. A species' position in its guild may help us determine whether we can develop a reconciled habitat to save it.

Some species never live outside reserves. For example, Little and Crowe (1994) showed that six species of South African birds were seen only within reserves of fynbos. We will probably never discover a compromise habitat that suits these species. They will always be *kulturmeider* and require relictual habitats. They will probably resist reconciliation ecology forever.

Which species are *kulturmeider*? The example of the fynbos birds tells us that taxonomy provides no clue. Species found only in reserves have close relatives that live elsewhere. But the lessons of optimal behavior studies for community organization may well provide a clue.

Tolerance-Intolerance Organization

Methods derived from foraging models (including optimal habitat selection models) have focused our attention on "shared preference" community organization (also termed "tolerance-intolerance" organization) (Rosenzweig 1991). In shared preference organization, species divide a quantitative niche axis; that is, an axis whose values differ merely because the measure of a single variable changes. All species do best at a certain value of this measure (frequently, the largest possible value). But species differ in their ability to dominate habitats representing the best portion of the axis. In the simplest case, one species, often because it is the more efficient forager, is more tolerant of the poorer habitats along the axis. The other, the intolerant species, requires the richer ones. There it can dominate, perhaps by aggressive behavior, perhaps by better dispersal or mobility, perhaps by some other means.

Because the intolerant species requires the best habitats, it often tends to have a smaller geographic range, less dense populations, and therefore, a greater risk of extinction. In contrast, species capable of profitably using the poorer habitats should be the behavioral opportunists, flexible enough to go wherever they have the chance to go. In one sense, they do not require the poorer habitats—they do even better as individuals when they get the opportunity to use the richer habitats. But in another sense, they do require the poorer habitats. Without them, they could not coexist with the intolerant species.

Consider the Amani sunbird, which is abundant within and restricted to about 5,000 ha of *Brachystegia* forest in coastal Kenya. Is its habitat its prison or its refuge? Joseph Oyugi (2005) answered this question with a foraging approach based on patch use, density-dependent habitat selection, and interspecific interactions.

The collared sunbird—a widespread *kulturfolger*—shares the forest with the Amani sunbird. But the collared sunbird uses all surrounding habitats, too. In his habitat selection studies, Oyugi discovered that these two sunbird species allocate foraging heights in *Brachystegia* trees. The Amani sunbird forages high, the collared sunbird forages low, and both species overlap in the mid-canopy.

When Oyugi examined foraging at the scale of patch use (individual branches), he found that the mid- and lower canopies offer better foraging opportunities than does the high canopy. Collared sunbirds interfere with the mild-mannered Amani sunbirds. The interference raises Amani foraging costs in the richer habitats and prevents them from using those habitats.

Overall, we see a case of shared preference habitat selection, with the less preferred but critical habitat being the crowns of mature *Brachystegia*. The Amani is actually the habitat generalist, and the crowns of *Brachystegia* its refuge. Lose these trees, and we lose the Amani sunbird to competitive exclusion by the collared sunbird.

The differences between the sunbirds remind me of the many cases of shared preference organization that I have seen in the field and the literature— beginning with the classic case of *Chthamalus* and *Balanus* in the intertidal zone of Scotland (Connell 1961). The story of *Bufo bufo* and *B. calamita* fits, too. The natterjack toad is the intolerant species, needing the warmer waters where its larvae can grow rapidly. But, in contrast to the unusual case of the sunbirds, the intolerant species (the natterjack) has the smaller population and narrower range.

I am speculating, of course, but I believe that studying shared preference organization may be the easiest way to predict which species can be fitted with a reconciled habitat. Dominant species in a shared preference system may be entirely incapable of succeeding in any but the richest, most pristine habitats. If they require relictual habitats, dominant species may be *kulturmeider* forever. Tolerant species, on the other hand, may be among those most likely to take advantage of new habitats. Tolerant species may provide the best targets for us to turn into *kulturfolger*.

But the reverse hypothesis also has merit. Human-dominated habitats often have an unnaturally steady supply of abundant resources. For example, in Tucson, total bird populations are about thirty times as large as they are in the surrounding national park (Emlen 1974). Whether accidentally or intentionally, Tucsonans supply water and food to those species that can stand to live alongside us. Under such circumstances, the intolerant species may be the most successful *kulturfolger*.

Regardless of which hypothesis succeeds in a given case, I suggest that a good way to recognize inveterate *kulturmeider* will be to examine the tolerance-intolerance organization of natural communities.

14.6   Management of Relictual and Novel Habitats

During the Homogocene, the hand of Man will be everywhere. We might as well admit that, although it may require getting over an emotional hurdle to do so. Intentionally or not, we are destined to manage life on this planet. We might as well try to do a good job, and what we learn about foraging behavior can help us. In fact, we can use behaviors as leading indicators of environmental quality and population change.

Monitoring Species and their Habitats by Measuring Behavior

Well-adapted behaviors are good indicators of both food quality and habitat quality. The forager that leaves a lot of food behind (has a high giving-up density) is telling us that it perceives a relatively rewarding set of habitats. In contrast, stressed animals faced with a poorer environment will extract almost all the available food from a patch (have a low giving-up density) in habitats they always use, may accept lesser habitat types, and may use habitats with an elevated threat of predation.

Because a well-adapted species expands its choice of habitats as its population grows, the variety of habitats it uses can help us monitor its population size (Rosenzweig 1987). Use of low-ranking habitats suggests large populations. The manager armed with a ranked list of habitats can census them in inverse quality order. When she knows the poorest habitat that the target species currently uses, she can infer the current population size. For species whose overabundance could pose a problem, a quick census of only a few inferior habitat types could reveal the threat quickly and cheaply. For exploited species, whose overabundance would represent an opportunity for a large yield, a census based on optimal habitat selection would allow an earlier and more efficient harvest. For species whose recent population sizes create a concern for their future, managers could start censusing in the poorest habitat in which observers last reported them. Then higher- or lower-ranking habitats would be censused, depending on whether the first census found any individuals. The census would proceed up-rank until individuals were detected or down-rank until they were no longer detected.

Sometimes we can use the foraging behavior of one species to monitor another. For example, although preservation efforts often target large carnivores, most large carnivores are scarce and difficult to observe. But their prey are not so scarce. And their prey are experts; they have been in the business of detecting predators for untold generations. So if we can learn to interpret their behavior as a reflection of predation threat, we can indirectly census the predators.

Changes in the foraging behavior of a potential prey individual seem straightforward (Brown 1999; Brown et al. 1999). With a predator nearby, the individual should spend more time being vigilant. It should also reject some riskier habitats entirely and exhibit higher giving-up densities in those it continues to use. Foraging time proportions should shift asymmetrically: away from more dangerous habitats and toward safer ones. Abundant evidence confirms such changes (e.g. Abramsky et al. 1996; Brown 1988; Brown and Alkon 1990; Dall et al. 2001; Dill and Fraser 1984; Fraser and Cerri 1982; Kotler, Brown, Slotow et al. 1993; Kotler et al. 1994; Lima 1985a; Milinski and Heller 1978; Nonacs and Dill 1990; Rosenzweig et al. 1997; Sih 1982; Werner et al. 1983).

So we know that foraging behavior changes predictably in response to predation threat. People studying rare and elusive carnivores in the field now apply this knowledge to their work. For example, mule deer (*Odocoileus hemionus*) foraging behavior signals the presence and threat of mountain lions (*Puma concolor*) nearby (Altendorf et al. 2001) (see chap. 13). The behavior of Nilgai antelope (*Boselaphus tragocamelus*) at water holes helps rangers in India to monitor tigers (*Panthera tigris*) (Brown, personal communication). And the behavior of blue sheep (*Pseudois nayaur*) in Nepal reveals the proximity of snow leopards (*Panthera uncia*) (Gurung 2003). Recently, the vigilance behaviors of Himalayan tahr helped Som Ale (unpublished data) to find and see two snow leopards in a span of weeks, providing the first confirmed records in 40 years of snow leopards in the region of his work.

Managing Reserves for *Kulturmeider*

Even after we successfully deploy reconciled habitats all over the earth, our reserves will retain great importance (Rosenzweig 2006). Although big fierce species sometimes find surprising places among us (Gaby et al. 1985; Dinerstein 2002), I guess that some never will. And our reserves will also provide the only habitats for inveterate *kulturmeider*. We shall probably wish to maximize the ability of those reserves to support those species that cannot find natural homes elsewhere (Rosenzweig 2005).

Suppose, for example, that careful work reveals the population of a *kulturmeider* (recall the Amani sunbird) to be suffering from competition with other members of its guild that are *kulturfolger* (like collared sunbirds). The manager may want to take steps to restrict the *kulturfolger* in the reserve. I doubt that he will find this easy to do. There may never be general rules to guide us. The job will require perceptive observation of behaviors and considerable inventiveness. And some well-meaning people may not understand. But the result will make the most of the environmental relicts that we save.

One might readily summarize my attitude toward enlightened management of reserves thus: Behaviors are the shadows of natural selection, of population dynamics, and of community processes (Rosenzweig 2001c). Optimality theories teach us how to ponder these shadows, revealing the state, the genesis, and the fundamental workings of the natural assemblages that have cast them. Because they give us some basic understanding of what is going on, they also provide other valuable services. Optimality theories suggest what we must do to achieve our conservation goals. They also give us some confidence that what we do will turn out as we intend. And they constitute an organizing system that will help us when, inevitably, we struggle to understand our mistakes and to correct them.

14.7   Coevolution in the Homogocene

Conservation ecology—both types, reservation and reconciliation—will supply a new set of ecological theaters for G. E. Hutchinson's evolutionary plays. That is, after all, the idea; the old theaters are vanishing so rapidly that if we do not supply new theaters, most of the players will vanish. The existing players may not know exactly how to act in the new theaters. That is to say, most or all will have to learn a new part. Yet at least the new theaters give them the chance to rehearse and improve. Natural selection being the consummate teacher that she is, we can expect them to improve a great deal. Much evidence indicates that evolution can occur quite rapidly in an anthropogenic environment (Ashley et al. 2003).

How will species change? What will happen to their niches and their behaviors? How will life in the new communities function? Optimal foraging was set up to answer exactly such questions (MacArthur and Pianka 1966). So far, progress in answering them has come from a growing body of theory with exciting potential.

One may tap into the literature of this work in several places (Abrams 2001; Cohen et al. 2001; Holt and Gomulkiewicz 1997b; Rosenzweig and Ziv 1999). This work asks a basic question: How can we predict the evolution of fundamental niches? Many subquestions arise, including the following: How does diversity affect the shape of niches (especially their breadths)? What prevents niches from evolving in response to environmental change? Does competition restrict the degree of specialization, and if so, how can we predict its upper limit? The work has only begun.

Nevertheless, we can depend on one aspect of change during the Homogocene. Not change directed by natural selection, but change elicited by the environments we provide for ourselves. If we continue down our current path, nature will wither and diminish. We will scarcely notice. Our baseline will decline each generation, and our disappointment will occupy a low priority in our lives. Sad Bayesians though we be, natural selection has made us Bayesians. We cannot be anything else.

But if we resolve to take advantage of what we already know, to learn more, to invent new environments and inject life into the sterility with which we now surround ourselves, then our baseline will shift upward. Reconciliation ecology will become the great environmental educator. Encompassing us in beauty, it will teach us what we can have and what to work for. That change in behavior will be most welcome.

14.8   Summary

Foraging and habitat selection theories provide a sound basis for conservation of species diversity. Optimality-grounded research can efficiently and rapidly monitor the population sizes of species. It can reveal underlying habitat needs and preferences as well as fundamental community organization. *Homo sapiens* is the single indispensable species whose optimal behaviors we must understand. People set the rules for managing biotic diversity, and those rules must not oppose the natural, evolved behaviors of our species, particularly those that judge—however subconsciously—the costs and benefits of what we do and might do. In addition, people will increasingly engage in reconciliation ecology, redesigning their own habitats to welcome more and more nonhuman species. Those redesigned habitats must do more than sustain us in health and comfort. We will deploy them only if they satisfy us aesthetically. Once we do, they will reverse the intergenerational decline in human environmental expectations known as the shifting baseline.

14.9   Suggested Readings

Rosenzweig (2005) provides an article that complements this chapter. Heerwagen and Orians (1993) discuss attempts to understand what people like. Penn (2003) speculates on the evolutionary roots of human-environment interactions. Rosenzweig (2003b) provides a manifesto for reconciliation ecology. Daily and Ellison (2002) explore the problem of bringing profit under the tent of conservation. McNeely and Scherr (2002) review the crucial task of combining farming with conservation in the world's tropics. Dinerstein (2002) reviews World Wildlife's massive and inspiring undertaking to integrate some quite dangerous mammals into productive human habitats.

Contributors

Melissa M. Adams-Hunt
Psychology Department
University of California at Berkeley
3210 Tolman Hall #1650
Berkeley, CA 94720-1650
USA
m_adams@berkeley.edu

Peter A. Bednekoff
Biology Department
Eastern Michigan University
Ypsilanti, MI 48197
USA
peter.bednekoff@emich.edu

Anders Brodin
Department of Theoretical Ecology
Lund University
Ecology Building
S-223 62 Lund,
Sweden
Anders.Brodin@teorekol.lu.se

Joel S. Brown
Department of Biological Sciences
University of Illinois at Chicago
845 West Taylor St
Chicago, Illinois 60607-7060
USA
squirrel@uic.edu

Thomas W. Castonguay
Nutrition and Food Science
University of Maryland
0112 Skinner Building
College Park, MD 20742-7521
USA
castong@umd.edu

Colin W. Clark
Department of Applied Mathematics
University of British Columbia
Vancouver, British Columbia V6T 1Z2
Canada
colin_clark@shaw.ca

Kristin L. Field
Department of Evolution, Ecology, and
    Organismal Biology
Ohio State University
318 W. 12th Avenue
Columbus, OH 43210-1293
USA
field.23@osu.edu

Ian M. Hamilton
Department of Evolution, Ecology and
    Organismal Biology
Ohio State University
318 W. 12th Avenue
Columbus, OH 43210-1293
USA
hamilton.598@osu.edu

Robert D. Holt
Department of Zoology
University of Florida
P.O. Box 118525
Gainesville, FL 32611-8525
USA
rdholt@zoo.ufl.edu

Alasdair I. Houston
Department of Biological Sciences
University of Bristol
Bristol, BS8 1UG
UK
a.i.houston@bristol.ac.uk

Lucia F. Jacobs
Psychology Department
University of California at Berkeley
3210 Tolman Hall #1650
Berkeley, CA 94720-1650
USA
jacobs@berkeley.edu

Tristan Kimbrell
Department of Zoology
University of Florida
P.O. Box 118525
Gainesville, FL 32611-8525
USA
kimbrell@ufl.edu

Burt P. Kotler
Mitrani Department of Desert Ecology
Ben-Gurion University of the Negev
Jacob Blaustein Institute for Desert
    Research,
Midreshet Ben-Gurion, 84990
Israel
kotler@bgu.ac.il

Åke Lindström
Department of Animal Ecology
Lund University
S-223 62 Lund,
Sweden
ake.lindstrom@zooekol.lu.se

Georgia Mason
Department of Animal and Poultry
    Science
University of Guelph
Guelph, Ontario N1G 2W1
Canada
gmason@uoguelph.ca

John M. McNamara
Department of Mathematics
University of Bristol
University Walk
Bristol, BS8 1TW
UK
John.McNamara@bristol.ac.uk

John B. Mitchell
Department of Social Science
Brescia University College
University of Western Ontario
London, Ontario N6G 1H2
Canada
jbmitche@uwo.ca

Jonathan A. Newman
Department of Environmental Biology
University of Guelph
Guelph, Ontario N1G 2W1
Canada
jnewma01@uoguelph.ca

Vladimir V. Pravosudov
University of Nevada, Reno
Biology Department m/s 314
Reno, NV 89557
vpravosu@unr.edu

Frederick D. Provenza
Department of Wildland Resources
Utah State University
Logan, UT 84322-5230
USA
stan@cc.usu.edu

David Raubenheimer
School of Biological Sciences and
    Liggins Institute
University of Auckland
Private Bag 92019
Auckland
New Zealand
d.raubenheimer@auckland.ac.nz

Michael L. Rosenzweig
Department of Ecology & Evolutionary
    Biology
University of Arizona
Tucson, AZ 85721
USA
scarab@u.arizona.edu

Kenneth A. Schmidt
Department of Biological Sciences,
    Texas Tech University
MS 3131
Lubbock, TX 79409
USA
Kenneth.Schmidt@ttu.edu

David F. Sherry
Department of Psychology, University
    of Western Ontario
London, Ontario N6A 5C2
Canada
sherry@uwo.ca

Peter Shizgal
Center for Studies in Behavioral
    Neurobiology
Concordia University
7141 Sherbrooke St. W.
Montreal, Quebec H4B 1 R6
Canada
shizgal@csbn.concordia.ca

David W Stephens
Ecology, Evolution and Behavior
University of Minnesota
1987 Upper Buford Circle
St. Paul, MN 55108
USA
dws@umn.edu

Thomas A. Waite
Department of Evolution, Ecology, and
    Organismal Biology
Ohio State University
318 W. 12th Avenue
Columbus, OH 43210-1293
USA
waite.1@osu.edu

Christopher J. Whelan
Illinois Natural History Survey
Midewin National Tall Grass Prairie
South State Highway 53
Wilmington, IL 60481
USA
virens@isp.com

Stephen C. Woods
Obesity Research Center
University of Cincinnati
2170 East Galbraith Road
Cincinnati, OH 45237
steve.woods@psychiatry.uc.edu

Ron Ydenberg
Department of Biological Sciences
Simon Fraser University
8888 University Drive
Burnaby, British Columbia V5A 1S6
Canada
ydenberg@sfu.ca

Literature Cited

Abe, T., and Higashi, M. 1991. Cellulose centered perspective on terrestrial community structure. *Oikos* 60:127–133.

Abijaoude, J. A., Morand-Fehr, P., Tessier, J., Schmidely, P., and Sauvant, D. 2000. Diet effect on the daily feeding behaviour, frequency and characteristics of males in dairy goats. *Livestock Production Science* 64:29–37.

Abrahams, M. V. 1986. Patch choice under perceptual constraints: A case for departures from an ideal free distribution. *Behavioral Ecology and Sociobiology* 19:409–415.

Abrahams, M. V., and Dill, L. M. 1989. A determination of the energetic equivalence of the risk of predation. *Ecology* 70:999–1007.

———. 1996. The value of titration experiments: A reply to Moody et al. (1996). *Behavioral Ecology and Sociobiology* 44:147–148.

Abrahams, M. V., and Sutterlin, A. 1999. The foraging and antipredator behaviour of growth-enhanced transgenic Atlantic salmon. *Animal Behaviour* 58:933–942.

Abrams, P. A. 1982. Functional responses of optimal foragers. *American Naturalist* 120:382–390.

———. 1984. Foraging time optimization and interactions within foodwebs. *American Naturalist* 124:80–96.

———. 1987. The functional responses of adaptive consumers of two resources. *Theoretical Population Biology* 32:262–288.

———. 1988. Resource productivity-consumer species diversity: Simple models of competition in spatially heterogeneous environments. *Ecology* 69:1418–1433.

———. 1992. Adaptive foraging by predators as a cause of predator-prey cycles. *Evolutionary Ecology* 6:56–72.

———. 1993. Indirect effects arising from optimal foraging. In *Mutualism and Community Organization*, ed. H. Kawanabe, J. Cohen, and K. Iwasake, 255–279. New York: Oxford University Press.

———. 1994. Should prey overestimate the risk of predation? *American Naturalist* 144:317–328.

———. 1995. Overestimation versus underestimation of predation risk: A reply to Bouskila et al. *American Naturalist* 145:1020–1024.

507

———. 1997. Anomalous predictions of ratio-dependent models of predation. *Oikos* 80:163–171.

———. 1999. The adaptive dynamics of consumer choice. *American Naturalist* 153:83–97.

———. 2000. The evolution of predator-prey interactions: Theory and evidence. *Annual Reviews of Ecology and Systematics* 31:79–105.

———. 2001. Adaptive dynamics: Neither F nor G. *Evolutionary Ecology Research* 3:369–372.

Abrams, P. A., and Kawecki, T. J. 1999. Adaptive host preference and the dynamics of host-parasitoid interactions. *Theoretical Population Biology* 56:307–324.

Abrams, P. A., and Schmitz, O. J. 1999. The effect of risk of mortality on the foraging behavior of animals faced with time and digestive capacity constraints. *Evolutionary Ecology Research* 1:285–301.

Abrams, T. W., Karl, K. A., and Kandel, E. R. 1991. Biochemical studies of stimulus convergence during classical conditioning in *Aplysia*: Dual regulation of adenylate cyclase by Ca^{2+}/calmodulin and transmitter. *Journal of Neuroscience* 11:2655–2665.

Abramsky, Z., Ovadia, O., and Rosenzweig, M. L. 1994. The shape of a *Gerbillus pyramidum* (Rodentia, Gerbillinae) isocline: An experimental field study. *Oikos* 69:318–326.

Abramsky, Z., and Pinshow, B. 1989. Changes in the foraging effort in 2 gerbil species correlate with habitat type. *Oikos* 56:43–53.

Abramsky, Z., Rosenzweig, M. L., and Pinshow, B. 1991. The shape of a gerbil isocline measured using principles of optimal habitat selection. *Ecology* 72:329–340.

Abramsky, Z., Rosenzweig, M. L., Pinshow, B., Brown, J. S., Kotler, B., and Mitchell, W. A. 1990. Habitat selection: An experimental field test with two gerbil species. *Ecology* 71:2358–2369.

Abramsky, Z., Rosenzweig, M. L., and Subach, A. 1992. The shape of a gerbil isocline—an experimental field study. *Oikos* 63:193–199.

———. 1998. Do gerbils care more about competition or predation? *Oikos* 83:75–84.

———. 2000. The energetic cost of competition: Gerbils as moneychangers. *Evolutionary Ecology Research* 2:279–292.

———. 2001. The cost of interspecific competition in gerbils. *Journal of Animal Ecology* 70:561–517.

———. 2002a. The costs of apprehensive foraging. *Ecology* 83:1330–1340.

———. 2002b. Measuring the benefit of habitat selection. *Behavioral Ecology* 13:497–502.

Abramsky, Z., Shachak, M., Subach, A., Brand, S., and Alfia, H. 1992. Predator-prey relationships: Rodent-snail interactions in the central Negev. *Oikos* 65:128–133.

Abramsky, Z., Strauss, E., Subach, A., Kotler, B. P., and Reichman, A. 1996. The effect of barn owls (*Tyto alba*) on the activity and microhabitat selection of *Gerbillus allenbyi* and *G. pyramidum*. *Oecologia* 105:313–319.

Adolph, E. F. 1947. Urges to eat and drink in rats. *American Journal of Physiology* 151:110–125.

Adriaensen, F., Dhondt, A. A., Dongen, S. V., Lens, L., and Matthysen, E. 1998. Stabilizing selection on blue tit fledging mass in the presence of sparrowhawks. *Proceedings of the Royal Society of London* B 265:1011–1016.

Afik, D., McWilliams, S. R., and Karasov, W. H. 1997. A test for passive absorption of glucose in yellow-rumped warblers and its ecological implications. *Physiological Zoology* 70:370–377.

Afik, D., Vidal, E. C., Martínez del Rio, C., and Karasov, W. H. 1995. Dietary modulation of intestinal hydrolytic enzymes in yellow-rumped warblers. *American Journal of Physiology* 38:R413–R420.

Ainslie, G. 1974. Impulse control in pigeons. *Journal of the Experimental Analysis of Behavior* 21:485–489.

Akcakaya, H. R., Arditi, R., and Ginzburg, L. R. 1995. Ratio-dependent predation: An abstraction that works. *Ecology* 76:995–1004.

Alerstam, T., and Lindström, Å. 1990. Optimal bird migration: The relative importance of time, energy and safety. In *Bird Migration: Physiology and Ecophysiology*, ed. E. Gwinner, 331–351. Heidelberg: Springer-Verlag.

Alexander, R. M. 1991. Optimization of gut structure and diet for higher vertebrate herbivores. *Philosophical Transactions of the Royal Society* B 333:249–255.

———. 1993. The relative merits of foregut and hindgut fermentation. *Journal of Zoology* 231:391–401.

———. 1994. Optimum gut structure for specified diets. In *The Digestive System in Mammals: Food, Form, and Function*, ed. D. J. Chivers and P. Langer, 54–62. Cambridge: Cambridge University Press.

Allden, W. G., and Whitaker, I. A. 1970. The determinants of herbage intake by grazing sheep: The interrelationship of factors influencing herbage intake and availability. *Australian Journal of Agricultural Research* 21:755–766.

Altendorf, K. B., Laundré, J. W., González, C. A. L., and Brown, J. S. 2001. Assessing effects of predation risk on foraging behavior of mule deer. *Journal of Mammalogy* 82:430–439.

Andersson, M., and Krebs, J. R. 1978. The evolution of hoarding behaviour. *Animal Behaviour* 26:707–711.

Andrewartha, H. G., and Birch, L. C. 1954. *The Distribution and Abundance of Animals*. Chicago: University of Chicago Press.

Anholt, B. R., Werner, E., and Skelly, D. K. 2000. Effect of food and predators on the activity of four larval ranid frogs. *Ecology* 81:3509–3521.

Anholt, R. R. H. 1994. Signal integration in the nervous system: Adenylate cyclases as molecular coincidence detectors. *Trends in Neurosciences* 17:37–41.

Appleby, M. C., and Lawrence, A. B. 1987. Food restriction as a cause of stereotypic behavior in tethered gilts. *Animal Production* 45:103–110.

Arcis, V., and Desor, D. 2003. Influence of environment structure and food availability on the foraging behaviour of the laboratory rat. *Behavioural Processes* 60:191–198.

Armstrong, R. A., and McGehee, R. 1980. Competitive exclusion. *American Naturalist* 115:151–170.

Arnold, G. W. 1975. Herbage intake and grazing behaviour in cows of four breeds at different physiological states. *Australian Journal of Agricultural Research* 26:1017–1024.

Arnold, S. J. 1978. The evolution of a special class of modifiable behaviors in relation to environmental pattern. *American Naturalist* 112:415–427.

Ashley, M. V., Willson, M. F., Pergams, O. R. W., O'Dowd, D. J., Gende, S. M., and Brown, J. S. 2003. Evolutionarily enlightened management. *Biological Conservation* 111:115–123.

Åstrand, P. O., and Rodahl, K. 1970. *Textbook of Work Physiology*. New York: McGraw-Hill.

Axelrod, R., and Hamilton, W. D. 1981. The evolution of cooperation. *Science* 211:1390–1396.

Bachevalier, J. M. M. 1986. Visual recognition impairment follows ventromedial but not dorsolateral prefrontal lesions in monkeys. *Behavioural Brain Research* 20:249–261.

Bacskai, B. J., Hochner, B., Mahaut-Smith, M., Adams, S. R., Kaang, B. K., Kandel, E. R., and Tsien, R. Y. 1993. Spatially resolved dynamics of cAMP and protein kinase A subunits in *Aplysia* sensory neurons. *Science* 260:222–226.

Baddeley, A. 1986. *Working Memory*. Oxford: Oxford University Press.

———. 1998. Recent developments in working memory. *Current Opinion in Neurobiology* 8:234–238.

Baharav, D., and Rosenzweig, M. 1985. Optimal foraging in Dorcas gazelles. *Journal of Arid Environments* 9:167–171.

Bailey, D. 1995. Daily selection of feeding areas by cattle in homogeneous and heterogeneous environments. *Applied Animal Behaviour Science* 45:183–199.

Baird, R. W., and Dill, L. M. 1996. Ecological and social determinants of group size in transient killer whales. *Behavioral Ecology* 7:408–416.

Bakker, R. T. 1983. The deer flees, the wolf pursues: Incongruencies in predator-prey co-evolution. In *Coevolution*, ed. D. J. Futuyma and M. Slatkin, 350–382. Sunderland, MA: Sinauer Associates.

Balas, M. T., and Adams, E. S. 1996. The dissolution of cooperative groups: Mechanisms of queen mortality in incipient fire ant colonies. *Behavioral Ecology and Sociobiology* 38:391–399.

Balda, R. P., and Kamil, A. C. 1989. A comparative study of cache recovery by three corvid species. *Animal Behaviour* 38:486–495.

———. 1992. Long-term spatial memory in Clark's nutcracker, *Nucifraga columbiana*. *Animal Behaviour* 44:761–769.

Banks, P. B. 2001. Predation-sensitive grouping and habitat use by eastern grey kangaroos: A field experiment. *Animal Behaviour* 61:1013–1021.

Bao, S., Chan, V. T., and Merzenich, M. M. 2001. Cortical remodelling induced by activity of ventral tegmental dopamine neurons. *Nature* 412.79 83.

Bar, Y., Abramsky, Z., and Gutterman, Y. 1984. Diet of gerbilline rodents of the Israeli desert. *Journal of Arid Environments* 7:371–376.

Barber, I., Nairn, D., and Huntingford, F. A. 2001. Nests as ornaments: Revealing construction by male sticklebacks. *Behavioral Ecology* 12:390–396.

Barbosa, A., Barluenga, M., and Moreno, E. 2000. Effects of body mass on the foraging behaviour of subordinate Coal Tits *Parus ater*. *Ibis* 142:428–434.

Barkley, C. L., and Jacobs, L. F. 1998. Visual environment and delay affect cache retrieval accuracy in a food-storing rodent. *Animal Learning & Behavior* 26:439–447.

Barluenga, M., Barbosa, A., and Moreno, E. 2003. Effect of daily body mass variation on the foraging behaviour of tit species (*Parus* spp.). *Ethology* 109:971–979.

Barnard, C. J., and Sibly, R. M. 1981. Producers and scroungers: A general model and its application to captive flocks of house sparrows. *Animal Behaviour* 29:543–550.

Barnea, A., and Nottebohm, F. 1994. Seasonal recruitment of hippocampal neurons in adult free-ranging black-capped chickadees. *Proceedings of the National Academy of Sciences U.S.A.* 91:11217–11221.

———. 1996. Recruitment and replacement of hippocampal neurons in young and adult chickadees: An addition to the theory of hippocampal learning. *Proceedings of the National Academy of Sciences U.S.A.* 93:714–718.

Barta, Z., McNamara, J. M., Houston, A. I., Welham, R. K., Hedenström, A., Weber, T. P., and Fero, O. 2006. Annual routines of non-migratory birds: Optimal moult strategies. *Oikos* 112:580–593.

Barton, R. A., and Dean, P. 1993. Comparative evidence indicating neural specialization for predatory behaviour in mammals. *Proceedings of the Royal Society of London* B 254:63–68.

Barton, R. A., and Harvey, P. H. 2000. Mosaic evolution of brain structure in mammals. *Nature* 405:1055–1058.

Bartsch, D., Ghirardi, M., Skehel, P. A., Karl, K. A., Herder, S. P., Chen, M., Bailey, C. H., and Kandel, E. R. 1995. *Aplysia* CREB2 represses long-term facilitation: Relief of repression converts transient facilitation into long-term functional and structural change. *Cell* 83:979–992.

Bashaw, M. J., Tarou, L. R., Maki, T. S., and Maple, T. L. 2001. A survey assessment of variables related to stereotypy in captive giraffe and okapi. *Applied Animal Behaviour Science* 73:235–247.

Basil, J. A., Kamil, A. C., Balda, R. P., and Fite, K. V. 1996. Differences in hippocampal volume among food storing corvids. *Brain Behavior and Evolution* 47:156–164.

Bateson, M., and Kacelnik, A. 1998. Risk-sensitive foraging: Decision making in variable environments. In *Cognitive Ecology: The Evolutionary Ecology of Information Processing and Decision Making*, ed. R. Dukas, 297–420. Chicago: University of Chicago Press.

Bauer, R. H., and Fuster, J. M. 1976. Delayed-matching and delayed-response deficit from cooling dorsolateral prefrontal cortex in monkeys. *Journal of Comparative and Physiological Psychology* 90:293–302.

Baumont, R. 1996. Palatability and feeding behaviour in ruminants. A review. *Annales de Zootechnie* 15:005 100.

Baumont, R., Prache, S., Meuret, M., and Morand-Fehr, P. 2000. How forage characteristics influence behaviour and intake in small ruminants: A review. *Livestock Production Science* 64:15–28.

Bautista, L. M., and Lane, S. L. 2001. Coal tits increase evening mass in response to tawny owl calls. *Acta Ethologica* 2:105–110.

Bautista, L. M., Tinbergen, J., Wiersma, P., and Kacelnik, A. 1998. Optimal foraging and beyond: How starlings cope with changes in food availability. *American Naturalist* 152:543–561.

Beach, F. A. 1950. The Snark was a Boojum. *American Psychologist* 5:115–124.

Beauchamp, G. 1992. Effects of energy requirements and worker mortality on colony growth and foraging in the honey bee. *Behavioral Ecology and Sociobiology* 31:123–132.

———. 1998. The effect of group size on mean food intake rate in birds. *Biological Reviews of the Cambridge Philosophical Society* 73:449–472.

Beauchamp, G., Bélisle, M., and Giraldeau, L.-A. 1997. Influence of conspecific attraction on the spatial distribution of learning foragers in a patchy habitat. *Journal of Animal Ecology* 66:671–682.

Beauchamp, G., Ens, B. J., and Kacelnik, A. 1991. A dynamic model of food allocation to starling (*Sturnus vulgaris*) nestlings. *Behavioral Ecology* 2:21–37.

Beauchamp, G., and Giraldeau, L.-A. 1996. Group foraging revisited: Information sharing or producer-scrounger game? *American Naturalist* 148:738–743.

———. 1997. Patch exploitation in a producer-scrounger system: Test of a hypothesis using flocks of spice finches (*Lonchura punctulata*). *Behavioral Ecology* 8:54–59.

Beauchamp, G., Guillemette, M., and Ydenberg, R. 1992. Prey selection while diving by common eiders, *Somateria mollissima*. *Animal Behaviour* 44:417–426.

Beckerman, A. P., Uriarte, M., and Schmitz, O. 1997. Experimental evidence for a behavior-mediated trophic cascade in a terrestrial food chain. *Proceedings of the National Academy of Sciences U.S.A.* 94:10735–10738.

Bednekoff, P. A. 1996. Translating mass dependent flight performance into predation risk: An extension of Metcalfe & Ure. *Proceedings of the Royal Society of London* B 263:887–889.

————. 1997. Mutualism among safe, selfish sentinels: A dynamic game. *American Naturalist* 150:373–392.

Bednekoff, P. A., Balda, R. P., Kamil, A. C., and Hile, A. G. 1997. Long-term spatial memory in four seed-caching corvid species. *Animal Behaviour* 53:335–341.

Bednekoff, P. A., Biebach, H., and Krebs, J. R. 1994. Great tit fat reserves under unpredictable temperatures. *Journal of Avian Biology* 25:156–160.

Bednekoff, P. A., and Houston, A. I. 1994a. Avian daily foraging patterns: Effects of digestive constraints and variability. *Evolutionary Ecology* 8:36–52.

————. 1994b. Optimizing fat reserves over the whole winter: A dynamic model. *Oikos* 71:408–415.

Bednekoff, P. A., and Krebs, J. R. 1995. Great tit reserves: Effects of changing and unpredictable feeding day length. *Functional Ecology* 9:457–462.

Bednekoff, P. A., and Lima, S. L. 1998a. Randomness, chaos and confusion in the study of antipredator vigilance. *Trends in Ecology and Evolution* 13:284–287.

————. 1998b. Re-examining safety in numbers: Interactions between risk dilution and collective detection depend upon predator targeting behaviour. *Proceedings of the Royal Society of London* B 265:2021–2026.

Beecham, J., and Farnsworth, K. 1998. Animal foraging from an individual perspective: An object orientated model. *Ecological Modelling* 113:141–156.

Behmer, S. T., Raubenheimer, D., and Simpson, S. J. 2001. Frequency-dependent food selection in locusts: A geometric analysis of the role of nutrient balancing. *Animal Behaviour* 61:995–1005.

Bekoff, M., Allen, C., and Burghardt, G. M. 2002. *The Cognitive Animal: Empirical and Theoretical Perspectives on Animal Cognition*. Cambridge, MA: MIT Press.

Bélisle, M. 1998. Foraging group size: Models and a test with jaegers kleptoparasitizing terns. *Ecology* 79:1922–1938.

Bell, R. H. V. 1970. The use of the herb layer by grazing ungulates in the Serengeti. In *Animal Populations in Relation to Their Food Resources*, ed. A. Watson, 111–123. New York: Blackwell Scientific Publications.

Belovsky, G. E. 1978. Diet optimization in a generalist herbivore: The moose. *Theoretical Population Biology* 14:105–134.

Ben-Natan, G., Abramsky, Z., Kotler, B. P., and Brown, J. S. 2004. Seed redistribution in sand dunes: A basis for coexistence of two rodent species. *Oikos* 105:325–335.

Benkman, C. 1988. Seed handling ability, bill structure, and the cost of specialization for crossbills. *Auk* 105:715–719.

Bennett, P. M., and Harvey, P. H. 1985. Relative brain size and ecology in birds. *Journal of Zoology* 207:151–169.

Benson, K., and Stephens, D. W. 1996. Interruptions, tradeoffs and temporal discounting. *American Zoologist* 36:506–517.

Berger, J., and Cunningham, C. 1994. Phenotypic alterations, evolutionarily significant structures, and rhino conservation. *Conservation Biology* 8:833–840.

Berger, J., Swenson, J. E., and Persson, I. L. 2001. Recolonizing carnivores and naive prey: Conservation lessons from Pleistocene extinctions. *Science* 291:1036–1039.

Bergeron, R., Badnell-Waters, A., Lambton, S., and Mason, G. 2006. Oral behaviors in captive ungulates: Foraging, frustration and gut health. In *Stereotypic Animal Behaviour: Fundamentals and Applications to Welfare*, 2nd ed., ed. G. Mason, and J. Rushen, 19–57. Wallingford, UK: CAB International.

Berkovitch, F. B., Hauser, M. D., and Jones, J. H. 1995. The endocrine stress response and alarm vocalizations in rhesus macaques. *Animal Behaviour* 49:1703–1706.

Bernays, E. A. 1991. Evolution of insect morphology in relation to plants. *Philosophical Transactions of the Royal Society* B 333:257–264.

Bernays, E. A., Augner, M., and Abbot, D. K. 1997. A behavioral mechanism for incorporating an unpalatable food in the diet of a generalist herbivore (Orthoptera: Acrididae). *Journal of Insect Behavior* 10:841–858.

Bernays, E. A., and Chapman, R. F. 1970a. Experiments to determine the basis of food selection by *Chorthippus parallelus* (Zetterstedt) (Orthoptera: Acrididae) in the field. *Journal of Animal Ecology* 39:761–776.

———. 1970b. Food selection by *Chorthippus parallelus* (Zetterstedt) (Orthoptera: Acrididae) in the field. *Journal of Animal Ecology* 39:383–394.

———. 1994. *Host-Plant Selection by Phytophagous Insects.* New York: Chapman & Hall.

Bernstein, C. 2000. Host-parasitoid models: The story of a successful failure. In *Parasitoid Population Biology,* ed. M. E. Hochberg and A. R. Ives, 41–57. Princeton, NJ: Princeton University Press.

Berteaux, D., Crete, M., Huot, J., Maltais, J., and Ouellet, J. P. 1998. Food choice by white-tailed deer in relation to protein and energy content of the diet. A field experiment. *Oecologia* 115:84–92.

Bertram, B. C. R. 1978. Living in groups: Predator and prey. In *Behavioral Ecology: An Evolutionary Approach,* ed. J. R. Krebs and N. B. Davies, 64–96. Oxford: Blackwell Scientific Publications.

Best, P. J., White, A. M., and Minai, A. 2001. Spatial processing in the brain: The activity of hippocampal place cells. *Annual Review of Neuroscience* 24:459–486.

Bhatt, R. S., Wasserman, E. A., Reynolds, W. F., and Knauss, K. S. 1988. Conceptual behavior in pigeons: Categorization of both familiar and novel examples from four classes of natural and artificial stimuli. *Journal of Experimental Psychology: Animal Behavior Processes* 14:219–234.

Bibby, C. J., and Green, R. E. 1980. Foraging behaviour of migrant pied flycatchers, *Ficedula hypoleuca,* on temporary territories. *Journal of Animal Ecology* 49:507–521.

Biebach, H. 1996. Energetics of winter and migratory fattening. In *Avian Energetics and Nutritional Ecology,* ed. C. Carey, 280–323. New York: Chapman & Hall.

Biegler, R., McGregor, A., Krebs, J. R., and Healy, S. D. 2001. A larger hippocampus is associated with longer-lasting spatial memory. *Proceedings of the National Academy of Sciences U.S.A.* 98:6941–6944.

Biegler, R., and Morris, R. G. M. 1993. Landmark stability is a prerequisite for spatial but not discrimination learning. *Nature* 361:631–633.

Bjorndal, K., and Bolten, A. B. 1993. Digestive efficiencies in herbivorous and omnivorous freshwater turtles on plant diets: Do herbivores have a nutritional advantage? *Physiological Zoology* 66:384–395.

Björnhag, G. 1994. Adaptations in the large intestine allowing small animals to eat fibrous foods. In *The Digestive System in Mammals: Food, Form, and Function,* ed. D. J. Chivers and P. Langer, 287–314. Cambridge: Cambridge University Press.

Bjornstad, O. N., and Grenfell, B. T. 2001. Noisy clockwork: Time series analysis of population fluctuations in animals. *Science* 293:638–643.

Black, J. L., and Kenney, P. A. 1984. Factors affecting diet selection by sheep. 2. Height and density of pasture. *Australian Journal of Agricultural Research* 35:565–578.

Blanckenhorn, W. U. 1991. Fitness consequences of food-based territoriality in water striders *Gerris remigis*. *Animal Behaviour* 42:147–149.

Blem, C. R. 1990. Avian energy storage. *Current Ornithology* 7:59–113.

Bliss, T. V. P., and Collingridge, G. L. 1993. A synaptic model of memory: Long-term potentiation in the hippocampus. *Nature* 361:31–39.

Bliss, T. V. P., and Lomo, T. 1973. Long-lasting potentiation of synaptic transmission in the dentate area of the anaesthetized rabbit following stimulation of the perforant path. *Journal of Physiology* 232:331–356.

Blood, D. C., and Radostits, O. M. 1989. *Veterinary Medicine: A Textbook of the Diseases of Cattle, Sheep, Pigs, Goats and Horses*. London: Balliere-Tindall.

Blough, D. S. 1992. Features of forms in pigeon perception. In *Cognitive Aspects of Stimulus Control*, ed. W. K. Honig, and J. Gregor Fetterman, 263–277. Hillsdale, NJ: Lawrence Erlbaum Associates.

Blumstein, D. T. 1992. Multivariate analysis of golden marmot maximum running speed: A new method to study MRS in the field. *Ecology* 73:1757–1767.

———. 1996. How much does social group size influence golden marmot vigilance? *Behaviour* 133:1133–1151.

———. 1998. Quantifying predation risk for refuging animals: A case study with golden marmots. *Ethology* 104:501–516.

———. 1999. Selfish sentinels. *Science* 284:1633–1634.

Blumstein, D. T., Evans, C. S., and Daniel, J. C. 1999. An experimental study of behavioural group size effects in tammar wallabies, *Macropus eugenii*. *Animal Behaviour* 58:351–360.

Bobisud, L. I., and Potratz, C. J. 1976. One-trial versus multi-trial learning for a predator encountering a model-mimic system. *American Naturalist* 110:121–128.

Bodner, M., Kroger, J., and Fuster, J. M. 1996. Auditory memory cells in dorsolateral prefrontal cortex. *NeuroReport* 7:1905–1908.

Boer, N. J. de. 1999. Pyrrolizidine alkaloid distribution in *Senecio jacobaea* rosettes minimises losses to generalist feeding. *Entomologia Experimentalis et Applicata* 91:169–173.

Boesch, C. 1991. Teaching among wild chimpanzees. *Animal Behaviour* 41:530–532.

Bolker, B., Holyoak, M., Krivan, V., Rowe, L., and Schmitz, O. 2003. Connecting theoretical and empirical studies of trait-mediated interactions. *Ecology* 84:1101–1114.

Bond, A. B., and Kamil, A. C. 1998. Apostatic selection by blue jays produces balanced polymorphism in virtual prey. *Nature* 395:594–596.

———. 1999. Searching image in blue jays: Facilitation and interference in sequential priming. *Animal Learning & Behavior* 27:461–471.

———. 2002. Visual predators select for crypticity and polymorphism in virtual prey. *Nature* 415:609–613.

Boness, D. J., and Bowen, W. D. 1996. The evolution of maternal care in pinnipeds. *BioScience* 46:645–654.

Bourtchuladze, R., Frenguelli, B., Blendy, J., Cioffi, D., Schutz, G., and Silva, A. J. 1994. Deficient long-term memory in mice with a targeted mutation of the cAMP-responsive element-binding protein. *Cell* 79:59–68.

Bouskila, A. 1995. Interactions between predation risk and competition: A field study with kangaroo rats and snakes. *Ecology* 76:165–178.

Bouskila, A., and Blumstein, D. T. 1992. Rules of thumb for predation hazard assessment: Predictions from a dynamic model. *American Naturalist* 139:161–176.

Bouskila, A., Blumstein, D. T., and Mangel, M. 1995. Prey under stochastic conditions should probably overestimate predation risk: A reply to Abrams. *American Naturalist* 145:1015–1019.

Boutin, S., Krebs, C. J., Boonstra, R., Dale, M. R. T., Hannon, S. J., Martin, K., and Sinclair, A. R. E. 1995. Population changes of the vertebrate community during a snowshoe hare cycle in Canada boreal forest. *Oikos* 74:69–80.

Bovet, D., and Vauclair, J. 1998. Functional categorization of objects and of their pictures in baboons (*Papio anubis*). *Learning and Motivation* 29:309–322.

Bowers, M., and Breland, B. 1996. Foraging of gray squirrels on an urban-rural gradient: Use of the GUD to assess anthropogenic impact. *Ecological Applications* 6:1135–1142.

Bowers, M. A., Jefferson, J. L., and Kuebler, M. 1993. Variation in giving-up densities of foraging chipmunks (*Tamias striatus*) and squirrels (*Sciurus carolinensis*). *Oikos* 66:229–236.

Bozinovic, F., and Torres, C. H. 1998. Does digestion rate affect diet selection? A study in *Octodon degus*, a generalist herbivorous rodent. *Acta Theriologica* 43:205–212.

Bradbury, J. W., and Vehrencamp, S. L. 1998. *Principles of Animal Communication.* Sunderland MA: Sinauer Associates.

Bradshaw, J. W. S., Goodwin, D., Legrand, D., V., and Nott, H. M. 1996. Food selection by the domestic cat, an obligate carnivore. *Comparative Biochemistry and Physiology* A 114:205–209.

Brandauer, N., and Wu, S. K. 1978. The freshwater mussels (family Unionidae). *Natural History Inventory of Colorado* 2:41–60.

Breland, K., and Breland, M. 1961. The misbehavior of organisms. *American Psychologist* 16:681–684.

Breuner, C. W., and Wingfield, J. C. 2000. Rapid behavioral response to corticosterone varies with photoperiod and dose. *Hormones and Behavior* 37:23–30.

Brodbeck, D. R. 1994. Memory for spatial and local cues: A comparison of a storing and a nonstoring species. *Animal Learning & Behavior* 22:119–133.

Brodin, A. 1992. Cache dispersion affects retrieval time in hoarding willow tits. *Ornis Scandinavica* 23:7–12.

———. 1994a. The disappearance of caches that have been stored by naturally foraging willow tits. *Animal Behaviour* 47:730–732.

———. 1994b. The role of naturally stored food supplies in the winter diet of the boreal willow tit, *Parus montanus. Ornis Svecica* 4:31–40.

———. 1994c. Separation of caches between individual willow tits hoarding under natural conditions. *Animal Behaviour* 47:1031–1035.

———. 2000. Why do hoarding birds gain fat in winter in the wrong way? Suggestions from a dynamic model. *Behavioral Ecology* 11:27–39.

———. 2001. Mass-dependent predation and metabolic expenditure in wintering birds: Is there a trade-off between different forms of predation? *Animal Behaviour* 62:993–999.

Brodin, A., and Clark, C. W. 1997. Long-term hoarding in the Paridae: A dynamic model. *Behavioral Ecology* 8:178–185.

Brodin, A., and Ekman, J. 1994. Benefits of food hoarding. *Nature* 372:510.

Brodin, A., and Kunz, C. 1997. An experimental study of cache recovery by hoarding willow tits after different retention intervals. *Behaviour* 134:881–890.

Brodin, A., Lundborg, K., and Clark, C. 2001. The effect of dominance on food hoarding: A game theoretical model. *American Naturalist* 157:66–75.

Brodin, A., Olsson, O., and Clark, C. W. 1998. Modeling the breeding cycle of long-lived birds: Why do king penguins try to breed late? *Auk* 115:767–771.

Brooke, M. D. 1981. How an adult wheatear (*Oenanthe oenanthe*) uses its territory when feeding nestlings. *Journal of Animal Ecology* 50:683–696.

Brough, C. N., and Dixon, A. F. G. 1989. Intraclonal trade-off between reproductive investment and size of body fat in the vetch aphid, *Megoura viciae* Buckton. *Functional Ecology* 3:747–751.

Brown, B. W., and Batzli, G. D. 1984. Habitat selection by fox and gray squirrels: A multivariate analysis. *Journal of Wildlife Management* 48:616–621.

Brown, J. H. 1975. Geographical ecology of desert rodents. In *Ecology and Evolution of Communities*, ed. M. L. Cody, and J. M. Diamond, 315–341. Cambridge, MA: Harvard University Press.

Brown, J. H., and Ernest, S. K. M. 2002. Rain and rodents: Complex dynamics of desert consumers. *BioScience* 52:979–987.

Brown, J. H., and Kurzius, M. A. 1987. Composition of desert rodent faunas: Combinations of coexisting species. *Annales Zoologici Fennici* 24:227–237.

Brown, J. L. 1983. Cooperation: A biologist's dilemma. *Advances in the Study of Behavior* 13:1–37.

Brown, J. S. 1988. Patch use as an indicator of habitat preference, predation risk, and competition. *Behavioral Ecology and Sociobiology* 22:37–47.

———. 1989a. Coexistence on a seasonal resource. *American Naturalist* 133:168–182.

———. 1989b. Desert rodent community structure: A test of four mechanisms of coexistence. *Ecological Monographs* 59:1–20.

———. 1992. Patch use under predation risk. I. Models and predictions. *Annales Zoologici Fennici* 29:301–309.

———. 1996. Coevolution and community organization in three habitats. *Oikos* 75:193–206.

———. 1998. Game theory and habitat selection. In *Game Theory and Animal Behavior*, ed. L. A. Dugatkin, and H. K. Reeve, 188–220. New York: Oxford University Press.

———. 1999. Vigilance, patch use and habitat selection: Foraging under predation risk. *Evolutionary Ecology Research* 1:49–71.

———. 2000. Foraging ecology of animals in response to heterogeneous environments. In *The Ecological Consequences of Environmental Heterogeneity*, ed. M. J. Hutchings, L. A. John, and A. J. A. Stewart, 181–215. Oxford: Blackwell Scientific.

Brown, J. S., and Alkon, P. A. 1990. Testing values of crested porcupine habitats by experimental food patches. *Oecologia* 83:512–518.

Brown, J. S., and Kotler, B. P. 2004. Hazardous-duty pay and the foraging cost of predation. *Ecology Letters* 7:999–1014.

Brown, J. S., Kotler, B. P., and Bouskila, A. 2001. Ecology of fear: Foraging games between predators and prey with pulsed resources. *Annales Zoologici Fennici* 38:71–87.

Brown, J. S., Kotler, B. P., and Knight, M. H. 1998. Patch use in the pygmy rock mouse (*Peromyscus collinus*). *Mammalia* 62:108–112.

Brown, J. S., Kotler, B. P., and Mitchell, W. A. 1994. Foraging theory, patch use, and the structure of a Negev Desert granivore community. *Ecology* 75:2286–2300.

———. 1997. Competition between birds and mammals: A comparison of giving-up densities between crested larks and gerbils. *Evolutionary Ecology* 11:757–771.

Brown, J. S., Kotler, B. P., Smith, R. J., and Wirtz, W. O. II. 1988. The effects of owl predation on the foraging behavior of heteromyid rodents. *Oecologia* 76:408–415.

Brown, J. S., Kotler, B. P., and Valone, T. J. 1994. Foraging under predation: A comparison of energetic and predation costs in a Negev and Sonoran Desert rodent community. *Australian Journal of Zoology* 42:435–448.

Brown, J. S., Laundré, J. W., and Gurung, M. 1999. The ecology of fear: Optimal foraging, game theory, and trophic interactions. *Journal of Mammalogy* 80:385–399.

Brown, J. S., and Mitchell, W. A. 1989. Diet selection on depletable resources. *Oikos* 54:33–43.

Brown, J. S., and Morgan, R. A. 1995. Effects of foraging behaviour and spatial scale on diet selectivity: A test with fox squirrels. *Oikos* 74:122–136.

Brown, J. S., Morgan, R. A., and Dow, B. D. 1992. Patch use under predation risk. II. A test with fox squirrels, *Sciurus niger*. *Annales Zoologici Fennici* 29:311–318.

Brown, J. S., and Pavlovic, N. B. 1992. Evolution in heterogeneous environments: Effects of migration on habitat specialization. *Evolutionary Ecology* 6:360–382.

Brown, L. C., and Yeager, L. E. 1945. Fox and gray squirrels in Illinois. *Illinois Natural History Survey* 23.419–436.

Brust, D. G. 1993. Maternal brood care by *Dendrobates pumilio*: A frog that feeds its young. *Journal of Herpetology* 27:96–98.

Bryant, D. M., and Newton, A. V. 1994. Metabolic costs of dominance in dippers, *Cinclus cinclus*. *Animal Behaviour* 48:447–455.

Bshary, R., and Noe, R. 1997. Red colobus and Diana monkeys provide mutual protection against predators. *Animal Behaviour* 54:1461–1474.

Buddington, R. K., Chen, W., and Diamond, J. M. 1991. Dietary regulation of intestinal brush-border sugar and amino-acid transport in carnivores. *American Journal of Physiology* 261:R793–R801.

Bugnyar, T., and Kotrschal, K. 2002. Observational learning and the raiding of food caches in ravens, *Corvus corax*: Is it "tactical" deception? *Animal Behaviour* 64:185–195.

Bulmer, M. 1994. *Theoretical Evolutionary Ecology*. Sunderland, MA: Sinauer Associates.

Bumann, D., Krause, J., and Rubenstein, D. 1997. Mortality risk of spatial positions in animal groups: The danger of being in the front. *Behaviour* 134:1063–1076.

Bunsey, M., and Eichenbaum, H. 1995. Selective damage to the hippocampal region blocks long-term retention of a natural and nonspatial stimulus-stimulus association. *Hippocampus* 5:546–556.

———. 1996. Conservation of hippocampal memory function in rats and humans. *Nature* 379:255–257.

Burd, M., Archer, D., Aranwela, N., and Stradling, D. J. 2002. Traffic dynamics of the leaf-cutting ant *Atta cephalotes*. *American Naturalist* 159:283–293.

Burlison, A. J., Hodgson, J., and Illius, A. W. 1991. Sward canopy structure and the bite dimensions and bite weight of grazing sheep. *Grass and Forage Science* 46:29–38.

Bush, R. R., and Mosteller, F. 1955. *Stochastic Models for Learning*. New York: John Wiley and Sons.

Cahan, S., and Julian, G. E. 1999. Fitness consequences of cooperative colony founding in the desert leaf-cutter ant *Acromyrmex versicolor*. *Behavioral Ecology* 10:585–591.

Calder, W. A. 1985. The comparative biology of longevity and lifetime energetics. *Experimental Gerontology* 20:161–170.

Caldwell, C. A., and Whiten, A. 2002. Evolutionary perspectives on imitation: Is a comparative psychology of social learning possible? *Animal Cognition* 5:193–208.

Canali, E., Ferrante, V., Mattiello, S., Gottardo, F., and Verga, M. 2001. Are oral stereotypies and abomasal lesions correlated in veal calves? In *Proceedings of the 35th Congress of the ISAE*, ed. J. P. Garner, J. A. Mench, and S. P. Heekin, 103. University of California–Davis.

Cano Lozano, V., Bonnard, E., Gauthier, M., and Richard, D. 1996. Mecamylamine-induced impairment of acquisition and retrieval of olfactory conditioning in the honeybee. *Behavioural Brain Research* 81:215–222.

Capaldi, E. A., Robinson, G. E., and Fahrbach, S. E. 1999. Neuroethology of spatial learning: The birds and the bees. *Annual Review of Psychology* 50:651–682.

Capaldi, E. J., and Miller, D. J. 1988. Counting in rats: Its functional significance and the independent cognitive processes that constitute it. *Journal of Experimental Psychology: Animal Behavior Processes* 14:3–17.

Cappuccino, N., and Price, P. W. 1995. *Population Dynamics: New Approaches and Synthesis*. San Diego: Academic Press.

Caraco, T. 1980. On foraging time allocation in a stochastic environment. *Ecology* 61:119–128.

———. 1981. Risk-sensitivity and foraging groups. *Ecology* 62:527–531.

———. 1987. Foraging games in a random environment. In *Foraging Behavior*, ed. A. C. Kamil, J. R. Krebs, and H. R. Pulliam, 389–414. New York: Plenum Press.

Caraco, T., and Giraldeau, L.-A. 1991. Social foraging: Producing and scrounging in a stochastic environment. *Journal of Theoretical Biology* 153:559–583.

Caraco, T., Martindale, S., and Whitham, T. S. 1980. An empirical demonstration of risk-sensitive foraging preferences. *Animal Behaviour* 28:820–830.

Carbone, C., and Gittleman, J. L. 2002. A common rule for the scaling of carnivore density. *Science* 295:2273–2276.

Carbone, C., Mace, G. M., Roberts, S. C., and MacDonald, D. W. 1999. Energetic constraints on the diet of terrestrial carnivores. *Nature* 402:286–288.

Carbone, C., Toit, J. T. D., and Gordon, I. J. 1997. Feeding success in African wild dogs: Does kleptoparasitism by spotted hyaenas influence hunting group size? *Journal of Animal Ecology* 66:318–326.

Caro, T. M., and Fitzgibbon, C. D. 1992. Large carnivores and their prey: The quick and the dead. In *Natural Enemies: The Population Biology of Predators, Parasites and Diseases*, ed. M. J. Crawley, 117–142. Oxford: Blackwell Scientific Publications.

Caro, T. M., and Hauser, M. D. 1992. Is there teaching in nonhuman animals? *Quarterly Review of Biology* 67:151–174.

Carpenter, F. L., and Hixon, M. A. 1988. A new function for torpor: Fat conservation in a wild migrant hummingbird. *Condor* 90:373–378.

Carrascal, L. M., and Polo, V. 1999. Coal tits, *Parus ater*, lose weight in response to chases by predators. *Animal Behaviour* 58:281–285.

Cartar, R. V. 1991. A test of risk-sensitive foraging in wild bumble bees. *Ecology* 72:888–895.

Caswell, H. 2001. *Matrix Population Models*. Sunderland, MA: Sinauer Associates.

Caton, J., Lawes, M., and Cunningham, C. 2000. Digestive strategy of the south-east African lesser bushbaby, *Galago moholi*. *Comparative Biochemistry and Physiology* A 127:39–49.

Chai, P., and Dudley, R. 1999. Maximum flight performance of hummingbirds: Capacities, constraints, and trade-offs. *American Naturalist* 153:398–411.

Chain, D. G., Schwartz, J. H., and Hegde, A. N. 2000. Ubiquitin-mediated proteolysis in learning and memory. *Molecular Neurobiology* 20:125–142.

Chambers, P. G., Simpson, S. J., and Raubenheimer, D. 1995. Behavioural mechanisms of nutrient balancing in *Locusta migratoria* nymphs. *Animal Behaviour* 50:1513–1523.

Chang, M. H., Chediack, J. G., Caviedes-Vidal, E., and Karasov, W. H. 2004. L-glucose absorption in house sparrows (*Passer domesticus*) is nonmediated. *Journal of Comparative Physiology* B 174:181–188.

Chapman, R., and Ascoli-Christensen, A. 1999. Sensory coding in the grasshopper (Orthoptera: Acrididae) gustatory system. *Annals of the Entomological Society of America* 92:873–879.

Chappell, J., and Kacelnik, A. 2002. Tool selectivity in a non-primate, the New Caledonian crow (*Corvus moneduloides*). *Animal Cognition* 5:71–78.

Charnov, E. L. 1976a. Optimal foraging: Attack strategy of a mantid. *American Naturalist* 110:141–151.

———. 1976b. Optimal foraging: The marginal value theorem. *Theoretical Population Biology* 9:129–136.

Charnov, E. L., and Orians, G. H. 1973. Optimal foraging: Some theoretical explorations. Unpublished manuscript.

Charnov, E. L., Orians, G. H., and Hyatt, K. 1976. Ecological implications of resource depletion. *American Naturalist* 110:247–259.

Chase, J. M. 1998. Central-place forager effects on food web dynamics and spatial pattern in northern California meadows. *Ecology* 79:1236–1245.

Chase, J. M., Wilson, W. G., and Richards, S. A. 2001. Foraging tradeoffs and resource patchiness: Theory and experiments with a freshwater snail community. *Ecology Letters* 4:304–312.

Chediack, J. G., Caviedes-Vidal, E., Fasulo, V., Yamin, L. J., and Karasov, W. H. 2003. Intestinal passive absorption of water-soluble compounds by sparrows: Effect of molecular size and luminal nutrients. *Journal of Comparative Physiology* B 173:187–197.

Chediack, J. G., Caviedes-Vidal, E., Karasov, W. H., and Pestchanker, M. 2001. Passive absorption of hydrophilic carbohydrate probes by the House Sparrow, *Passer domesticus*. *Journal of Experimental Biology* 204:723–731.

Cheng, K. 1994. The determination of direction in landmark-based spatial search in pigeons: A further test of the vector sum model. *Animal Learning & Behavior* 22:291–301.

Chesson, P. 1990. MacArthur's consumer resource model. *Theoretical Population Biology* 37:26–38.

———. 2000. Mechanisms of maintenance of species diversity. *Annual Review of Ecology and Systematics* 31:343–366.

Cheverton, J., Kacelnik, A., and Krebs, J. R. 1985. Optimal foraging: Constraints and currencies. In *Experimental Behavioral Ecology and Sociobiology*, ed. B. Hölldobler and M. Lindauer, 109–126. Sunderland, MA: Sinauer Associates.

Chittka, L., and Thomson, J. D. 1997. Sensori-motor learning and its relevance for task specialization in bumble bees. *Behavioral Ecology and Sociobiology* 41:385–398.

Chittka, L., Thomson, J. D., and Waser, N. M. 1999. Flower constancy, insect psychology, and plant evolution. *Naturwissenschaften* 86:361–377.

Chivers, D., and Langer, P., eds. 1994. *The Digestive System in Mammals: Food, Form, and Function*. Cambridge: Cambridge University Press.

Choe, J. C., and Perlman, D. L. 1997. Social conflict and cooperation among founding queens in ants (Hymenoptera: Formicidae). In *The Evolution of Social Behavior in Insects and Arachnids*, ed. J. C. Choe and B. J. Crespi, 392–406. New York: Cambridge University Press.

Choong, M. F., Lucas, P. W., Ong, J. S. Y., Pereira, B., Tan, H. T. W., and Turner, I. M. 1992. Leaf fracture-toughness and sclerophylly—their correlations and ecological implications. *New Phytologist* 121:597–610.

Church, D. 1988. *The Ruminant Animal: Digestive Physiology and Nutrition*. Englewood Cliffs, NJ: Prentice Hall.

Ciavarella, T., Simpson, R., Dove, H., Leury, B., and Sims, I. 2000. Diurnal changes in the concentration of water-soluble carbohydrates in *Phalaris aquatica* L. pasture in spring, and the effect of short-term shading. *Australian Journal of Agricultural Research* 51: 749–756.

Claessen, D., Van Oss, C., de Roos, A. M., and Persson, L. 2002. The impact of size-dependent predation on population dynamics and individual life history. *Ecology* 83:1660–1675.

Clark, C. W. 1994. Antipredator behavior and the asset protection principle. *Behavioral Ecology* 5:159–170.

Clark, C. W., and Dukas, R. 1994. Balancing foraging and antipredator demands: An advantage of sociality. *American Naturalist* 144:542–548.

———. 2000. Winter survival strategies for small birds: Managing energy expenditure through hypothermia. *Evolutionary Ecology Research* 2:473–491.

Clark, C. W., and Ekman, J. 1995. Dominant and subordinate fattening strategies: A dynamic game. *Oikos* 72:205–212.

Clark, C. W., and Mangel, M. 1984. Foraging and flocking strategies: Information in an uncertain environment. *American Naturalist* 123:626–641.

———. 1986. The evolutionary advantage of group foraging. *Theoretical Population Biology* 30:45–75.

———. 2000. *Dynamic State Variable Models in Ecology: Methods and Applications*. Oxford: Oxford University Press.

Clarke, M. F., and Kramer, D. L. 1994. Scatter-hoarding by a larder-hoarding rodent: Intraspecific variation in the hoarding behaviour of the eastern chipmunk, *Tamias striatus*. *Animal Behaviour* 48:299–308.

Clayton, N. S., and Dickinson, A. 1998. Episodic-like memory during cache recovery by scrub jays. *Nature* 395:272–274.

———. 1999. Scrub jays (*Aphelocoma coerulescens*) remember the relative time of caching as well as the location and content of their caches. *Journal of Comparative Psychology* 113:403–416.

Clayton, N. S., and Krebs, J. R. 1994. Memory for spatial and object-specific cues in food-storing and non-storing birds. *Journal of Comparative Physiology* A 174:371–379.

Clements, K. C., and Stephens, D. W. 1995. Testing models of non-kin cooperation: Mutualism and the Prisoner's Dilemma. *Animal Behaviour* 50:527–535.

Clements, K. D., and Choat, J. H. 1995. Fermentation in tropical marine herbivorous fishes. *Physiological Zoology* 68:355–378.

Clements, K. D., and Raubenheimer, D. 2005. Feeding and nutrition. In *The Physiology of Fishes*, ed. D. H. Evans, and J. B. Clairborne, 47–82. Boca Raton, FL: CRC Press.

Clubb, R., and Vickery, S. 2006. The motivational basis of pacing in caged carnivores. In *Stereotypic Animal Behaviour: Fundamentals and Applications to Welfare*, 2nd ed., ed. G. Mason, and J. Rushen, 58–85. Wallingford, UK: CAB International.

Clutton-Brock, T. H. 1991. *The Evolution of Parental Care*. Princeton, NJ: Princeton University Press.

Clutton-Brock, T. H., Iason, G. R., Albon, S. D., and Guinness, F. E. 1982. The effects of lactation on feeding behaviour and habitat use of wild red deer hinds. *Journal of Zoology* 198:227–236.

Clutton-Brock, T. H., O'Riain, M. J., Brotherton, P. N. M., Gaynor, D., Kansky, R., Griffin, A. S., and Manser, M. 1999. Selfish sentinels in cooperative mammals. *Science* 284:1640–1644.

Cochran, P. 1987. Optimal digestion in a batch-reactor gut: The analogy to partial prey. *Oikos* 50:268–270.

Cockrem, J. F., and Silverin, B. 2002. Sight of a predator can stimulate a corticosterone response in the great tit (*Parus major*). *General and Comparative Endocrinology* 125:248–255.

Cohen, A. 1995. Extra-oral digestion in predaceous terrestrial arthropods. *Annual Review of Entomology* 40:85–103.

Cohen, Y., Vincent, T. L., and Brown, J. S. 1999. A G-function approach to fitness minima, fitness maxima, evolutionarily stable strategies and adaptive landscapes. *Evolutionary Ecology Research* 1:923–942.

———. 2001. Does the G-function deserve an F? *Evolutionary Ecology Research* 3:375–377.

Colinvaux, P. A. 1978. *Why Big Fierce Animals Are Rare: An Ecologist's Perspective*. Princeton, NJ: Princeton University Press.

Collett, T. S. 1996. Insect navigation en route to the goal: Multiple strategies for the use of landmarks. *Journal of Experimental Biology* 199:227–235.

Collett, T. S., Cartwright, B. A., and Smith, B. A. 1986. Landmark learning and visuo-spatial memories in gerbils. *Journal of Comparative Physiology* A 158:835–51.

Commons, M. L., Nevin, J. A., and Davison, M. 1991. *Signal Detection: Mechanisms, Models, and Applications*. Hillsdale, NJ: Lawrence Erlbaum Associates.

Connell, J. H. 1961. The influence of interspecific competition and other factors on the distribution of the barnacle *Chthamalus stellatus*. *Ecology* 42:710–723.

Conover, K. L., and Shizgal, P. 1994. Competition and summation between rewarding effects of sucrose and lateral hypothalamic stimulation in the rat. *Behavioral Neuroscience* 108:537–548.

Conradt, L. 2000. Use of a seaweed habitat by red deer (*Cervus elaphus* L.). *Journal of Zoology* 250:541–549.

Cook, R. G. 1992. Dimensional organization and texture discrimination in pigeons. *Journal of Experimental Psychology: Animal Behavior Processes* 18:354–363.

Cook, R. G., Brown, M. F., and Riley, D. A. 1985. Flexible memory processing by rats: Use of prospective and retrospective information in the radial maze. *Journal of Experimental Psychology: Animal Behavior Processes* 11:453–469.

Coolen, I., Giraldeau, L.-A., and LaVoie, M. 2001. Head position as an indicator of producer and scrounger tactics in a ground-feeding bird. *Animal Behaviour* 61:895–903.

Cooper, J., Gordon, I. J., and Pike, A. W. 2000. Strategies for the avoidance of faeces by grazing sheep. *Applied Animal Behaviour Science* 69:15–33.

Cooper, S. D. B., Kyriazakis, I., and Oldham, J. D. 1996. The effects of physical form of feed, carbohydrate source, and inclusion of sodium bicarbonate on the diet selections of sheep. *Journal of Animal Science* 74:1240–1251.

Cooper, W. E. 2000. Tradeoffs between predation risk and feeding in a lizard, the broad-headed skink (*Eumeces laticeps*). *Behaviour* 137:1175–1189.

Coppedge, B., and Shaw, J. 1998. Bison grazing patterns on seasonally burned tallgrass prairie. *Journal of Range Management* 51:258–264.

Cork, S. J. 1994. Digestive constraints on dietary scope in small and moderately small mammals: How much do we really understand? In *The Digestive System in Mammals: Food, Form, and Function*, ed. D. J. Chivers and P. Langer, 337–369. Cambridge: Cambridge University Press.

Cork, S. J., and Foley, W. J. 1991. Digestive and metabolic strategies of arboreal mammalian folivores in relation to chemical defenses in temperate and tropical forests. In *Plant Defences against Mammalian Herbivory*, ed. R. T. Palo, and C. T. Robbins, 133–166. Boca Raton, FL: CRC Press.

Cosgrove, G. P., and Niezen, J. H. 2000. Intake and selection for white clover by grazing lambs in response to gastrointestinal parasitism. *Applied Animal Behaviour Science* 66: 71–85.

Courtney, S., and Sallabanks, R. 1992. It takes guts to handle fruits. *Oikos* 65:163–166.

Covich, A. P. 1976. Analyzing shapes of foraging areas: Some ecological and economic theories. *Annual Review of Ecology and Systematics* 7:235–257.

Cowie, R. J. 1977. Optimal foraging in great tits (*Parus major*). *Nature* 268:137–139.

Crane, K., Smith, M., and Reynolds, D. 1997. Habitat selection patterns of feral horses in southcentral Wyoming. *Journal of Range Management* 50:374–380.

Crespi, B. J. 2000. The evolution of maladaptation. *Heredity* 84:623–629.

———. 2001. The evolution of social behavior in microorganisms. *Trends in Ecology and Evolution* 16:178–183.

Cresswell, W. 1994. Flocking is an effective anti-predation strategy in redshanks, *Tringa totanus*. *Animal Behaviour* 47:433–442.

———. 1996. Surprise as a winter hunting strategy in Sparrowhawks *Accipiter nisus*, Peregrines *Falco peregrinus* and Merlins *F. columbarius*. *Ibis* 138:684–692.

Cresswell, W., Hilton, G. M., and Ruxton, G. D. 2000. Evidence for a rule governing the avoidance of superfluous escape flights. *Proceedings of the Royal Society of London* B 267: 733–737.

Cresswell, W., and Quinn, J. L. 2004. Faced with a choice, sparrowhawks more often attack the more vulnerable prey group. *Oikos* 104:71–76.

Cummings, D. E., Purnell, J. Q., Frayo, R. S., Schmidova, K., Wisse, B. E., and Weigle, D. S. 2001. A preprandial rise in plasma ghrelin levels suggests a role in meal initiation in humans. *Diabetes* 50:1714–1719.

Cuthill, I. C., Haccou, P., and Kacelnik, A. 1994. Starlings (*Sturnus vulgaris*) exploiting patches: Response to long-term changes in travel time. *Behavioral Ecology* 5:81–90.

Cuthill, I. C., Haccou, P., Kacelnik, A., Krebs, J. R., and Iwasa, Y. 1990. Starlings exploiting patches: The effect of recent experience on foraging decisions. *Animal Behaviour* 40:625–640.

Cuthill, I. C., and Houston, A. I. 1997. Managing time and energy. In *Behavioral Ecology: An Evolutionary Approach*, 4th ed., ed. J. R. Krebs and N. B. Davies, 97–120. Oxford: Blackwell Science.

Cuthill, I. C., Maddocks, S. A., Weall, C. V., and Jones, E. K. M. 2000. Body mass regulation in response to changes in feeding predictability and overnight energy expenditure. *Behavioral Ecology* 11:189–195.

Daan, S., Deerenberg, C., and Dijkstra, C. 1996. Increased daily work precipitates natural death in the kestrel. *Journal of Animal Ecology* 65:539–544.

Daan, S., Masman, D., and Groenewold, A. 1990. Avian basal metabolic rates: Their association with body composition and energy expenditure in nature. *American Journal of Physiology* 259:R333–R340.

Dade, W., Jumars, P. A., and Penry, D. L. 1990. Supply-side optimization: Maximizing absorptive rates. In *Behavioural Mechanisms of Food Selection*, ed. R. N. Hughes, 531–556. Berlin: Springer-Verlag.

Dailey, J. W., and McGlone, J. J. 1997. Oral/nasal/facial and other behaviors of sows kept individually outdoors on pasture, soil or indoors in gestation crates. *Applied Animal Behaviour Science* 52:25–43.

Daily, G. C., and Ellison, K. 2002. *The New Economy of Nature: The Quest to Make Conservation Profitable*. Washington, DC: Island Press.

Dall, S. R. X., Giraldeau, L.-A., Olsson, O., McNamara, J. M., and Stephens, D. W. 2005. Information and its use in evolutionary ecology. *Trends in Ecology and Evolution* 20:187–193.

Dall, S. R. X., and Johnstone, R. A. 2002. Managing uncertainty: Information and insurance under the risk of starvation. *Philosophical Transactions of the Royal Society* B 357:1519–1526.

Dall, S. R. A., Kotler, B. P., and Bouskila, A. 2001. Attention, "apprehension" and gerbils searching in patches. *Annales Zoologici Fennici* 38:15–23.

Dallal, N. L., and Meck, W. H. 1990. Hierarchical structures: Chunking by food type facilitates spatial memory. *Journal of Experimental Psychology: Animal Behavior Processes* 16:69–84.

Daly, M., Behrends, P. R., Wilson, M. I., and Jacobs, L. F. 1992. Behavioral modulations of predation risk: Moonlight avoidance and crepuscular compensation in a nocturnal desert rodent. *Animal Behaviour* 44:1–9.

Dänhardt, J., and Lindström, Å. 2001. Optimal departure decisions of songbirds from an experimental stopover site and the significance of weather. *Animal Behaviour* 62:235–243.

Darwin, C. 1876. *On the Effects of Cross and Self-Fertilization in the Vegetable Kingdom*. London: John Murray.

Davidson, D. W. 1997. Resource imbalances in the evolutionary ecology of tropical arboreal ants. *Biological Journal of the Linnean Society* 61:153–181.

Davies, N. B., and Houston, A. I. 1981. Owners and satellites: The economics of territory defence in the pied wagtail, *Motacilla alba*. *Journal of Animal Ecology* 50:157–180.

Davis, H. 1996. Underestimating the rat's intelligence. *Cognitive Brain Research* 3:291–298.

Davoren, G., and Burger, A. E. 1999. Differences in prey selection and behaviour during self-feeding and chick provisioning in rhinoceros auklets. *Animal Behavior* 58:853–863.

Dawkins, R. 1976. *The Selfish Gene*. New York: Oxford University Press.

Day, T., and Detling, J. 1990. Grassland patch dynamics and herbivore grazing preference following urine deposition. *Ecology* 71:180–188.

Day, T., Pritchard, J., and Schluter, D. 1994. A comparison of two sticklebacks. *Evolution* 48:1723–1734.

Dayan, P., and Abbott, L. F. 2001. *Theoretical Neuroscience: Computational Modeling of Neural Systems*. Cambridge, MA: MIT Press.

Dearing, M. D. 1996. Disparate determinants of summer and winter diet selection of a generalist herbivore, *Ochotona princeps*. *Oecologia* 108:467–478.

Dearing, M. D., and Schall, J. J. 1992. Testing models of optimal diet assembly by the generalist herbivorous lizard *Cnemidophorus murinus*. *Ecology* 73:845–858.

de Belle, J. S., and Heisenberg, M. 1994. Associative odor learning in *Drosophila* abolished by chemical ablation of mushroom bodies. *Science* 263:692–695.

de Bruin, J. P. C., van Oyen, H. G. M., and van de Poll, N. 1983. Behavioural changes following lesions of the orbital prefrontal cortex in male rats. *Behavioural Brain Research* 10:209–232.

de Castro, J. M. 1988. Physiological, environmental and subjective determinants of food intake in humans: A meal pattern analysis. *Physiology and Behavior* 44:651–659.

DeGroot, M. H. 1970. *Optimal Statistical Decisions*. New York: McGraw-Hill.

Delestrade, A. 1999. Foraging strategy in a social bird, the alpine chough: Effect of variation in quantity and distribution of food. *Animal Behaviour* 57:299–305.

Del Moral, R. 1984. The impact of the Olympic marmot on subalpine vegetation structure. *American Journal of Botany* 71:1228–36.

Demment, M. W., and Van Soest, P. J. 1985. A nutritional explanation for body-size patterns of ruminant and nonruminant herbivores. *American Naturalist* 125:641–672.

Denbow, D. 2000. Gastrointestinal anatomy and physiology. In *Sturkie's Avian Physiology*, ed. G. Whittow, 299–325. New York: Academic Press.

Denton, J. S., Hitchings, S. P., Beebee, T. J. C., and Gent, A. 1997. A recovery program for the natterjack toad (*Bufo calamita*) in Britain. *Conservation Biology* 11:1329–1338.

Devenport, L. D., and Devenport, J. A. 1994. Time-dependent averaging of foraging information in least chipmunks and golden-mantled ground squirrels. *Animal Behaviour* 47:787–802.

Devenport, L. D., Hill, T., Wilson, M., and Ogden, E. 1997. Tracking and averaging in variable environments: A transition rule. *Journal of Experimental Psychology: Animal Behavior Processes* 23:450–460.

Devereux, C. 1998. The fiscal shrike. *Africa Birds & Birding* 3:52–57.

Diamond, D. M., Bennett, M. C., Fleshner, M., and Rose, G. M. 1992. Inverted-U relationship between the level of peripheral corticosterone and the magnitude of hippocampal primed burst potentiation. *Hippocampus* 2:421–430.

Diamond, J. 1991. Evolutionary design of intestinal nutrient absorption: Enough but not too much. *News in Physiological Sciences* 6:92–96.

Diamond, J., Karasov, W. H., Phan, D., and Carpenter, F. L. 1986. Digestive physiology is a determinant of foraging bout frequency in hummingbirds. *Nature* 320:62–63.

Dill, L. M., and Fraser, A. H. G. 1984. Risk of predation and the feeding behavior of juvenile coho salmon (*Oncorhynchus kisutch*). *Behavioral Ecology and Sociobiology* 16:65–71.

Dinerstein, E. 2002. *The Return of the Unicorns: A Success Story in the Conservation of Asian Rhinoceros*. New York: Columbia University Press.

Distel, R. A., Laca, E. A., Griggs, T. C., and Demment, M. W. 1995. Patch selection by cattle: Maximization of intake rate in horizontally heterogeneous pastures. *Applied Animal Behaviour Science* 45:11–21.

Dolby, A. S., and Grubb, T. C., Jr. 2000. Social context affects risk taking by a satellite species in a mixed-species foraging group. *Behavioral Ecology* 11:110–114.

Domjan, M. 1998. *The Principles of Learning and Behavior*. 4th ed. Bellmont, CA: Wadsworth.

———. 2005. Pavlovian conditioning: A functional perspective. *Annual Review of Psychology* 56:179–206.

Dornhaus, A., and Chittka, L. 2004. Why do honey bees dance? *Behavioral Ecology and Sociobiology* 55:395–401.

Doucet, C. M., Adams, I. T., and Fryxell, J. M. 1994. Beaver dam and cache composition: Are woody species used differently? *Ecoscience* 1:268–270.

Dougherty, C. T., Bradley, N. W., Cornelius, P. L., and Lauriault, L. M. 1989. Short-term fasts and the ingestive behaviour of grazing cattle. *Grass and Forage Science* 44:295–302.

Dougherty, C. T., Lauriault, L., Bradley, N., Gay, N., and Cornelius, P. 1991. Induction of tall fescue toxicosis in heat-stressed cattle and its alleviation with thiamin. *Journal of Animal Science* 69:1008–1018.

Dowling, J. E., and Dubin, M. W. 1984. The vertebrate retina. In *Handbook of Physiology*, section 1, *The Nervous System*, vol. 3, *Sensory Processes*, ed. V. B. Mountcastle and I. Darian-Smith, 317–339. Bethesda, MD: American Physiological Society.

Downes, S., and Hoefer, A. M. 2004. Antipredatory behaviour in lizards: Interactions between group size and predation risk. *Animal Behaviour* 67:485–492.

Drent, R. H., and Daan, S. 1980. The prudent parent: Energetic adjustments in avian breeding. *Ardea* 68:225–252.

Dubnau, J., and Tully, T. 1998. Gene discovery in *Drosophila*: New insights for learning and memory. *Annual Review of Neuroscience* 21:407–444.

Dugatkin, L. A. 1997. *Cooperation among Animals: An Evolutionary Perspective*, New York: Oxford University Press.

———. 1998. Game theory and cooperation. In *Game Theory and Animal Behavior*, ed. L. A. Dugatkin, and H. K. Reeve, 38–63. New York: Oxford University Press.

Dugatkin, L. A., and Reeve, H. K. 1998. *Game Theory and Animal Behavior*. New York: Oxford University Press.

Dukas, R., ed. 1998a. *Cognitive Ecology: The Evolutionary Ecology of Information Processing and Decision Making*. Chicago: University of Chicago Press.

———. 1998b. Constraints on information processing and their effects on behavior. In *Cognitive Ecology: The Evolutionary Ecology of Information Processing and Decision Making*, ed. R. Dukas, 89–127. Chicago: University of Chicago Press.

———. 1998c. Evolutionary ecology of learning. In *Cognitive Ecology: The Evolutionary Ecology of Information Processing and Decision Making*, ed. R. Dukas, 129–174. Chicago: University of Chicago Press.

———. 1999. Costs of memory: Ideas and predictions. *Journal of Theoretical Biology* 197:41–50.

———. 2001. Effects of predation risk on pollinators and plants. In *Cognitive Ecology of Pollination*, ed. L. Chittka and J. Thomson, 214–236. Cambridge: Cambridge University Press.

———. 2004a. Evolutionary biology of animal cognition. *Annual Review of Ecology Evolution and Systematics* 35:347–374.

———. 2004b. Male fruit flies learn to avoid interspecific courtship. *Behavioral Ecology* 15:695–698.

———. 2005a. Experience improves courtship in male fruit flies. *Animal Behaviour* 69:1203–1209.

———. 2005b. Learning affects mate choice in female fruit flies. *Behavioral Ecology* 16:800–804.

Dukas, R., and Kamil, A. C. 2000. The cost of limited attention in blue jays. *Behavioral Ecology* 11:502–506.

Dumont, B. 1995. Dietary choice determinism for grazing herbivores: Principal theories. *Productions Animales* 8:285–292.

Dumont, B., and Boissy, A. 1999. Impact of social on grazing behaviour in herbivores. *Productions Animales* 12:3–10.

————. 2000. Grazing behaviour of sheep in a situation of conflict between feeding and social motivations. *Behavioural Processes* 49:121–138.

Dumont, B., Dutrone, A., and Petit, M. 1998. How readily will sheep walk for a preferred forage. *Journal of Animal Science* 76:965–971.

Dumont, B., Maillard, J. F., and Petit, M. 2000. The effect of the spatial distribution of plant species within the sward on the searching success of sheep when grazing. *Grass and Forage Science* 55:138–145.

Dumont, B., and Petit, M. 1998. Spatial memory of sheep at pasture. *Applied Animal Behaviour Science* 60:43–53.

Duncan, A. J., and Gordon, I. J. 1999. Habitat selection according to the ability of animals to eat, digest and detoxify foods. *Proceedings of the Nutrition Society* 58:799–805.

Duncan, P., and Cowtan, P. 1980. An unusual choice of habitat helps Camargue horses to avoid blood-sucking horse-flies. *Biology of Behaviour* 5:55–60.

Duriez, O., Weimerskirsch, H., and Fritz, H. 2000. Regulation of chick provisioning in the thin-billed prion: An interannual comparison and manipulation of parents. *Canadian Journal of Zoology* 78:1275–1283.

Dusenbery, D. B. 1992. *Sensory Ecology*. New York: W. H. Freeman.

Dyer, F. C. 1996. Spatial memory and navigation by honeybees on the scale of the foraging range. *Journal of Experimental Biology* 199:147–154.

————. 1998. Cognitive ecology of navigation. In *Cognitive Ecology: The Evolutionary Ecology of Information Processing and Decision Making*, ed. R. Dukas, 201–260. Chicago: University of Chicago Press.

Dyer, F. C., and Seeley, T. D. 1991. Nesting behavior and the evolution of worker tempo in four honey bee species. *Ecology* 72:156–170.

Earn, D. J. D., and Johnstone, R. A. 1997. A systematic error in tests of ideal free theory. *Proceedings of the Royal Society of London* B 264:1671–1675.

Eckert, C. D., Winston, M. L., and Ydenberg, R. C. 1994. The relationship between population size, amount of brood and individual foraging effort in the honey bee *Apis mellifera*. *Oecologia* 97:248–255.

Edwards, G. R., Newman, J. A., Parsons, A. J., and Krebs, J. R. 1994. Effects of the scale and spatial distribution of the food resource and animal state on diet selection: An example with sheep. *Journal of Animal Ecology* 63:816–826.

————. 1996a. Effects of the total, vertical and horizontal availability of the food resource on diet selection and intake of sheep. *Journal of Agricultural Science* 127:555–562.

————. 1996b. The use of spatial memory by grazing animals to locate food patches in spatially heterogeneous environments: An example with sheep. *Applied Animal Behaviour Science* 50:147–160.

————. 1997. Use of cues by grazing animals to locate food patches: An example with sheep. *Applied Animal Behaviour Science* 51:59–68.

Egan, J. P. 1975. *Signal Detection Theory and ROC-Analysis*. New York: Academic Press.

Ehlinger, T. 1990. Phenotype-limited feeding efficiency and habitat choice in bluegill: Individual differences and trophic polymorphism. *Ecology* 71:886–896.

Eichenbaum, H. 2000. A cortical-hippocampal system for declarative memory. *Nature Neuroscience Reviews* 1:41–50.

Eilam, D., Dayan, T., Ben-Eliyahu, S., Schulman, I., Shefer, G., and Hendrie, C. A. 1999. Differential behavioral and hormonal responses of voles and spiny mice to owl calls. *Animal Behaviour* 58:1085–1093.

Ekman, J. B., and Lilliendahl, K. 1993. Using priority to food access: Fattening strategies in dominance-structured willow tit (*Parus montanus*) flocks. *Behavioral Ecology* 4:232–238.

Ekman, J. B, and Rosander, B. 1992. Survival enhancement through food sharing: A means for parental control of natal dispersal. *Theoretical Population Biology* 42:117–129.

Elgar, M. A. 1989. Predator vigilance and group size in mammals and birds: A critical review of the empirical evidence. *Biological Review* 64:13–33.

Elgar, M. A., and Crespi, B. J. 1992. *Cannibalism: Ecology and Evolution among Diverse Taxa.* Oxford: Oxford University Press.

Elgar, M. A., McKay, H., and Woon, P. 1986. Scanning, pecking and alarm flights in house sparrows. *Animal Behaviour* 34:1892–1894.

Elman, J. L., Bates, E. A., Johnson, M. H., Karmiloff-Smith, A., Parisi, D., and Plunkett, K. 1996. *Rethinking Innateness: A Connectionist Perspective on Development.* London: Bradford Books.

Emlen, J. M. 1966. The role of time and energy in food preference. *American Naturalist* 100:611–617.

Emlen, J. T. 1974. An urban bird community in Tucson, Arizona: Derivation, structure, regulation. *Condor* 76:184–197.

Emmans, G. C., and Kyriazakis, I. 1995. The idea of optimisation in animals: Uses and dangers. *Livestock Production Science* 44:189–197.

Ennaceur, A., and Delacour, J. 1988. A new one-trial test for neurobiological studies of memory in rats. I. Behavioral data. *Behavioural Brain Research* 31:47–59.

Ennaceur, A., and Meliani, K. 1992. A new one-trial test for neurobiological studies of memory in rats. III. Spatial vs. non-spatial working memory. *Behavioural Brain Research* 51:83–92.

Epstein, R., Kirshnit, C. E., Lanza, R. P., and Rubin, L. C. 1984. "Insight" in the pigeon: Antecedents and determinants of an intelligent performance. *Nature* 308: 61–62.

Erber, J., Homberg, U., and Gronenberg, W. 1987. Functional roles of the mushroom bodies in insects. In *Arthropod Brain: Its Evolution, Development, Structure, and Functions*, ed. A. P. Gupta, 485–511. New York: Wiley.

Erber, J., Masuhr, T., and Menzel, R. 1980. Localization of short-term memory in the brain of the bee, *Apis mellifera*. *Physiological Entomology* 5:343–358.

Evans, P. D., and Robb, S. 1993. Octopamine receptor subtypes and their modes of action. *Neurochemical Research* 18:869–874.

Evans, P. R. 1969. Winter fat deposition and overnight survival of yellow buntings (*Emberiza citrinella*). *Journal of Animal Ecology* 38:415–423.

Faaborg, J., Parker, P. G., DeLay, L., de Vries, T., Bednarz, J. C., Paz, S. M., Naranjo, J., and Waite, T. A. 1995. Confirmation of cooperative polyandry in the Galapagos hawk (*Buteo galapagoensis*) using DNA fingerprinting. *Behavioral Ecology and Sociobiology* 36:83–90.

Farmer, C. G. 2000. Parental care: The key to understanding endothermy and other convergent features in birds and mammals. *American Naturalist* 155:326–334.

Farnsworth, K. D., and Beecham, J. A. 1999. How do grazers achieve their distribution? A continuum of models from random diffusion to the ideal free distribution using biased random walks. *American Naturalist* 153:509–526.

Farnsworth, K. D., and Illius, A. W. 1996. Large grazers back in the fold: Generalizing the prey model to incorporate mammalian herbivores. *Functional Ecology* 10:678–680.

———. 1998. Optimal diet choice for large herbivores: An extended contingency model. *Functional Ecology* 12:74–81.

Feinsinger, P. 1976. Organization of a tropical guild of nectarivorous birds. *Ecological Monographs* 46:257–291.

Feinsinger, P., and Colwell, R. K. 1978. Community organization among nectar-feeding neotropical birds. *American Zoologist* 19:779–795.

Fereres, A., Kampmeier, G., and Irwin, M. 1999. Aphid attraction and preference for soybean and pepper plants infected with Potyviridae. *Annals of the Entomological Society of America* 92:542–548.

Ferguson, N. S., Nelson, L., and Gous, R. M. 1999. Diet selection in pigs: Choices made by growing pigs when given foods differing in nutrient density. *Animal Science* 68:691–699.

Ferraris, R., and Diamond, J. 1997. Regulation of intestinal sugar transport. *Physiological Reviews* 77:257–306.

Fewell, J. H., and Winston, M. L. 1992. Colony state and regulation of foraging in the honey bee, *Apis mellifera* L. *Behavioral Ecology and Sociobiology* 30:387–393.

Fiala, A., Müller, U., and Menzel, R. 1999. Reversible downregulation of protein kinase A during olfactory learning using antisense technique impairs long-term memory formation in the honeybee, *Apis mellifera*. *Journal of Neuroscience* 19:10125–10134.

Fierer, N., and Kotler, B. P. 2000. Evidence for micropatch partitioning and effects of boundaries on patch use in two species of gerbils. *Functional Ecology* 14:176–182.

Flanagan, D., and Mercer, A. R. 1989. An atlas and 3-D reconstruction of the antennal lobes in the worker honey bee, *Apis mellifera* L. (Hymenoptera: Apidae). *International Journal of Insect Morphology and Embryology* 18·145–159.

Floresco, S. B., Seamans, J. K., and Phillips, A. G. 1997. Selective roles for hippocampal, prefrontal cortical, and ventral striatal circuits in radial-arm maze tasks with or without a delay. *Journal of Neuroscience* 17:1880–1890.

Focardi, S., and Marcellini, P. 1995. A mathematical framework for optimal foraging of herbivores. *Journal of Mathematical Biology* 33:365–387.

Foley, W., and Cork, S. J. 1992. Use of fibrous diets by small herbivores: How far can the rules be "bent"? *Trends in Ecology and Evolution* 7:159–162.

Forbes, J. 2001. Consequences of feeding for future feeding. *Comparative Biochemistry and Physiology* A 128:463–470.

Forchhammer, M. C., and Boomsma, J. J. 1995. Foraging strategies and seasonal diet optimization of muskoxen in West Greenland. *Oecologia* 104:169–180.

Fortin, D. 2001. An adjustment of the extended contingency model of Farnsworth and Illius (1998). *Functional Ecology* 15:138–139.

Fransson, T., and Weber, T. P. 1997. Migratory fuelling in blackcaps (*Sylvia atricapilla*) under perceived risk of predation. *Behavioral Ecology and Sociobiology* 41:75–80.

Fraser, D. F., and Cerri, R. D. 1982. Experimental evaluation of predator-prey relationships in a patchy environment: Consequences for habitat use patterns in minnows. *Ecology* 63:307–313.

Fraser, D. F., Gilliam, J. F., Akkara, J. T., Albanese, B. W., and Snider, S. B. 2004. Night feeding by guppies under predator release: Effects on growth and daytime courtship. *Ecology* 85:312–319.

Fretwell, S. D. 1972. *Populations in a Seasonal Environment*. Princeton, NJ: Princeton University Press.

Fretwell, S. D., and Lucas, H. L. 1969. On territorial behavior and other factors influencing habitat distribution in birds. *Acta Biotheoretica* 19:16–36.

Frey-Roos, F., Brodmann, P. A., and Reyer, H.-U. 1995. Relationships between food resources, foraging patterns, and reproductive success in the water pipit, *Anthus sp. spinoletta*. *Behavioral Ecology* 6:287–295.

Friedman, H. R., and Goldman-Rakic, P. S. 1994. Coactivation of prefrontal cortex and inferior parietal cortex in working memory tasks revealed by 2DG functional mapping in the rhesus monkey. *Journal of Neuroscience* 14:2775–2788.

Fryxell, J. M., and Lundberg, P. 1994. Diet choice and predator-prey dynamics. *Evolutionary Ecology* 8:407–421.

———. 1998. *Individual Behavior and Community Dynamics*. London: Chapman & Hall.

Fryxell, J. M., Wilmshurst, J. F., Sinclair, A. R. E., Haydon, D. T., Holt, R. D., and Abrams, P. A. 2005. Landscape scale, heterogeneity, and the viability of Serengeti grazers. *Ecology Letters* 8:328–335.

Funmilayo, O. 1979. Food consumption, preferences and storage in the mole. *Acta Theriologica* 24:379–389.

Fuster, J. M. 1973. Unit activity in prefrontal cortex during delayed-response performance: Neuronal correlates of transient memory. *Journal of Neurophysiology* 36:61–78.

———. 1985. The prefrontal cortex and temporal integration. In *Cerebral Cortex*, vol. 4, *Association and Auditory Cortices*, ed. A. Peters and E. G. Jones, 151–177. New York: Plenum Press.

———. 1991. The prefrontal cortex and its relation to behavior. In *Progress in Brain Research*, vol. 87, *Role of the Forebrain in Sensation and Behavior*, ed. G. Holstege, 201–211. Amsterdam: Elsevier.

———. 1997. *The Prefrontal Cortex: Anatomy, Physiology, and Neuropsychology of the Frontal Lobe*. Philadelphia: Lippincott-Raven.

Fuster, J. M., and Alexander, G. E. 1971. Neuron activity related to short-term memory. *Science* 173:652–654.

Fuster, J. M., and Bauer, R. H. 1974. Visual short-term memory deficit from hypothermia of frontal cortex. *Brain Research* 81:393–400.

Gaby, R., McMahon, M. P., Mazzoti, F. J., Gillies, W. N., and Wilcox, J. R. 1985. Ecology of a population of *Crocodylus acutus* at a power plant site in Florida. *Journal of Herpetology* 19:189–198.

Gadgil, M., and Seshagiri Rao, P. R. 1995. Designing incentives to conserve India's biodiversity. In *Property Rights in a Social and Ecological Context: Case Studies and Design Applications*, ed. S. Hanna and M. Munasinghe, 53–62. Washington, DC: The Beijer International Institute of Ecological Economics & The World Bank.

Galef, B. G., Jr. 1988. Imitation in animals: History, definition, and interpretation of data for the psychological laboratory. In *Social Learning: Psychological and Biological Perspectives*, ed. T. R. Zentall, and B. G. Galef, Jr., 3–28. Hillsdale, NJ: Lawrence Erlbaum Associates.

———. 1989. Enduring social enhancement of rats' preferences for the palatable and the piquant. *Appetite* 13:81–92.

———. 1991. Information centres of Norway rats: Sites for information exchange and information parasitism. *Animal Behaviour* 41:295–302.

———. 2004. Approaches to the study of traditional behaviors of free-living animals. *Learning and Behavior* 32:53–61.

Galef, B. G., Jr., and Giraldeau, L.-A. 2001. Social influences on foraging in vertebrates: Causal mechanisms and adaptive functions. *Animal Behaviour* 61:3–15.

Galef, B. G., Jr., Mason, J. R., Preti, G., and Bean, N. J. 1988. Carbon disulphide: A semio-chemical mediating socially-induced diet choice in rats. *Physiology and Behavior* 42:119–124.

Galef, B. G., Jr., and Stein, M. 1985. Demonstrator influence on observer diet preference: Analyses of critical social interactions and olfactory signals. *Animal Learning & Behavior* 13:31–38.

Gallistel, C. R. 1990. *The Organization of Learning*. Cambridge, MA: MIT Press.

———. 1994. Foraging for brain stimulation: Toward a neurobiology of computation. *Cognition* 50:151–170.

Gallistel, C. R., and Gibbon, J. 2000. Time, rate and conditioning. *Psychological Review* 107:289–344.

Gallistel, C. R., and Leon, M. 1991. Measuring the subjective magnitude of brain stimulation reward by titration with rate of reward. *Behavioral Neuroscience* 105:913–925.

Gallistel, C. R., Mark, T. A., King, A. P., and Latham, P. E. 2001. The rat approximates an ideal detector of changes in rates of reward: Implications for the law of effect. *Journal of Experimental Psychology: Animal Behavior Processes* 27:354–372.

Ganslosser, U., and Brunner, C. 1997. Influence of food distribution on behavior in captive bongos, *Taurotragus euryceros*: An experimental investigation. *Zoo Biology* 16:237–245.

Garcia, J., and Koelling, R. A. 1966. Relation of cue to consequence in avoidance learning. *Psychonomic Science* 4:123–124.

Gardiner, M. 1972. *The Biology of Invertebrates*. New York: McGraw-Hill Book Company.

Gardner, S. N., and Agrawal, A. A. 2002. Induced plant defense and the evolution of counter-defenses in herbivores. *Evolutionary Ecology Research* 4:1131–1151.

Garland, T., Jr., Dickerman, A. W., Janis, C. M., and Jones, J. A. 1993. Phylogenetic analysis of covariance by computer-simulation. *Systematic Biology* 42:265–292.

Garris, P. A., Kilpatrick, M., Bunin, M. A., Michael, D., Walker, Q. D., and Wightman, R. M. 1999. Dissociation of dopamine release in the nucleus accumbens from intracranial self-stimulation. *Nature* 398:67–69.

Gass, C. L., and Garrison, J. S. E. 1999. Energy regulation by traplining hummingbirds. *Functional Ecology* 13:483–492.

Gaston, A. J. 1992. *The Ancient Murrelet*. London: Poyser.

Gaulin, S. J. C., and FitzGerald, R. W. 1986. Sex differences in spatial ability: An evolutionary hypothesis and test. *American Naturalist* 127:74–88.

———. 1988. Home-range size as a predictor of mating systems in *Microtus*. *Journal of Mammalogy* 69:311–319.

———. 1989. Sexual selection for spatial-learning ability. *Animal Behavior* 37:322–331.

Gauthier, M., Cano-Lozano, V., Zaoujal, A., and Richard, D. 1994. Effects of intracranial injections of scopolamine on olfactory conditioning retrieval in the honeybee. *Behavioural Brain Research* 63:145–149.

Gegear, R. J., and Laverty, T. M. 1998. How many flower types can bumble bees work at the same time? *Canadian Journal of Zoology* 76:1358–1365.

Gentle, L. K., and Gosler, A. G. 2001. Fat reserves and perceived predation risk in the great tit, *Parus major*. *Proceedings of the Royal Society of London* B 268:487–491.

Gentry, R. L., and Kooyman, G. L. 1986. *Fur Seals: Maternal Strategies for Breeding on Land and at Sea*. Princeton, NJ: Princeton University Press.

Gescheider, G. A. 1985. *Psychophysics: Method, Theory, and Application*. Hillsdale, NJ: Lawrence Erlbaum Associates.

Getty, T. 1985. Discriminability and the sigmoid functional response: How optimal foragers could stabilize model-mimic complexes. *American Naturalist* 125:239–256.

———. 1995. Search, discrimination and selection—mate choice by Pied Flycatchter. *American Naturalist* 145:146–154.

Getty, T., Kamil, A. C., and Real, P. G. 1987. Signal detection theory and foraging for cryptic and mimetic prey. In *Foraging Behavior*, ed. A. C. Kamil, J. R. Krebs, and H. R. Pulliam, 525–548. New York: Plenum Press.

Ghirlanda, S., and Enquist, M. 2003. A century of generalization. *Animal Behaviour* 66:15–36.

Ghosh, A., and Greenberg, M. E. 1995. Calcium signaling in neurons: Molecular mechanisms and cellular consequences. *Science* 268:239–247.

Gibb, J. A. 1966. Tit predation and the abundance of *Ernarmonia conicolana* (Hey) on Weeting Heath, Norfolk, 1962–63. *Journal of Animal Ecology* 35:43–54.

Gibb, M. J., Huckle, C. A., Nuthall, R., and Rook, A. J. 1997. Effect of sward surface height on intake and grazing behaviour by lactating Holstein Friesian cows. *Grass and Forage Science* 52:309–321.

Gibbon, J., Church, R. M., Fairhust, S., and Kacelnik, A. 1988. Scalar expectancy theory and choice between delayed rewards. *Psychological Review* 95:102–114.

Gibbon, J., Malapani, C., Dale, C. I., and Gallistel, C. R. 1997. Toward a neurobiology of temporal cognition: Advances and challenges. *Current Opinion in Neurobiology* 7:170–184.

Gidenne, T. 1992. Rate of passage of fibre particles of different size in the rabbit: Effect of dietary lignin level. *The Journal of Applied Rabbit Research* 15:1175–1182.

Gigerenzer, G., and Selten, R. 2001. *Bounded Rationality: The Adaptive Toolbox.* Cambridge, MA: MIT Press.

Gillham, S. B., Dodman, N. H., Shuster, L., Kream, R., and Rand, W. 1994. The effect of diet on cribbing behavior and plasma beta-endorphin in horses. *Applied Animal Behaviour Science* 41:147–153.

Gilliam, J. F. 1982. Habitat use and competitive bottlenecks in size-structured fish populations. Ph.D. dissertation, Michigan State University.

———. 1990. Hunting by the hunted: Optimal prey selection by foragers under predation hazard. In *Behavioural Mechanisms of Food Selection*, ed. R. N. Hughes, 797–818. Berlin: Springer-Verlag.

Gilliam, J. F., and Fraser, D. F. 1987. Habitat selection under predation hazard: A test of a model with foraging minnows. *Ecology* 68:1856–1862.

Gilpin, M. E. 1975. *Group Selection in Predator-Prey Communities.* Princeton, NJ: Princeton University Press.

Ginnett, T., and Demment, M. 1999. Sexual segregation by Masai giraffes at two spatial scales. *African Journal of Ecology* 37:93–106.

Giraldeau, L.-A. 1988. The stable group and determinants of foraging group size. In *The Ecology of Social Behavior*, ed. C. N. Slobodchikoff, 33–53. New York: Academic Press.

———. 1997. The ecology of information use. In *Behavioral Ecology: An Evolutionary Approach*, 4th ed., ed. J. R. Krebs and N. B. Davies, 42–68. Oxford: Blackwell Science.

Giraldeau, L.-A., and Beauchamp, G. 1999. Food exploitation: Searching for the optimal joining policy. *Trends in Ecology and Evolution* 14:102–106.

Giraldeau, L.-A., and Caraco, T. 1993. Genetic relatedness and group size in an aggregation economy. *Evolutionary Ecology* 7:429–438.

———. 2000. *Social Foraging Theory.* Princeton, NJ: Princeton University Press.

Giraldeau, L.-A., Hogan, J. A., and Clinchy, M. J. 1990. The payoffs to producing and scrounging: What happens when patches are divisible? *Ethology* 85:132–146.

Giraldeau, L.-A., and Kramer, D. L. 1982. The marginal value theorem: A quantitative test using load size variation in a central place forager, the eastern chipmunk *Tamias striatus*. *Animal Behaviour* 30:1036–1042.

Giraldeau, L.-A., and Lefebvre, L. 1987. Scrounging prevents cultural transmission of food-finding behaviour in pigeons. *Animal Behaviour* 35:387–394.

Giraldeau, L.-A., and Livoreil, B. 1998. Game theory and social foraging. In *Game Theory and Animal Behavior*, ed. L. A. Dugatkin, and H. K. Reeve, 16–37. New York: Oxford University Press.

Giraldeau, L.-A., Soos, C., and Beauchamp, G. 1994. A test of the producer-scrounger foraging game in captive flocks of spice finches, *Lonchura punctulata*. *Behavioral Ecology and Sociobiology* 34:251–256.

Gittleman, J. L., and Harvey, P. H. 1980. Why are distasteful prey not cryptic? *Nature* 286:149–150.

Giurfa, M. 2003. Cognitive neuroethology: Dissecting non-elemental learning in a honeybee brain. *Current Opinion in Neurobiology* 13:726–735.

Gleeson, S. K., and Wilson, D. S. 1986. Equilibrium diet—optimal foraging and prey coexistence. *Oikos* 46:139–144.

Goldman-Rakic, P. S. 1990. Cellular and circuit basis of working memory in prefrontal cortex of nonhuman primates. In *Progress in Brain Research*, vol. 85, *The Prefrontal Cortex: Its Structure, Function and Pathology*, ed. H. B. M. Uylings, C. G. Van Eden, J. P. C. De Bruin, M. A. Corner, and M. G. P. Feenstra, 325–336. Amsterdam: Elsevier.

Goldman-Rakic, P. S., Selemon, L. D., and Schwartz, M. L. 1984. Dual pathways connecting the dorsolateral prefrontal cortex with the hippocampal formation and parahippocampal cortex in the rhesus monkey. *Neuroscience* 12:719–743.

Goodenough, J. E., McGuire, B., and Wallace, R. A. 1993. *Perspectives on Animal Behavior*. New York: John Wiley & Sons.

———. 2001. Mechanisms of orientation. In *Perspectives on Animal Behavior*, 2nd ed., 217–263. New York: John Wiley & Sons.

Goodridge, J. P., and Taube, J. S. 1995. Preferential use of the landmark navigation system by head direction cells in rats. *Behavioral Neuroscience* 109:49–61.

Gorman, M. L., Mills, M. G., Raath, J. P., and Speakman, J. R. 1998. High hunting costs make African wild dogs vulnerable to kleptoparasitism by hyaenas. *Nature* 391:479–481.

Gosler, A. G., Greenwood, J. J. D., and Perrins, C. M. 1995. Predation risk and the cost of being fat. *Nature* 377:621–623.

Gould, E., Tanapat, P., Rydel, T., and Hastings, N. 2000. Regulation of hippocampal neurogenesis in adulthood. *Biological Psychiatry* 48:715–720.

Gould, J. P. 1974. Risk, stochastic preference, and the value of information. *Journal of Economic Theory* 8:64–84.

Gould, S. J., and Lewontin, R. C. 1979. The spandrels of San Marco and the Panglossian paradigm: A critique of the adaptationist programme. *Proceedings of the Royal Society of London* B 205:581–598.

Grafen, A. 2000. Developments of the Price equation and natural selection under uncertainty. *Proceedings of the Royal Society of London* B 267:1223–1227.

Grajal, A. 1995. Structure and function of the digestive tract of the hoatzin (*Opisthocomus hoazin*): A folivorous bird with foregut fermentation. *Auk* 112:20–28.

Grajal, A., Strahl, S. D., Parra, R., Dominguez, M. G., and Neher, A. 1989. Foregut fermentation in the hoatzin, a neotropical leaf-eating bird. *Science* 245:1236–1238.

Grand, T. C. 2002. Foraging-predation risk trade-offs, habitat selection, and the coexistence of competitors. *American Naturalist* 159:106–112.

Grand, T. C., and Dill, L. M. 1997. The energetic equivalence of cover to juvenile coho salmon (*Oncorhynchus kisutch*): Ideal free distribution theory applied. *Behavioral Ecology* 8:437–444.

———. 1999. Predation risk, unequal competitors and the ideal free distribution. *Evolutionary Ecology Research* 1:389–409.

Grant, P. R. 1986. *Ecology and Evolution of Darwin's Finches*. Princeton, NJ: Princeton University Press.

Gray, J. M., Yarian, D., and Ramenofsky, M. 1990. Corticosterone, foraging behavior, and metabolism in dark-eyed juncos, *Junco hyemalis*. *General and Comparative Endocrinology* 79:375–384.

Green, D. M., and Swets, J. A. 1966. *Signal Detection Theory and Psychophysics*. New York: Wiley.

Green, L., Fisher, E. B., Perlow, S., and Sherman, L. 1981. Preference reversal and self-control: Choice as a function of reward amount and delay. *Behavior Analysis Letters* 1:244–256.

Green, L., and Rachlin, H. 1991. Economic substitutability of electrical brain stimulation, food and water. *Journal of the Experimental Analysis of Behavior* 55:133–143.

Green, R. F. 1980. Bayesian birds: A simple example of Oaten's stochastic model of optimal foraging. *Theoretical Population Biology* 18:244–256.

Greenstone, M. H. 1979. Spider feeding behaviour optimises dietary essential amino acid composition. *Nature* 282:501–503.

Greenwood, G. B., and Demment, M. W. 1988. The effects of fasting on short-term cattle grazing behaviour. *Grass and Forage Science* 43:377–386.

Greenwood, M. F. D., and Metcalfe, N. B. 1998. Minnows become nocturnal at low temperatures. *Journal of Fish Biology* 53:25–32.

Greggers, U., and Menzel, R. 1993. Memory dynamics and foraging strategies of honeybees. *Behavioral Ecology and Sociobiology* 32:17–29.

Griffin, D. R. 2001. *Animal Minds: Beyond Cognition to Consciousness*. Chicago: University of Chicago Press.

Grimm, V., and Railsback, S. F. 2005. *Individual-Based Modeling and Ecology*. Princeton, NJ: Princeton University Press.

Gros-Louis, J. 2004. Responses of white-faced capuchins (*Cebus capucinus*) to naturalistic and experimentally presented food-associated calls. *Journal of Comparative Psychology* 118:396–402.

Grunbaum, D. 1998. Using spatially explicit models to characterize foraging performance in heterogeneous landscapes. *American Naturalist* 151:97–115.

Grünbaum, L., and Müller, U. 1998. Induction of a specific olfactory memory leads to a long-lasting activation of protein kinase C in the antennal lobe of the honeybee. *Journal of Neuroscience* 18:4384–4392.

Gudmundsson, G. A., Lindström, Å., and Alerstam, T. 1991. Optimal fat loads and long-distance flights by migrating knots, *Calidris canutus*, sanderlings, *C. alba*, and turnstones, *Arenaria interpres*. *Ibis* 133:140–152.

Guerra, B., and Vickery, W. L. 1998. How do red squirrels, *Tamiasciurus hudsonicus*, and eastern chipmunks, *Tamias striatus*, coexist? *Oikos* 83:139–144.

Guerrieri, E., Poppy, G. M., Powell, W., Tremblay, E., and Pennacchio, F. 1999. Induction and systemic release of herbivore-induced plant volatiles mediating in-flight orientation of *Aphidius ervi. Journal of Chemical Ecology* 25:1247–1261.

Guglielmo, C. G., Karasov, W. H., and Jakubas, W. J. 1996. Nutritional costs of a plant secondary metabolite explain selective foraging by ruffed grouse. *Ecology* 77:1103–1115.

Guillemette, M., Ydenberg, R. C., and Himmelman, J. H. 1993. The role of energy intake and habitat selection of common eiders *Somateria mollissima* in winter: A risk-sensitive interpretation. *Journal of Animal Ecology* 61:599–610.

Guo, P. Z., Mueller, L. D., and Ayala, F. J. 1991. Evolution of behavior by density-dependent natural selection. *Proceedings of the National Academy of Sciences U.S.A.* 88:10905–10906.

Gurung, M. 2003. Ecology of fear: A case study of blue sheep and snow leopards in the Nepal Himalayas. Ph.D. dissertation, University of Illinois at Chicago.

Gustaffson, L., and Pärt, T. 1990. Acceleration of senescence in the collared flycatcher *Ficedula albicollis. Nature* 347:279–281.

Gustaffson, L., and Sutherland, W. J. 1988. The costs of reproduction in the collared flycatcher *Ficedula albicollis. Nature* 335:813–815.

Gustafsson, A., Herrmann, A., and Huber, F. 2003. *Conjoint Measurement: Methods and Applications.* 3rd ed. Heidelberg: Springer-Verlag.

Gustafsson, L., Nordling, D., Andersson, M. S., Sheldon, B. C., and Qvarnstrom, A. 1994. Infectious diseases, reproductive effort and the cost of reproduction in birds. *Philosophical Transactions of the Royal Society* B 346:323–331.

Gustafsson, M., Jensen, P., de Jonge, P. F. H., and Schuurman, T. 1999. Domestication effects on foraging strategies in pigs (*Sus scrofa*). *Applied Animal Behaviour Science* 62:305–317.

Haftorn, S. 1956. *Contribution to the food biology of tits especially about storing of surplus food.* Part 4. A comparative analysis of *Parus atricapillus* L, *P. cristatus* L., and *P. ater* L. Det Kongelige Norske Videnskabers Selskabs Skrifter.

———. 1959. The proportion of spruce seeds removed by the tits in a Norwegian spruce forest in 1954–55. *Kongelige Norske Videnskabers Selskabs Forhandlinger* 32:121–125.

———. 1972. Hypothermia of tits in the arctic winter. *Ornis Scandinavica* 3:153–166.

———. 1992. The diurnal body weight circle in titmice *Parus* spp. *Ornis Scandinavica* 23:435–443.

Hainsworth, F. R. 1981. Energy regulation in hummingbirds. *American Scientist* 69:420–429.

Hairston, N. G., Smith, F. E., and Slobodkin, L. B. 1960. Community structure, population control and competition. *American Naturalist* 94:421–425.

Hamback, P. A. 1998. Seasonality, optimal foraging, and prey coexistence. *American Naturalist* 152:881–895.

Hamilton, I. M. 2000. Recruiters and joiners: Using optimal skew theory to predict group size and the division of resources within groups of social foragers. *American Naturalist* 155:684–695.

———. 2002. Kleptoparasitism and the distribution of unequal competitors. *Behavioral Ecology* 13:260–267.

Hamilton, W. D. 1964. The genetical evolution of social behavior, I & II. *Journal of Theoretical Biology* 7:1–52.

Hammer, M. 1993. An identified neuron mediates the unconditioned stimulus in associative olfactory learning in honeybees. *Nature* 366:59–63.

Hammer, M., and Menzel, R. 1995. Learning and memory in the honeybee. *Journal of Neuroscience* 15:1617–1630.

———. 1998. Multiple sites of associative odor learning as revealed by local brain microinjections of octopamine in honeybees. *Learning and Memory* 5:146–156.

Hammond, K. A., and Diamond, J. 1997. Maximal sustained energy budgets in humans and animals. *Nature* 386:457–462.

Hampton, R. R., Healy, S. D., Shettleworth, S. J., and Kamil, A. C. 2002. "Neuroecologists" are not made of straw. *Trends in Cognitive Science* 6:6–7.

Hampton, R. R., Sherry, D. F., Shettleworth, S. J., Khurgel, M., and Ivy, G. 1995. Hippocampal volume and food-storing behavior are related in parids. *Brain Behavior and Evolution* 45:54–61.

Hampton, R. R., and Shettleworth, S. J. 1996. Hippocampal lesions impair memory for location but not color in passerine birds. *Behavioral Neuroscience* 110:831–835.

Hampton, R. R., Shettleworth, S. J., and Westwood, R. P. 1998. Proactive interference, recency, and associative strength: Comparisons of black-capped chickadees and dark-eyed juncos. *Animal Learning & Behavior* 26:475–485.

Hanski, I. 1999. *Metapopulation Ecology*. Oxford: Oxford University Press.

Hanski, I., Henttonen, H., Korpimaki, E., Oksanen, L., and Turchin, P. 2001. Small-rodent dynamics and predation. *Ecology* 82:1505–1520.

Hanson, H. M. 1959. Effects of discrimination training on stimulus generalization. *Journal of Experimental Psychology* 58:321–334.

Hardin, G. 1968. The tragedy of the commons. *Science* 162:1243–1248.

Harfenist, A. E., and Ydenberg, R. C. 1995. Parental provisioning and predation risk in rhinoceros auklets (*Cerorhinca monocerata*): Effects on nestling growth and fledging. *Behavioral Ecology* 6:82–86.

Harley, C. B. 1981. Learning the evolutionarily stable strategy. *Journal of Theoretical Biology* 89:611–633.

Harrison, J. 1986. Caste-specific changes in honeybee flight capacity. *Physiological Zoology* 59:175–187.

Harvey, A., Parsons, A. J., Rook, A. J., Penning, P. D., and Orr, R. J. 2000. Dietary preference of sheep for perennial ryegrass and white clover at contrasting sward surface heights. *Grass and Forage Science* 55:242–252.

Harvey, P. H., and Gittleman, J. L. 1992. Correlates of carnivory: Approaches and answers. In *Natural Enemies: The Population Biology of Predators, Parasites and Diseases*, ed. M. J. Crawley, 26–39. Oxford: Blackwell Scientific Publications.

Harvey, P. H., and Greenwood, P. J. 1978. Anti-predator defence strategies: Some evolutionary problems. In *Behavioral Ecology: An Evolutionary Approach*, ed. J. R. Krebs and N. B. Davies, 129–158. Oxford: Blackwell Scientific Publications.

Hassell, M. P. 1978. *The Dynamics of Arthropod Predator-Prey Systems*. Princeton, NJ: Princeton University Press.

———. 2000. *The Spatial and Temporal Dynamics of Host-Parasitoid Interactions*. Oxford: Oxford University Press.

Hassell, M. P., and Comins, H. N. 1978. Sigmoid functional responses and population stability. *Theoretical Population Biology* 14:62–67.

Hassell, M. P., and May, R. M. 1973. Stability in insect host-parasite models. *Journal of Animal Ecology* 42:693–726.

Hassell, M. P., and Varley, G. C. 1969. New inductive population model for insect parasites and its bearing on biological control. *Nature* 223:1133–1137.

Hatfield, T., and Schluter, D. 1999. Ecological speciation in sticklebacks: Environment dependent hybrid fitness. *Evolution* 53:866–873.

Hauser, M. D. 1997. Artifactual kinds and functional design features: What a primate understands without language. *Cognition* 64:285–308.

Healy, S. D., and Hurly, T. A. 1998. Rufous hummingbirds' (*Selasphorus rufus*) memory for flowers: Patterns or actual spatial locations? *Journal of Experimental Psychology: Animal Behavior Processes* 24:396–404.

Healy, S. D., and Krebs, J. R. 1992. Food storing and the hippocampus in corvids: Amount and volume are correlated. *Proceedings of the Royal Society of London* B 248:241–245.

Hebb, D. O. 1949. *The Organization of Behavior: A Neuropsychological Theory*. New York: Wiley & Sons.

Hedenström, A. 1992. Flight performance in relation to fuel load in birds. *Journal of Theoretical Biology* 158:535–537.

Hedenström, A., and Alerstam, T. 1992. Climbing performance of migrating birds as a basis for estimating limits for fuel-carrying capacity and muscle work. *Journal of Experimental Biology* 164:19–38.

———. 1995. Optimal flight speeds of birds. *Philosophical Transactions of the Royal Society* B 348:471–487.

———. 1997. Optimum fuel loads in migratory birds: Distinguishing between time and energy minimization. *Journal of Theoretical Biology* 189:227–234.

Hedenström, A., and Sunada, S. 1999. On the aerodynamics of moult gaps in birds. *Journal of Experimental Biology* 202:67–76.

Heerwagen, J. H., and Orians, G. H. 1993. Humans, habitats, and aesthetics. In *The Biophilia Hypothesis*, ed. S. R. Kellert and E. O. Wilson, 138–172. Washington, DC: Island Press.

Hegde, A. N., Inokuchi, K., Pei, W., Casadio, A., Ghirardi, M., Chain, D. G., Martin, K. C., Kandel, E. R., and Schwartz, J. H. 1997. Ubiquitin C-terminal hydrolase is an immediate-early gene essential for long-term facilitation in *Aplysia*. *Cell* 89:115–126.

Heinrich, B. 1995. An experimental investigation of insight in Common Ravens (*Corvus corax*). *Auk* 112:994–1003.

Heisenberg, M. 1998. What do the mushroom bodies do for the insect brain? An introduction. *Learning and Memory* 5:1–10.

Heisenberg, M., Borst, A., Wagner, S., and Byers, D. 1985. *Drosophila* mushroom body mutants are deficient in olfactory learning. *Journal of Neurogenetics* 2:1–30.

Heithaus, M. R. 2001. Habitat selection by predators and prey in communities with asymmetrical intraguild predation. *Oikos* 92:542–554.

Herman, C. S., and Valone, T. J. 2000. The effect of mammalian predator scent on the foraging behavior of *Dipodomys merriami*. *Oikos* 91:139–145.

Heyes, C. M., and Galef, B. G., Jr., eds. 1996. *Social Learning in Animals: The Roots of Culture*. New York: Academic Press.

Hibbard, B., Peters, J. P., Chester, S. T., Robinson, J. A., Kotarski, S. F., Croom, W. J., Jr., and Hagler, W. M., Jr. 1995. The effect of slaframine on salivary output and subacute and acute acidosis in growing beef steers. *Journal of Animal Science* 73:516–525.

Higgs, P., and Fox, B. J. 1993. Interspecific competition: A mechanism for succession after fire in wet heathland. *Oikos* 67:358–370.

Hildebrandt, H., and Müller, U. 1995a. Octopamine mediates rapid stimulation of Protein Kinase A in the antennal lobe of honeybees. *Journal of Neurobiology* 27:44–50.

———. 1995b. PKA activity in the antennal lobe of honeybees is regulated by chemosensory stimulation in vivo. *Brain Research* 679:281–288.

Hill, C., Quigley, M. A., Cavaletto, J. F., and Gordon, W. 1992. Seasonal changes in lipid content and composition in the benthic amphipods *Monoporeia affinis* and *Pontoporeia femorata*. *Limnology and Oceanography* 37:1280–1289.

Hilton, G. M., Cresswell, W., and Ruxton, G. D. 1999. Intraflock variation in the speed of escape-flight response on attack by an avian predator. *Behavioral Ecology* 10:391–395.

Hintz, H. F., Sedgewick, C. J., and Schryver, H. F. 1976. Some observations on digestion of pelleted diet by ruminants and non-ruminants. *International Zoo Yearbook* 16:54–57.

Hipfner, J. M., Gaston, A. J., and Storey, A. E. 2001. Nest-site safety predicts the relative investment made in first and replacement eggs by two long-lived seabirds. *Oecologia* 129:234–242.

Hirakawa, H. 1995. Diet optimization with a nutrient or toxin constraint. *Theoretical Population Biology* 47:1–16.

———. 1997a. Digestion-constrained optimal foraging in generalist mammalian herbivores. *Oikos* 78:37–47.

———. 1997b. How important is digestive quality? A correction of Verlinden and Wiley's digestive rate model. *Evolutionary Ecology* 11:249–252.

Hirvonen, H., Ranta, E., Rita, H., and Peuhkuri, N. 1999. Significance of memory properties in prey choice decisions. *Ecological Modelling* 115:177–189.

Hissa, R. 1997. Physiology of the European brown bear (*Ursus arctos arctos*). *Annales Zoologici Fennici* 34:267–287.

Hitchcock, C. L., and Houston, A. I. 1994. The value of a hoard: Not just energy. *Behavioral Ecology* 5:202–205.

Hobbs, N. T. 1990. Diet selection by generalist herbivores: A test of the linear programming model. In *Behavioral Mechanisms of Food Selection*, ed. R. N. Hughes, 395–422. Berlin: Springer-Verlag.

Hodgson, J., and Illius, A. W., eds. 1996. *The Ecology and Management of Grazing Systems*. Wallingford, UK: CAB International.

Hofer, H., and East, M. L. 1993. The commuting system of Serengeti spotted hyaenas: How a predator copes with migratory prey. III. Attendance and maternal care. *Animal Behavior* 46:575–589.

Hogstad, O. 1987. It is expensive to be dominant. *Auk* 104:333–336.

———. 1988. Rank-related resource access in winter flocks of willow tit *Parus montanus*. *Ornis Scandinavica* 19:169–174.

Holekamp, K. E., and Smale, L. 1990. Provisioning and food-sharing by lactating spotted hyenas (*Crocuta crocuta*). *Ethology* 86:191–202.

Holling, C. S. 1959a. The components of predation as revealed by a study of small mammal predation of the European Pine Sawfly. *Canadian Entomologist* 91:234–261.

———. 1959b. Some characteristics of simple types of predation and parasitism. *Canadian Entomologist* 91:385–398.

———. 1965. The functional response of predators to prey density and its role in mimicry and population regulation. *Memoirs of the Entomological Society of Canada* 45:1–60.

———. 1966. The functional response of invertebrate predators to prey density. *Memoirs of the Entomological Society of Canada* 48:1–86.

Holmes, W. G. 1984. Predation risk and foraging behavior of the hoary marmot in Alaska. *Behavioral Ecology and Sociobiology* 15:293–301.

Holmgren, N. 1995. The ideal free distribution of unequal individuals: Predictions from a behaviour-based functional response. *Journal of Animal Ecology* 64:197–212.

Holt, R. D. 1977. Predation, apparent competition, and structure of prey communities. *Theoretical Population Biology* 12:197–229.

———. 1983. Optimal foraging and the form of the predator isocline. *American Naturalist* 122:521–541.

———. 1984. Spatial heterogeneity, indirect interactions, and the coexistence of prey species. *American Naturalist* 124:377–406.

———. 1985. Population dynamics in 2-patch environments—some anomalous consequences of an optimal habitat distribution. *Theoretical Population Biology* 28:181–208.

———. 1987a. Population dynamics and evolutionary processes: The manifold effects of habitat selection. *Evolutionary Ecology* 1:331–347.

———. 1987b. Prey communities in patchy environments. *Oikos* 50:276–290.

———. 1996a. Demographic constraints in evolution: Towards unifying the evolutionary theories of senescence and niche conservatism. *Evolutionary Ecology* 10:1–11.

———. 1996b. Rarity and evolution: Some theoretical considerations. In *The Biology of Rarity*, ed. W. E. Kunin, and K. J. Gaston, 209–234. London: Chapman & Hall.

———. 1997. Community modules. In *Multitrophic Interactions in Terrestrial Ecosystems*, ed. A. C. Gange and V. K. Brown, 333–349. Oxford: Blackwell Science.

Holt, R. D., Barfield, M., and Gomulkiewicz, R. 2005. Theories of niche conservatism and evolution: Could exotic species be potential tests? In *Species Invasions: Insights into Ecology, Evolution, and Biogeography*, ed. D. Sax, J. Stachowicz, and S. D. Gaines, 259–290. Sunderland, MA: Sinauer Associates.

Holt, R. D., and Gomulkiewicz, R. 1997a. The evolution of species niches: A population dynamic perspective. In *Case Studies in Mathematical Modelling: Ecology, Physiology and Cell Biology*, ed. H. G. Othmer, F. R. Adler, M. A. Lewis, and J. C. Dallon, 25–50. Upper Saddle River, NJ: Prentice Hall.

———. 1997b. How does immigration influence local adaptation? A reexamination of a familiar paradigm. *American Naturalist* 149:563–572.

Holt, R. D., Grover, J., and Tilman, D. 1994. Simple rules for interspecific dominance in systems with exploitative and apparent competition. *American Naturalist* 144:741–771.

Holt, R. D., and Kotler, B. P. 1987. Short-term apparent competition. *American Naturalist* 130:412–430.

Holt, R. D., and Lawton, J. H. 1994. The ecological consequences of shared natural enemies. *Annual Review of Ecology and Systematics* 25:495–520.

Hopkins, W. A., Roe, J. H., Philippi, T., and Congdon, J. D. 2004. Standard and digestive metabolism in the banded water snake, *Nerodia fasciata fasciata*. *Comparative Biochemistry and Physiology* A 137:141–149.

Horn, M. H. 1989. Biology of marine herbivorous fishes. *Oceanography and Marine Biology: An Annual Review* 27:167–272.

Horn, M. H., and Messer, K. S. 1992. Fish guts as chemical reactors—a model of the alimentary canals of marine herbivorous fishes. *Marine Biology* 113:527–535.

Houston, A. I. 1987. Optimal foraging by parent birds feeding dependent young. *Journal of Theoretical Biology* 124:251–274.

————. 1993. The efficiency of mass loss in breeding birds. *Proceedings of the Royal Society of London* B 254:221–225.

————. 1998. Models of optimal avian migration: State, time and predation. *Journal of Avian Biology* 29:395–404.

Houston, A. I., and McNamara, J. M. 1982. A sequential approach to risk-taking. *Animal Behaviour* 30:1260–1261.

————. 1985. A general theory of central-place foraging for single prey loaders. *Theoretical Population Biology* 28:233–262.

————. 1993. A theoretical investigation of the fat reserves and mortality levels of small birds in winter. *Ornis Scandinavica* 24:205–219.

————. 1999. *Models of Adaptive Behaviour*. Cambridge: Cambridge University Press.

Houston, A. I., McNamara, J. M., and Hutchinson, J. M. C. 1993. General results concerning the trade-off between gaining energy and avoiding predation. *Philosophical Transactions of the Royal Society* B 341:375–397.

Houston, A. I., Thompson, W. A., and Gaston, A. J. 1996. The use of a time and energy budget model of a parent bird to investigate limits to fledging mass in the thick-billed murre. *Functional Ecology* 10:432–439.

Houston, A. I., Welton, N., and McNamara, J. M. 1997. Acquisition and maintenance costs in the long-term regulation of avian fat reserves. *Oikos* 78:331–340.

Houtman, R., and Dill, L. M. 1998. The influence of predation risk on diet selectivity: A theoretical analysis. *Evolutionary Ecology* 12:251–262.

Howery, L. D., Bailey, D. W., Ruyle, G. B., and Renken, W. J. 2000. Cattle use visual cues to track food locations. *Applied Animal Behaviour Science* 67:1–14.

Hughes, A. 1971. Topographical relationships between the anatomy and physiology of the rabbit visual system. *Documenta Ophthalmologica* 30:33–159.

Hughes, J. J., and Ward, D. 1993. Predation risk and distance to cover affect foraging behavior in Namib desert gerbils. *Animal Behaviour* 46:1243–1245.

Hughes, J. J., Ward, D., and Perrin, M. R. 1995. Effects of substrate on foraging decisions by a Namib desert gerbil. *Journal of Mammalogy* 67:638–645.

Hughes, N. F., and Grand, T. C. 2000. Physiological ecology meets the ideal-free distribution: Predicting the distribution of size-structured fish populations across temperature gradients. *Environmental Biology of Fishes* 59:285–298.

Hugie, D. M., and Dill, L. M. 1994. Fish and game: A game theoretic approach to habitat selection by predators and prey. *Journal of Fish Biology* 45:151–169.

Hugie, D. M., and Grand, T. C. 1998. Movement between patches, unequal competitors and the ideal free distribution. *Evolutionary Ecology* 12:1–19.

————. 2003. Movement between habitats used by unequal competitors: Effects of finite population size on ideal free distributions. *Evolutionary Ecology Research* 5:131–153.

Hume, I. D. 1989. Optimal digestive strategies in mammalian herbivores. *Physiological Zoology* 62:1145–1163.

Hume, I. D., and Sakaguchi, E. 1991. Patterns of digesta flow and digestion in foregut and hindgut fermenters. In *Physiological Aspects of Digestion and Metabolism in Ruminants*, ed. T. Tsuda, Y. Sasaki, and R. Kwawshima, 427–451. San Diego: Academic Press.

Hunt, R. R. 1996. Manufacture and use of hook-tools by New Caledonian crows. *Nature* 379:249–251.

Huntly, N. J. 1987. Influence of refuging consumers (Pikas: *Ochotona princeps*) on subalpine meadow vegetation. *Ecology* 68:274–283.

Hutchings, M. R., Gordon, I. J., Robertson, E., Kyriazakis, I., and Jackson, F. 2000. Effects of parasitic status and level of feeding motivation on the diet selected by sheep grazing grass/clover swards. *Journal of Agricultural Science* 135:65–75.

Hutchings, M. R., Kyriazakis, I., Anderson, D. H., Gordon, I. J., and Coop, R. L. 1998. Behavioural strategies used by parasitized and non-parasitized sheep to avoid ingestion of gastro-intestinal nematodes associated with faeces. *Animal Science* 67:97–106.

Hutchings, M. R., Kyriazakis, I., and Gordon, I. J. 2001. Herbivore physiological state affects foraging trade-off decisions between nutrient intake and parasite avoidance. *Ecology* 82:1138–1150.

Hutchings, M. R., Kyriazakis, I., Gordon, I. J., and Jackson, F. 1999. Trade-offs between nutrient intake and faecal avoidance in herbivore foraging decisions: The effect of animal parasitic status, level of feeding motivation and sward nitrogen content. *Journal of Animal Ecology* 68:310–323.

Hutchings, M. R., Kyriazakis, I., Papachristou, T. G., Gordon, I. J., and Jackson, F. 2000. The herbivores' dilemma: Trade-offs between nutrition and parasitism in foraging decisions. *Oecologia* 124:242–251.

Hutchings, N. J., and Gordon, I. J. 2001. A dynamic model of herbivore-plant interactions on grasslands. *Ecological Modelling* 136:209–222.

Iason, G. R., Mantecon, A. R., Sim, D. A., Gonzalez, J., Foreman, E., Bermudez, F. F., and Elston, D. A. 1999. Can grazing sheep compensate for a daily foraging time constraint? *Journal of Animal Ecology* 68:87–93.

Iason, G. R., and Van Wieren, S. E. 1999. Digestive and ingestive adaptations of mammalian herbivores to low-quality forage. In *Herbivores: Between Plants and Predators*, ed. H. Olff, V. K. Brown, and R. H. Drent, 337–369. Oxford: Blackwell Science.

Illius, A. W. 2006. Linking function responses and foraging behavior to population dynamics. In *Large Herbivore Ecology: Ecosystem Dynamics and Conservation*, ed. K. Danell, R. Bergström, P. Duncan, and J. Pastor, 71–96. Cambridge: Cambridge University Press.

Illius, A. W., and FitzGibbon, C. 1994. Costs of vigilance in foraging ungulates. *Animal Behaviour* 47:481–484.

Illius, A. W., and Gordon, I. J. 1987. The allometry of food intake in grazing ruminants. *Journal of Animal Ecology* 56:989–999.

———. 1990. Constraints on diet selection and foraging behaviour in mammalian herbivores. In *Behavioural Mechanisms of Food Selection*, ed. R. N. Hughes, 367–392. Berlin: Springer-Verlag.

———. 1997. Scaling up from functional response to numerical response in vertebrate herbivores. In *Herbivores: Between Plants and Predators*, ed. H. Olff, V. K. Brown, and R. H. Drent, 397–425. Oxford: Blackwell Science.

Illius, A. W., Gordon, I. J., Elston, D. A., and Milne, J. D. 1999. Diet selection in goats: A test of intake-rate maximization. *Ecology* 80:1008–1018.

Illius, A. W., Jessop, N. S., and Gill, M. 2000. Mathematical models of food intake and metabolism in ruminants. In *Ruminant Physiology: Digestion, Metabolism, Growth and Reproduction*, ed. P. Cronjé and E. A. Boomker, 21–40. Wallingford, UK: CAB International.

Inman, A. J. 1990. Foraging decisions: The effects of conspecifics and environmental stochasticity. Ph.D. dissertation, Oxford University.

Irving, L., West, G. C., and Peyton, L. J. 1967. Winter feeding program of Alaska willow ptarmigan shown by crop contents. *Condor* 69:69–71.

Irwin, L. 1975. Deer-moose relationships on a burn in Northeastern Minnesota. *Journal of Wildlife Management* 39:653–662.

Iverson, S. J., Bowen, W. D., Boness, D. J., and Oftedal, O. T. 1993. The effect of maternal size and milk energy output on pup growth in grey seals (*Halichoerus grypus*). *Physiological Zoology* 66:61–88.

Iwasa, Y., Higashi, M., and Yamamura, N. 1981. Prey distribution as a factor determining the choice of optimal foraging strategy. *American Naturalist* 117:710–723.

Jackson, J. B. C. 2001. What was natural in the coastal oceans? *Proceedings of the National Academy of Sciences U.S.A.* 98:5411–5418.

Jacobs, L. F. 1995. The ecology of spatial cognition: Adaptive patterns of hippocampal size and space use in wild rodents. In *Studies of the Brain in Naturalistic Settings*, ed. E. Alleva, A. Fasolo, H.-P. Lipp, and L. Nadel, 301–322. Dordrecht: Kluwer Academic Publishers.

Jacobs, L. F., Gaulin, S. J. C., Sherry, D. F., and Hoffman, G. E. 1990. Evolution of spatial cognition: Sex-specific patterns of spatial behavior predict hippocampal size. *Proceedings of the National Academy of Sciences U.S.A.* 87:6349–6352.

Jacobs, L. F., and Schenk, F. 2003. Unpacking the cognitive map: The parallel map theory of hippocampal function. *Psychological Review* 110:285–315.

Jacobs, L. F., and Shiflett, M. W. 1999. Spatial orientation on a vertical maze in free-ranging fox squirrels (*Sciurus niger*). *Journal of Comparative Psychology* 113:116–127.

Jacobsen, C. F. 1936. Studies of cerebral function in primates. I. The functions of the frontal association areas in monkeys. *Comparative Psychology Monographs* 13:1–60.

Jacobsen, C. F., and Nissen, H. W. 1937. Studies of cerebral function in primates. IV. The effects of frontal lobe lesions on the delayed alternation habit in monkeys. *Journal of Comparative Psychology* 23:101–112.

James, P. C., and Verbeek, N. A. M. 1984. Temporal and energetic aspects of food storage in northwestern crows. *Ardea* 72:207–215.

James, W. 1890. *The Principles of Psychology*. New York: Holt.

Janis, I. L. 1972. *Victims of Groupthink: A Psychological Study of Foreign-Policy Decisions and Fiascoes*. Boston: Houghton Mifflin.

Janmaat, A. F., Winston, M. L., and Ydenberg, R. C. 2000. Condition-dependent response to changes in pollen stores by honey bee (*Apis mellifera*) colonies with different parasite loads. *Behavioral Ecology and Sociobiology* 47:171–179.

Janson, C. H. 1998. Experimental evidence for spatial memory in wild brown capuchin monkey (*Cebus apella*). *Animal Behaviour* 55:1129–1143.

Jansson, C. J., Ekman, J., and von Bromssen, A. 1981. Winter mortality and food supply in tits *Parus* spp. *Oikos* 37:313–322.

Jarman, P. J. 1974. The social organisation of antelope in relation to their ecology. *Behaviour* 48:215–267.

Jarrard, L. E. 1993. On the role of the hippocampus in learning and memory in the rat. *Behavioral and Neural Biology* 60:9–26.

———. 1995. What does the hippocampus really do? *Behavioural Brain Research* 71:1–10.

Jenni, L., and Jenni-Eiermann, S. 1998. Fuel supply and metabolic constraints in migrating birds. *Journal of Avian Biology* 29:521–528.

Jeschke, J. M., Kopp, M., and Tollrian, R. 2002. Predator functional responses: Discriminating between handling and digesting prey. *Ecological Monographs* 72:95–112.

Jiang, Z., and Hudson, R. 1993. Optimal grazing of wapiti (*Cervus elaphus*) on grassland: Patch and feeding station departure rules. *Evolutionary Ecology* 7:488–498.

Johnson, K. G., Tyrrell, J., Rowe, J. B., and Pethick, D. W. 1998. Behavioural changes in stabled horses given nontherapeutic levels of virginiamycin. *Equine Veterinary Journal* 30:139–143.

Johnsson, J. I., Petersson, E., Jonsson, E., Bjornsson, B. T., and Jarvi, T. 1996. Domestication and growth hormone alter antipredator behaviour and growth patterns in juvenile brown trout, *Salmo trutta*. *Canadian Journal of Fisheries and Aquatic Sciences* 53:1546–1554.

Jones, G., and Rydell, J. 1994. Foraging strategy and predation risk as factors influencing emergence time in echolocating bats. *Philosophical Transactions of the Royal Society* B 346:445–455.

Jönsson, K. I., and Rebecchi, L. 2002. Experimentally induced anhydrobiosis in the tardigrade *Richtersius coronifer*: Phenotypic factors affecting survival. *Journal of Experimental Zoology* 293:578–584.

Jullien, M., and Clobert, J. 2000. The survival value of flocking in neotropical birds: Reality or fiction? *Ecology* 81:3416–3430.

Jumars, P. A. 2000a. Animal guts as ideal chemical reactors: Maximizing absorption rates. *American Naturalist* 155:527–543.

———. 2000b. Animal guts as nonideal chemical reactors: Partial mixing and axial variation in absorption kinetics. *American Naturalist* 155:544–555.

Jumars, P. A., and Martínez del Rio, C. 1999. The tau of continuous feeding on simple foods. *Physiological and Biochemical Zoology* 72:633–641.

Kacelnik, A. 1984. Central place foraging in starlings (*Sturnus vulgaris*) I. Patch residence time. *Journal of Animal Ecology* 53:283–299

———. 1993. Leaf-cutting ants tease optimal foraging theorists. *Trends in Ecology and Evolution* 8:346–348.

Kacelnik, A., and Bateson, M. 1997. Risk-sensitivity: Crossroads for theories of decision-making. *Trends in Cognitive Sciences* 1:304–309.

Kacelnik, A., Brunner, D., and Gibbon, J. 1990. Timing mechanisms in optimal foraging: Some applications of scalar expectancy theory. In *Behavioral Mechanisms in Food Selection*, ed. R. N. Hughes, 61–82. Berlin: Springer Verlag.

Kacelnik, A., Krebs, J. R., and Bernstein, C. 1992. The ideal free distribution and predator-prey populations. *Trends in Ecology and Evolution* 7:50–55.

Kacelnik, A., and Todd, I. A. 1992. Psychological mechanisms and the marginal value theorem: Effect of variability in travel time on patch exploitation. *Animal Behaviour* 43:313–322.

Kallander, H., and Smith, H. G. 1990. Food storing in birds: An evolutionary perspective. *Current Ornithology* 7:147–207.

Kamil, A. C. 1978. Systematic foraging by a nectar-feeding bird, the amakihi (*Loxops virens*). *Journal of Comparative and Physiological Psychology* 92:388–396.

———. 1994. A synthetic approach to the study of animal intelligence. In *Behavioral Mechanisms in Evolutionary Ecology*, ed. L. A. Real, 11–45. Chicago: University of Chicago Press.

Kamil, A. C., and Balda, R. P. 1990. Spatial memory in seed-caching corvids. In *The Psychology of Learning and Motivation*, ed. G. H. Bower, 1–25. San Diego: Academic Press.

Kamil, A. C., Balda, R. P., and Olson, D. J. 1994. Performance of four seed-caching corvid species in the radial-arm maze analog. *Journal of Comparative Psychology* 108:385–393.

Kamil, A. C., Krebs, J. R., and Pulliam, H. R., ed. 1987. *Foraging Behavior*. New York: Plenum Press.

Kamil, A. C., Misthal, R. L., and Stephens, D. W. 1993. Failure of simple optimal foraging models to predict residence time when patch quality is uncertain. *Behavioral Ecology* 4:350–363.

Kamil, A. C., and Sargent, T. D., eds. 1981. *Foraging Behaviour: Ecological, Ethological and Psychological Approaches*. New York: Garland STPM Press.

Kamin, L. J. 1969. Predictability, surprise, attention, and conditioning. In *Punishment and Aversive Behavior*, ed. B. A. Campbell and R. M. Church, 279–296. New York: Appleton-Century-Crofts.

Karasov, W. H. 1990. Digestion in birds: Chemical and physiological determinants and ecological implications. In *Avian Foraging: Theory, Methodology and Applications*, ed. M. L. Morrison, C. J. Ralph, J. Verner, and J. R. Jehl, 391–415. Los Angeles, CA: Cooper Ornithological Society.

———. 1992. Tests of the adaptive modulation hypothesis for dietary control of intestinal nutrient transport. *American Journal of Physiology* 263:R496–R502.

———. 1996. Digestive plasticity in avian energetics and feeding ecology. In *Avian Energetics and Nutritional Ecology*, ed. C. Carey, 61–84. New York: Chapman & Hall.

Karasov, W. H., and Cork, S. J. 1994. Glucose absorption by a nectarivorous bird: The passive pathway is paramount. *American Journal of Physiology* 267:G18–G26.

———. 1996. Test of a reactor-based digestion optimization model for nectar-eating rainbow lorikeets. *Physiological Zoology* 69:117–138.

Karasov, W. H., and Diamond, J. M. 1983. Adaptive regulation of sugar and amino acid transport by vertebrate intestine. *American Journal of Physiology* 245:G443–G462.

———. 1985. Digestive adaptations for fueling the cost of endothermy. *Science* 228:202–204.

———. 1988. Interplay between physiology and ecology in digestion. *BioScience* 38:602–611.

Karasov, W. H., and Hume, I. D. 1997. The vertebrate gastrointestinal system. In *Handbook of Physiology*, section 13, *Comparative Physiology*, vol. 1, ed. W. H. Dantzler, 407–480. New York: Oxford University Press.

Karasov, W. H., Pinshow, B., Starck, J. M., and Afik, D. 2004. Anatomical and histological changes in the alimentary tract of migrating blackcaps (*Sylvia atricapilla*): A comparison among fed, fasted, food-restricted, and refed birds. *Physiological and Biochemical Zoology* 77:149–160.

Kareiva, P., Morse, D. H., and Eccleston, J. 1989. Stochastic prey arrivals and crab spider performance using two simple "rules of thumb." *Oecologia* 78:542–549.

Kasuya, E. 1982. Central place water collection in a Japanese paper wasp (*Polistes chinensis antennalis*). *Animal Behaviour* 30:1010–1014.

Kats, L. B., and Dill, L. M. 1998. The scent of death: Chemosensory assessment of predation risk by prey animals. *Ecoscience* 5:361–394.

Kawecki, T. J. 1995. Demography of source-sink populations and the evolution of ecological niches. *Evolutionary Ecology* 9:38–44.

Kay, A. 2004. The relative availabilities of complementary resources affect the feeding preferences of ant colonies. *Behavioral Ecology* 15:63–70.

Keddy, P. A. 1981. Experimental demography of the sand-dune annual, *Cakile edentula*, growing along an environmental gradient in Nova Scotia. *Journal of Ecology* 69:615–630.

Keeling, M. J., Wilson, H. B., and Pacala, S. W. 2000. Reinterpreting space, time lags, and functional responses in ecological models. *Science* 290:1758–1761.

Keith, L. B. 1963. *Wildlife's Ten-Year Cycle*. Madison: University of Wisconsin Press.

Kellert, S. R., and Wilson, E. O., eds. 1993. *The Biophilia Hypothesis.* Washington, DC: Island Press.

Kemp, A. C. 1995. *The Hornbills.* Oxford: Oxford University Press.

Kendall, B. E., Briggs, C. J., Murdoch, W. W., Turchin, P., Ellner, S. P., McCauley, E., Nisbet, R. M., and Wood, S. N. 1999. Why do populations cycle? A synthesis of statistical and mechanistic modeling approaches. *Ecology* 80:1789–1805.

Kennedy, M., and Gray, R. D. 1993. Can ecological theory predict the distribution of foraging animals? A critical evaluation of experiments on the ideal free distribution. *Oikos* 68:158–166.

Kennedy, P. 1995. Intake and digestion in swamp buffaloes and cattle. 4. Particle size and buoyancy in relation to voluntary intake. *Journal of Agricultural Science* 124:277–287.

Kenney, P. A., and Black, J. L. 1984. Factors affecting diet selection by sheep. 1. Potential intake rate and acceptability of feed. *Australian Journal of Agricultural Research* 35:551–563.

Keogh, R. 1975. Grazing behaviour of sheep during summer and autumn in relation to facial eczema. *Proceedings of the New Zealand Society of Animal Production* 35:198–203.

Kesner, R. P., and DiMattia, B. V. 1987. Neurobiology of an attribute model of memory. *Progress in Psychobiology and Physiological Psychology* 12:207–277.

Kesner, R. P., and Holbrook, T. 1987. Dissociation of item and order spatial memory in rats following medial prefrontal cortex lesions. *Neuropsychologia* 25:653–664.

Ketelaars, J. J. M. H., and Tolkamp, B. J. 1991. Toward a new theory of feed intake regulation in ruminants. 1. Causes of differences in voluntary feed intake: Critique of current views. *Livestock Production Science* 30:269–296.

———. 1992. Toward a new theory of feed intake regulation in ruminants. 3. Optimum feed intake: In search of a physiological background. *Livestock Production Science* 31:235.

Keunen, J. E., Plaizner, J. C., Kyriazakis, I., Duffield, T. F., Widowski, T. M., Lindinger, M. I., and McBride, B. W. 2002. Effects of a subacute ruminal acidosis model on the diet selection of dairy cows. *Journal of Dairy Science* 85:3304–3313.

Khallad, Y. 2004. Conceptualization in the pigeon: What do we know? *International Journal of Psychology* 39:73–94.

Kie, J. G. 1999. Optimal foraging and risk of predation: Effects on behavior and social structure in ungulates. *Journal of Mammalogy* 80:1114–1129.

Kihlstrom, J. F. 1987. The cognitive unconscious. *Science* 237:1445–1452.

Kiley-Worthington, M. 1983. Stereotypies in horses. *Equine Practice* 5:34–40.

Kilpatrick, A. 2003. The impact of thermoregulatory costs on foraging behaviour: A test with American Crows (*Corvus brachyrhynchos*) and eastern grey squirrels (*Sciurus carolinensis*). *Evolutionary Ecology Research* 5:781–786.

Kimbrell, T., and Holt, R. D. 2004. On the interplay of predator switching and prey evasion in determining the stability of predator-prey dynamics. *Israel Journal of Zoology* 50:187–205.

———. 2005. Individual behaviour, space and predator evolution promote persistence in a two-patch system with predator switching. *Evolutionary Ecology Research* 7:53–71.

Kirkwood, J. R. 1983. A limit to metabolisable energy intake in mammals and birds. *Comparative Biochemistry and Physiology* A 75:1–3.

Kirsch, I., Lynn, S. J., Vigorito, M., and Miller, R. R. 2004. The role of cognition in classical and operant conditioning. *Journal of Clinical Psychology* 60:369–392.

Kitaysky, A. S., Wingfield, J. C., and Piatt, J. F. 1999. Dynamics of food availability, body condition and physiological stress response in breeding black-legged kittiwakes. *Functional Ecology* 13:577–584.

Koenig, W. D., and Mumme, R. L. 1987. *Population Ecology of the Cooperatively Breeding Acorn Woodpecker*. Princeton, NJ: Princeton University Press.

Kohler, W. 1925. *The Mentality of Apes*. London: K. Paul, Trench, Trubner & Co.; New York: Harcourt, Brace.

Kohlmann, S. G., and Risenhoover, K. L. 1996. Using artificial food patches to evaluate habitat quality for granivorous birds: An application of foraging theory. *Condor* 98:854–857.

Koivula, K., Orell, M., Rytkonen, S., and Lahti, K. 1995. Fatness, sex and dominance: Seasonal and daily body mass changes in Willow Tits. *Journal of Avian Biology* 26:209–216.

Kojima, S., and Goldman-Rakic, P. S. 1982. Delay-related activity of prefrontal neurons in rhesus monkeys performing delayed response. *Brain Research* 248:43–49.

———. 1984. Functional analysis of spatially discriminative neurons in prefrontal cortex of rhesus monkey. *Brain Research* 291:229–240.

Kokko, H., and Ruxton, G. D. 2000. Breeding suppression and predator-prey dynamics. *Ecology* 81:252–260.

Kolb, B. 1974. Prefrontal lesions alter eating and hoarding behavior in rats. *Physiology and Behavior* 12:507–511.

Kolb, B., and Whishaw, I. Q. 1983. Dissociation of the contributions of the prefrontal, motor, and parietal cortex to the control of movement in the rat: An experimental review. *Canadian Journal of Psychology* 37:211–232.

Koops, M. A., and Abrahams, M. V. 1998. Life history and the fitness consequences of imperfect information. *Evolutionary Ecology* 12:601–613.

Koops, M. A., and Giraldeau, L.-A. 1996. Producer-scrounger foraging games in starlings: A test of rate-maximizing and risk-sensitive models. *Animal Behaviour* 51:773–783.

Korb, J., and Linsenmair, K. E. 2002. Evaluation of predation risk in the collectively foraging termite *Macrotermes bellicosus*. *Insectes Sociaux* 49:264–269.

Kotler, B. P. 1984. Predation risk and the structure of desert rodent communities. *Ecology* 65:689–701.

———. 1997. Patch use by gerbils in a risky environment: Manipulating food and safety to test four models. *Oikos* 78:274–282.

Kotler, B. P., and Blaustein, L. 1995. Titrating food and safety in a heterogeneous environment: When are the risky and safe patches of equal value? *Oikos* 74:251–258.

Kotler, B. P., Blaustein, L., and Brown, J. S. 1992. Predator facilitation: The combined effects of snakes and owls on the foraging behaviour of gerbils. *Annales Zoologici Fennici* 29:199–206.

Kotler, B. P., and Brown, J. S. 1988. Environmental heterogeneity and the coexistence of desert rodents. *Annual Review of Ecology and Systematics* 19:281–307.

———. 1990. Harvest rates of two species of gerbilline rodents. *Journal of Mammalogy* 71:591–596.

Kotler, B. P., Brown, J. S., and Bouskila, A. 2004. Apprehension and time allocation in gerbils: The effects of predatory risk and energetic state. *Ecology* 85:917–922.

Kotler, B. P., Brown, J. S., Bouskila, A., Mukherjee, S., and Goldberg, T. 2004. Foraging games between gerbils and their predators: Seasonal changes in schedules of activity and apprehension. *Israel Journal of Zoology* 50:256–271.

Kotler, B. P., Brown, J. S., Dall, S. R. X., Gresser, S., Ganey, D., and Bouskila, A. 2002. Foraging games between gerbils and their predators: Temporal dynamics of resource depletion and apprehension in gerbils. *Evolutionary Ecology Research* 4:495–518.

Kotler, B. P., Brown, J. S., and Hasson, O. 1991. Factors affecting gerbil foraging behavior and rates of owl predation. *Ecology* 72:2249–2260.

Kotler, B. P., Brown, J. S., and Knight, M. H. 1999. Patch use in hyraxes: There's no place like home? *Ecology Letters* 2:82–88.

Kotler, B. P., J. S. Brown, and W. A. Mitchell. 1993. Environmental factors affecting patch use in two species of gerbilline rodents. *Journal of Mammalogy* 74:614–620.

Kotler, B. P., Brown, J. S., Slotow, R. H., Goodfriend, W., and Strauss, M. 1993. The influence of snakes on the foraging behavior of gerbils. *Oikos* 67:309–316.

Kotler, B. P., Brown, J. S., Smith, R. J., and Wirtz, W. O. I. 1988. The effects of morphology and body size on rate of owl predation on desert rodents. *Oikos* 53:145–152.

Kotler, B. P., Brown, J. S., and Subach, A. 1993. Mechanisms of species coexistence of optimal foragers: Temporal partitioning by two species of sand dune gerbils. *Oikos* 67:548–556.

Kotler, B. P., Gross, J. E., and Mitchell, W. A. 1994. Applying patch use to assess aspects of foraging behavior in Nubian ibex. *Journal of Wildlife Management* 58:299–307.

Kotler, B. P., and Holt, R. D. 1989. Predation and competition—the interaction of two types of species interactions. *Oikos* 54:256–260.

Kraemer, P. J., and Spear, N. E. 1993. Retrieval processes and conditioning. In *Animal Cognition: A Tribute to Donald A. Riley*, ed. T. R. Zentall, 87–107. Hillsdale, NJ. Lawrence Erlbaum Associates.

Kramer, D. L. 1988. The behavioral ecology of air breathing by aquatic animals. *Canadian Journal of Zoology* 66:89–94.

———. 1995. Intestine length in the fishes of a tropical stream. 2. Relationships to diet—the long and short of a convoluted issue. *Environmental Biology of Fishes* 42:129–141.

Krams, I. A. 1996. Predation risk and shifts of foraging sites in mixed Willow and Crested Tit flocks. *Journal of Avian Biology* 27:153–156.

———. 2000. Length of feeding day and body weight of great tits in a single-and a two-predator environment. *Behavioral Ecology and Sociobiology* 48:147–153.

Krause, J. 1993. The effect of "Schreckstoff" on the shoaling behaviour of the minnow: A test of Hamilton's selfish herd theory. *Animal Behaviour* 45:1019–1024.

Krause, J., and Godin, J. G. J. 1995. Predator preferences for attacking particular prey group sizes: Consequences for predator hunting success and prey predation risk. *Animal Behaviour* 50:465–473.

———. 1996. Influence of prey foraging posture on flight behavior and predation risk: Predators take advantage of unwary prey. *Behavioral Ecology* 7:264–271.

Krause, J., and Ruxton, G. D. 2002. *Living in Groups*. Oxford: Oxford University Press.

Krebs, C. J. 2001. *Ecology: The Experimental Analysis of Distribution and Abundance*. San Francisco: Benjamin Cummings.

Krebs, C. J., Boutin, S., Boonstra, R., Sinclair, A. R. E., Smith, J. N. M., Dale, M. R. T., Martin, K., and Turkington, R. 1995. Impact of food and predation on the snowshoe hare cycle. *Science* 269:1112–1115.

Krebs, J. R. 1978. Optimal foraging: Decision rules for predators. In *Behavioral Ecology: An Evolutionary Approach*, ed. J. R. Krebs and N. B. Davies, 23–63. Oxford: Blackwell Scientific Publications.

Krebs, J. R., and Avery, M. I. 1984. Chick growth and prey quality in the European Bee-Eater (*Merops apiaster*). *Oecologia* 64:363–368.

———. 1985. Central place foraging in the European bee-eater *Merops apiaster*. *Journal of Avian Ecology* 54:459–472.

Krebs, J. R., Clayton, N. S., Healy, S. D., Cristol, D. A., Patel, S. N., and Jolliffe, A. R. 1996. The ecology of the avian brain: Food-storing memory and the hippocampus. *Ibis* 138:34–46.

Krebs, J. R., and Davies, N. B. 1978. *Behavioral Ecology: An Evolutionary Approach*. Oxford: Blackwell Scientific Publications.

———. 1981. *Introduction to Behavioral Ecology*. Oxford: Blackwell.

Krebs, J. R., and Harvey, P. H. 1986. Busy doing nothing—efficiently. *Nature* 320:18–19.

Krebs, J. R., and Inman, A. J. 1992. Learning and foraging: Individuals, groups and populations. *American Naturalist* 140:S63–S84.

Krebs, J. R., Kacelnik, A., and Taylor, P. 1978. Test of optimal sampling by foraging great tits. *Nature* 275:27–31.

Krebs, J. R., MacRoberts, M., and Cullen, J. 1972. Flocking and feeding in the great tit *Parus major*: An experimental study. *Ibis* 114:507–530.

Krebs, J. R., Sherry, D. F., Healy, S. D., Perry, V. H., and Vaccarino, A. L. 1989. Hippocampal specialization of food-storing birds. *Proceedings of the National Academy of Sciences U.S.A.* 86:1388–1392.

Krebs, J. R., Stephens, D. W., and Sutherland, W. J. 1983. Perspectives in optimal foraging. In *Perspectives in Ornithology*, ed. A. H. Brush and G. A. Clark, 165–216. New York: Cambridge University Press.

Kreissl, S., Eichmüller, S., Bicker, G., Rapus, J., and Eckert, M. 1994. Octopamine-like immunoreactivity in the brain and subesophageal ganglion of the honeybee. *Journal of Comparative Neurology* 348:583–595.

Krivan, V. 1996. Optimal foraging and predator-prey dynamics. *Theoretical Population Biology* 49:265–290.

———. 1997. Dynamic ideal free distribution: Effects of optimal patch choice on predator-prey dynamics. *American Naturalist* 149:164–178.

Krivan, V., and Sikder, A. 1999. Optimal foraging and predator-prey dynamics, II. *Theoretical Population Biology* 55:111–126.

Kullberg, C. 1995. Strategy of the pygmy owl while hunting avian and mammalian prey. *Ornis Fennica* 72:72–78.

———. 1998a. Does diurnal variation in body mass affect take-off ability in wintering willow tits? *Animal Behaviour* 56:227–233.

———. 1998b. Spatial niche dynamics under predation risk in the Willow Tit *Parus montanus*. *Journal of Avian Biology* 29:235–240.

Kullberg, C., Fransson, T., and Jakobsson, S. 1996. Impaired predator evasion in fat blackcaps (*Sylvia atricapilla*). *Proceedings of the Royal Society of London* B 263:1671–1675.

Kullberg, C., Houston, D. C., and Metcalfe, N. B. 2002. Impaired flight ability—a cost of reproduction in female blue tits. *Behavioral Ecology* 13:575–579.

Kullberg, C., Jakobsson, S., and Fransson, T. 2000. High migratory fuel loads impair predator evasion in Sedge Warblers. *Auk* 117:1034–1038.

———. 1998. Predator-induced take-off strategy in great tits (*Parus major*). *Proceedings of the Royal Society of London* B 265:1659–1664.

Kullberg, C., Metcalfe, N. B., and Houston, D. C. 2002. Impaired flight ability during incubation in the pied flycatcher. *Journal of Avian Biology* 33:179–183.

Kvist, A., and Lindström, Å. 2000. Maximum daily energy intake: It takes time to lift the metabolic ceiling. *Physiological and Biochemical Zoology* 73:30–36.

———. 2003. Gluttony in migratory waders—unprecedented energy assimilation rates in vertebrates. *Oikos* 103:397–403.

Labandeira, C. C. 1997. Insect mouthparts: Ascertaining the paleobiology of insect feeding strategies. *Annual Review of Ecology and Systematics* 28:153–193.

Laca, E. A. 1998. Spatial memory and food searching mechanisms of cattle. *Journal of Range Management* 51:370–378.

Laca, E. A., and Demment, M. W. Foraging strategies of grazing animals. In *The Ecology and Management of Grazing Systems*, ed. J. Hodgson, and A. W. Illius, 137–158. Wallingford, UK: CAB International.

Laca, E. A., Distel, R., Griggs, T., and Demment, M. 1994. Effects of canopy structure on patch depression by grazers. *Ecology* 75:706–716.

Laca, E. A., Ungar, E. D., Seligman, N., and Demment, M. W. 1992. Effects of sward height and bulk density on bite dimensions of cattle grazing homogeneous swards. *Grass and Forage Science* 47:91–102.

Lachnit, H., Giurfa, M., and Menzel, R. 2004. Odor processing in honeybees: Is the whole equal to, more than, or different from the sum of its parts? *Advances in the Study of Behavior* 34:241–264.

Lack, D. 1968. *Ecological Adaptations for Breeding in Birds*. London: Methuen.

Lair, H., Kramer, D. L., and Giraldeau, L.-A. 1994. Interference competition in central place foragers: The effect of imposed waiting on patch use decisions of eastern chipmunks. *Behavioral Ecology* 5:237–244.

Laland, K. N., and Williams, K. 1998. Social transmission of maladaptive information in the guppy. *Behavioral Ecology* 9:493–499.

Lamprecht, R., Hazvi, S., and Dudai, Y. 1997. cAMP response element-binding protein in the amygdala is required for long- but not short-term conditioned taste aversion memory. *Journal of Neuroscience* 17:8443–8450.

Land, M. F. 1985. The eye: Optics. In *Comprehensive Insect Physiology, Biochemistry and Pharmacology*, ed. G. A. Kerut and L. I. Gilbert, 225–275. Oxford: Pergamon Press.

Lane, J. S., Whang, E. E., Rigberg, D. A., Hins, O. J., Kwan, D., Zinner, M. J., McFadden, D. W., Diamond, J., and Ashley, S. W. 1999. Paracellular glucose transport plays a minor role in the unanesthetized dog. *American Journal of Physiology* 276:G789–G794.

Langer, P. 1994. Food and digestion of Cenozoic mammals in Europe. In *The Digestive System in Mammals: Food, Form, and Function*, ed. D. J. Chivers and P. Langer, 9–24. Cambridge: Cambridge University Press.

Langhans, W. 1996. Metabolic and glucostatic control of feeding. *Proceedings of the Nutrition Society* 55:497–515.

Langley, C. M., Riley, D. A., Bond, A. B., and Goel, N. 1996. Visual search for natural grains in pigeons (*Columba livia*): Search images and selective attention. *Journal of Experimental Psychology: Animal Behavior Processes* 22:139–151.

Lanham, C. R. 1999. Mechanisms of coexistence in urban fox squirrels and gray squirrels. Master's thesis, University of Illinois at Chicago.

Lank, D. B., and Ydenberg, R. C. 2003. Death and danger at migratory stopovers: Problems with "predation risk." *Journal of Avian Biology* 34:225–228.

Laughlin, S. B. 2001. Energy as a constraint on the coding and processing of sensory information. *Current Opinion in Neurobiology* 11:475–480.

Launchbaugh, K. L. 1996. Biochemical aspects of grazing behaviour. In *The Ecology and Management of Grazing Systems*, ed. J. Hodgson and A. W. Illius, 159–184. Wallingford, UK: CAB International.

Laundré, J. W., Hernandez, L., and Altendorf, K. B. 2001. Wolves, elk, and bison: Reestablishing the "landscape of fear" in Yellowstone National Park, USA. *Canadian Journal of Zoology* 79:1401–1409.

Lauriault, L., Dougherty, C., Bradley, N., and Cornelius, P. 1990. Thiamin supplementation and the ingestive behavior of beef cattle grazing endophyte-infected tall fescue. *Journal of Animal Science* 68:1245–1253.

Laverty, G., and Skadhauge, E. 1999. Physiological roles and regulation of transport activities in the avian lower intestine. *Journal of Experimental Zoology* 283:480–494.

Lawton, J. H. 1992. There are not 10 million kinds of population dynamics. *Oikos* 63:337–338.

Leader, N., and Yom-Tov, Y. 1998. The possible function of stone ramparts at the nest entrance of the blackstart. *Animal Behavior* 56:207–217.

Lee, E. A., Weiss, S. L., Lam, M., Torres, R., and Diamond, J. 1998. A method for assaying intestinal brush-border sucrase in an intact intestinal preparation. *Proceedings of the National Academy of Sciences U.S.A.* 95:2111–2116.

Lee, S. J., Witter, M. S., Cuthill, I. C., and Goldsmith, A. R. 1996. Reduction in escape performance as a cost of reproduction in gravid starlings, *Sturnus vulgaris*. *Proceedings of the Royal Society of London* B 263:619–624.

Lehikoinen, E. 1987. Seasonality of daily weight cycle in wintering passerines and its consequences. *Ornis Scandinavica* 18:216–226.

Lemen, C. A. 1978. Seed size selection in heteromyids: A second look. *Oecologia* 35:13–19.

Lenormand, T. 2002. Gene flow and the limits to natural selection. *Trends in Ecology and Evolution* 17:183–189.

Lentle, R. G., Hume, I. D., Stafford, K. J., Kennedy, M., Springett, B. P., Browne, R., and Haslett, S. 2004. Temporal aspects of feeding events in tammar (*Macropus eugenii*) and parma (*Macropus parma*) wallabies. I. Food acquisition and oral processing. *Australian Journal of Zoology* 52:81–95.

Leon, M. I., and Gallistel, C. R. 1998. Self-stimulating rats combine subjective reward magnitude and subjective reward rate multiplicatively. *Journal of Experimental Psychology: Animal Behavior Processes* 24:265–277.

Leonard, B., and McNaughton, B. L. 1990. Spatial representation in the rat: Conceptual, behavioral, and neurophysiological perspectives. In *Neurobiology of Comparative Cognition*, ed. R. P. Kesner and D. S. Olton, 363–422. Hillsdale, NJ: Lawrence Erlbaum Associates.

Lessells, C. M. 1991. The evolution of life histories. In *Behavioral Ecology: An Evolutionary Approach*, 3rd ed., ed. J. R. Krebs and N. B. Davies, 32–68. Oxford: Blackwell Scientific Publications.

———. 1995. Putting resource dynamics into continuous input ideal free distribution models. *Animal Behaviour* 49:487–494.

Lessells, C. M., and Stephens, D. W. 1983. Central place foraging: Single prey loaders again. *Animal Behavior* 31:238–243.

Levey, D. J. 1987. Seed size and fruit-handling techniques of birds. *American Naturalist* 129:471–485.

Levey, D. J., and Cipollini, M. L. 1996. Is most glucose absorbed passively in northern bobwhite? *Comparative Biochemistry and Physiology* A 113:225–231.

Levey, D. J., and Karasov, W. H. 1989. Digestive responses of temperate birds switched to fruit or insect diets. *Auk* 106:675–686.

———. 1992. Digestive modulation in a seasonal frugivore, the American Robin (*Turdus migratorius*). *American Journal of Physiology* 262:G711–G718.

Levey, D. J., and Martínez del Rio, C. 1999. Test, rejection, and reformulation of a chemical reactor-based model of gut function in a fruit-eating bird. *Physiological and Biochemical Zoology* 72:369–383.

———. 2001. It takes guts (and more) to eat fruit: Lessons from avian nutritional ecology. *Auk* 118:819–831.

Levri, E. P. 1998. Perceived predation risk, parasitism, and the foraging behavior of a freshwater snail (*Potamopyrgus antipodarum*). *Canadian Journal of Zoology* 76:1878–1884.

Lic, V. 2001. Applying foraging theory to wildlife conservation: An application with the raccoon (*Procyon lotor*). Master's thesis, University of Illinois at Chicago.

Lilliendahl, K. 1997. The effect of predator presence on body mass in captive greenfinches. *Animal Behaviour* 53:75–81.

———. 1998. Yellowhammers get fatter in the presence of a predator. *Animal Behaviour* 55:1335–1340.

———. 2000. Daily accumulation of body reserves under increased predation risk in captive Greenfinches *Carduelis chloris*. *Ibis* 142:587–595.

Lima, S. L. 1983. Downy woodpecker foraging behavior: Efficient sampling in simple stochastic environments. *Ecology* 65:166–174.

———. 1985a. Maximizing feeding efficiency and minimizing time exposed to predators: A trade-off in the black-capped chickadee. *Oecologia* 66:60–67.

———. 1985b. Sampling behavior of starlings foraging in simple patch environments. *Behavioral Ecology and Sociobiology* 16:135–142.

———. 1986. Predation risk and unpredictable foraging conditions: Determinants of body mass in birds. *Ecology* 67:377–385.

———. 1988a. Initiation and termination of daily feeding in dark-eyed juncos: Influences of predation risk and energy reserves. *Oikos* 53:3–11.

———. 1988b. Vigilance and diet selection: The classical diet model reconsidered. *Journal of Theoretical Biology* 132:127–143.

———. 1988c. Vigilance during the initiation of daily feeding in dark-eyed juncos. *Oikos* 53:12–16.

———. 1991. Energy, predators and the behavior of feeding hummingbirds. *Evolutionary Ecology* 5:220–230.

———. 1994a. Collective detection of predatory attack by birds in the absence of alarm signals. *Journal of Avian Biology* 25:319–326.

———. 1994b. On the personal benefits of anti-predatory vigilance. *Animal Behaviour* 48:734–736.

———. 1995a. Back to the basics of anti-predatory vigilance: The group size effect. *Animal Behaviour* 49:11–20.

———. 1995b. Collective detection of predatory attack by social foragers: Fraught with ambiguity? *Animal Behaviour* 50:1097–1108.

———. 1998. Stress and decision making under the risk of predation: Recent developments from behavioral, reproductive, and ecological perspectives. *Advances in the Study of Behavior* 27:215–290.

————. 2002. Putting predators back into behavioral predator-prey interactions. *Trends in Ecology and Evolution* 17:70–75.

Lima, S. L., and Bednekoff, P. A. 1999a. Back to the basics of antipredatory vigilance: Can nonvigilant animals detect attack? *Animal Behaviour* 58:537–543.

————. 1999b. Temporal variation in danger drives antipredator behavior: The predation risk allocation hypothesis. *American Naturalist* 153:649–659.

Lima, S. L., and Dill, L. M. 1990. Behavioral decisions made under the risk of predation: A review and prospectus. *Canadian Journal of Zoology* 68:619–640.

Lima, S. L., and Zollner, P. A. 1996. Anti-predatory vigilance and the limits to collective detection: Visual and spatial separation between foragers. *Behavioral Ecology and Sociobiology* 38:355–363.

Lin, S., Binder, B., and Hart, E. 1998a. Chemical ecology of cottonwood leaf beetle adult feeding preferences on *Populus*. *Journal of Chemical Ecology* 24:1791–1802.

————. 1998b. Insect feeding stimulants from the leaf surface of *Populus*. *Journal of Chemical Ecology* 24:1781–1790.

Lind, J., Fransson, T., Jakobsson, S., and Kullberg, C. 1999. Reduced take-off ability in robins (*Erithacus rubecula*) due to migratory fuel load. *Behavioral Ecology and Sociobiology* 46:65–70.

Lind, J., Jakobsson, S., and Kullberg, C. Impaired predator evasion in the life-history of birds: Behavioral and physiological adaptations to reduced flight ability. *Current Ornithology* 17. In press.

Lindén, M., and Møller, A. P. 1989. Cost of reproduction and covariation of life history traits in birds. *Trends in Ecology and Evolution* 4:367–371.

Linder, Y. 1987. Seasonal differences in thermal regulation in *Gerbillus allenbyi* and *Gerbillus pyramidum* and their contributions to energy budgets. Beer Sheva, Israel: Ben-Gurion University of the Negev.

Lindley, D. V. 1985. *Making Decisions*. London; New York: Wiley.

Lindström, Å. 2003. Fuel deposition rates in migrating birds: Causes, constraints and consequences. In *Avian Migration*, ed. P. Berthold, E. Gwinner, and E. Sonnenschein, 307–320. Berlin: Springer-Verlag.

Lindström, Å., and Alerstam, T. 1992. Optimal fat loads in migrating birds: A test of the time-minimization hypothesis. *American Naturalist* 140:477–491.

Lindström, Å., Kvist, A., Piersma, T., Dekinga, A., and Dietz, M. W. 2000. Avian pectoral muscle size rapidly tracks body mass changes during flight, fasting and fuelling. *Journal of Experimental Biology* 203:913–919.

Little, R. M., and Crowe, T. M. 1994. Conservation implications of deciduous fruit farming on birds in the Elgin district, Western Cape Province, South Africa. *Transactions of the Royal Society of South Africa* 49:185–198.

Liu, H. P., Mitton, J. B., and Herrmann, S. J. 1996. Genetic differentiation in and management recommendations for the freshwater mussel, *Pyganodon grandis* (Say, 1829). *American Malacological Bulletin* 13:117–124.

Livoreil, B., and Giraldeau, L.-A. 1997. Patch departure decisions by spice finches foraging singly or in groups. *Animal Behaviour* 54:967–977.

Loft, E., Menke, J., and Kie, J. 1991. Habitat shifts by mule deer: The influence of cattle grazing. *Journal of Wildlife Management* 55:16–26.

Longland, W. S., and Price, M. V. 1991. Direct observations of owls and heteromyid rodents: Can predation risk explain microhabitat use? *Ecology* 72:2261–2273.

Lopez-Calleja, M. V., and Bozinovic, F. 2000. Energetics and nutritional ecology of small herbivorous birds. *Revista Chilena de Historia Natural* 73:411–420.

Lopez-Calleja, M. V., Bozinovic, F., and Martínez del Rio, C. 1997. Effects of sugar concentration on hummingbird feeding and energy use. *Comparative Biochemistry and Physiology* A 118:1291–1299.

Louviere, J. J. 1988. *Analyzing Decision Making: Metric Conjoint Analysis.* Newbury Park, CA: Sage Publications.

Lucas, J. R., Brodin, A., de Kort, S. R., and Clayton, N. S. 2004. Does hippocampal size correlate with the degree of caching specialization? *Proceedings of the Royal Society of London* B 271:2432–2429.

Lucas, J. R., Pravosudov, V. V., and Zielinski, D. L. 2001. A reevaluation of the logic of pilferage effects, predation risk, and environmental variability on avian energy regulation: The critical role of time budgets. *Behavioral Ecology* 12:246–260.

Lucas, J. R., and Walter, L. R. 1991. When should chickadees hoard food? Theory and experimental results. *Animal Behaviour* 41:579–601.

Lucas, P. W. 1994. Categorisation of food items relevant to oral processing. In *The Digestive Systems in Mammals: Food, Form, and Function,* ed. D. J. Chivers and P. Langer, 197–218. Cambridge: Cambridge University Press.

Ludwig, D., and Rowe, L. 1990. Life-history strategies for energy gain and predator avoidance under time constraints. *American Naturalist* 135:686–707.

Lundberg, P., and Danell, K. 1990. Functional response of browsers: Tree exploitation by moose. *Oikos* 58.078–004.

Luo, J., and Fox, B. J. 1994. Diet of the eastern chestnut mouse (*Pseudomys gracilicaudatus*). II. Seasonal and successional patterns. *Wildlife Research* 21:419–431.

Lutge, B., Hatch, G., and Hardy, M. 1995. The influence of urine and dung deposition on patch grazing patterns of cattle and sheep in the Southern Tall Grassveld. *African Journal of Range and Forage Science* 12:104–110.

Luttbeg, B., and Schmitz, O. J. 2000. Predator and prey models with flexible individual behavior and imperfect information. *American Naturalist* 155:669–683.

Lythgoe, J. N. 1979. *The Ecology of Vision.* Oxford: Clarendon Press.

MacArthur, R. H. 1972. *Geographical Ecology.* New York: Harper and Row.

MacArthur, R. H., and Pianka, E. R. 1966. On optimal use of a patchy environment. *American Naturalist* 100:603–609.

Mace, G. M., Harvey, P. H., and Clutton-Brock, T. H. 1981. Brain size and ecology in small mammals. *Journal of Zoology* 193:333–354.

Macleod, R., Barnett, P., Clark, J. A., and Cresswell, W. 2005. Body mass change strategies in blackbirds *Turdus merula*: The starvation-predation risk trade-off. *Journal of Animal Ecology* 74:292–302.

Macphail, E. M., and Bolhuis, J. J. 2001. The evolution of intelligence: Adaptive specializations versus general process. *Biological Reviews of the Cambridge Philosophical Society* 76:341–364.

Magnhagen, C. 1991. Predation risk as a cost of reproduction. *Trends in Ecology and Evolution* 6:183–186.

Magnhagen, C., and Heibo, E. 2001. Gape size allometry in pike reflects variation between lakes in prey availability and relative body depth. *Functional Ecology* 15:754–762.

Maguire, E. A., Gadian, D. G., Johnsrude, I. S., Good, C. D., Ashburner, J., Frackowiak, R. S. J., and Frith, C. D. 2000. Navigation-related structural change in the hippocampi of taxi drivers. *Proceedings of the National Academy of Sciences U.S.A.* 97:4398–4403.

Mainguy, S. K., and Thomas, V. G. 1985. Comparisons of body reserve buildup and use in several groups of Canada geese. *Canadian Journal of Zoology* 63:1765–1772.

Maki, W. S. 1979. Pigeons' short-term memories for surprising vs. expected reinforcement and nonreinforcement. *Animal Learning & Behavior* 7:31–37.

Malenka, R. C., and Nicoll, R. A. 1999. Long-term potentiation—a decade of progress? *Science* 285:1870–1874.

Mandelik, Y., Jones, M., and Dayan, T. 2003. Structurally complex habitat and sensory adaptations mediate the behavioural responses of a desert rodent to an indirect cue for increased predation risk. *Evolutionary Ecology Research* 5:501–515.

Mangel, M., and Clark, C. W. 1988. *Dynamic Modeling in Behavioral Ecology*. Princeton, NJ: Princeton University Press.

Manser, M. B., and Brotherton, P. N. M. 1995. Environmental constraints on the foraging behavior of a dwarf antelope (*Madoqua kirkii*). *Oecologia* 102:404–412.

Markman, S., Pinshow, B., and Wright, J. 1999. Orange-tufted sunbirds do not feed nectar to their chicks. *Auk* 116:257–259.

———. 2002. The manipulation of food resources reveals sex-specific trade-offs between parental self-feeding and offspring care. *Proceedings of the Royal Society of London* B 269:1931–1938.

Marquis, R. J., and Whelan, C. J. 1994. Insectivorous birds increase growth of white oak through consumption of leaf-chewing insects. *Ecology* 75:2007–2014.

———. 1998. Revelations and limitations of the experimental approach for the study of plant-animal interactions. In *The State of Experimental Ecology*, ed. J. Bernardo and W. Resetarits, Jr., 416–436. Oxford: Oxford University Press.

Marra, P. P., and Holberton, R. L. 1998. Corticosterone levels as indicators of habitat quality: Effects of habitat segregation in a migratory bird during the non-breeding season. *Oecologia* 116:284–292.

Martin, A. C., Zim, H. S., and Nelson, A. L. 1951. *American Wildlife and Plants: A Guide to Wildlife Food Habits*. New York: McGraw-Hill.

Martin, J., and Lopez, P. 1999. An experimental test of the costs of antipredatory refuge use in the wall lizard, *Podarcis muralis*. *Oikos* 84:499–505.

Martin, M. M. 1991. The evolution of cellulose digestion in insects. *Philosophical Transactions of the Royal Society* B 333:281–288.

Martin, R. D., Chivers, D. J., Maclarnon, A. M., and Hladik, C. M. 1985. Gastrointestinal allometry in primates and other mammals. In *Size and Scaling in Primate Biology*, ed. W. L. Jungers, 61–89. New York: Plenum Press.

Martindale, S. 1982. Nest defense and central place foraging: A model and experiment. *Behavioral Ecology and Sociobiology* 10:85–89.

Martinez, F. A., and Marschall, E. A. 1999. A dynamic model of group-size choice in the coral reef fish *Dascyllus albisella*. *Behavioral Ecology* 10:572–577.

Martínez del Rio, C. 1992. Great shrike-tyrant predation on a green-backed firecrown. *Wilson Bulletin* 104:368–369.

Martínez del Rio, C., Baker, H. G., and Baker, I. 1992. Ecological and evolutionary implications of digestive processes: Bird preferences and the sugar constituents of floral nectar and fruit pulp. *Experientia* 48:544–551.

Martínez del Rio, C., Cork, S., and Karasov, W. H. 1994. Engineering and digestion: Modeling gut function. In *The Digestive System in Mammals: Food, Form, and Function*, ed. D. J. Chivers and P. Langer, 25–53. Cambridge: Cambridge University Press.

Martínez del Rio, C., and Karasov, W. H. 1990. Digestion strategies in nectar-eating and fruit-eating birds and the sugar composition of plant rewards. *American Naturalist* 136:618–637.

Martínez del Rio, C., and Stevens, B. R. 1989. Physiological constraint on feeding-behavior: Intestinal-membrane disaccharidases of the starling. *Science* 243:794–796.

Martins, T. L. F., and Wright, J. 1993a. Brood reduction in response to manipulated brood sizes in the common swift (*Apus apus*). *Behavioral Ecology and Sociobiology* 32:61–70.

———. 1993b. Cost of reproduction and allocation of food between parent and young in the swift (*Apus apus*). *Behavioral Ecology* 4:213–223.

Mason, G. J. 1991. Stereotypies: A critical review. *Animal Behaviour* 41:1015–1037.

Mason, G. J., Clubb, R., Latham, N., and Vickery, S. 2006. Why and how should we use environmental enrichments to tackle stereotypic behaviours? *Applied Animal Behaviour Science* 102:163–188.

Mason, G. J., Cooper, J., and Clarebrough, C. 2001. Frustrations of fur-farmed mink. *Nature* 410:35–36.

Mason, G. J., and Mendl, M. 1997. Do the stereotypies of pigs, chickens and mink reflect adaptive species differences in the control of foraging? *Applied Animal Behaviour Science* 53:45–58.

Mathison, G. W., Okine, E. K., Vaage, A. S., Kaske, M., and Milligan, L. P. 1995. Current understanding of the contribution of the propulsive activities in the forestomach to the flow of digesta. In *Ruminant Physiology: Digestion, Metabolism, Growth and Reproduction.*, ed. W. V. Englehardt, S. Leonhard-Marek, G. Breves, and D. Gieseck, 23–41. Stuttgart: Ferdinand Enke Verlag.

May, R. 1973. *Stability and Complexity in Model Ecosystems*. Princeton, NJ: Princeton University Press.

Mayer, J. 1955. Regulation of energy intake and the body weight: The glucostatic and lipostatic hypothesis. *Annals of the New York Academy of Sciences* 63:14–42.

Maynard Smith, J. 1982. *Evolution and the Theory of Games*. Cambridge: Cambridge University Press.

Maynard Smith, J., and Price, G. 1973. The logic of animal conflict. *Nature* 246:15–18.

Mayntz, D., Raubenheimer, D., Salomon, M., Toft, S., and Simpson, S. J. 2005. Nutrient-specific foraging in invertebrate predators. *Science* 307:111–113.

Mazur, J. E., and Logue, A. W. 1978. Choice in a "self-control" paradigm: Effects of a fading procedure. *Journal of the Experimental Analysis of Behavior* 30:11–17.

McCleery, R. H. 1978. Optimal behaviour sequences and decision making. In *Behavioral Ecology: An Evolutionary Approach*, ed. J. R. Krebs and N. B. Davies, 377–410. Oxford: Blackwell Scientific Publications.

McConnachie, S., and Alexander, G. J. 2004. The effect of temperature on digestive and assimilation efficiency, gut passage time and appetite in an ambush foraging lizard, *Cordylus melanotus melanotus*. *Journal of Comparative Physiology* B 174:99–105.

McEwen, B. S. 2000. The neurobiology of stress: From serendipity to clinical relevance. *Brain Research* 886:172–189.

McEwen, B. S., and Sapolsky, R. M. 1995. Stress and cognitive function. *Current Opinion in Neurobiology* 5:205–216.

McGinley, M. A. 1984. Central place foraging for nonfood items: Determination of the stick-size value relationship of house building materials collected by eastern woodrats. *American Naturalist* 123:841–853.

McGreevy, P. D., Cripps, P. J., French, N. P., Green, L. E., and Nicol, C. J. 1995. Management factors associated with stereotypic and redirected behaviour in the Thoroughbred horse. *Equine Veterinary Journal* 27:86–91.

McGreevy, P. D., Hawson, L. A., Habermann, T. C., and Cattle, S. R. 2001. Geophagia in horses: A short note on 13 cases. *Applied Animal Behaviour Science* 71:119–125.

McLaughlin, R. L., and Montgomerie, R. D. 1989. Brood dispersal and multiple central place foraging by Lapland longspur parents. *Behavioral Ecology and Sociobiology* 25:207–215.

McLinn, C. M., and Stephens, D. W. 2006. What makes information valuable: Signal reliability and environmental certainty. *Animal Behaviour* 71:1119–1129.

McNab, B. K. 1963. Bioenergetics and the determination of home range size. *American Naturalist* 97:133–140.

McNamara, J. M. 1982. Optimal patch use in a stochastic environment. *Theoretical Population Biology* 21:269–288.

———. 1990. The policy which maximizes long-term survival of an animal faced with the risks of starvation and predation. *Advances in Applied Probability* 22:295–308.

———. 1995. Implicit frequency dependence and kin selection in fluctuating environments. *Evolutionary Ecology* 9:185–203.

———. 1996. Risk-prone behaviour under rules which have evolved in a changing environment. *American Zoologist* 36:484–495.

———. 2001. The effect of adaptive behaviour on the stability of population dynamics. *Annales Zoologici Fennici* 38:25–36.

McNamara, J. M., Barta, Z., Houston, A. I., and Race, P. 2005. A theoretical investigation of the effect of predators on foraging behaviour and energy reserves. *Proceedings of the Royal Society of London* B 272:929–934.

McNamara, J. M., and Houston, A. I. 1982. Short-term behaviour and life-time fitness. In *Functional Ontogeny*, ed. D. J. McFarland, 60–87. London: Pitman.

———. 1986. The common currency for behavioral decisions. *American Naturalist* 127:358–378.

———. 1987a. Memory and the efficient use of information. *Journal of Theoretical Biology* 125:385–395.

———. 1987b. Starvation and predation as factors limiting population size. *Ecology* 68:1515–1519.

———. 1990. The value of fat reserves and the tradeoff between starvation and predation. *Acta Biotheoretica* 38:37–61.

———. 1992. Evolutionarily stable levels of vigilance as a function of group size. *Animal Behavior* 43:641–658.

———. 1994. The effect of a change in foraging options on intake rate and predation rate. *American Naturalist* 144:978–1000.

———. 1997. Currencies for foraging based on energetic gain. *American Naturalist* 150:603–617.

McNamara, J. M., Houston, A. I., and Krebs, J. R. 1990. Why hoard? The economy of food hoarding in tits, *Parus* spp. *Behavioral Ecology* 1:12–23.

McNamara, J. M., Houston, A. I., and Lima, S. L. 1994. Foraging routines of small birds in winter: A theoretical investigation. *Journal of Avian Biology* 25:287–302.

McNamara, J. M., Welham, R. K., and Houston, A. I. 1998. The timing of migration within the context of an annual routine. *Journal of Avian Biology* 29:416–423.

McNeely, J. A., and Scherr, S. J. 2002. *Ecoagriculture: Strategies to Feed the World and Save Wild Biodiversity*. Washington, DC: Island Press.

McPeek, M. A., Rodenhouse, N. L., Holmes, R. T., and Sherry, T. W. 2001. A general model of site-dependent population regulation: Population-level regulation without individual-level interactions. *Oikos* 94:417–424.

McWhorter, T. J., and Martínez del Rio, C. 2000. Does gut function limit hummingbird food intake? *Physiological and Biochemical Zoology* 73:313–324.

Menzel, R., Heyne, A., Kinzel, C., Gerber, B., and Fiala, A. 1999. Pharmacological dissociation between the reinforcing, sensitizing, and response-releasing functions of reward in honeybee classical conditioning. *Behavioral Neuroscience* 113:744–754.

Menzel, R., and Müller, U. 1996. Learning and memory in honeybees: From behavior to neural substrates. *Annual Review of Neuroscience* 19:379–404.

Merilaita, S., and Jormalainen, V. 2000. Different roles of feeding protection in diel microhabitat choice of sexes in *Idotea baltica*. *Oecologia* 122:445–451.

Mery, F., and Kawecki, T. J. 2002. Experimental evolution of learning ability in fruit flies. *Proceedings of the National Academy of Sciences U.S.A.* 99:14274–14279.

———. 2004. The effect of learning on experimental evolution of resource preference in *Drosophila melanogaster*. *Evolution* 58:757–767.

Mesterton-Gibbons, M. 1991. *An Introduction to Game-Theoretic Modelling*. Redwood City, CA: Addison-Wesley.

Metcalfe, N. B., Fraser, N. H. C., and Burns, M. D. 1999. Food availability and the nocturnal vs. diurnal foraging trade-off in juvenile salmon. *Journal of Animal Ecology* 68:371–381.

Metcalfe, N. B., and Ure, S. E. 1995. Diurnal variation in flight performance and hence potential predation risk in small birds. *Proceedings of the Royal Society of London* B 261:395–400.

Meyer, D. B. 1977. The avian eye and its adaptations. In *Handbook of Sensory Physiology*, vol. 7/5, ed. F. Crescitelli., 549–611. New York: Springer Verlag.

Meyer, M. D., and Valone, T. J. 1999. Foraging under multiple costs: The importance of predation, energetic, and assessment error costs to a desert forager. *Oikos* 87:571–579.

Micheau, J., and Riedel, G. 1999. Protein kinases: Which one is the memory molecule? *Cellular and Molecular Life Sciences* 55:534–548.

Michener, C. D. 1974. *The Social Behavior of the Bees: A Comparative Study*. Cambridge: Cambridge University Press.

Midford, P. E., Hailman, J. P., and Woolfenden, G. E. 2000. Social learning of a novel foraging patch in families of free-living Florida scrub-jays. *Animal Behaviour* 59:1199–1207.

Milinski, M., and Heller, R. 1978. Influence of a predator on the optimal foraging behaviour of sticklebacks (*Gasterosteus aculeatus* L.). *Nature* 275:642–644.

Milinski, M., and Parker, G. A. 1991. Competition for resources. In *Behavioural Ecology: An Evolutionary Approach*, 3rd ed., ed. J. R. Krebs and N. B. Davies, 137–168. Oxford: Blackwell Scientific.

Milinski, M., Semmann, D., and Krambeck, H.-J. 2002. Reputation helps solve the "tragedy of the commons." *Nature* 415:424–426.

Milinski, M., Semmann, D., Krambeck, H.-J., and Marotzke, J. 2006. Stabilizing the Earth's climate is not a losing game: Supporting evidence from public goods experiments. *Proceedings of the National Academy of Sciences U.S.A.* 103:3994–3998.

Miller, R. R., and Escobar, M. 2001. Contrasting acquisition-focused and performance-focused models of acquired behavior. *Current Directions in Psychological Science* 10:141–145.

Miller, T. E., and Kerfoot, W. C. 1987. Redefining indirect effects. In *Predation: Direct and Indirect Impacts on Aquatic Communities*, ed. W. C. Kerfoot and A. Sih, 33–37. Hanover, NH: University Press of New England.

Milner, B., Petrides, M., and Smith, M. L. 1985. Frontal lobes and the temporal organization of memory. *Human Neurobiology* 4:137–142.

Mintzer, A. C. 1987. Primary polygyny in the ant *Atta texana*: Number and weight of females and colony success in the laboratory. *Insectes Sociaux* 34:108–117.

Mischel, W., Shoda, Y., and Rodriguez, M. L. 1989. Delay of gratification in children. *Science* 244:933–938.

Mitchell, J. B., and Laiacona, J. 1998. The medial frontal cortex and temporal memory: Tests using spontaneous exploratory behaviour in the rat. *Behavioural Brain Research* 97:107–113.

Mitchell, W. A. 1989. Informational constraints on optimally foraging hummingbirds. *Oikos* 55:145–154.

———. 1990. An optimal-control theory of diet selection—the effects of resource depletion and exploitative competition. *Oikos* 58:16–24.

———. 2000. Limits to species richness in a continuum of habitat heterogeneity: An ESS approach. *Evolutionary Ecology Research* 2:293–316.

Mitchell, W. A., Z. Abramsky, B. P. Kotler, B. P. Pinshow, and J. S. Brown. 1990. The effect of competition on foraging activity in desert rodents: Theory and experiments. *Ecology* 71:844–854.

Mitchell, W. A., and Valone, T. J. 1990. The optimization research program: Studying adaptations by their function. *Quarterly Review of Biology* 65:43–52.

Mittelbach, G. G. 1981. Foraging efficiency and body size: A study of optimal diet and habitat use by bluegills. *Ecology* 62:1370–1386.

Mittelbach, G. G., Osenberg, C. W., and Wainwright, P. C. 1999. Variation in feeding morphology between pumpkinseed populations: Phenotypic plasticity or evolution? *Evolutionary Ecology Research* 1:111–128.

Mobbs, P. G. 1982. The brain of the honeybee *Apis mellifera*. I. The connections and spatial organization of the mushroom bodies. *Philosophical Transactions of the Royal Society* B 298:309–354.

Mohr, K., Vibe-Petersen, S., Jeppesen, L. L., Bildsoe, M., and Leirs, H. 2003. Foraging of multimammate mice, *Mastomys natalensis*, under different predation pressure: Cover, patch-dependent decisions and density-dependent GUDs. *Oikos* 100:459–468.

Moir, R. J. 1994. The "carnivorous" herbivores. In *The Digestive System in Mammals: Food, Form, and Function*, ed. D. J. Chivers and P. Langer, 87–102. Cambridge: Cambridge University Press.

Montague, P. R., Dayan, P., Person, C., and Sejnowski, T. J. 1995. Bee foraging in uncertain environments using predictive hebbian learning. *Nature* 377:725–728.

Montague, P. R., Dayan, P., and Sejnowski, T. J. 1996. A framework for mesencephalic dopamine systems based on predictive hebbian learning. *Journal of Neuroscience* 16:1936–1947.

Montminy, M. R., and Bilezikjian, L. M. 1987. Binding of a nuclear protein to the cyclic-AMP response element of the somatostatin gene. *Nature* 328:175–178.

Moody, A. L., and Houston, A. I. 1995. Interference and the ideal free distribution. *Animal Behaviour* 49:1065–1072.

Moody, A. L., Houston, A. I., and McNamara, J. M. 1996. Ideal free distribution under predation risk. *Behavioral Ecology and Sociobiology* 38:131–143.

Mooney, H. A., and Cleland, E. E. 2001. The evolutionary impact of invasive species. *Proceedings of the National Academy of Sciences U.S.A.* 98:5446–5451.

Moore, D. J. 2002. The provisioning tactics of parent common terns (*Sterna hirundo*) in relation to brood energy requirement. Ph.D. dissertation, Simon Fraser University.

Moreno, J. 1989. Strategies of mass change in breeding birds. *Biological Journal of the Linnean Society* 37:297–310.

Moreno, J., Lundberg, A., and Carlson, A. 1981. Hoarding of individual nuthatches *Sitta europaea* and marsh tits *Parus palustris*. *Holarctic Ecology* 4:263–269.

Morgan, R. A. 1994. Using giving-up densities to investigate search images, functional responses and habitat selection. Ph.D. dissertation, University of Illinois at Chicago.

Morgan, R. A., and Brown, J. S. 1996. Using giving-up densities to detect search images. *American Naturalist* 148:1059–1074.

Morgan, R. A., Brown, J. S., and Thorson, J. M. 1997. The effect of spatial scale on the functional response of fox squirrels. *Ecology* 78:1087–1097.

Morris, D. W. 1987. Tests of density-dependent habitat selection in a patchy environment. *Ecological Monographs* 57:269–281.

———. 1988. Habitat-dependent population regulation and community structure. *Evolutionary Ecology* 2:253–269.

———. 1996. Coexistence of specialist and generalist rodents via habitat selection. *Ecology* 77:2352–2364.

———. 1997. Optimally foraging deer mice in prairie mosaics: A test of habitat theory and absence of landscape effects. *Oikos* 80:31–42.

Morris, D. W., and Davidson, D. L. 2000. Optimally foraging mice match patch use with habitat differences in fitness. *Ecology* 81:2061–2066.

Morris, D. W., Fox, B. J., Luo, J., and Monamy, V. 2000. Habitat dependent competition and the coexistence of Australian heathland rodents. *Oikos* 91:294–306.

Morris, D. W., and Kingston, S. R. 2002. Predicting future threats to biodiversity from habitat selection by humans. *Evolutionary Ecology Research* 4:787–810.

Morris, R. G. M. 1981. Spatial localization does not require the presence of local cues. *Learning and Motivation* 12:239–260.

Morse, D. H., and Stephens, E. G. 1996. The consequences of adult foraging success on the components of lifetime fitness in a semelparous, sit and wait predator. *Evolutionary Ecology* 10:361–373.

Mottley, K., and Giraldeau, L.-A. 2000. Experimental evidence that group foragers can converge on predicted producer-scrounger equilibria. *Animal Behaviour* 60:341–350.

Mougeot, F., and Bretagnolle, V. 2000. Predation as a cost of sexual communication in nocturnal seabirds: An experimental approach using acoustic signals. *Animal Behaviour* 60:647–656.

Mountfort, D., Campbell, J., and Clements, K. D. 2002. Hindgut fermentation in three species of New Zealand marine herbivorous fish. *Applied and Environmental Microbiology* 68:1374–1380.

Muchapondwa, E. 2002a. *The Contingent Valuation of the Elephant in Mudzi: An Econometric Analysis*. Department of Economics, Goteborg University, Sweden.

———. 2002b. *An Overview of Community-Based Wildlife Conservation in Zimbabwe*. Department of Economics, Goteborg University, Sweden.

Müller, U. 1997. Neuronal cAMP-dependent protein kinase type II is concentrated in mushroom bodies of *Drosophila melanogaster* and the honeybee *Apis mellifera*. *Journal of Neurobiology* 33:33–44.

Munz, F. W., and McFarland, W. N. 1977. Evolutionary adaptations of fishes to the photic environment. In *Handbook of Sensory Physiology*, vol. 7/5, ed. F. Crescitelli, 193–274. New York: Springer Verlag.

Murdoch, W. W. 1977. Stabilizing effects of spatial heterogeneity in predator-prey systems. *Theoretical Population Biology* 11:252–273.

Murdoch, W. W., and Oaten, A. 1975. Predation and population stability. *Advances in Ecological Research* 9:1–131.

Murphy, S. M., and Linhart, Y. B. 1999. Comparative morphology of the gastrointestinal tract in the feeding specialist *Sciurus aberti* and several generalist congeners. *Journal of Mammalogy* 80:1325–1330.

Murray, M. G., and Illius, A. W. Multispecies grazing in the Serengeti. In *The Ecology and Management of Grazing Systems*, ed. J. Hodgson, and A. W. Illius, 247–272. Wallingford, UK: CAB International.

Myers, N., Mittermeier, R. A., Mittermeier, C. G., da Fonseca, G. A. B., and Kent, J. 2000. Biodiversity hotspots for conservation priorities. *Nature* 403:853–858.

Mysterud, A., Larsen, P., Ims, R., and Ostbye, E. 1999. Habitat selection by roe deer and sheep: Does habitat ranking reflect resource availability. *Canadian Journal of Zoology* 77:776–783.

Nakamura, H., and Wako, Y. 1988. Food storing behaviour of willow tit *Parus montanus*. *Journal of Yamashina Institute of Ornithology* 20:21–36.

Nakaoka, M. 2000. Nonlethal effects of predators on prey populations: Predator-mediated change in bivalve growth. *Ecology* 81:1031–1045.

Newman, J. A., Parsons, A. J., and Harvey, A. 1992. Not all sheep prefer clover: Diet selection revisited. *Journal of Agricultural Science* 119:275–283.

Newman, J. A., Parsons, A. J., and Penning, P. D. 1994. A note on the behavioral strategies used by grazing animals to alter their intake rates. *Grass and Forage Science* 49:502–505.

Newman, J. A., Parsons, A. J., Thornley, J. H. M., Penning, P. D., and Krebs, J. R. 1995. Optimal diet selection by a generalist grazing herbivore. *Functional Ecology* 9:255–268.

Newman, J. A., Penning, P. D., Parsons, A. J., Harvey, A., and Orr, R. J. 1994. Fasting affects intake behavior and diet preference of grazing sheep. *Animal Behaviour* 47:185–193.

Newman, J. A., Recer, G. M., Zwicker, S. M., and Caraco, T. 1988. Effects of predation hazard on foraging "constraints": Patch-use strategies in gray squirrels. *Oikos* 53:93–97.

Nicol, C. J. 2000. Equine stereotypies. In *Recent Advances in Companion Animal Behavior Problems*, ed. K. A. Houpt. Ithaca, NY: International Veterinary Information Service.

———. 2004. Development, direction, and damage limitation: Social learning in domestic fowl. *Learning and Behavior* 32:72–81.

Nicol, C. J., Wilson, A. D., Waters, A. J., Harris, P. A., and Davidson, H. P. B. 2001. Crib-biting in foals is associated with gastric ulceration and mucosal inflammation. In *Proceedings of the 35th Congress of the ISAE*, ed. J. P. Garner, J. A. Mench, and S. P. Heekin, 40. University of California–Davis.

Nixon, C. M., Worley, D. M., and McClain, M. W. 1968. Food habitats of squirrels in southeast Ohio. *Journal of Wildlife Management* 62:294–305.

Nolet, B. A. 2002. Efficiency as a foraging currency in animals attaining a gain below the energetic ceiling. *Behavioral Ecology* 13:571–574.

Nolet, B. A., and Klaassen, M. 2005. Time and energy constraints in demanding phases of the annual cycle: An example of time limitation in refuelling migratory swans. *Oikos* 111:302–310.

Nolet, B. A., Van Der Veer, P. J., Evers, E. G. J., and Ottenheim, M. M. 1995. A linear programming model of diet choice of free-living beavers. *Netherlands Journal of Zoology* 45:315–337.

Nonacs, P. 1990. Size and kinship affect success of co-founding *Lasius pallitarsis* queens. *Psyche* 97:217–228.

———. 1993. The economics of brood raiding and nest consolidation during ant colony founding. *Evolutionary Ecology* 7:623–633.

———. 2001. State dependent behavior and the Marginal Value Theorem. *Behavioral Ecology* 12:71–83.

Nonacs, P., and Dill, L. M. 1990. Mortality risk vs. food quality trade-offs in a common currency: Ant patch preferences. *Ecology* 71:1886–1892.

Norberg, R. A. 1981. Optimal flight speed in birds when feeding young. *Journal of Animal Ecology* 50:473–477.

Nores, A. I., and Nores, M. 1994. Nest building and nesting behavior of the brown cacholote. *Wilson Bulletin* 106:106–120.

Norris, K., and Evans, M. R. 2000. Ecological immunology: Life history trade-offs and immune defense in birds. *Behavioral Ecology* 11:19–26.

Nowak, M. A., Page, K. M., and Sigmund, K. 2000. Fairness versus reason in the ultimatum game. *Science* 289:1773–1775.

Nowak, M. A., and Sigmund, K. 1998. Evolution of indirect reciprocity by image scoring. *Nature* 393:573–577.

Noy-Meir, I. 1975. Stability of grazing systems: An application of predator-prey graphs. *Journal of Ecology* 63:459–481.

Nur, N. 1987. Parents, nestlings, and feeding frequency: A model of optimal parental investment and implications for reproductive strategies. In *Foraging Behavior*, ed. A. C. Kamil, J. R. Krebs, and H. R. Pulliam, 457–475. New York: Plenum Press.

Oaten, A. 1977. Optimal foraging in patches: A case for stochasticity. *Theoretical Population Biology* 12:263–285.

O'Connor, M., and Spash, C. L. 1999. *Valuation and the Environment*. Cheltenham: Edward Elgar Publishing.

O'Donoghue, M., Boutin, S., Krebs, C. J., and Hofer, E. J. 1997. Numerical responses of coyotes and lynx to the snowshoe hare cycle. *Oikos* 80:150–162.

O'Donoghue, M., Boutin, S., Krebs, C. J., Zuleta, G., Murray, D. L., and Hofer, E. J. 1998. Functional responses of coyotes and lynx to the snowshoe hare cycle. *Ecology* 79:1193–1208.

Ohtani, T., and Fukuda, H. 1977. Factors governing the spatial distribution of adult drone honey bees in the hive. *Journal of Apicultural Research* 16:14–26.

O'Keefe, J., and Burgess, N. 1996. Geometric determinants of the place fields of hippocampal neurons. *Nature* 381:425–428.

O'Keefe, J., and Dostrovsky, J. 1971. The hippocampus as a spatial map: Preliminary evidence from unit activity in the freely-moving rat. *Brain Research* 34:171–175.

Oksanen, L. 1990. Exploitation ecosystems in seasonal environments. *Oikos* 57:14–24.

Oksanen, L., Fretwell, S. D., Arruda, D., and Niemela, P. 1981. Exploitation ecosystems in gradients of primary productivity. *American Naturalist* 118:240–261.

Oksanen, L., and Oksanen, T. 1999. The logic and realism of the hypothesis of exploitation ecosystems. *American Naturalist* 155:703–723.

Olff, H., Brown, V. K., and Drent, R. H., eds. 1999. *Herbivores: Between Plants and Predators*. Oxford: Blackwell Science.

Olson, D. J. 1991. Species differences in spatial memory among Clark's nutcrackers, scrub jays, and pigeons. *Journal of Experimental Psychology: Animal Behavior Processes* 17:363–376.

Olson, D. J., Kamil, A. C., Balda, R. P., and Nims, P. J. 1995. Performance of four seed-caching corvid species in operant tests of spatial and nonspatial memory. *Journal of Comparative Psychology* 109:173–181.

Olsson, O., Brown, J. S., and Smith, H. G. 2001. Gain curves in depletable food patches: A test of five models with European starlings. *Evolutionary Ecology Research* 3:285–310.

———. 2002. Long- and short-term state-dependent foraging under predation risk: An indication of habitat quality. *Animal Behavior* 63:981–989.

Olsson, O., and Holmgren, N. M. A. 1998. The survival-rate-maximizing policy for Bayesian foragers: Wait for good news. *Behavioral Ecology* 9:345–353.

Olsson, O., Wiktander, U., Holmgren, N. M. A., and Nilsson, S. G. 1999. Gaining ecological information about Bayesian foragers through their behaviour. II. A field test with woodpeckers. *Oikos* 87:264–276.

Olsson, O., Wiktander, U., and Nilsson, S. G. 2000. Daily foraging routines and feeding effort of a small bird feeding on a predictable resource. *Proceedings of the Royal Society of London* B 267:1457–1461.

Olton, D. S., Becker, J. T., and Handelmann, G. E. 1979. Hippocampus, space, and memory. *Behavioral and Brain Sciences* 2:313–365.

Olupot, W., and Waser, P. M. 2001. Activity patterns, habitat use and mortality risks of mangabey males living outside social groups. *Animal Behaviour* 61:1227–1235.

Omura, H., Honda, K., and Hayashi, N. 2000. Identification of feeding attractants in oak sap for adults of two nymphalid butterflies, *Kaniska canace* and *Vanessa indica*. *Physiological Entomology* 25:281–287.

Orians, G. H. 1998. Human behavioral ecology: 140 years without Darwin is too long. *Bulletin of the Ecological Society of America* 79:15–28.

Orians, G. H., and Pearson, N. E. 1979. On the theory of central place foraging. In *Analysis of Ecological Systems*, ed. D. J. Horn, R. D. Mitchell, and R. D. Stairs, 154–177. Columbus: Ohio State University Press.

Ormerod, T. C., MacGregor, J. N., and Chronicle, E. P. 2002. Dynamics and constraints in insight problem solving. *Journal of Experimental Psychology: Learning, Memory and Cognition* 28:791–799.

Orr, R. J., Penning, P. D., Harvey, A., and Champion, R. A. 1997. Diurnal patterns of intake rate by sheep grazing monocultures of ryegrass or white clover. *Applied Animal Behaviour Science* 52:65–77.

Osenberg, C. W., and Mittelbach, G. G. 1989. Effects of body size on the predator-prey interaction between pumpkinseed sunfish and gastropods. *Ecological Monographs* 59:405–432.

Oster, G. F., and Wilson, E. O. 1978. *Caste and Ecology in the Social Insects*. Princeton, NJ: Princeton University Press.

Ovadia, O., and Abramsky, Z. 1995. Density-dependent habitat selection: Evaluation of the isodar method. *Oikos* 73:86–94.

Ovadia, O., and H. Z. Dohna. 2003. The effect of intra- and interspecific aggression on patch residence time in Negev Desert gerbils: A competing risk analysis. *Behavioral Ecology* 14:583–591.

Ovadia, O., and Schmitz, O. J. 2002. Linking individuals with ecosystems: Experimentally identifying the relevant organizational scale for predicting trophic abundances. *Proceedings of the National Academy of Sciences U.S.A.* 99:12927–12931.

Ovadia, O., Ziv, Y., Abramsky, Z., Pinshow, B., and Kotler, B. P. 2001. Harvest rates and foraging strategies in Negev Desert gerbils. *Behavioral Ecology* 12:219–226.

Overgaard, J., Andersen, J. B., and Wang, T. 2002. The effects of fasting duration on the metabolic response to feeding in *Python molurus*: An evaluation of the energetic costs associated with gastrointestinal growth and upregulation. *Physiological and Biochemical Zoology* 75:360–368.

Owen, J. 1980. *Feeding Strategy. Survival in the Wild*. Chicago: University of Chicago Press.

Owen-Smith, N. 1993. Evaluating optimal diet models for an African browsing ruminant, the kudu: How constraining are the assumed constraints? *Evolutionary Ecology* 7:499–524.

———. 1996. Circularity in linear programming models of optimal diet. *Oecologia* 108:259–261.

———. 1997. How successful was Edwards' linear programming model for marmots? *Oecologia* 112:331–332.

Oyugi, J. 2005. Community ecology and conservation of the Amani Sunbird in Arabuko-Sokoke Forest, Kenya. Ph.D. dissertation, University of Illinois at Chicago.

Oyugi, J. O., and Brown, J. S. 2003. Giving-up densities and habitat preferences of European Starlings and American Robins. *Condor* 105:130–135.

Packer, C., Scheel, D., and Pusey, A. E. 1990. Why lions form groups: Food is not enough. *American Naturalist* 136:1–19.

Palmer, S. E. 1999. *Vision Science: From Photons to Phenomenology*. Cambridge, MA: MIT Press.

Pappenheimer, J. R. 1993. On the coupling of membrane digestion with intestinal absorption of sugars and amino-acids. *American Journal of Physiology* 265:G409–G417.

———. 2001. Intestinal absorption of hexoses and amino acids: From apical cytosol to villus capillaries. *Journal of Membrane Biology* 184:233–239.

Pappenheimer, J. R., Dahl, C. E., Karnovsky, M. L., and Maggio, J. E. 1994. Intestinal absorption and excretion of octapeptides composed of D-amino acids. *Proceedings of the National Academy of Sciences U.S.A.* 91:1942–1945.

Pappenheimer, J. R., and Michel, C. C. 2003. Role of villus microcirculation in intestinal absorption of glucose: Coupling of epithelial with endothelial transport. *Journal of Physiology* 553:561–574.

Pappenheimer, J. R., and Reiss, K. Z. 1987. Contribution of solvent drag through intercellular junctions to absorption of nutrients by the small intestine of the rat. *Journal of Membrane Biology* 100:123–136.

Parker, G. A. 1978. Searching for mates. In *Behavioural Ecology: An Evolutionary Approach*, ed. J. R. Krebs and N. B. Davies, 214–244. Oxford: Blackwell Scientific Publications.

Parker, G. A., and Sutherland, W. J. 1986. Ideal free distributions when animals differ in competitive ability: Phenotype limited ideal free models. *Animal Behaviour* 34:1222–1242.

Parsons, A. J., and Chapman, D. F. 2000. The principles of pasture growth and utilization. In *Grass: Its Production and Utilization*, ed. A. Hopkins, 31–89. Oxford: Blackwell Science.

Parsons, A. J., Newman, J. A., Penning, P. D., Harvey, A., and Orr, R. J. 1994. Diet preference of sheep: Effects of recent diet, physiological state and species abundance. *Journal of Animal Ecology* 63:465–478.

Parsons, A. J., Schwinning, S., and Carrere, P. 2001. Plant growth functions and possible spatial and temporal scaling errors in models of herbivory. *Grass and Forage Science* 56:21–34.

Parsons, A. J., Thornley, J. H. M., Newman, J., and Penning, P. D. 1994. A mechanistic model of some physical determinants of intake rate and diet selection in a two-species temperate grassland sward. *Functional Ecology* 8:187–204.

Pasanen, S., and Koskela, P. 1974. Seasonal and age variation in the metabolism of the common frog *Rana temporaria* in northern Finland. *Comparative Biochemistry and Physiology* A 47:635–654.

Pastor, J., Standke, K., Farnsworth, K., Moen, R., and Cohen, Y. 1999. Further development of the Spalinger-Hobbs mechanistic foraging model for free-ranging moose. *Canadian Journal of Zoology* 77:1505–1512.

Paszkowski, C. A., Penttinen, O. P., Holopainen, I. J., and Tonn, W. M. 1996. Predation risk and feeding patterns of crucian carp. *Journal of Fish Biology* 48:818–828.

Patterson, D. M., McGilloway, D. A., Cushnahan, A., Mayne, C. S., and Laidlaw, A. S. 1998. Effect of duration of fasting period on short-term intake rates of lactating dairy cows. *Animal Science* 66:299–305.

Pauly, D. 1995. Anecdotes and the shifting baseline syndrome of fisheries. *Trends in Ecology and Evolution* 10:430.

Peacor, S. D., and Werner, E. E. 1997. Trait-mediated indirect interactions in a simple aquatic food web. *Ecology* 78:1146–1156.

———. 2000. Predator effects on an assemblage of consumers through induced changes in consumer foraging behavior. *Ecology* 81:1998–2010.

Pels, B., de Roos, A. M., and Sabelis, M. W. 2002. Evolutionary dynamics of prey exploitation in a metapopulation of predators. *American Naturalist* 159:172–189.

Penn, D. J. 2003. The evolutionary roots of our environmental problems: Toward a Darwinian ecology. *Quarterly Review of Biology* 78:275–301.

Penning, P. D. 1986. Some effects of sward conditions on grazing behaviour and intake by sheep. In *Grazing Research at Northern Latitudes*, ed. O. Gudmundsson, 219–226. London: Plenum.

Penning, P. D., Parsons, A. J., Newman, J. A., Orr, R. J., and Harvey, A. 1993. The effects of group size on grazing time in sheep. *Applied Animal Behaviour Science* 37:101–109.

Penning, P. D., Parsons, A. J., Orr, R. J., Harvey, A., and Champion, R. A. 1995. Intake and behavior responses by sheep, in different physiological states, when grazing monocultures of grass or white clover. *Applied Animal Behaviour Science* 45:63–78.

Penning, P. D., Parsons, A. J., Orr, R. J., and Treacher, T. T. 1991. Intake and behavior responses by sheep to changes in sward characteristics under continuous stocking. *Grass and Forage Science* 46:15–28.

Pennycuick, C. 1992. *Newton Rules Biology: A Physical Approach to Biological Problems*. Oxford: Oxford University Press.

Penry, D. L. 1993. Digestive constraints on diet choice. In *Diet Selection. An Interdisciplinary Approach to Foraging Behaviour*, ed. R. Hughes, 32–55. Oxford: Blackwell Scientific Publications.

Penry, D. L., and Jumars, P. A. 1986. Chemical reactor analysis and optimal digestion. *BioScience* 36:310–315.

———. 1987. Modeling animal guts as chemical reactors. *American Naturalist* 129:69–96.

———. 1990. Gut architecture, digestive constraints and feeding ecology of deposit-feeding and carnivorous polychaetes. *Oecologia* 82:1–11.

Pepperberg, I. M. 1994. Numerical competence in an African gray parrot (*Psittacus erithacus*). *Journal of Comparative Psychology* 108:36–44.

Perez-Barberia, F. J., and Gordon, I. J. 1999. Body size dimorphism and sexual segregation in polygynous ungulates: An experimental test with Soay sheep. *Oecologia* 120:258–267.

Perry, G., and Pianka, E. 1997. Animal foraging: Past, present and future. *Trends in Ecology and Evolution* 12:360–364.

Pettifor, R. A., Perrins, C. M., and McCleery, R. H. 1988. Individual optimization of clutch size in great tits. *Nature* 336:160–162.

Pfeffer, K., Fritz, J., and Kotrschal, K. 2002. Hormonal correlates of being an innovative greylag goose, *Anser anser*. *Animal Behaviour* 63:687–695.

Pfennig, D. W. 1995. Absence of joint nesting advantage in desert seed harvester ants: Evidence from a field experiment. *Animal Behaviour* 49:567–575.

Pfister, J. A., Provenza, F. D., Manners, G. D., Gardner, D. R., and Ralphs, M. H. 1997. Tall larkspur ingestion: Can cattle regulate intake below toxic levels? *Journal of Chemical Ecology* 23:759–777.

Pierce, G. J., and Ollason, J. G. 1987. Eight reasons why optimal foraging theory is a complete waste of time. *Oikos* 49:111–118.

Piersma, T., and Lindström, Å. 1997. Rapid reversible changes in organ size as a component of adaptive behavior. *Trends in Ecology and Evolution* 12:134–138.

Pietrewicz, A. T., and Kamil, A. C. 1979. Search image formation in the blue jay (*Cyanocitta cristata*). *Science* 204:1332–1333.

Pilastro, A., and A. Magnani. 1997. Weather conditions and fat accumulation dynamics in pre-migratory roosting barn swallows, *Hirundo rustica*. *Journal of Avian Biology* 28:338–344.

Pinto-Hamuy, T., and Linck, P. 1965. Effect of frontal lesions on performance of sequential tasks by monkeys. *Experimental Neurology* 12:96–107.

Pitt, W. C. 1999. Effects of multiple vertebrate predators on grasshopper habitat selection: Trade-offs due to predation risk, foraging, and thermoregulation. *Evolutionary Ecology* 13:499–515.

Polis, G. A. 1981. The evolution and dynamics of intraspecific predation. *Annual Review of Ecology and Systematics* 12:225–251.

Pond, C. M. 1981. Storage. In *Physiological Ecology: An Evolutionary Approach to Resource Use*, ed. C. R. Townsend and P. Calow, 190–219. Oxford: Blackwell Science.

Porte, D. J., Baskin, D. G., and Schwartz, M. W. 2005. Insulin signaling in the central nervous system: A critical role in metabolic homeostasis and disease from *C. elegans* to humans. *Diabetes* 54:1264–1276.

Powlesland, R. G. 1980. Food storing behavior of the South Island robin *Petroica australis australis*. *Mauri Ora* 8:11–20.

Prache, S. 1997. Intake rate, intake per bite and time per bite of lactating ewes on vegetative and reproductive swards. *Applied Animal Behaviour Science* 52:53–64.

Prache, S., and Peyraud, J. L. 1997. Sward prehensibility in cattle and sheep. *Productions Animales* 10:377–390.

Pravosudov, V. V. 1985. Search for and storage of food by *Parus cinctus lapponicus* and *P montanus borealis*. *Zoologicheski Zhurnal* 64:1036–1043.

———. 2003. Long-term moderate elevation in corticosterone facilitates avian food caching behavior and enhances spatial memory. *Proceedings of the Royal Society of London* B 270:2599–2604.

Pravosudov, V. V., and Clayton, N. S. 2001. Effects of demanding foraging conditions on cache retrieval accuracy in food-caching mountain chickadees (*Poecile gambeli*). *Proceedings of the Royal Society of London* B 268:363–368.

Pravosudov, V. V., and Grubb, T. C., Jr. 1997. Energy management in passerine birds during the non-breeding season: A review. *Current Ornithology* 14:189–234.

————. 1998. Management of fat reserves in tufted titmice *Baelophus bicolor* in relation to risk of predation. *Animal Behaviour* 56:49–54.

Pravosudov, V. V., Kitaysky, A. S., Wingfield, J. C., and Clayton, N. S. 2001. Long-term unpredictable foraging conditions and physiological stress response in mountain chickadees (*Poecile gambeli*). *General and Comparative Endocrinology* 123:324–331.

Pravosudov, V. V., and Lucas, J. R. 2000. The costs of being cool: A dynamic model of nocturnal hypothermia by small food-caching birds in winter. *Journal of Avian Biology* 31:463–472.

————. 2001a. Daily patterns of energy storage in food-caching birds under variable daily predation risk: A dynamic state variable model. *Behavioral Ecology and Sociobiology* 50:239–250.

————. 2001b. A dynamic model of short-term energy management in small food-caching and non-caching birds. *Behavioral Ecology* 12:207–218.

Prins, H. H. T., Ydenberg, R. C., and Drent, R. H. 1980. The interaction of brant geese *Branta bernicla* and sea plantain *Plantago maritima* during spring staging: Field observations and experiments. *Acta Botanica Neerlandica* 29:585–596.

Prinz, J. F., and Lucas, P. W. 1997. An optimization model for mastication and swallowing in mammals. *Proceedings of the Royal Society of London* B 264:1715–1721.

Proctor, C. J., Broom, M., and Ruxton, G. D. 2001. Modelling antipredator vigilance and flight response in group foragers when warning signals are ambiguous. *Journal of Theoretical Biology* 211:409–417.

Provenza, F. D. 1995a. Postingestive feedback as an elementary determinant of food preference and intake in ruminants. *Journal of Range Management* 48:2–17.

————. 1995b. Tracking variable environments: There is more than one kind of memory. *Journal of Chemical Ecology* 21:911–923.

————. 1996. Acquired aversions as the basis for varied diets of ruminants foraging on rangelands. *Journal of Animal Science* 74:2010–2020.

————. 2003. *Foraging Behavior: Managing to Survive in a World of Change.* Logan, UT: Utah State University Press.

Provenza, F. D., Bowns, J. E., Urness, P. J., Malechek, J. C., and Butcher, J. E. 1983. Biological manipulation of blackbrush by goat browsing. *Journal of Range Management* 36:513–518.

Provenza, F. D., Scott, C. B., Phy, T. S., and Lynch, J. J. 1996. Preference of sheep for foods varying in flavors and nutrients. *Journal of Animal Science* 74:2355–2361.

Provenza, F. D., and Villalba, J. J. 2006. Foraging in domestic vertebrates: Linking the internal and external milieu. In *Feeding in Domestic Vertebrates: From Structure to Function*, ed. V. L. Bels. Oxfordshire, UK: CAB International. In press.

Provenza, F. D., Villalba, J. J., and Bryant, J. P. 2003. Foraging by herbivores: Linking the biochemical diversity of plants with herbivore culture and landscape diversity. In *Landscape Ecology and Resource Management: Linking Theory with Practice*, ed. J. A. Bissonette and I. Storch, 387–421. New York: Island Press.

Provenza, F. D., Villalba, J. J., Dziba, L. E., Atwood, S. B., and Banner, R. E. 2003. Linking herbivore experience, varied diets, and plant biochemical diversity. *Small Ruminant Research* 49:257–274.

Pulliam, H. R. 1973. On the advantages of flocking. *Journal of Theoretical Biology* 38:419–422.

————. 1974. On the theory of optimal diets. *American Naturalist* 108:59–75.

————. 1976. The principle of optimal behavior and the theory of communities. *Perspectives in Ethology* 2:311–332.

Pusenius, J., and Ostfeld, R. S. 2002. Mammalian predator scent, vegetation cover and tree seedling predation by meadow voles. *Ecography* 25:481–487.

Pyke, G. H., Pulliam, H. R., and Charnov, E. L. 1977. Optimal foraging: A selective review of theory and tests. *Quarterly Review of Biology* 52:137–154.

Queller, D. C. 1995. The spaniels of St. Marx and the Panglossian paradox: A critique of a rhetorical programme. *Quarterly Review of Biology* 70:485–489.

Rachlin, H. 2000. *The Science of Self-Control.* Cambridge, MA: Harvard University Press.

Rachlin, H., and Green, L. 1972. Commitment, choice and self-control. *Journal of the Experimental Analysis of Behavior* 17:15–22.

Randolph, J. C., and Cameron, G. N. 2001. Consequences of diet choice by a small generalist herbivore. *Ecological Monographs* 71:117–136.

Rands, S. A., and Cuthill, I. C. 2001. Separating the effects of predation risk and interrupted foraging upon mass changes in the blue tit *Parus caeruleus. Proceedings of the Royal Society of London* B 268:1783–1790.

Rands, S. A., Houston, A. I., and Gasson, C. E. 2000. Prey processing in central place foragers. *Journal of Theoretical Biology* 202:161–174.

Ranta, E., Peuhkuri, N., Hirvonen, H., and Barnard, C. J. 1998. Producers, scroungers and the price of a free meal. *Animal Behaviour* 55:737–744.

Ranta, E., Peuhkuri, N., Laurila, A., Rita, H., and Metcalfe, N. B. 1996. Producers, scroungers and foraging group structure. *Animal Behaviour* 51:171–175.

Ranta, E., Rita, H., and Lindström, K. 1993. Competition versus cooperation: Success of individuals foraging alone and in groups. *American Naturalist* 142:42–58.

Raubenheimer, D. 1995. Problems with ratio analysis in nutritional studies. *Functional Ecology* 9:21–29.

Raubenheimer, D., and Jones, S. A. 2006. Nutritional imbalance in an extreme generalist omnivore: Tolerance and recovery through complementary food selection. *Animal Behaviour* 71:1253–1262.

Raubenheimer, D., and Simpson, S. J. 1993. The geometry of compensatory feeding in the locust. *Animal Behaviour* 45:953–964.

————. 1994. The analysis of nutrient budgets. *Functional Ecology* 8:783–791.

————. 1995. Constructing nutrient budgets. *Entomologia Experimentalis et Applicata* 77: 99–104.

————. 1997. Integrative models of nutrient balancing: Application to insects and vertebrates. *Nutrition Research Reviews* 10:151–179.

————. 1998. Nutrient transfer functions: The site of integration between feeding behaviour and nutritional physiology. *Chemoecology* 8:61–68.

Raupp, M. J., and Sadof, C. S. 1991. Responses of leaf beetles to injury-related changes in their salicaceous hosts. In *Phytochemical Induction by Herbivores*, ed. D. W. Tallamy and M. J. Raupp, 183–203. New York: Wiley Interscience.

Rebar, C. E. 1995. Ability of *Dipodomys merriami* and *Chaetodipus intermedius* to locate resource distributions. *Journal of Mammalogy* 76:437–447.

Rebhun, W. C., Dill, S. G., and Power, H. T. 1982. Gastric ulcers in foals. *Journal of American Veterinary Association* 180:404–407.

Reboreda, J. C., Clayton, N. S., and Kacelnik, A. 1996. Species and sex differences in hippocampus size in parasitic and non-parasitic cowbirds. *Neuroreport* 7:505–508.

Redbo, I. 1992. The influence of restraint on the occurrence of oral stereotypies in dairy cows. *Applied Animal Behaviour Science* 35:115–123.

Redbo, I., Redbo-Torstensson, P., Odberg, F. O., Hadendahl, A., and Holm, J. 1998. Factors affecting behavioural disturbances in race-horses. *Journal of Animal Science* 66:475–481.

Reeve, H. K. 1998. Game theory, reproductive skew, and nepotism. In *Game Theory and Animal Behavior*, ed. L. A. Dugatkin, and H. K. Reeve, 118–148. New York: Oxford University Press.

Rehder, V. 1989. Sensory pathways and motoneurons of the proboscis reflex in the suboesophageal ganglion of the honey bee. *Journal of Comparative Neurology* 279:499–513.

Rehkämper, G., Haase, E., and Frahm, H. D. 1988. Allometric comparison of brain weight and brain structure volumes in different breeds of the domestic pigeon, *Columba livia* f.d. (fantails, homing pigeons, strassers). *Brain Behavior and Evolution* 31:141–149.

Reinertsen, R. E. 1996. Physiological and ecological aspects of hypothermia. In *Avian Energetics and Nutritional Ecology*, ed. C. Carey, 125–127. London: Chapman & Hall.

Reinertsen, R. E., and Haftorn, S. 1983. Nocturnal hypothermia and metabolism in the willow tit *Parus montanus* at 63″N. *Journal of Comparative Physiology* 151:109–118.

Reinhardt, U. G. 2002. Asset protection in juvenile salmon: How adding biological realism changes a dynamic foraging model. *Behavioral Ecology* 13:94 100.

Reinhardt, U. G., and Healey, M. C. 1999. Season- and size-dependent risk taking in juvenile coho salmon: Experimental evaluation of asset protection. *Animal Behaviour* 57:923–933.

Relyea, R. A., and Auld, J. R. 2004. Having the guts to compete: How intestinal plasticity explains costs of inducible defences. *Ecology Letters* 7:869–875.

Reneerkens, J., Piersma, T., and Ramenofsky, M. 2002. An experimental test of the relationship between temporal variability of feeding opportunities and baseline levels of corticosterone in a shorebird. *Journal of Experimental Zoology* 293:81–88.

Renshaw, E. 1991. *Modelling Biological Populations in Space and Time.* Cambridge: Cambridge University Press.

Rescorla, R. A. 1988. Pavlovian conditioning: It's not what you think it is. *American Psychologist* 43:151–160.

Rescorla, R. A., and Wagner, A. R. 1972. A theory of Pavlovian conditioning: Variations in the effectiveness of reinforcement and non-reinforcement. In *Classical Conditioning II: Current Research and Theory*, ed. A. Black and W. R. Prokasy, 64–99. New York: Appleton-Century-Crofts.

Reuss, L. 2000. The mechanism of glucose absorption in the intestine. *Annual Review of Physiology* 62:939–946.

Reyer, H., and Westerterp, K. 1985. Parental energy expenditure: A proximate cause of helper recruitment in the pied kingfisher (*Ceryle rudis*). *Behavioral Ecology and Sociobiology* 17:363–369.

Reynoldson, T. B. 1983. The population biology of Turbellaria with special reference to the freshwater triclads of the British Isles. *Advances in Ecological Research* 13:237–326.

Richards, S. A., and de Roos, A. M. 2001. When is habitat assessment an advantage when foraging? *Animal Behaviour* 61:1101–1112.

Richardson, N. R., and Gratton, A. 1996. Behavior-relevant changes in nucleus accumbens dopamine transmission elicited by food reinforcement: An electrochemical study in rat. *Journal of Neuroscience* 16:8160–8169.

Richner, H., Christie, P., and Oppliger, A. 1995. Paternal investment affects prevalence of malaria. *Proceedings of the National Academy of Sciences U.S.A.* 92:1192–1194.

Ricklefs, R. 1996. Avian energetics, ecology and evolution. In *Avian Energetics and Nutritional Ecology*, ed. C. Carey, 1–30. New York: Chapman & Hall.

Riley, J. R., Greggers, U., Smith, A. D., Reynolds, D. R., and Menzel, R. 2005. The flight paths of honeybees recruited by the waggle dance. *Nature* 435:205–207.

Rind, M. I., and Phillips, C. J. C. 1999. The effects of group size on the ingestive and social behaviour of grazing dairy cows. *Animal Science* 68:589–596.

Rinkevich, B., and Shapira, M. 1999. Multi-partner urochordate chimeras outperform two-partner chimerical entities. *Oikos* 87:315–320.

Rinkevich, B., and Weissman, I. L. 1987. A long-term study of fused subclones of a compound ascidian: The resorption phenomenon. *Journal of Zoology* 213:717–733.

———. 1992. Chimeras vs. genetically homogeneous individuals: Potential fitness costs and benefits. *Oikos* 63:119–124.

Rissing, S. W., and Pollack, G. B. 1987. Queen aggression, pleometrotic advantage and brood raiding in the ant *Veromessor pergandei* (Hymenoptera: Formicidae). *Animal Behaviour* 34:226–233.

———. 1991. An experimental analysis of pleometrotic advantage in the desert seed-harvester ant *Messor pergandei* (Hymenoptera: Formicidae). *Insectes Sociaux* 38:205–211.

Rissing, S. W., Pollack, G. B., Higgins, M. R., Hagen, R. H., and Smith, D. R. 1989. Foraging specialization without relatedness or dominance among co-founding ant queens. *Nature* 338:420–422.

Rita, H., and Ranta, E. 1998. Competition in a group of equal foragers. *American Naturalist* 152:71–81.

Rita, H., Ranta, E., and Peuhkuri, N. 1997. Group foraging, patch exploitation time and the finder's advantage. *Behavioral Ecology and Sociobiology* 40:35–39.

Robb, S., Cheek, T. R., Hannan, F. L., Hall, L. M., Midgley, J. M., and Evans, P. D. 1994. Agonist-specific coupling of a cloned *Drosophila* octopamine/tyramine receptor to multiple second messenger systems. *EMBO Journal* 13:1325–1330.

Robbins, C. T. 1993. *Wildlife Feeding and Nutrition*. New York: Academic Press.

Roberts, G. 1996. Why individual vigilance declines as group size increases. *Animal Behaviour* 51:1077–1086.

Roberts, S., and Church, R. M. 1978. Control of an internal clock. *Journal of Experimental Psychology: Animal Behavior Processes* 4:318–337.

Roberts, W. A. 1998. *Principles of Animal Cognition*. Boston: McGraw Hill.

Roces, F., and Nuñez, J. A. 1993. Information about food quality influences load-size selection in recruited leaf-cutting ants. *Animal Behaviour* 45:135–143.

Roessingh, P., Bernays, E. A., and Lewis, A. C. 1985. Physiological factors influencing preference for wet or dry food in *Schistocera gregaria* nymphs. *Entomologia Experimentalis et Applicata* 37:89–94.

Rogers, C. M. 1987. Predation risk and fasting capacity: Do wintering birds maintain optimal body mass? *Ecology* 68:1051–1061.

Rogers, C. M., and Rogers, C. J. 1990. Seasonal variation in daily mass amplitude and minimum body mass: A test of a recent model. *Ornis Scandinavica* 21:105–114.

Romey, W. L. 1995. Position preferences within groups: Do whirligigs select positions which balance feeding opportunities with predator avoidance? *Behavioral Ecology and Sociobiology* 37:195–200.

Romoser, W. 1973. *Science of Entomology*. New York: Macmillan.

Rook, A. J., and Huckle, C. A. 1995. Synchronization of ingestive behavior by grazing dairy-cows. *Animal Science* 60:25–30.

Rook, A. J., and Penning, P. D. 1991. Synchronisation of eating, ruminating and idling activity by grazing sheep. *Applied Animal Behaviour Science* 32:157–166.

Roper, K. L., Kaiser, D. H., and Zentall, T. R. 1995. True directed forgetting in pigeons may occur only when alternative working memory is required on forget-cue-trials. *Animal Learning & Behavior* 23:280–285.

Rosenheim, J. A., Nonacs, P., and Mangel, M. 1996. Sex ratios and multifaceted parental investment. *American Naturalist* 148:501–535.

Rosenkilde, C. E. 1979. Functional heterogeneity of the prefrontal cortex in the monkey: A review. *Behavioral and Neural Biology* 25:301–345.

Rosenthal, G. A., and Berenbaum, M. 1992. *Herbivores: Their Interaction with Secondary Plant Metabolites*. London: Academic Press.

Rosenzweig, M. L. 1971. The paradox of enrichment: The destabilization of exploitation ecosystems in ecological time. *Science* 171:385–387.

———. 1974. On the optimal above-ground activity of bannertail kangaroo rats, *Journal of Mammalogy* 55:193–199.

———. 1981. A theory of habitat selection. *Ecology* 62:327–335.

———. 1987. Density-dependent habitat selection: A tool for more effective population management. In *Modeling and Management of Resources under Uncertainty*, ed. T. Vincent, Y. Cohen, W. J. Grantham, G. P. Kirkwood, and J. M. Skowronski, 98–111. Berlin: Springer-Verlag.

———. 1991. Habitat selection and population interactions: The search for mechanism. *American Naturalist* 137:S5–S28.

———. 2001a. The four questions: What does the introduction of exotic species do to diversity? *Evolutionary Ecology Research* 3:361–367.

———. 2001b. Loss of speciation rate will impoverish future diversity. *Proceedings of the National Academy of Sciences U.S.A.* 98:5404–5410.

———. 2001c. Optimality—the biologist's tricorder. *Annales Zoologici Fennici* 38:1–3.

———. 2003a. Reconciliation ecology and the future of species diversity. *Oryx* 37:194–205.

———. 2003b. *Win-Win Ecology: How the Earth's Species Can Survive in the Midst of Human Enterprise*. New York: Oxford University Press.

———. 2005. Avoiding mass extinction: Basic and applied challenges. *American Midland Naturalist* 153:195–208.

———. 2006. Beyond set-asides. In *The Endangered Species Act at Thirty: Renewing the Conservation Promise*, ed. D. D. Goble, J. M. Scott, and F. W. Davis. Washington, DC: Island Press.

Rosenzweig, M. L., and Abramsky, Z. 1986. Centrifugal community organization. *Oikos* 45:79–88.

———. 1997. Two gerbils of the Negev: A long-term investigation of optimal habitat selection and its consequences. *Evolutionary Ecology* 11:733–756.

Rosenzweig, M. L., Abramsky, Z., and Subach, A. 1997. Safety in numbers: Sophisticated vigilance by Allenby's gerbil. *Proceedings of the National Academy of Sciences U.S.A.* 94:5713–5715.

Rosenzweig, M. L., and MacArthur, R. H. 1963. Graphical representation and stability of predator-prey interactions. *American Naturalist* 97:209–223.

Rosenzweig, M. L., and Winakur, J. 1969. Population ecology of desert rodent communities: Habitat and environmental complexity. *Ecology* 50:558–572.

Rosenzweig, M. L., and Ziv, Y. 1999. The echo pattern in species diversity: Pattern and process. *Ecography* 22:614–628.

Rothley, K. D., Schmitz, O. J., and Cohon, J. L. 1997. Foraging to balance conflicting demands: Novel insights from grasshoppers under predation risk. *Behavioural Ecology* 8:551–559.

Roughgarden, J., and Feldman, M. 1975. Species packing and predation pressure. *Ecology* 56:489–492.

Royama, T. 1971. A comparative study of models of predation and parasitism. *Researches in Population Ecology* 1 (suppl): 1–91.

———. 1992. *Analytical Population Dynamics*. London: Chapman & Hall.

Ruel, J. J., and Ayres, M. P. 1999. Jensen's inequality predicts effects of environmental variation. *Trends in Ecology and Evolution* 14:361–366.

Rushen, J., Lawrence, A. B., and Terlouw, E. M. C. 1993. The motivational basis of stereotypies. In *Stereotypic Animal Behaviour: Fundamentals and Applications to Welfare*, ed. A. B. Lawrence and J. Rushen, 41–64. Wallingford, UK: CAB International.

Ruxton, G. D., Hall, S. J., and Gurney, W. S. C. 1995. Attraction towards feeding conspecifics when food patches are exhaustible. *American Naturalist* 145:653–660.

Ruxton, G. D., and Moody, A. L. 1997. The ideal free distribution with kleptoparasitism. *Journal of Theoretical Biology* 186:449–458.

Sambraus, H. H. 1985. Mouth-based anomalous syndromes. In *Ethology of Farm Animals*, ed. A. F. Fraser, 391–422. Amsterdam: Elsevier.

Sapolsky, R. M. 1992. Neuroendocrinology of the stress response. In *Behavioral Endocrinology*, ed. J. B. Becker, S. M. Breedlove, and D. Crews, 287–324. Cambridge, MA: MIT Press.

Sasaki, K., Satoh, T., and Obara, Y. 1996. Cooperative foundation of colonies by unrelated foundresses in the ant *Polyrhachis moesta*. *Insectes Sociaux* 43:217–226.

Sassone-Corsi, P. 1995. Transcription factors responsive to cAMP. *Annual Review of Cell and Developmental Biology* 11:355–377.

Sato, S., Kubo, T., and Abe, M. 1992. Factors influencing tongue-rolling and the relationship between tongue-rolling and production traits in fattening cattle. *Journal of Animal Science* 70:157.

Sauvant, D., Meschy, F., and Mertens, D. 1999. Components of ruminal acidosis and acidogenic effects of diets. *Productions Animales* 12:49–60.

Schacter, D. L. 1996. *Searching for Memory*. New York: Basic Books.

Schaller, G. B. 1975. *The Serengeti Lion: A Study of Predator-Prey Relationships*. Chicago: University of Chicago Press.

Schettini, M., Prigge, E., and Nestor, E. 1999. Influence of mass and volume of ruminal contents on voluntary intake and digesta passage of a forage diet in steers. *Journal of Animal Science* 77:1896–1904.

Schluter, D. 1981. Does the theory of optimal diets apply in complex environments? *American Naturalist* 118:139–147.

———. 1993. Adaptive radiation in sticklebacks: Size, shape, and habitat use efficiency. *Ecology* 74:699–709.

———. 1995. Adaptive radiation in sticklebacks: Trade-offs in feeding performance and growth. *Ecology* 76:82–90.

Schluter, D., and Grant, P. R. 1984. Determinants of morphological patterns in communities of Darwin's finches. *American Naturalist* 123:175–196.

Schluter, D., and McPhail, J. D. 1992. Ecological character displacement and speciation in sticklebacks. *American Naturalist* 140:85–108.

Schluter, D., Price, T. D., and Grant, P. R. 1985. Character displacement in Darwin's finches. *Science* 227:1056–1059.

Schmaranzer, S., and Stabenheiner, A. 1988. Variability of the thermal behavior of honeybees on a feeding place. *Journal of Comparative Physiology* B 158:135–141.

Schmid-Hempel, P., Kacelnik, A., and Houston, A. I. 1985. Honeybees maximize efficiency by not filling their crop. *Behavioral Ecology and Sociobiology* 17:61–66.

Schmid-Hempel, P., Winston, M. L., and Ydenberg, R. C. 1993. The foraging behavior of individual workers in relation to colony state in the social Hymenoptera. *Canadian Entomologist* 125:129–160.

Schmidt, K. A. 1998. The consequences of partially directed search effort. *Evolutionary Ecology* 12:263–277.

Schmidt, K. A., and Brown, J. S. 1996. Patch assessment in fox squirrels: The role of resource density, patch size, and patch boundaries. *American Naturalist* 147:360–380.

Schmidt, K. A., Brown, J. S., and Morgan, R. A. 1998. Plant defenses as complementary resources: A test with squirrels. *Oikos* 81:130–142.

Schmidt, P. A., and Mech, L. D. 1997. Wolf pack size and food acquisition. *American Naturalist* 150:513–517.

Schmidt-Nielson, K. 1997. *Animal Physiology: Adaptation and Environment*. Cambridge: Cambridge University Press.

Schmitz, O. J. 1998. Direct and indirect effects of predation and predation risk in oldfield interaction webs. *American Naturalist* 151:327–342.

———. 2001. From interesting details to dynamical relevance: Toward more effective use of empirical insights in theory construction. *Oikos* 94:39–50.

Schmitz, O. J., Beckerman, A. P., and Litman, S. 1997. Functional responses of adaptive consumers and community stability with emphasis on the dynamics of plant-herbivore systems. *Evolutionary Ecology* 11:773–784.

Schmitz, O. J., Beckerman, A. P., and O'Brien, K. M. 1997. Behaviorally mediated trophic cascades: Effects of predation risk on food webs. *Ecology* 78:1388–1399.

Schmitz, O. J., and Suttle, K. B. 2001. Effects of top predator species on direct and indirect interactions in a food web. *Ecology* 82:2072–2081.

Schön, M. 1998. Conservation measures and implementation for the great grey shrike (*Lanius excubitor*) in the southwestern Schwäbische Alb of southwestern Germany. In *Shrikes of the World* II: *Conservation Implementation*, ed. R. Yosef and F. Lohrer, 68–73. Eilat: International Birdwatching Center in Eilat.

Schoener, T. W. 1968. Sizes of feeding territories among birds. *Ecology* 49:704–726.

———. 1969. Models of optimal size for solitary predators. *American Naturalist* 103:277–313.

———. 1971. Theory of feeding strategies. *Annual Review of Ecology and Systematics* 2:369–404.

Schooley, R. L., Sharpe, P. B., and Horne, B. V. 1996. Can shrub cover increase predation risk for a desert rodent? *Canadian Journal of Zoology* 74:157–163.

Schorger, A. W. 1955. *The Passenger Pigeon: Its Natural History and Extinction*. Madison: University of Wisconsin Press.

Schultz, W. 1998. Predictive reward signal of dopamine neurons. *Journal of Neurophysiology* 80:1–27.

———. 2000. Multiple reward signals in the brain. *Nature Reviews. Neuroscience* 1:199–207.

Schultz, W., Dayan, P., and Montague, P. R. 1997. A neural substrate of prediction and reward. *Science* 275:1593–1599.

Schultz, W., and Dickinson, A. 2000. Neuronal coding of prediction errors. *Annual Review of Neuroscience* 23:473–500.

Schwartz, M. W., Woods, S. C., Porte, D., Seeley, R. J., and Baskin, D. G. 2000. Central nervous system control of food intake. *Nature* 404:661–671.

Schwegler, H., and Lipp, H. P. 1995. Variations in the morphology of the septo-hippocampal complex and maze learning in rodents: Correlation between morphology and behavior. In *Behavioural Brain Research in Naturalistic and Semi-naturalistic Settings: Possibilities and Perspectives*, ed. E. Alleva, A. Fasolo, H. P. Lipp, L. Nadel, and L. Ricceri, 259–276. Dordrecht: Kluwer Academic Publishers.

Schwinning, S., and Parsons, A. J. 1999. The stability of grazing systems revisited: Spatial models and the role of heterogeneity. *Functional Ecology* 13:737–747.

Scrivener, A. M., and Slaytor, M. 1994. Cellulose digestion in *Panesthia cribrata* Saussure: Dose fungal cellulase play a role? *Comparative Biochemistry and Physiology* B 107:309–315.

Seamans, J. K., Floresco, S. B., and Phillips, A. G. 1995. Functional differences between the prelimbic and anterior cingulate regions of rat prefrontal cortex. *Behavioral Neuroscience* 109:1063–1073.

Secor, S. M. 2003. Gastric function and its contribution to the postprandial metabolic response of the Burmese python *Python molurus*. *Journal of Experimental Biology* 206:1621–1630.

Secor, S. M., and Diamond, J. M. 1995. Adaptive responses to feeding in Burmese pythons—Pay before pumping. *Journal of Experimental Biology* 198:1313–1325.

———. 1998. A vertebrate model of extreme physiological regulation. *Nature* 395:659–662.

———. 2000. Evolution of regulatory responses to feeding in snakes. *Physiological and Biochemical Zoology* 73:123–141.

Seeley, T. D. 1985. *Honeybee Ecology*. Princeton, NJ: Princeton University Press.

Seger, J., and Brockmann, H. J. 1987. What is bet-hedging? *Oxford Surveys in Evolutionary Biology* 4:181–211.

Sevi, A., Casamassima, D., and Muscio, A. 1999. Group size effects on grazing behaviour and efficiency in sheep. *Journal of Range Management* 52:327–331.

Sheldon, B. C., and Verhulst, S. 1996. Ecological immunology: Costly parasite defences and trade-offs in evolutionary ecology. *Trends in Ecology and Evolution* 11:317–321.

Sherman, P. W., Jarvis, J., and Alexander, R. 1991. *The Biology of the Naked Mole-Rat*. Princeton, NJ: Princeton University Press.

Sherman, P. W., Lacey, E. A., Reeve, H. K., and Keller, L. 1995. The eusociality continuum. *Behavioral Ecology* 6:102–108.

Sherry, D. F. 1984. Food storage by black-capped chickadees: Memory for the location and contents of caches. *Animal Behaviour* 32:451–464.

Sherry, D. F., and Duff, S. J. 1996. Behavioral and neural bases of orientation in food-storing birds. *Journal of Experimental Biology* 199:165–172.

Sherry, D. F., Forbes, M. R. L., Khurgel, M., and Ivy, G. O. 1993. Females have a larger hippocampus than males in the brood-parasitic brown-headed cowbird. *Proceedings of the National Academy of Sciences U.S.A.* 90:7839–7843.

Sherry, D. F., and Galef, B. G., Jr. 1990. Social learning without imitation: More about milk bottle opening by birds. *Animal Behaviour* 40:987–989.

Sherry, D. E., Krebs, J. R., and Cowie, R. J. 1981. Memory for the location of stored food in marsh tits. *Animal Behaviour* 29:1260–1266.

Sherry, D. F., and Vaccarino, A. L. 1989. Hippocampus and memory for food caches in black-capped chickadees. *Behavioral Neuroscience* 103:308–318.

Sherry, D. F., Vaccarino, A. L., Buckenham, K., and Herz, R. S. 1989. The hippocampal complex of food-storing birds. *Brain Behavior and Evolution* 34:308–317.

Shettleworth, S. J. 1995. Comparative studies of memory in food storing birds: From the field to the Skinner box. In *Behavioural Brain Research in Naturalistic and Semi-naturalistic Settings: Possibilities and Perspectives*, ed. E. Alleva, A. Fasolo, H. P. Lipp, L. Nadel, and L. Ricceri, 159–192. Dordrecht: Kluwer Academic Publishers.

———. 1998. *Cognition, Evolution and Behavior.* New York: Oxford University Press.

———. 2003. Memory and hippocampal specialization in food-storing birds: Challenges for research on comparative cognition. *Brain Behavior and Evolution* 62:108–116.

Shettleworth, S. J., Krebs, J. R., Stephens, D. W., and Gibbon, J. 1988. Tracking a fluctuating environment: A study of sampling. *Animal Behaviour* 36:87–105.

Shettleworth, S. J., and Plowright, C. M. S. 1992. How pigeons estimate rates of prey encounter. *Journal of Experimental Psychology: Animal Behavior Processes* 18:219–235.

Shiflett, M. W., Tomaszycki, M. L., Rankin, A. Z., and DeVoogd, T. J. 2004. Long-term memory for spatial locations in a food-storing bird (*Poecile atricapilla*) requires activation of NMDA receptors in the hippocampal formation during learning. *Behavioral Neuroscience* 118:121–130.

Shizgal, P., and Murray, B. 1989. Neuronal basis of intracranial self-stimulation. In *The Neuropharmacological Basis of Reward*, ed. S. J. Cooper, 106–163. Cambridge, MA: MIT Press.

Sibly, R. M. 1981. Strategies of digestion and defecation. In *Physiological Ecology: An Evolutionary Approach to Resource Use*, ed. C. R. Townsend, and P. Calow, 109–139. Oxford: Blackwell Science.

Sih, A. 1980. Optimal behavior: Can foragers balance two conflicting demands? *Science* 210:1041–1042.

———. 1982. Foraging strategies and the avoidance of predation by an aquatic insect, *Notonecta hoffmanni. Ecology* 63:786–796.

———. 1984. The behavioral response race between predator and prey. *American Naturalist* 123:143–150.

———. 1992. Prey uncertainty and the balancing of antipredator and feeding needs. *American Naturalist* 139:1052–69.

———. 1998. Game theory and predator-prey response races. In *Game Theory and Animal Behavior*, ed. L. A. Dugatkin, and H. K. Reeve, 221–238. New York: Oxford University Press.

Sih, A., and Christensen, B. 2001. Optimal diet theory: When does it work, and when and why does it fail? *Animal Behaviour* 61:379–390.

Sih, A., Enlund, G., and Wooster, D. 1998. Emergent impacts of multiple predators on prey. *Trends in Ecology and Evolution* 13:350–355.

Siikamäki, P., Haimi, J., Hovi, M., and Rätti, O. 1998. Properties of food loads delivered to nestlings in the pied flycatcher: Effects of clutch size manipulation, year, and sex. *Oecologia* 115:579–585.

Silanikove, N., Perevolotsky, A., and Provenza, F. D. 2001. Use of tannin-binding chemicals to assay for tannins and their negative postingestive effects in ruminants. *Animal Feed Science and Technology* 91:69–81.

Silverin, B. 1986. Corticosterone-binding proteins and behavioral effects of high plasma levels of corticosterone during the breeding period in pied flycatchers, *Ficedula hypoleuca*. *General and Comparative Endocrinology* 64:67–74.

———. 1998. The stress response in birds. *Poultry and Avian Biology Review* 9:151–168.

Simon, H. A. 1956. Rational choice and the structure of the environment. *Psychological Review* 63:129–138.

Simpson, S. J., and Raubenheimer, D. 1993a. The central role of the hemolymph in the regulation of nutrient intake in insects. *Physiological Entomology* 18:395–403.

———. 1993b. A multilevel analysis of feeding behavior—the geometry of nutritional decisions. *Philosophical Transactions of the Royal Society* B 342:381–402.

———. 1995. The geometric analysis of feeding and nutrition: A user's guide. *Journal of Insect Physiology* 41:545–553.

———. 1996. Feeding behaviour, sensory physiology and nutrient feedback: A unifying model. *Entomologica Experimentalis et Applicata* 80:55–64.

———. 2000. The hungry locust. *Advances in the Study of Behavior* 29:1–44.

———. 2001. The geometric analysis of nutrient-allelochemical interactions: A case study using locusts. *Ecology* 82:422–439.

Sinclair, A. R. E., and Arcese, P. 1995. Population consequences of predation-sensitive foraging: The Serengeti wildebeest. *Ecology* 76:882–891.

Sjerps, M., and Haccou, P. 1994. Effects of competition on optimal patch leaving: A war of attrition. *Theoretical Population Biology* 46:300–318.

Skalski, G. T., and Gilliam, J. F. 2002. Feeding under predation hazard: Testing models of adaptive behavior with stream fish. *American Naturalist* 160:158–172.

Skelly, D. K. 1994. Activity level and the susceptibility of anuran larvae to predation. *Animal Behaviour* 47:465–468.

Skoczen, S. 1961. On food storage of the mole *Talpa europaeae* Linnaeus 1758. *Acta Theriologica* 2:23–43.

Slaytor, M. 1992. Cellulose digestion in termites and cockroaches: What role do symbionts play? *Comparative Biochemistry and Physiology* B 103:775–784.

Smith, B. H., and Menzel, R. 1989. An analysis of variability in the feeding motor program of the honey bee: The role of learning in releasing a modal action pattern. *Ethology* 82:68–81.

Smith, D. F., and Litvaitis, J. A. 2000. Foraging strategies of sympatric lagomorphs: Implications for differential success in fragmented landscapes. *Canadian Journal of Zoology* 78:2134–2141.

Smith, G. P., and Gibbs, J. 1998. The satiating effect of cholecystokinin and bombesin-like peptides. In *Satiation: From Gut to Brain*, ed. G. P. Smith, 97–125. New York: Oxford University Press.

Smith, J. W., Benkman, C. W., and Coffey, K. 1999. The use and misuse of public information by foraging red crossbills. *Behavioral Ecology* 10:54–62.

Smith, R. M., Young, M. R., and Marquiss, M. 2001. Bryophyte use by an insect herbivore: Does the crane-fly *Tipula montana* select food to maximise growth? *Ecological Entomology* 26:83–90.

Smith, T. B., and Skulason, S. 1996. Evolutionary significance of resource polymorphisms in fishes, amphibians, and birds. *Annual Review of Ecology and Systematics* 27:111–133.

Smulders, T. V. 1998. A game theoretical model of the evolution of food hoarding: Applications to the Paridae. *American Naturalist* 151:356–366.

Smulders, T. V., and DeVoogd, T. J. 2000. The avian hippocampal formation and memory for hoarded food: Spatial learning out in the real world. In *Brain, Perception, Memory*, ed. J. J. Bolhuis, 127–148. Oxford: Oxford University Press.

Smulders, T. V., and Dhondt, A. A. 1997. How much memory do tits need? *Trends in Ecology and Evolution* 12:417–418.

Smulders, T. V., Sasson, A. D., and DeVoogd, T. J. 1995. Seasonal variation in hippocampal volume in a food-storing bird, the black-capped chickadee. *Journal of Neurobiology* 27:15–25.

Sober, E., and Wilson, D. S. 1998. *Unto Others: The Evolution and Psychology of Unselfish Behavior*. Cambridge, MA: Harvard University Press.

Sokolowski, M. B., Pereira, H. S., and Hughes, K. 1997. Evolution of foraging behavior in *Drosophila* by density-dependent selection. *Proceedings of the National Academy of Sciences U.S.A.* 94:7373–1377.

Soler, M., Soler, J. J., Møller, A. P., Moreno, J., and Lindell, M. 1996. The functional significance of sexual display: Stone carrying in the black wheatear. *Animal Behavior* 51:247–254.

Sonerud, G. 1989. Allocation of prey between self-consumption and transport in two different-sized central place foragers. *Ornis Scandinavica* 20:69–71.

Sorensen, J. S., Turnbull, C. A., and Dearing, M. D. 2004. A specialist herbivore (*Neotoma stephensi*) absorbs fewer plant toxins than does a generalist (*Neotoma albigula*). *Physiological and Biochemical Zoology* 77:139–148.

Spalinger, D. E., and Hobbs, N. T. 1992. Mechanisms of foraging in mammalian herbivores: New models of functional response. *American Naturalist* 140:325–348.

Springer, A. M., Estes, J. A., van Vliet, G. B., Williams, T. M., Doak, D. F., Danner, E. M., Forney, K. A., and Pfister, B. 2003. Sequential megafaunal collapse in the North Pacific Ocean: An ongoing legacy of industrial whaling? *Proceedings of the National Academy of Sciences U.S.A.* 100:12223–12228.

Spritzer, M. D., Meikle, D. B., and Solomon, N. G. 2005. Female choice based on male spatial ability and aggressiveness among meadow voles. *Animal Behaviour* 69:1121–1130.

Spritzer, M. D., Solomon, N. G., and Meikle, D. B. 2005. Influence of scramble competition for mates upon the spatial ability of male meadow voles. *Animal Behaviour* 69:375–386.

Squire, L. R. 1992. Memory and the hippocampus: A synthesis from findings with rats, monkeys, and humans. *Psychological Review* 99:195–231.

Stamm, J. S. 1969. Electrical stimulation of the monkey's prefrontal cortex during delayed-response performance. *Journal of Comparative and Physiological Psychology* 67:535–546.

Stapanian, M. A., and Smith, C. C. 1984. Density-dependent survival of scatter-hoarded nuts: An experimental approach. *Ecology* 65:1387–1396.

Starck, J. M. 2003. Shaping up: How vertebrates adjust their digestive system to changing environmental conditions. *Animal Biology* 53:245–257.

Starck, J. M., and Beese, K. 2001. Structural flexibility of the intestine of Burmese python in response to feeding. *Journal of Experimental Biology* 204:325–335.

———. 2002. Structural flexibility of the small intestine and liver of garter snakes in response to feeding and fasting. *Journal of Experimental Biology* 205:1377–1388.

Starck, J. M., Moser, P., Werner, R. A., and Linke, P. 2004. Pythons metabolize prey to fuel the response to feeding. *Proceedings of the Royal Society of London* B 271:903–908.

Starck, J. M., and Wang, T., eds. 2005. *Physiological and Ecological Adaptations to Feeding in Vertebrates.* Enfield, NH: Science Publishers.

Stearns, S. C. 1992. *The Evolution of Life Histories.* Oxford: Oxford University Press.

Steele, M. A., and Wiegl, P. D. 1992. Energetics of patch use in the fox squirrel *Sciurus niger*: Responses to variability in prey profitability and patch density. *American Midland Naturalist* 128:156–167.

Stephens, D. W. 1981. The logic of risk-sensitive foraging preferences. *Animal Behavior* 29: 628–629.

———. 1987. On economically tracking a variable environment. *Theoretical Population Biology* 32:15–25.

———. 1989. Variance and the value of information. *American Naturalist* 134:128–140.

———. 1991. Change, regularity and value in the evolution of animal learning. *Behavioral Ecology* 2:77–89.

———. 2002. Discrimination, discounting and impulsivity: A role for an informational constraint. *Philosophical Transactions of the Royal Society* 357:1527–1537.

Stephens, D. W., and Anderson, D. 2001. The adaptive value of preference for immediacy: When shortsighted rules have farsighted consequences. *Behavioral Ecology* 12:330–339.

Stephens, D. W., and Charnov, E. L. 1982. Optimal foraging: Some simple stochastic models. *Behavioral Ecology and Sociobiology* 10:251–263.

Stephens, D. W., and Krebs, J. R. 1986. *Foraging Theory.* Princeton, NJ: Princeton University Press.

Stephens, D. W., and McLinn, C. M. 2003. Choice and context: Testing a simple short-term choice rule. *Animal Behaviour* 66:59–70.

Stephens, D. W., McLinn, C. M., and Stevens, J. R. 2002. Discounting and reciprocity in an iterated prisoner's dilemma. *Science* 298:2216–2218.

Stevens, C. E., and Hume, I. D. 1995. *Comparative Physiology of the Vertebrate Digestive System.* Cambridge: Cambridge University Press.

Stokke, S. 1999. Sex differences in feeding-patch choice in a megaherbivore: Elephants in Chobe National Park, Botswana. *Canadian Journal of Zoology* 77:1723–1732.

Stonehouse, B. 1960. The king penguin *Aptenodytes patagonica* of South Georgia. I. Breeding behaviour and development. *Falkland Islands Dependencies Survey, Scientific Reports,* no. 23.

Strausfeld, N. J., Hansen, L., Yongsheng, L., Gomez, R. S., and Ito, K. 1998. Evolution, discovery, and interpretations of arthropod mushroom bodies. *Learning and Memory* 5: 11–37.

Striedter, G. F. 2005. *Principles of Brain Evolution.* Sunderland, MA: Sinauer Associates.

Stuartsmith, A. K., and Boutin, S. 1995. Predation on red squirrels during a snowshoe hare decline. *Canadian Journal of Zoology* 73:713–722.

Suarez, R. 1996. Upper limits to mass-specific metabolic rates. *Annual Review of Physiology* 58:583–605.

———. 1998. Oxygen and the upper limits to animal design and performance. *Journal of Experimental Biology* 201:1065–1072.

Subiaul, F., Cantalon, J. F., Halloway, R. L., and Terrace, H. S. 2004. Cognitive imitation in rhesus macaques. *Science* 305:407–410.

Sutherland, R. D., Betteridge, K., Fordham, R. A., Stafford, K. J., and Costall, D. A. 2000. Rearing conditions for lambs may increase tansy ragwort grazing. *Journal of Range Management* 53:432–436.

Sutherland, W. J. 1996. *From Individual Behaviour to Population Ecology*. New York: Oxford University Press.

Sutherland, W. J., and Parker, G. A. 1985. Distribution of unequal competitors. In *Behavioural Ecology: Ecological Consequences of Adaptive Behaviour*, ed. R. M. Sibly and R. H. Smith, 255–274. Oxford: Blackwell Scientific.

Sutton, R. S., and Barto, A. G. 1981. Toward a modern theory of adaptive networks: Expectation and prediction. *Psychological Review* 88:135–170.

———. 1998. *Reinforcement Learning: An Introduction*. Cambridge, MA: MIT Press.

Swaddle, J. P., and Biewener, A. A. 2000. Exercise and reduced muscle mass in starlings. *Nature* 406:585–586.

Swanberg, P. O. 1951. Food storage, territory, and song in the Thick-billed Nutcracker. In *Proceedings of the 10th International Ornithological Congress 1950, Uppsala*, 545–554.

Swaney, W., Kendal, J., Capon, H., Brown, C., and Laland, K. N. 2001. Familiarity facilitates social learning of foraging behaviour in the guppy. *Animal Behaviour* 62:591–598.

Swanson, L. W. 1982. A direct projection from Ammon's horn to prefrontal cortex in the rat. *Brain Research* 217:150–154.

Sweitzer, R. A., and Berger, J. 1992. Size related effects of predation on habitat use and behavior of porcupines (*Erethizon dorsatum*). *Ecology* 73:867–875.

Swets, J. A. 1996. *Signal Detection Theory and ROC Analysis in Psychology and Diagnostics: Collected Papers*. Mahwah, NJ: Lawrence Erlbaum Associates.

Taborsky, M. 1988. Kiwis and dog predation: Observations in Waitangi State Forest. *Notornis* 35:197–202.

Takeda, K. 1961. Classical conditioning response in the honey bee. *Journal of Insect Physiology* 6:168–179.

Tamm, S. 1987. Tracking varying environments: Sampling by hummingbirds. *Animal Behaviour* 35:1725–1734.

Tamura, N., Hashimoto, Y., and Hayashi, F. 1999. Optimal distances for squirrels to transport and hoard walnuts. *Animal Behaviour* 58:635–642.

Taube, J. S., and Burton, H. L. 1995. Head direction cell activity monitored in a novel environment and during a cue conflict situation. *Journal of Neurophysiology* 74:1953–1971.

Taube, J. S., Goodridge, J. P., Golob, E. J., Dudchenko, P. A., and Stackman, R. W. 1996. Processing the head direction cell signal: A review and commentary. *Brain Research Bulletin* 40:477–486.

Temple, S. A. 1987. Do predators always capture substandard individuals disproportionately from prey populations? *Ecology* 68:669–674.

Templeton, J. J., and Giraldeau, L.-A. 1995. Patch assessment in foraging flocks of European starlings: Evidence for public information use. *Behavioral Ecology* 6:65–72.

———. 1996. Vicarious sampling: The use of personal and public information by starlings in a simple patchy environment. *Behavioral Ecology and Sociobiology* 38:105–113.

Terlouw, E. M. C., Lawrence, A. B., and Illius, A. W. 1991. Influences of feeding level and physical restriction on development of stereotypies in sows. *Animal Behaviour* 42:981–992.

Théry, M. 1992. The evolution of leks through female choice: Differential clustering and space utilization in six sympatric manakins. *Behavioral Ecology and Sociobiology* 30:227–237.

Thiollay, J. M. 1999. Frequency of mixed species flocking in tropical forest birds and correlates of predation risk: An intertropical comparison. *Journal of Avian Biology* 30:282–294.

Thomas, R. J. 2000. Strategic diel regulation of body mass in European robins. *Animal Behaviour* 59:787–791.

Thompson, M. C. 1974. Migratory patterns of ruddy turnstones in the central Pacific region. *Living Bird* 12:5–23.

Thomson, J. D., Slatkin, M., and Thomson, B. A. 1997. Trapline foraging by bumble bees. II. Definition and detection from sequence data. *Behavioral Ecology* 8:199–210.

Thorndike, E. L. 1911/2000. *Animal Intelligence: Experimental Studies*. New Brunswick, NJ: Transaction Publishers.

Thornley, J. H. M. 1998. *Grassland Dynamics: An Ecosystem Simulation Model*. Wallingford, UK: CAB International.

Thornley, J. H. M., Parsons, A. J., Newman, J., and Penning, P. D. 1994. A cost-benefit model of grazing intake and diet selection in a two-species temperate grassland sward. *Functional Ecology* 8:5–16.

Thorpe, W. H. 1963. *Learning and Instinct in Animals*. Cambridge, MA: Harvard University Press.

Thorson, J. M., Morgan, R. A., Brown, J. S., and Norman, J. E. 1998. Direct and indirect cues of predatory risk and patch use by fox squirrels and thirteen-lined ground squirrels. *Behavioral Ecology* 9:151–157.

Tibbets, T. M., and Faeth, S. H. 1999. *Neotyphodium* endophytes in grasses: Deterrents or promoters of herbivory by leaf-cutting ants? *Oecologia* 118:297–305.

Tilman, D. 1980. Resources: A graphical-mechanistic approach to competition and predation. *American Naturalist* 116:362–393.

———. 1982. *Resource Competition and Community Structure*. Princeton, NJ: Princeton University Press.

Tilman, D., and Pacala, S. W. 1994. Limiting similarity in mechanistic and spatial models of plant competition in heterogeneous environments. *American Naturalist* 143:222–257.

Tinbergen, L. 1960. The natural control of insects in pinewoods. I. Factors influencing the intensity of predation by songbirds. *Archives Neerlandaises de Zoologie* 13:265–343.

Tinbergen, N. 1963. On aims and methods of ethology. *Zeitschrift für Tierpsychologie* 20:410–433.

Todd, I. A., and Cowie, R. J. 1990. Measuring the risk of-predation in an energy currency: Field experiments with foraging blue tits, *Parus caeruleus*. *Animal Behaviour* 40:112–117.

Tolkamp, B. J., and Kyriazakis, I. 1997. Measuring diet selection in dairy cows: Effect of training on choice of dietary protein level. *Animal Science* 64:197–207.

Tolman, E. C. 1948. Cognitive maps in rats and men. *Psychological Review* 55:189–208.

Tomback, D. 1977. Foraging strategies of Clark's nutcracker. *Living Bird* 16:123–161.

Torres, C. H., and Bozinovic, F. 1997a. Food selection in an herbivorous rodent: Balancing nutrition with thermoregulation. *Ecology* 78:2230–2237.

———. 1997b. Foraging strategy in an herbivorous small mammal in central Chile: Time minimizer or energy maximizer? *Revista Chilena de Historia Natural* 70:577–585.

Tregenza, T. 1995. Building on the ideal free distribution. *Advances in Ecological Research* 26:253–307.

Tregenza, T., Parker, G. A., and Thompson, D. J. 1996. Interference and the ideal free distribution: Models and tests. *Behavioral Ecology* 7:379–386.

Treisman, A. M., and Gelade, G. 1980. A feature-integration theory of attention. *Cognitive Psychology* 12:97–136.

Treisman, M. 1975. Predation and the evolution of gregariousness. I. Models for concealment and evasion. *Animal Behaviour* 23:779–800.

Treves, A. 1998. The influence of group size and neighbors on vigilance in two species of arboreal monkeys. *Behaviour* 135:453–481.

———. 2000. Theory and method in studies of vigilance and aggregation. *Animal Behaviour* 60:711–722.

Trivers, R. L. 1971. The evolution of reciprocal altruism. *Quarterly Review of Biology* 46:189–226.

Trunzer, B., Heinze, J., and Hölldobler, B. 1998. Cooperative colony founding and experimental primary polygyny in the ponerine ant *Pachycondyla villosa*. *Insectes Sociaux* 45:267–276.

Tullock, G. 1971. The coat tit as a careful shopper. *American Naturalist* 105:77–80.

Turchin, P. 1995. Population regulation: Old arguments and a new synthesis. In *Population Dynamics: New Approaches and Synthesis*, ed. N. Cappuccino and P. W. Price, 19–40. San Diego: Academic Press.

———. 1999. Population regulation: A synthetic view. *Oikos* 84:153–159.

———. 2001. Does population ecology have general laws? *Oikos* 94:17–26.

Turchin, P., and Kareiva, P. 1989. Aggregation of *Aphis varians*. An effective strategy for reducing predation risk. *Ecology* 70:1008–1016.

Turner, M. G., Wu, Y., Wallace, L. L., Romme, W. H., and Brenkert, A. 1994. Simulating winter interactions among ungulates, vegetation, and fire in Northern Yellowstone Park. *Ecological Applications* 4:472–496.

Tversky, A., and Kahneman, D. 1974. Judgement under uncertainty: Heuristics and biases. *Science* 185:1124–1131.

Uchmanski, J., and Grimm, V. 1996. Individual-based modelling in ecology: What makes the difference? *Trends in Ecology and Evolution* 11:437–441.

Uetz, G. W., Bischoff, J., and Raver, J. 1992. Survivorship of wolf spiders (Lycosidae) reared on different diets. *Journal of Arachnology* 20:207–211.

Underwood, R. 1982. Vigilance behaviour in grazing antelopes. *Behaviour* 79:79–107.

Ungar, E. D. 1996. Ingestive behaviour. In *The Ecology and Management of Grazing Systems*, ed. J. G. Hodgson, and A. W. Illius, 185–218. Wallingford, UK: CAB International.

Valone, T. J. 1989. Group foraging, public information and patch estimation. *Oikos* 56:357–363.

———. 1993. Patch information and estimation: A cost of group foraging. *Oikos* 68:258–266.

Valone, T. J., and Giraldeau, L.-A. 1993. Patch estimation in group foragers: What information is used? *Animal Behaviour* 45:721–728.

van Baalen, M., Krivan, V., van Rijn, P. C. J., and Sabelis, M. W. 2001. Alternative food, switching predators, and the persistence of predator-prey systems. *American Naturalist* 157:512–524.

van Balaan, M., and Sabelis, M. W. 1993. Coevolution of patch strategies of predator and prey and the consequences for ecological stability. *American Naturalist* 142:646–670.

———. 1999. Nonequilibrium population dynamics of "ideal and free" predator and prey. *American Naturalist* 154:69–88.

Van der Meer, J., and Ens, B. J. 1997. Models of interference and their consequences for the distribution of ideal and free predators. *Journal of Animal Ecology* 66:846–858.

Van der Merwe, M. 2004. Foraging ecology of squirrels. Ph.D. dissertation, University of Illinois at Chicago.

Van der Veen, I. T. 1999. Effects of predation risk on diurnal mass dynamics and foraging routines of yellowhammers (*Emberiza citrinella*). *Behavioral Ecology* 10:545–551.

Van der Veen, I. T., and Lindström, K. M. 2000. Escape flights of yellowhammers and green-finches: More than just physics. *Animal Behaviour* 59:593–601.

Van der Veen, I. T., and Sivars, L. E. 2000. Causes and consequences of mass loss upon predator encounter: Feeding interruption, stress or fit-for-flight? *Functional Ecology* 14:638–644.

Van der Wal, R., Madan, N., Lieshout, S., van, Dormann, C., Langvatn, R., and Albon, S. 2000. Trading forage quality for quantity? Plant phenology and patch choice by Svalbard reindeer. *Oecologia* 123:108–115.

Vander Wall, S. B. 1988. Foraging strategies of Clark's nutcracker on rapidly changing pine seed resources. *Condor* 90:621–631.

———. 1990. *Food Hoarding in Animals*. Chicago: University of Chicago Press.

van Gils, J. A., Schenk. I. V., Bos, O., and Piersma, T. 2003. Incompletely informed shore-birds that face a digestive constraint maximize net energy gain when exploiting patches. *American Naturalist* 161:777–791.

van Nerum, K., and Buelens, H. 1997. Hypoxia-controlled winter metabolism in honeybees (*Apis mellifera*). *Comparative Biochemistry and Physiology* A 117:445–455.

van Nieuwenhuyse, D. 1998. Conservation opportunities for the red-backed shrike (*Lanius collurio*). In *Shrikes of the World* II: *Conservation Implementation*, ed. R. Yosef and F. Lohrer, 79–82. Eilat: International Birdwatching Center in Eilat.

Van Wieren, S. E. 1996. Do large herbivores select a diet that maximizes short-term energy intake? *Forest Ecology and Management* 88:149–156.

Vasquez, R. A. 1994. Assessment of predation risk via illumination level: Facultative central place foraging in the cricetid rodent *Phyllotis darwini*. *Behavioral Ecology and Sociobiology* 34:375–381.

Vasquez, R. A., Ebensperger, L. A., and Bozinovic, F. 2002. The influence of habitat on travel speed, intermittent locomotion, and vigilance in a diurnal rodent. *Behavioral Ecology* 13:182–187.

Vasquez, R. A., and Kacelnik, A. 2000. Foraging rate versus sociality in the starling *Sturnus vulgaris*. *Proceedings of the Royal Society of London* B 267:157–164.

Veasey, J. S., Houston, D. C., and Metcalfe, N. B. 2001. A hidden cost of reproduction: The trade-off between clutch size and escape take-off speed in female zebra finches. *Journal of Animal Ecology* 70:20–24.

Veasey, J. S., Metcalfe, N. B., and Houston, D. C. 1998. A reassessment of the effect of body mass upon flight speed and predation risk in birds. *Animal Behaviour* 56:883–889.

Verhulst, S., and Hogstad, O. 1996. Social dominance and energy reserves in flocks of Willow tits. *Journal of Avian Biology* 27:203–208.

Verlinden, C., and Wiley, R. H. 1989. The constraints of digestive rate: An alternative model of diet selection. *Evolutionary Ecology* 3:264–273.

———. 1997. Digestive constraints on foraging: A further comment. *Evolutionary Ecology* 11:251–252.

Vickery, W. L., Giraldeau, L.-A., Templeton, J. J., Kramer, D. L., and Chapman, C. A. 1991. Producers, scroungers, and group foraging. *American Naturalist* 137:847–863.

Villalba, J. J., and Provenza, F. D. 2000. Postingestive feedback from starch influences the ingestive behaviour of sheep consuming wheat straw. *Applied Animal Behaviour Science* 66:49–63.

Vincent, T. L., and Brown, J. S. 1988. The evolution of ESS theory. *Annual Review of Ecology and Systematics* 19:423–443.

———. 2004. *Evolutionary Game Theory, Natural Selection and Darwinian Dynamics*. Cambridge: Cambridge University Press.

Vincent, T. L. S., Scheel, D., Brown, J. S., and Vincent, T. L. 1996. Trade-offs and coexistence in consumer-resource models: It all depends on what and where you eat. *American Naturalist* 148:1038–1058.

Vine, I. 1973. Detection of prey flocks by predators. *Journal of Theoretical Biology* 40:207–210.

Visalberghi, E., and Limongelli, L. 1994. Lack of comprehension of cause-effect relations in tool-using capuchin monkeys (*Cebus apella*). *Journal of Comparative Psychology* 108:15–22.

———. 1996. Acting and understanding: Tool use revisited through the minds of capuchin monkeys. In *Reaching into Thought: The Minds of the Great Apes*, ed. A. E. Russon, K. A. Bard, and S. T. Parker, 57–79. New York: Cambridge University Press.

Visalberghi, E., and Trinca, L. 1989. Tool use in capuchin monkeys: Distinguishing between performing and understanding. *Primates* 30:511–521.

Vitousek, P. M., Mooney, H. A., Lubchenco, J., and Melillo, J. M. 1997. Human domination of Earth's ecosystems. *Science* 277:494–499.

von Frisch, K. 1950. *Bees: Their Vision, Chemical Senses, and Language*. Ithaca, NY: Cornell University Press.

von Frisch, K., and Lindauer, M. 1955. Uber die Fluggeschwindigkeit der Bienen und uber ihre Richtungweisung bei Seitenwind. *Naturwissenschaften* 42:377–385.

von Uexküll, J. 1957. A stroll through the world of animals and men: A picture book of invisible worlds. In *Instinctive Behavior: The Development of a Modern Concept*, ed. C. H. Schiller, 5–80. New York: International Universities Press.

Vucetich, J. A., Peterson, R. O., and Waite, T. A. 2004. Raven scavenging favours group foraging in wolves. *Animal Behaviour* 67:1117–1126.

Wackernagel, M., and Rees, W. 1998. *Our Ecological Footprint: Reducing Human Impact on Earth*. Gabriola Island, BC: New Society Publishers.

Waddell, S., and Quinn, W. G. 2001. Flies, genes, and learning. *Annual Review of Neuroscience* 24:1283–1309.

Wagner, A. R., Rudy, J. W., and Whitlow, J. W. 1973. Rehearsal in animal conditioning. *Journal of Experimental Psychology* 97:407–426.

Wahungu, G. M., Catterall, C. P., and Olsen, M. F. 2001. Predator avoidance, feeding and habitat use in the red-necked pademelon, *Thylogale thetis*, at rainforest edges. *Australian Journal of Zoology* 49:45–58.

Waite, T. A., and Ydenberg, R. C. 1994a. Shift towards efficiency maximizing by grey jays hoarding in winter. *Animal Behaviour* 48:1466–1468.

———. 1994b. What currency do scatter-hoarding grey jays maximize? *Behavioral Ecology and Sociobiology* 34:43–49.

Wallis de Vries, M. 1996. Effects of resource distribution patterns on ungulate foraging behaviour: A modelling approach. *Forest Ecology and Management* 88:167–177.

Wallis de Vries, M., and Daleboudt, C. 1994. Foraging strategy of cattle in patchy grassland. *Oecologia* 100:98–106.

Wallis de Vries, M., Laca, E., and Demment, M. 1999. The importance of scale of patchiness for selectivity in grazing herbivores. *Oecologia* 121:355–363.

Wallis de Vries, M., and Schippers, P. 1994. Foraging in a landscape mosaic: Selection for energy and minerals in free-ranging cattle. *Oecologia* 100:107–117.

Wallraff, H. G. 1996. Seven theses on pigeon homing deduced from empirical findings. *Journal of Experimental Biology* 199:105–111.

Ward, D. 1992. The role of satisficing in foraging theory. *Oikos* 63:312–317.

Ward, P., and Zahavi, A. 1973. The importance of certain assemblages of birds as "information centres" for food finding. *Ibis* 115:517–534.

Warkentin, K. M. 1995. Adaptive plasticity in hatching age: A response to predation risk trade-offs. *Proceedings of the National Academy of Sciences U.S.A.* 92:3507–3510.

Warner, A. C. I. 1981. Rate of passage of digesta through the gut of mammals and birds. *Nutrition Abstracts and Reviews* 51B:789–820.

Wasserman, E. A., DeVolder, C. L., and Coppage, D. J. 1992. Non-similarity-based conceptualization in pigeons via secondary or mediated generalization. *Psychological Science* 3:374–378.

Wasserman, E. A., Kiedinger, R. E., and Bhatt, R. S. 1988. Conceptual behavior in pigeons: Categories, subcategories, and pseudocategories. *Journal of Experimental Psychology: Animal Behavior Processes* 14:235–246.

Watanabe, H., and Tokuda, G. 2001. Animal cellulases. *Cellular and Molecular Life Sciences* 58:1167–1178.

Watanabe, M. 1986. Prefrontal unit activity during delayed conditional go/no-go discrimination in the monkey. I. Relation to the stimulus. *Brain Research* 382:1–14.

Watanabe, S. 1993. Object-picture equivalence in the pigeon: An analysis with natural concept and pseudoconcept discriminations. *Behavioural Processes* 30:225–231

Watson, L., and Owen-Smith, N. 2000. Diet composition and habitat selection of eland in semi-arid shrubland. *African Journal of Ecology* 38:130–137.

Watt, P. J., Nottingham, S. F., and Young, S. 1997. Toad tadpole aggregation behaviour: Evidence for a predator avoidance function. *Animal Behaviour* 54:865–872.

Weber, T. P., Fransson, T., and Houston, A. I. 1999. Should I stay or should I go? Testing optimality models of stopover decisions in migrating birds. *Behavioral Ecology and Sociobiology* 46:280–286.

Wedekind, C., and Milinski, M. 2000. Cooperation through image scoring in humans. *Science* 288:850–852.

Weir, A. A. S., Chappell, J., and Kacelnik, A. 2002. Shaping of hooks in New Caledonian crows. *Science* 297:981.

Weis-Fogh, T. 1967. Respiration and tracheal ventilation in locusts and other flying insects. *Journal of Experimental Biology* 47:561–587.

Weiss, S. L., Lee, E. A., and Diamond, J. 1998. Evolutionary matches of enzyme and transporter capacities to dietary substrate loads in the intestinal brush border. *Proceedings of the National Academy of Sciences U.S.A.* 95:2117–2121.

Weissman, I. L., Saito, Y., and Rinkevich, B. 1990. Allorecognition histocompatibility in a protochordate species: Is the relationship to MHC semantic or structural? *Immunological Reviews* 113:227–241.

Weitzman, M. L. 2001. Gamma discounting. *American Economic Review* 91:260–271.

Wells, K. D., and Bevier, C. R. 1997. Contrasting patterns of energy substrate use in two species of frogs that breed in cold weather. *Herpetology* 53:70–80.

Welton, N. J., Houston, A. I., Ekman, J., and McNamara, J. M. 2002. A dynamic model of hypothermia as an adaptive response by small birds to winter conditions. *Acta Biotheoretica* 50:39–56.

Welty, J. 1975. *The Life of Birds*. Philadelphia: W. B. Saunders.

Werner, E. E. 1992. Individual behavior and higher order species interactions. *American Naturalist* 140:S5–S32.

Werner, E. E., and Anholt, B. R. 1993. Ecological consequences of the trade-off between growth and mortality rates mediated by foraging activity. *American Naturalist* 142:242–272.

———. 1996. Predator-induced behavioral indirect effects: Consequences to competitive interactions in anuran larvae. *Ecology* 77:157–169.

Werner, E. E., and Gilliam, J. F. 1984. The ontogenetic niche and species interactions in size-structured populations. *Annual Review of Ecology and Systematics* 15:393–425.

Werner, E. E., Gilliam, J. F., Hall, D. L., and Mittelbach, G. G. 1983. An experimental test of the effects of predation risk on habitat use in fish. *Ecology* 64:1540–1548.

Werner, E. E., and Hall, D. J. 1988. Ontogenetic habitat shifts in bluegill: The foraging rate-predation risk trade-off. *Ecology* 69:1352–1366.

Western, D. 2001. Human-modified ecosystems and future evolution. *Proceedings of the National Academy of Sciences U.S.A.* 98:5458–5465.

Westheimer, G. 1994. The Ferrier Lecture, 1992. Seeing depth with two eyes: Stereopsis. *Proceedings of the Royal Society of L____ B 257:205–214*

Wheelwright, N. T. 1986. The diet of American robins: An analysis of U.S. Biological Survey records. *Auk* 103:710–725.

———. 1988. Seasonal changes in food preferences of American robins in captivity. *Auk* 105:374–378.

Whelan, C. J., and Brown, J. S. 2005. Optimal foraging and gut constraints: Reconciling two schools of thought. *Oikos* 110:481–496.

Whelan, C. J., Brown, J. S., and Maina, G. 2003. Search biases, frequency-dependent predation and species co-existence. *Evolutionary Ecology Research* 5:329–343.

Whelan, C. J., Brown, J. S., Schmidt, K. A., Steele, B. B., and Willson, M. F. 2000. Linking consumer-resource theory and digestive physiology: Application to diet shifts. *Evolutionary Ecology Research* 2:911–934.

White, C. M., and West, G. C. 1977. The annual cycle and feeding behavior of Alaskan redpolls. *Oecologia* 27:227–238.

Whitham, T. G. 1977. Coevolution of foraging in *Bombus* and nectar dispensing in *Chilopsis*: A last dregs theory. *Science* 197:593–596.

Whittingham, L. A., and Robertson, R. J. 1993. Nestling hunger and parental care in red-winged blackbirds. *Auk* 110:240–246.

Whybrow, J., Cooper, J., Haskell, M., and Lewis, R. 1995. Feed quality and abnormal oral behaviour in lambs housed individually on unbedded slats. In *Proceedings of the 29th Congress of the ISAE*, ed. S. M. Rutter, J. Rushen, H. D. Randle, and J. C. Eddison, 251–252. Potters Bar: UFAW.

Wiepkema, P. R., Van Hellemond, K. K., Roessingh, P., and Romberg, H. 1987. Behavior and abomasal damage in individual veal calves. *Applied Animal Behaviour Science* 18:257–268.

Wilcoxon, H. C., Dragoin, W. B., and Kral, P. A. 1971. Illness-induced aversions in rat and quail: Relative salience of visual and gustatory cues. *Science* 171:826–828.

Wilde, J. E., Linton, S. M., and Greenaway, P. 2004. Dietary assimilation and the digestive strategy of the omnivorous anomuran land crab *Birgus latro* (Coenobitidae). *Journal of Comparative Physiology* B 174:299–308.

Wiley, R. H. 1994. Errors, exaggeration and deception in animal communication. In *Behavioral Mechanisms in Evolutionary Ecology*, ed. L. A. Real, 157–189. Chicago: University of Chicago Press.

Williams, G. C. 1966. *Adaptation and Natural Selection*. Princeton, NJ: Princeton University Press.

Williams, N. M., and Tepedino, V. J. 2003. Consistent mixing of near and distant resources in foraging bouts by the solitary mason bee *Osmia lignaria. Behavioral Ecology* 14:141–149.

Williams, N. M., and Thomson, J. D. 1998. Trapline foraging by bumble bees: Temporal patterns of visitation and foraging success at single plants. *Behavioral Ecology* 9:612–621.

Williams, T. D., and Vézina, F. 2001. Reproductive energy expenditure, intraspecific variation and fitness in birds. *Current Ornithology* 16:355–406.

Williamson, M. H. 1972. *The Analysis of Biological Populations*. London: Arnold.

Willson, M. F., and Whelan, C. J. 1993. Variation of dispersal phenology in a bird-dispersed shrub, *Cornus drummondii. Ecological Monographs* 63:151–172.

Wilson, D.S. 1976. Deducing the energy available in the environment: An application of optimal foraging theory. *Biotropica* 8:96–103.

———. 1977. Structured demes and the evolution of group-advantageous traits. *American Naturalist* 111:157–185.

———. 1990. Weak altruism, strong group selection. *Oikos* 59:135–140.

Wilson, E. O. 1971. *The Insect Societies*. Cambridge, MA: Belknap Press of Harvard University Press.

———. 1984. *Biophilia: The Human Bond with Other Species*. Cambridge, MA: Harvard University Press.

———. 2002. *The Future of Life*. New York: Alfred A. Knopf.

Wilson, J., and Kennedy, P. 1996. Plant and animal constraints to voluntary feed intake associated with fibre characteristics and particle breakdown and passage in ruminants. *Australian Journal of Agricultural Research* 47:199–225.

Wilson, W. G., and Richards, S. A. 2000. Evolutionarily stable strategies for consuming a structured resource. *American Naturalist* 155:83–100.

Wiltschko, W., and Wiltschko, R. 1996. Magnetic orientation in birds. *Journal of Experimental Biology* 199:29–38.

Wingfield, J. C., Breuner, C., and Jacobs, J. 1997. Corticosterone and behavioral responses to unpredictable events. In *Perspectives in Avian Endocrinology*, ed. S. Harvey and R. J. Etches, 267–278. Bristol: Journal of Endocrinology Ltd.

Wingfield, J. C., Maney, D. L., Breuner, C. W., Jacobs, J. D., Lynn, S., Ramenofsky, M., and Richardson, R. D. 1998. Ecological bases of hormone-behavior interactions: The "emergency life history stage." *American Zoology* 38:191–206.

Wingfield, J. C., and Silverin, B. 1986. Effects of corticosterone on territorial behavior of free-living male song sparrow *Melospiza melodia. Hormones and Behavior* 20:405–417.

Winston, M. L. 1987. *The Biology of the Honey Bee*. Cambridge, MA: Harvard University Press.

Wise, R. A. 1996. Addictive drugs and brain stimulation reward. *Annual Review of Neuroscience* 19:319–340.

Wisenden, B. D. 2000. Olfactory assessment of predation risk in the aquatic environment. *Philosophical Transactions of the Royal Society* B 355:1205–1208.

Witter, M. S., and Cuthill, I. C. 1993. The ecological costs of avian fat storage. *Philosophical Transactions of the Royal Society* B 340:73–92.

Witter, M. S., Cuthill, I. C., and Bonser, R. H. C. 1994. Experimental investigations of mass-dependent predation risk in the European starling, *Sturnus vulgaris. Animal Behaviour* 48:201–222.

Wolf, T., and Schmid-Hempel, P. 1990. On the integration of individual foraging strategies with colony ergonomics in social insects: Nectar collection in honeybees. *Behavioural Ecology and Sociobiology* 27:103–111.

Woods, H. A., and Kingsolver, J. G. 1999. Feeding rate and the structure of protein digestion and absorption in lepidopteran midguts. *Archives of Insect Biochemistry and Physiology* 42:74–87.

Woods, S. C. 1991. The eating paradox: How we tolerate food. *Psychological Reviews* 98:488–505.

Woods, S. C., Schwartz, M. W., Baskin, D. G., and Seeley, R. J. 2000. Food intake and the regulation of body weight. *Annual Review of Psychology* 51:255–277.

Woods, S. C., Seeley, R. J., Porte, D., and Schwartz, M. W. 1998. Signals that regulate food intake and energy homeostasis. *Science* 280:1378–1383.

Wootton, J. T. 1992. Indirect effects, prey susceptibility, and habitat selection impacts of birds on limpets and algae. *Ecology* 73:981–991.

———. 1993. Indirect effects and habitat use in an intertidal community: Interaction chains and interaction modifications. *American Naturalist* 141:71–89.

Wright, J., Both, C., Cotton, P. A., and Bryant, D. 1998. Quality vs. quantity: Energetic and nutritional trade-offs in parental provisioning strategies. *Journal of Animal Ecology* 67:620–634.

Wright, J., Maklakov, A. A., and Khazin, V. 2001. State-dependent sentinels: An experimental study in the Arabian babbler. *Proceedings of the Royal Society of London* B 268:821–826.

Wright, S. H., and Ahearn, G. A. 1997. Nutrient absorption in invertebrates. In *Handbook of Physiology*, section 13, *Comparative Physiology*, vol. 2, ed. W. H. Dantzler., 1137–1205. New York: Oxford University Press.

Wüstenberg, D., Gerber, B., and Menzel, R. 1998. Long- but not medium-term retention of olfactory memories in honeybees is impaired by actinomycin D and anisomycin. *European Journal of Neuroscience* 10:2742–2745.

Wynne, C. D. L. 2001. *Animal Cognition: The Mental Lives of Animals*. New York: Palgrave.

Wynne-Edwards, V. C. 1962. *Animal Dispersion in Relation to Social Behaviour*. Edinburgh: Oliver and Boyd.

Yang, Y., and Joern, A. 1994a. Gut size changes in relation to variable food quality and body size in grasshoppers. *Functional Ecology* 8:36–45.

———. 1994b. Influence of diet quality, developmental stage, and temperature on food residence time in the grasshopper *Melanopus differentialis*. *Physiological Zoology* 67:598–616.

Ydenberg, R. C. 1988. Foraging by diving birds. *Proceedings Congressus Internationalis Ornithologici* 19:1831–1842.

———. 1994. The behavioral ecology of provisioning in birds. *Ecoscience* 1:1–14.

———. 1998. Behavioral decisions about foraging and predator avoidance. In *Cognitive Ecology: The Evolutionary Ecology of Information Processing and Decision Making*, ed. R. Dukas, 343–378. Chicago: University of Chicago Press.

Ydenberg, R. C., and Bertram, D. F. 1988. Lack's clutch size hypothesis and brood enlargement studies on colonial seabirds. *Colonial Waterbirds* 12:134–137.

Ydenberg, R. C., and Clark, C. W. 1989. Aerobiosis and anaerobiosis during diving by western grebes: An optimal foraging approach. *Journal of Theoretical Biology* 139:437–449.

Ydenberg, R. C., and Dill, L. M. 1986. The economics of fleeing from predators. *Advances in the Study of Behavior* 16:229–249.

Ydenberg, R. C., Giraldeau, L.-A., and Kramer, D. L. 1986. Interference competition, payoff asymmetries, and the social relationships of central place foragers. *Theoretical Population Biology* 30:26–44.

Ydenberg, R. C., and Hurd, P. 1998. Simple models of feeding with time and energy constraints. *Behavioral Ecology* 9:49–53.

Ydenberg, R. C., and Krebs, J. R. 1987. The tradeoff between territorial defense and foraging in the great tit (*Parus major*). *American Zoologist* 27:337–346.

Ydenberg, R. C., and Schmid-Hempel, P. 1994. Modelling social insect foraging. *Trends in Ecology and Evolution* 9:491–493.

Yin, J. C. P., Wallach, J. S., Del Vecchio, M., Wilder, E. L., Zhou, H., Quinn, W. G., and Tully, T. 1994. Induction of a dominant negative CREB transgene specifically blocks long-term memory in *Drosophila*. *Cell* 79:49–58.

Yoganarasimha, D., and Knierim, J. J. 2005. Coupling between place cells and head direction cells during relative translations and rotations of distal landmarks. *Experimental Brain Research* 160:344–359.

Yosef, R., and Grubb, T. C., Jr. 1992. Territory size influences nutritional condition in nonbreeding loggerhead shrikes (*Lanius ludovicianus*): A ptilochronology approach. *Conservation Biology* 6:447–449.

———. 1994. Resource dependence and territory size in loggerhead shrikes (*Lanius ludovicianus*). *Auk* 111:465–469.

Yosef, R., and Lohrer, F. E., eds. 1995. *Shrikes of the World: Biology and Conservation. Proceedings of the Western Foundation of Vertebrate Zoology* 6:1–343.

Yunger, J. A., Meserve, P. L, and Gutierrez, J. R. 2002. Small-mammal foraging behavior: Mechanisms for coexistence and implication for population dynamics. *Ecological Monographs* 72:561–577.

Yuval, B., Holliday-Hanson, M. L., and Washino, R. K. 1994. Energy budget of swarming male mosquitoes. *Ecological Entomology* 19:74–78.

Zach, R., and Smith, J. N. M. 1981. Optimal foraging in wild birds? In *Foraging Behaviour: Ecological, Ethological and Psychological Approaches*, ed. A. C. Kamil and T. D. Sargent, 95–113. New York: Garland STPM Press.

Zentall, T. R., ed. 1993. *Animal Cognition: A Tribute to Donald A. Riley*. Hillsdale, NJ: Lawrence Erlbaum Associates.

———. 2004. Action imitation in birds. *Learning and Behavior* 32:15–23.

Zentall, T. R., and Wasserman, E., A. 2006. *Comparative Cognition: Experimental Explorations of Animal Intelligence*. New York: Oxford University Press.

Ziv, Y., Abramsky, Z., Kotler, B. P., and Subach, A. 1993. Interference competition and temporal and habitat partitioning in two gerbil species. *Oikos* 66:237–246.

Ziv, Y., and Smallwood, J. A. 2000. Gerbils and heteromyids: Interspecific competition and the spatio-temporal niche. In *Activity Patterns in Small Mammals: A Comparative Ecological Approach*, ed. H. S. Halle and N. C. Stenseth, 159–176. Berlin: Springer-Verlag.

Index